Design and Modeling of Low Power VLSI Systems

Manoj Sharma
BVC, India

Ruchi Gautam
MyResearch Labs, Gr Noida, India

Mohammad Ayoub Khan
Sharda University, India

A volume in the Advances in Computer and Electrical Engineering (ACEE) Book Series

ENGINEERING SCIENCE REFERENCE
An Imprint of IGI Global

Published in the United States of America by
Engineering Science Reference (an imprint of IGI Global)
701 E. Chocolate Avenue
Hershey PA, USA 17033
Tel: 717-533-8845
Fax: 717-533-8661
E-mail: cust@igi-global.com
Web site: http://www.igi-global.com

Library of Congress Cataloging-in-Publication Data

Names: Sharma, Manoj, 1981- editor. | Gautam, Ruchi, 1984- editor. | Khan, Mohammad Ayoub, 1980- editor.
Title: Design and modeling of low power VLSI systems / Manoj Sharma, Ruchi Gautam, and Mohammad Ayoub Khan, editors.
Description: Hershey, PA : Engineering Science Reference, 2016. | Includes bibliographical references and index.
Identifiers: LCCN 2016003311| ISBN 9781522501909 (hardcover) | ISBN 9781522501916 (ebook)
Subjects: LCSH: Low voltage integrated circuits--Design and construction. | Integrated circuits--Very large scale integration--Design and construction.
Classification: LCC TK7874.66 .D47 2016 | DDC 621.39/5--dc23 LC record available at http://lccn.loc.gov/2016003311

This book is published in the IGI Global book series Advances in Computer and Electrical Engineering (ACEE) (ISSN: 2327-039X; eISSN: 2327-0403)

British Cataloguing in Publication Data
A Cataloguing in Publication record for this book is available from the British Library.

All work contributed to this book is new, previously-unpublished material. The views expressed in this book are those of the authors, but not necessarily of the publisher.

For electronic access to this publication, please contact: eresources@igi-global.com.

Advances in Computer and Electrical Engineering (ACEE) Book Series

Srikanta Patnaik
SOA University, India

ISSN: 2327-039X
EISSN: 2327-0403

Mission

The fields of computer engineering and electrical engineering encompass a broad range of interdisciplinary topics allowing for expansive research developments across multiple fields. Research in these areas continues to develop and become increasingly important as computer and electrical systems have become an integral part of everyday life.

The **Advances in Computer and Electrical Engineering (ACEE) Book Series** aims to publish research on diverse topics pertaining to computer engineering and electrical engineering. **ACEE** encourages scholarly discourse on the latest applications, tools, and methodologies being implemented in the field for the design and development of computer and electrical systems.

Coverage

- Computer Hardware
- Programming
- Electrical Power Conversion
- Optical Electronics
- Algorithms
- Applied Electromagnetics
- Power Electronics
- Microprocessor Design
- Chip Design
- Computer science

IGI Global is currently accepting manuscripts for publication within this series. To submit a proposal for a volume in this series, please contact our Acquisition Editors at Acquisitions@igi-global.com or visit: http://www.igi-global.com/publish/.

The Advances in Computer and Electrical Engineering (ACEE) Book Series (ISSN 2327-039X) is published by IGI Global, 701 E. Chocolate Avenue, Hershey, PA 17033-1240, USA, www.igi-global.com. This series is composed of titles available for purchase individually; each title is edited to be contextually exclusive from any other title within the series. For pricing and ordering information please visit http://www.igi-global.com/book-series/advances-computer-electrical-engineering/73675. Postmaster: Send all address changes to above address.

Titles in this Series

For a list of additional titles in this series, please visit: www.igi-global.com

Reliability in Power Electronics and Electrical Machines Industrial Applications and Performance Models
Shahriyar Kaboli (Sharif University of Technology, Iran) and Hashem Oraee (Sharif University of Technology, Iran)
Engineering Science Reference • copyright 2016 • 481pp • H/C (ISBN: 9781466694293) • US $255.00 (our price)

Handbook of Research on Emerging Technologies for Electrical Power Planning, Analysis, and Optimization
Smita Shandilya (Sagar Institute of Research Technology & Science, India) Shishir Shandilya (Bansal Institute of Research & Technology, India) Tripta Thakur (Maulana Azad National Institute of Technology, India) and Atulya K. Nagar (Liverpool Hope University, UK)
Engineering Science Reference • copyright 2016 • 410pp • H/C (ISBN: 9781466699113) • US $310.00 (our price)

Sustaining Power Resources through Energy Optimization and Engineering
Pandian Vasant (Universiti Teknologi PETRONAS, Malaysia) and Nikolai Voropai (Energy Systems Institute SB RAS, Russia)
Engineering Science Reference • copyright 2016 • 494pp • H/C (ISBN: 9781466697553) • US $215.00 (our price)

Environmental Impacts on Underground Power Distribution
Osama El-Sayed Gouda (Cairo University, Egypt)
Engineering Science Reference • copyright 2016 • 405pp • H/C (ISBN: 9781466665095) • US $225.00 (our price)

Handbook of Research on Advanced Computational Techniques for Simulation-Based Engineering
Pijush Samui (Centre for Disaster Mitigation and Management, VIT University, India)
Engineering Science Reference • copyright 2016 • 616pp • H/C (ISBN: 9781466694798) • US $310.00 (our price)

Operation, Construction, and Functionality of Direct Current Machines
Muhammad Amin (COMSATS Institute of Information Technology, Wah Cantt, Pakistan) and Mubashir Husain Rehmani (COMSATS Institute of Information Technology, Wah Cantt, Pakistan)
Engineering Science Reference • copyright 2015 • 404pp • H/C (ISBN: 9781466684416) • US $200.00 (our price)

Performance Optimization Techniques in Analog, Mixed-Signal, and Radio-Frequency Circuit Design
Mourad Fakhfakh (University of Sfax, Tunisia) Esteban Tlelo-Cuautle (INAOE, Mexico) and Maria Helena Fino (New University of Lisbon, Portugal)
Engineering Science Reference • copyright 2015 • 464pp • H/C (ISBN: 9781466666276) • US $235.00 (our price)

IGI GLOBAL
DISSEMINATOR OF KNOWLEDGE
www.igi-global.com

701 E. Chocolate Ave., Hershey, PA 17033
Order online at www.igi-global.com or call 717-533-8845 x100
To place a standing order for titles released in this series, contact: cust@igi-global.com
Mon-Fri 8:00 am - 5:00 pm (est) or fax 24 hours a day 717-533-8661

Editorial Advisory Board

List of Reviewers

Table of Contents

Detailed Table of Contents

In contemporary world the technology has kept its vast identity in developing ultra NANO devices to give up the compact device utilities, in VLSI, Metal Oxide Semiconductor device plays an key role in power dissipation product, in terms of MOS theory characteristics it is predefined that a MOS transistor can conduct easily with low voltage which gives low power but in DSM technology there is a likelihood to achieve ultra low power, so this can be achieved due to the rapid shrinking of gate length, here the chapter deals with challenges and limitations of low power techniques. The predominant way to generate low power is to start with the fundamental principles that are defined in the existing technologies that it gives low power with less leakage current. Apart from this parameter consideration is also required to achieve this. The successful and the major parameter in generating low power is that the shrinking of supply voltage. To go through this, upcoming sections gives the brief idea about the different techniques that are utilized to generate low power with less leakage.

Ever increasing demand for portable and battery-operated systems has lead to aggressive scaling. While technology scaling facilitates faster and high performance devices, at the same time it causes excessive power dissipation. Leakage power dissipation is now a dominating component of total power consumption in such portable devices. So there is a tremendous need to limit the power dissipation in high density chips which has initiated many innovative techniques to develop in the design of low power circuits and systems. Today's nano-scaled VLSI chips have ultra-thin gate oxide, very low threshold voltage and having short channels. As such leakage power dissipation has emerged as the most challenging issue in VLSI circuit and systems. This Chapter review and compare the state of the art circuit techniques for leakage minimization. It also conceptually classifies the different techniques of leakage minimization. Moreover, a detailed comparison based on trading-offs with other design parameters is also given along with leakage minimization.

This chapter describes three contemporary low power design approaches; a resistor-less bandgap reference circuit, a hybrid voltage level shifter with a diode connected NMOS and a modified dynamic comparator, each design with the objective to demonstrate the feasibility of contemporary approaches in achieving lower power VLSI design. All three designs are simulated in 0.18 μm CMOS technology using industrial simulation tool and the results are based on performance parameters defined in the chapter.

With the rapidly evolving silicon technology, the power density becomes increasingly high. Quadratically related to power, the voltage scaling offers a means of minimizing energy. However, power supply scaling demands less threshold voltage, which rises leakage current. Several low power techniques have been devised. This chapter deals with the non-conventional low power design solutions, based on adiabatic switching theory. In such circuits, the energy rather than getting dissipated during every cycle, is transferred back and forth between the logic and power-clock sources. A brief discussion on the reversible logic circuits will be presented followed by the fully adiabatic and quasi-adiabatic circuits. The use of power-clock sources for operating the adiabatic circuits will also be introduced. The generalized energetics of an adiabatic circuit followed by the typical loss models of the adiabatic families are presented. Some of the adiabatic circuits employing CMOS transistors are introduced in the chapter. A short comparison for the adiabatic circuit leakage models follows.

This chapter presents a comprehensive overview of the conventional fully static master slave flip-flops used in low power VLSI systems where power budget is critical. In addition, the chapter also presents alternative realization of fully static master-slave flip-flops utilizing a modified feedback strategy. The flip-flops designed on the basis of modified architecture have been explained in detail and compared with state-of-the-art master slave flip-flop designs available in the literature. Extensive capacitance calculations have been performed in terms of clock load and capacitance at internal nodes has also been estimated for all the flip-flop configurations. This is executed in order to compare their relative power and delay characteristics which are well supported by simulation results.

The mission of this chapter is to introduce the reader the recent developments in the design of ultra-Low Power ADCs for Wearable and Implantable Medical Devices (WIMDs). The focus of this chapter will be on Signal-Adaptive Successive Approximation Register (SAR) ADC architectures and their derivatives, since the majority of the ULP medical devices rely on these architectures. The proposed chapter first provides an overview of the WIMDs, and electrophysiological signals. Then, basic SAR ADCs are introduced followed by the study of adaptive SAR ADCs. The chapter concludes with a brief summary of the other prevalent ADC architecture for WIMDs, namely the Level-Crossing ADCs.

In this chapter, an ultra low power CMOS Common Gate LNA (CGLNA) with a Capacitive Cross-Coupled (CCC) gm boosting scheme is designed and analysed. The technique described has been employed in literature to reduce the Noise Figure (NF) and power dissipation. In this work we have extended the concept for low voltage operation along with improving NF and also for significant reduction in current consumption. A gm boosted CCC-CGLNA is implemented in 90nm CMOS technology. It has a gain of 9.9dB and a noise figure of 0.87dB at 2.4GHz ISM band and consumes less power (0.5mw) from 0.6V supply voltage. The designed gm boosted CCC-CGLNA is suitable for low power application in CMOS technologies.

The objective of this chapter is to describe the various designed arithmetic circuit for an application of multimedia circuit that can be used in a high-performance or mobile microprocessor with a particular set of optimisation criteria. The aim of this chapter is to describe the design method of binary arithmetic especially using by CMOS and Pass Transistor Logic technique. The pass transistor techniques are reduced the noise margin for small circuit, which can be explained in this chapter. This chapter further describe the types of arithmetic and its techniques. The technique design principle procedure should make the following decisions: circuit family (complementary static CMOS, pass-transistor, or Shannon Theorem based); type of arithmetic to be used. The decisions on the designed logic level significantly affect the propagation delay, area and power dissipation.

Integrated circuits have been predominantly designed and developed by large firms and manufacturers; nowadays, any electronic engineer should be able to develop specific and innovative low-power designs using available open cores. This chapter presents the design process for a specific chip, beginning with a definition of its function, design considerations, power analysis, performance optimization, and chip optimization. The hardware and software for this circuit were developed for low-power implementation: it includes a processor, memory blocks, ports, buses, and a proposed application program, so it can be used as a starting point for other low-power very-large-scale integration (VLSI) circuits. The chip uses frequency synthesis and configuration parameters to deliver electric signals on a variety of waveforms and patterns. This design can be used in many research fields and application areas, where experiments or portable devices need low-power, programmable, and configurable electric signal generators.

Preface

For analog circuits the power analysis and optimization used to be less critical as compared to be digital circuits design. Both technical and economical needs drive the low power industry which has boomed with rapid market share. Historically fundamental implementation platform for circuits are silicon wafers based transistors. The advances in associated technologies have revolutionalised the integration levels and hence the functionalities implemented per chips. Unlike the vacuum tube which need large voltage and power the silicon based transistors have implemented large number of functionalities on small area and lesser power compared. Along with the increase in complexity, the issue of power dissipation has become very critical design parameters. In order to continue the integration trends on chips for more functionality and improved performance parameters the transistor size has to be reduced well supported by the fabrication unit capabilities. Gradually the technology approached their physical limitations. Simultaneously the power per unit area has increased many folds.

The power dissipated per unit area has reached to limiting values in the range of temperature in nuclear reactor and sun's surface many times. Handling this enormous power on a small chip is becoming exigent. The required heat removal mechanisms proved meager in this range and the associated cost rise rapidly. Even the reduced voltage swings could not provide comprehensive solutions to these extragent power values. The increased power figures also effect on circuit reliability.

Today's high demanding applications like portable multimedia, speech recognition, images processing, OCR, etc need large space and power. So power minimization is now very critical design parameter which needs conscious efforts across all design steps, domains and abstract levels. Hence the need of power optimization is forcing engineers to bring new strategies, techniques and principles aiming to bring the power equations within limits without compromising the integrated functionalities.

There is no single reason for power dissipation. The power optimization process is multi-dimensional needing to account all the different components involved.

Providing readers a knowledge and basic understanding of these different set of components is the primary aim of the book. Through the medium of this book, it is tried to collate the different aspects associated to the power dissipation. It is tried to capture both power analysis and power optimization aspects as an essential components of VLSI circuit low power design problem. Through power analysis the estimation of different power parameters are made and these are optimized using low power techniques.

It is further tried to highlight the need of understanding the power aware system components, requirements and optimize-able methodologies from engineering perspectives and methods for the problems to tackle in the multidimensional environment considering the integrated approach in all levels of design cycle.

The book presents practical as well as conceptual knowledge of the techniques and methodologies of low power integrated circuits design aspects. Each chapter presents the reader with an in-depth investigation regarding the actual or potential role of the techniques concerned. The book also presents state-of-the art and future perspectives in the domain of power aware design, where experts, researchers, and academicians had shared ideas and experiences covering frontier technologies, breakthrough and innovative solutions and applications.

The significance of low power design can be seen via development of numousouse applications/ products addressing one or the other issues. The smart phones, palm computing, high speed switches for IOTs, interlinked networks etc, can be developed because of the advances in low power circuit design capabilities.

The different chapters in the book are prepared by researchers and practitioners in low power domain from academia and industry. Some authors have extensively provided the comprehensive description and reported the results for their implementations as well. Throughout the book it is tried to present the content in a way understandable to readers easily.

ORGANIZATION OF THE BOOK

The book is organized into three parts and altogether 14 chapters. Part 1, titled "Fundamental of Power aware VLSI Design", contains Chapters 1-5. Part 2, named "Non Traditional Low Power Methadologies", contains Chapters 6-7. Part 3, titled "Low Power applications and studies", contains Chapters 8-14. A brief description of each of the chapters follows:

Chapter 1, "Power dissipation fundamentals" provide fundamental concepts of power dissipation. It briefly discusses the reasons of power dissipation focusing the modeling of different parameters in details. The chapter provides evaluation methodology for different main culprits contributing major role in power equations.

Chapter 2, "Low Power Design Techniques: Classical and Beyond CMOS Era" highlights the importance and need of power aware design techniques. The power limitation wrt high-performance system, portable applications, battery limitations are discussed. Further in this chapter, there is an attempt to discuss types of power dissipation, strategies and techniques for low-power design, CAD tools for low-power design. Lastly aspects wrt various nanoelectronic alternatives as possible candidates for MOSFET replacement in futuristic ultra-low power designs are considered.

Chapter 3, "Low power strategies for beyond Moore's law era: Low power device technologies and materials", gives a general overview of the various technologies; materials and architectures researchers are concentrating to continue the technology beyond Moore's law with low power consumption. The problems of semiconductor industries wrt chip design with less than 10nm technology are illustrated. The aspects of Moore's law forcing researchers to look upon devices in another aspect are highlighted. Further the various architectures and materials invented to increase the reliability, speed and most importantly low power operation without increasing the size of devices are discussed. The on-set of nanotechnology and nano-science leading to unconventional 3D structure devices to and 0-dimensional structures are also detailed.

Chapter 4, "Challenges and limitations of low power techniques: Low power methodologies in analog and digital circuits" presents the low power design issues wrt system design. In contemporary world the technology has kept its vast identity in developing ultra NANO devices to give up the Compaq device utilities. in VLSI, Metal Oxide Semiconductor device plays an key role in power dissipation product, in terms of MOS theory characteristics it is predefined that a MOS transistor can conduct easily with low voltage which gives low power but in DSM technology there is a likelihood to achieve ultra low power, so this can be achieved due to the rapid shrinking of gate length, here the chapter deals with challenges and limitations of low power techniques. The predominant way to generate low power is to start with the fundamental principles that are defined in the existing technologies that it gives low power with less leakage current. Apart from this parameter consideration is also required to achieve this. The successful and the major parameter in generating low power is that the shrinking of supply voltage. To go through this, upcoming sections gives the brief idea about the different techniques that are utilized to generate low power with less leakage.

Chapter 5, "Leakage Minimization in CMOS VLSI Circuits-A Brief Review", highlights the concern of Leakage power dissipation with nanoscaled devices. Aggressive scaling has resulted in high density chips with increased functionalities, but at the same time it is power hungry due to high performance. So it is absolutely necessary to adopt low power design strategies to minimize power. Moreover, sub-threshold conduction and band to band tunneling has become too much prominent with reduced device dimensions. Again, the life of a battery is also limited. So for portable battery operated systems, the only way to minimize power dissipation in standby mode (especially, the leakage) is by adopting some innovative circuit techniques. The chapter introduces first, the present leakage scenario, the trend according to ITRS and the design needs for leakage minimization. Next, it describes different types of leakage currents and their mechanisms. The chapter then describes the most widely used circuit level leakage minimization techniques in brief along with their merits/demerits. Further, a broad classification on leakage minimization techniques is presented based on the low power strategies being adopted by each such techniques. Finally, a conclusion is given at end of the chapter. This chapter is well organized and written in a simple language. It will certainly benefit researchers, PG students and Engineers working in the field of low power VLSI design.

Chapter 6, "Contemporary Low Power Design Approaches", described the contemporary low power design approaches that are still being used in basic building blocks of VLSI systems today focusing the reasons of still using these and the approaches that fit into today's technology scaled process. This Chapter explains three contemporary low power design approaches; a resistor-less bandgap reference (BGR) circuit, a hybrid voltage level shifter (VLS) with a diode connected NMOS and a modified dynamic comparator, each design with the objective to demonstrate the feasibility of contemporary approaches in achieving lower power VLSI design. The BGR yield a precise DC voltage with optimized output reference voltage of 910 mV with maximum temperature coefficient (TC) at 13 ppm/°C. In the second design, an extensive hand analysis towards achieving an optimized VLS relating transistor sizing with performance parameters are included. The final approach uses modified latch utilizing NMOS transistor has been applied in a dynamic comparator. The t_{latch} of the proposed comparator shows improvement of approximately 23% compared to conventional comparator. All three designs are simulated in 0.18 μm CMOS technology using industrial simulation tool and the results are based on performance parameters defined in the chapter.

Chapter 7, "Low Power VLSI Circuit Design using Energy Recovery Techniques", deals with the non-conventional low power design solutions, based on adiabatic switching theory. In such circuits, the energy rather than getting dissipated during every cycle, is transferred back and forth between the logic and power-clock sources. A brief discussion on typical reversible logic circuits is presented followed by the fully adiabatic and quasi-adiabatic circuits. The use of power-clock sources for operating the adiabatic circuits has also been introduced. The generalized energetics of an adiabatic circuit followed by the typical loss models of the adiabatic families found widely in the literature are presented. The quasi-adiabatic circuits employing CMOS transistors, namely, 2N-2P, 2N-2N2P, PFAL, DCPAL and PSAL are introduced in the chapter. Performance comparison for the adiabatic circuits has been made along with the leakage models of the quasi-adiabatic circuits. The influence of technology scaling on the energy recovery performance is analyzed using 350nm and 32nm technology nodes. In summary, this chapter can help an electronics circuit designer interested in adiabatic circuit in learning the nuances of the energy recovery property of the circuits.

Chapter 8, "State-of-the-art master slave flip-flop designs for low power VLSI systems" presents a detailed explanation for transistor-level implementation of state-of-the-art master-slave flip-flops has been presented and possibilities for reduced transistor count implementation using modified feedback structure have been explored. it is ensured that the static functionality is not deviated while optimization of power-delay product and power-delay-area product has also been emphasized simultaneously for the proposed designs. Technology calibration has been carried out so that the reader achieves a good level of understanding in terms of constraints in the digital IC design process (design space).

Chapter 9, "Signal-Adaptive Analog-to-Digital Converters for ULP wearable and implantable medical devices: A Survey", introduces the reader with recent developments in the design of ultra-low power ADCs for wearable and implantable medical devices (WIMDs). More specifically, power consumption of the Successive Approximation Register (SAR) ADCs and their derivatives targeting wearable and implantable devices is discussed. In particular, reconfigurable, and signal-adaptive SAR ADCs are reviewed and evaluated for their power consumption. An overview of the WIMDs, and electrophysiological signals is also provided in this chapter.

It is shown that there is room for improvement to optimize the power consumption of SAR ADCs for digitizing electrophysiological signals. Signal-aware SAR ADCs can be made to consume very low-power, especially when the needs for the target application is taken into account. Significant power reduction is shown when the ADC parameters such as resolution and sampling rate are modulated based on the requirements of the target application, and these results are reinforced with simulations against electrophysiological data. More specifically, it is shown that up to 67% power saving can be achieved for monitoring the QRS complex in EKG.

Chapter 10, "The Design of Ultra Low Power RF CMOS LNA in Nanometer Technology", discusses an ultra low power CMOS LNA and has designed it using a g_m boosting technique to reduce the power consumption and improve the NF of a CGLNA is described. A capacitor cross-coupled LNA is presented as an implementation of the general technique. The fully integrated differential CCC-CGLNA consumes 0.83mA from 0.6V voltage supply and attains a measured NF of 0.875dB at 2.4GHz the amplifier were implemented in a standard 90nm CMOS process. The designed technique makes the common gate topology attractive for low-power and high-frequency fully integrated designs.

A general g_m boosted CCC-CGLNA topology is detailed that exhibits lower noise figure and consumes less power than conventional CGLNA. It also preserves the advantages of insensitivity to parasitic input capacitance, higher reverse isolation and better stability, which makes it attractive for emerging high frequency applications. Implementation of the gm boosted CCC-CGLNA scheme is also discussed and simulations shown good agreement between analysis and simulated results. The gm boosted topology provides new opportunities for circuit innovation implementations using other coupling techniques are under development.

Chapter 11, "Low Power Arithmetic Circuit Design for multimedia applications: Adder design", describes the various designed arithmetic circuit for an application of multimedia circuit that can be used in a high-performance or mobile microprocessor with a particular set of optimisation criteria. The aim of this chapter is to describe the design method of binary arithmetic especially using by CMOS and Pass Transistor Logic technique. The pass transistor techniques are reduced the noise margin for small circuit, which is explained in this chapter. This chapter further describe the types of arithmetic and its techniques. The technique design principle procedure should make the following decisions: (a) Circuit family (complementary static CMOS, pass-transistor, or Shannon Theorem based); (b) Type of arithmetic to be used. The decisions on the designed logic level significantly affect the propagation delay, area and power dissipation.

Chapter 12, "Case study: System on a Chip for Electric Stimulation", presents a detailed description and procedures to develop an application specific integrated circuit: a configurable system on a chip which delivers multiple signal patterns. This system is based on open cores and it is intended to operate as a standalone module or to interconnect with other devices. Throughout the design process a low-power goal was pursued, both from a hardware and software perspective. This system uses open source cores only: an Open RISC 1200 processor, program memory, data memory, external address bus, data bus, communication port, interrupt controller, clock, reset and GPIO ports. The application program is optimized for minimum power and memory access. This system can be used as it is or can be adapted for other applications with little or no modifications. A simple user interface is also explained, so this system can be used without understanding the integrated circuit theory. This system can be the foundation to develop other systems, and also can be used in other applications such as particle manipulation, pollution detection, water and food quality, cell detection, nano-manipulation, and biomedical analysis. Generally speaking, any application that needs electric signals on different waveforms, frequencies and patterns, can use this system.

Chapter 13, "Low Power Design of High Speed Communication System Using IO Standard Technique over 28 nm VLSI Chip", proposes the low power design for VLSI based high speed communication designed over 28 nm VLSI chip packed in UltraScale Field Programming Gate Array (FPGA). The low power design for VLSI based high speed communication is achieved using IO Standard technique. The developed system is tested for 500 GHZ, 900 GHz, 10 THz and 17 THz carrier frequencies for developed low power design technique for VLSI based high speed communication. It is determined that using the High Speed Unterminated Logic IO Standard more than 96% power reduction is achieved for all frequencies. Furthermore, the designed system consumes 19% less chip power compared to existing VLSI based high-speed communication systems. The designed VLSI based high-speed communication system over 28 nm VLSI chip provides the low power consumption for designed system over chip for different high frequencies.

WHO AND HOW TO READ THIS BOOK

The reader is presumed to have knowledge of basic circuit design concepts and techniques involved. The book avoids details relating to circuit design aspects but focus on proved low power optimizations methods that have been in use today's electronics products, which have made it possible to integagate large complexities on mega chips impacting the global economy as a whole directly of indirectly.

This book has three groups of people as its potential audience, (i) undergraduate students and post-graduate students conducting research in the areas of Integrated circuit design and power low power equations. (ii) Researchers at universities and other institutions working in these fields; and (iii) early practitioners in the R&D departments of industrial settings for providing concerned information at a place. This book differs from other books that have comprehensive case study and real data from experiments. The book has focus on collating the different techniques involved.

We hope that this book appeals to the readers/designers focusing the low power VLSI design aspects and would inspire them to pursue their works further. We also hope that this book would certainly assist them in meeting the power requirements for their circuits on time and hence reduce the design cycle time.

SIGNIFICANCE OF BOOK IN TODAY'S ECOSYSTEM

The low power VLSI design methodology has now evolved into a separate engineering involving un-accountable techniques, methodologies and their combinations to minimize the power dissipation in VLSI circuits. Including the complete set of these is not feasible. Presently in this book different basic approaches are targeted and streamlined wrt fundamental advancements reported in the hardware design and related technologies. Efforts would be extended to incorporate the in-depth, comprehensive power estimation and optimization technologies subsequently. These collected efforts are useful for academics, researchers, and practitioners seeking the latest practices and knowledge in this field.

The proposed book aims collating the different low power techniques for IC design classifying them as traditional and modern techniques. It further aims to collate the limiting factors associated with the traditional low power techniques which lead to evolution of new non-traditional methodologies for optimizing the power equations.

Through the efforts readers would be able to appreciate the technicalities involved in low power VLSI hardware development process cycle. They would be able to understand the advances in these technologies and their application boundaries limitations. With advances in new capabilities in related domains, new design mechanism have been evolved for ASIC, Semicustom, Custom and other classified design flows along with associated verification process targeting optimized solution on power, performance and area parameters.

Readers would be able to appreciate the advancement of related CAD tools associated leading to much needed optimized products. They would be able to understand the basic limiting principles which are proving to be hindrances for further optimizing processes leading to evolvement of an era where new fusion of principles and concepts from different sciences are made and applied to optimize power parameters in VLSI hardware development, design, implementation and verification steps.

New trends in power and speed optimization would be illustrated enabling the readers to take advances from these researches and innovative solution approaches.

Hence readers would be benefited through the efforts, covering the development and generation of low power and high performance VLSI hardware modeling and verification techniques.

Manoj Sharma
BVC, India

Ruchi Gautam
MyResearch Labs, India

Mohammad Ayoub Khan
Sharda University, India

Introduction

Digital circuit design based upon MOS technologies has revolutionalized every sphere of life. The advances associated with circuit design is the fundamental reason of advances in capabilities of different inventions and applications which lead human make possible which were considered to be impossible in yester-years.

The increased performance capabilities of the devices were primarily evaluated on speed, power and area (SPA) parameters. These parameters for some functionalities are dependent upon the fabrication technologies. The target node plays critical role in this aspect. The same functionality became success on one node with specific tradeoffs of design parameters and capabilities. On the other hand these parameters become un-acceptable with other technological nodes. Speed changes with load, type of input and transistor size. Power depends upon time, switching frequencies, within that time, input values, slew rate, load capacitance and gate width. Area is purely a fabrication technology dependent scaled with the numbers of switches implemented.

Out of these primary parameters in SPA, SP has become most critical. The speed or delay associated can be predicted with Logical Effort (LE). Extrapolating LE fundamentals power estimation can also be madled which is called Logical Power. Vladimir Stcjanovic has given set of rules for consistent estimation of performance and power for flip flops designs. These set of rules can play important role for other circuit considerations as well.

Loss of power supply happens because of inertia of the circuit that needs to be overcome by the applied power supply. This opposing force depends upon the resistive and capacitive attributes of the device and their interconnections. Being a voltage controlled resistor, MOSFET has specific resistance (R_{on}) which causes power dissipation when a current flows across this. The resistive power loss is inversely proportional to the size of the transistor. Larger the size of the MOSFET, lower would be its R_{DS}(ON). The power dissipated because of capacitive attribute is generally referred as switching power. The element of power loss depends upon the switching operation of the MOSFET as the intrinsic parasitic capacitance stores and dissipates energy during these switching activities.

Power budgeting involves detailed and careful analysis of power supply sizing, current requirements, cooling/heatsink requirements, maximum reliable operating frequencies, device selections. So at every sphere low power circuit design aspects is very critical. Maximizing the battery life, cost effective packaging, and reliability issues highlights the urgencies of reducing the energy consumption in circuits. Predominantly SoCs have to deal with dynamic power and leakage power is issue in other circuits.

ITRS have extrapolated the power trends for dynamic power and static power pertaining to memory blocks and logic blocks as depicted in Figure 1. As evident the high gap in the requirements and trend needs to be addressed with innovative engineering solutions.

Figure 1. Power trends are outstripping power requirements

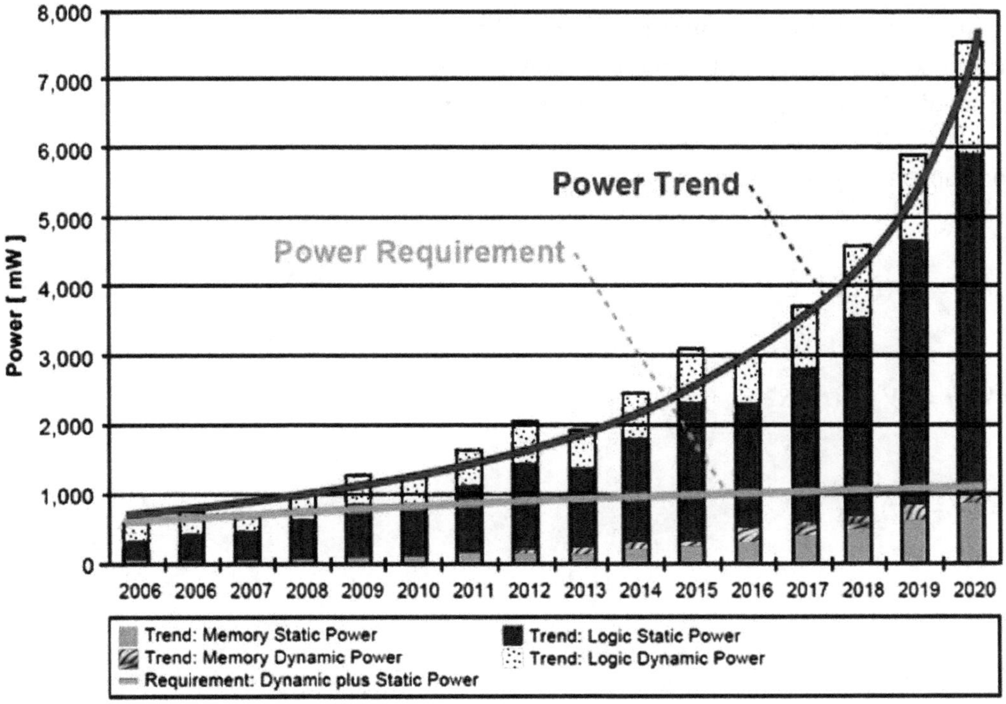

The heat generated by the system is not the only concerned but the investment and efforts required to take off that heat from the chip is also very expensive and un-contributory to the logic, shifting the design equations in negative side.

Power dissipation in MOS circuits constitutes static components viz static; sub-thresh hold conduction through OFF transistor, tunneling currents through gate oxide, leakage through reverse biased diodes, contention currents in ratioed circuits, and dynamic components due to charging and discharging of load capacitance, short circuit current. These transitions can further be considered as essential transitions/ useful transitions for logic evolution and unnecessary transition because of glitches and hazards. Therefore dynamic power have two sub-components namely because of necessary signal transitions and glitch power. The glitch power is totally waist and designers need to take extra care to have circuit implementation for glitch free different data evaluation. Therefore power equation can be depicted as

$$P_{Total} = P_{Static} + P_{Dynamic} \cdots \tag{1}$$

and

$$P_{Total} = P_{Static} + P_{Dynamic} + P_{Tran} + P_{SC} \tag{2}$$

$$P_{Total} = \left(0.5 * C_L * V_{DD}^2 * f\right) + \left(I_{SC} + I_{Leakage}\right) * V_{DD} \tag{3}$$

Transition Power (P_{tran})

Consider the basic inverter structure as sown in Figure 2. The charging and discharging of the load capacitor can be analyzed as a RC circuit as shown in Figure 3.

Charging of a Capacitor with a constant voltage source V can be analyzed as follows:

Charge on capacitor, $q(t) = Cv(t)$ (4)

Current, $i(t) = dq(t)/dt = C^{dv(t)}/dt$ (5)

$i(t) = C^{dv(t)}/dt = [V - v(t)]/R$ (6)

$dv(t)/dt = [V - v(t)]/RC$ (7)

Figure 2. Charge transfer in load capacitor

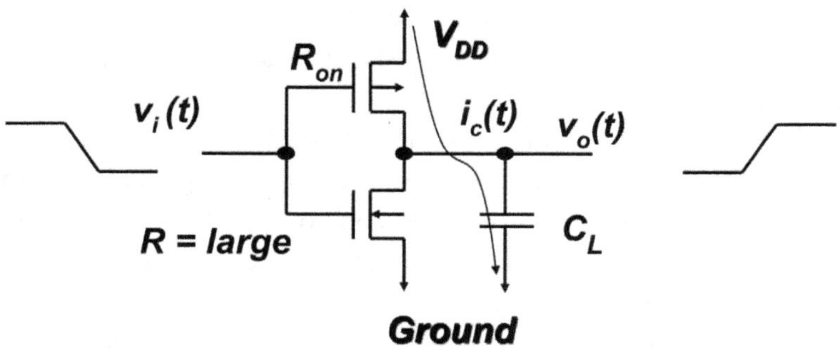

Figure 3. Equivalent RC model for charge transfer

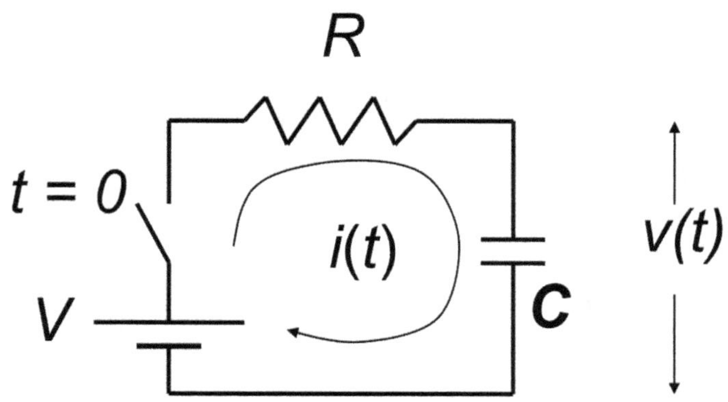

$$\int \frac{dv(t)}{dt} = \int \frac{dt}{RC} \tag{8}$$

$$ln\left[V - v(t)\right] = \frac{-t}{RC} + A \tag{9}$$

$$v(t) = V\left[1 - e^{\left(\frac{-t}{RC}\right)}\right] \tag{10}$$

$$i(t) = C\frac{dv(t)}{dt} = \frac{V}{R}e^{\left(\frac{-t}{RC}\right)} \tag{11}$$

Total Energy per Charging Transition from Power Supply is

$$E_{Trans} = \int_{0}^{infinite} Vi(t)\,dt = \int_{0}^{infinite} \frac{V^2}{R}e^{\left(\frac{-t}{RC}\right)dt} = CV^2 \tag{12}$$

Energy Dissipated per Transition in Resistance

$$R\int_{0}^{infinite} i^2(t)\,dt = R\frac{V^2}{R^2}\int_{0}^{infinite} e^{\left(\frac{-2t}{RC}\right)}dt = \frac{1}{2}CV^2 \tag{13}$$

Energy Stored in Charged Capacitor

$$\int_{0}^{infinite} v(t)\,i(t)\,dt = \int_{0}^{infinite} V\left[1 - e^{\left(\frac{-t}{RC}\right)}\right]\frac{V}{R}e^{\left(\frac{-t}{RC}\right)} = \frac{1}{2}CV^2 \tag{14}$$

So the Transition Power can be seen independently for high rising out and for falling output voltage.

- **Gate Output Rising Transition:** Total Energy dissipated in pMOS transistor is $CV^2/2$ and the same amount of energy is stored in the capacitor.
- **Gate Output Falling Transition:** The energy dissipated in nMOS transistor is $CV^2/2$ and the energy dissipated per transition = $CV^2/2$

Also total 'Transition Power' dissipation can be depicted as a function depicted in Equation 15.

$P_{Trans} = E_{Trans}\alpha fc_k = \alpha fck_{CV}2/^2$ where α = activity factor (15)

Transistor Power Dissipation is Data Dependent Function of S*witching Activity. Let the probability of transition from out put logic '0' to logic '1' is* P0→1 . Then,

Power = Energy/Transition * Transition rate (16)

$= CL * V_{DD}^2 * f_{0\to1} \qquad = CL * V_{DD}^2 * P_{0\to1} * f \qquad = CEFF * V_{DD}^2 * fP;$

where $CEFF$ = effective Capactance = $CL*P_{0\to1}$

The transition probabilities of some of the basic gates based on static CMOS implementation are depicted in Table 1.

Consider an example of a Static 2 Input NOR gate as shown in Figure 4

Assume P_A=1 = 1/2P_B=1 = ½. $P_{(Out=1)}$ = 1/4 (this is the *signal probability*)
$P_{(0\to1)}$= 3/4×1/4 = 3/16 (this is the *transition probability*) = $P_{(Out = 0)} \cdot P_{(Out = 1)}$CEFF= 3/16 CL

Hence Power Consumption is Data Dependent.

In the case of dependent signals the simple case is not sufficient. Consider a re-convergent fan-out circuit configuration as shown in Figure 5, the $P_{Z=1} = P_{B=1}P_{(X=1|B=1)} = P_{B=1}$

Another factor that is very critical is capacitance. There should be clear distinction between the parasitic capacitive elements and capacitive elements of intrinsic transistor. Parasitics include capacitances between the gate electrode and the source and drain regions (COS and COD), Junction capacitances between

Table 1. Transition probabilities

Gate	$P_{0\to1}$
AND	$(1 - P_A P_B)P_A P_B$
OR	$(1\text{-}P_A)(1P_B)(1\text{-}(1P_A)(1\text{-}P_B))$
XOR	$(1(P_A+P_B 2P_A P_B)(P_A+P_B\text{-}2P_A P_B)$

Table 2. Truth table of 2 input Nor gate

A	B	Out
0	0	1
0	1	0
1	0	0
1	1	0

Figure 4. Input NOR gate

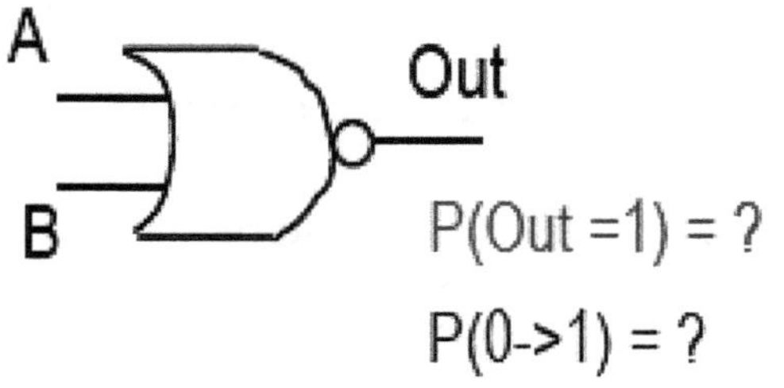

Figure 5. A re-convergent fan-out circuit

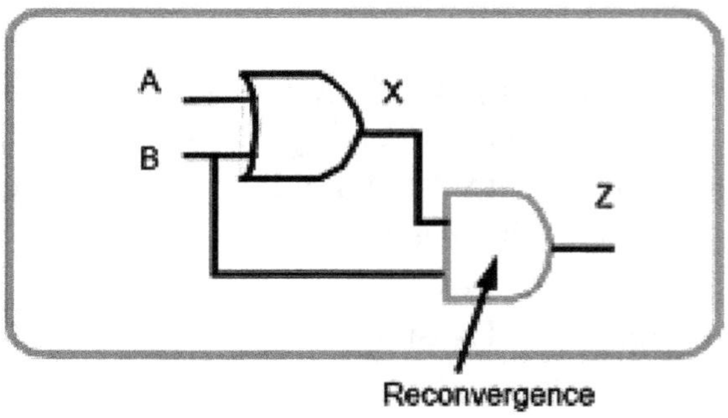

substrate and the source/drain regions (Cjs and Cjd), and capacitance between metal and source/drain/gate. Accurate calculation of the capacitance involved is equally important for correct power estimation. The aspects of device physics are considered in both intrinsic and extrinsic capacitance models in BSIM models. For intrinsic and overlap capacitances, proper settings should be made considering the following:

- Separate effective channel length and width used for capacitance
- Use of piece-wise equations
- Use of single equation models
- Continuousness of charge and capacitance over the regions
- Consistence of threshold voltage
- Consideration pertaining to body bias, short/narrow channel and DIBL effects
- Effective gate source voltages

The overlap capacitance has two components viz gate bias-independent and gate bias-dependent. The former component includes the total capacitance between gate and the heavily doped source/drain terminals. The bias independent fringing capacitance is also considered here. The second gate bias-dependent component considers the capacitance between gate and the lightly doped regions of the source and drain terminals.

For accuracy of the capacitance calculation consideration of terminal charges is always better as compared to terminal voltage levels. Taking the gate, drain, source, bulk charge as Q_g, Q_d, Q_s, Q_b and channel charge, accumulation charge, substrate depletion charge as Q_{inv}, Q_{acc}, Q_{sub}, then

$$Q_g = -(Q_{inv} + Q_{acc} + Q_{sub}) \tag{17}$$

$$Q_b = (Q_{acc} + Q_{sub}) \tag{18}$$

$$Q_{inv} = (Q_s + Q_d) \tag{19}$$

Also

$$Q_{sub} = (\delta Q_{sub0} + \delta Q_{sub}) \tag{20}$$

Therefore

$$Q_g = -(Q_{inv} + Q_{acc} + Q_{sub0} + \delta Q_{sub}) \tag{21}$$

The total charge can be find by integrating the charge along the channel. The 16 charge components can be determined using Equation 22 where i and j denote the transistor terminals.

$$C_{ij} = \frac{\delta Q_i}{\delta V_j} \tag{22}$$

Generally SPICE models overestimate the intrinsic capacitances which need serious consideration for transistors with thinner oxide thickness. The charge thickness (CTM) for capacitance estimation resolve this by considering a series capacitance with C_{ox} as depicted in Figure 6.

The effective capacitance can be evaluated using relation shown in Equation 23, 24, where X_{DC} is DC charge thickness.

$$C_{oxeff} = \frac{C_{oxp} * C_{cen}}{C_{oxp} + C_{cen}} \tag{23}$$

$$C_{ceb} = \varepsilon s_i / XD_C \tag{24}$$

Fringing capacitance has two components, a bias-independent outer fringing capacitance and a bias-dependent inner fringing capacitance. The inner fringing is not considered generally. The bias-independent outer fringing capacitance (CF) is estimated as per the relation in Equation 25; where EPSROX is Gate dielectric constant relative to vacuum having a typical value of 3.9 for SIO_2. TOXE is Electrical gate equivalent oxide thickness having a typical value of 3.0e-9m.

$$CF = \frac{2 * EPSROX * \varepsilon_0}{\Pi} * log\left(1 + \frac{4e^{-7}}{TOXE}\right) \tag{25}$$

As the capacitance of the transistor is amplified by its gain at the drain side the accurate estimation of overlap capacitance is very critical. The experimental data proves the dependency of overlap capacitance on gate-source and gate-drain voltages. The effect of series layout parameters like isolated, shared, merged source drain and multi-finger layout parameters should also be considered properly for correct estimation of the capacitances associated.

Figure 6. Charge thickness cap in CTM

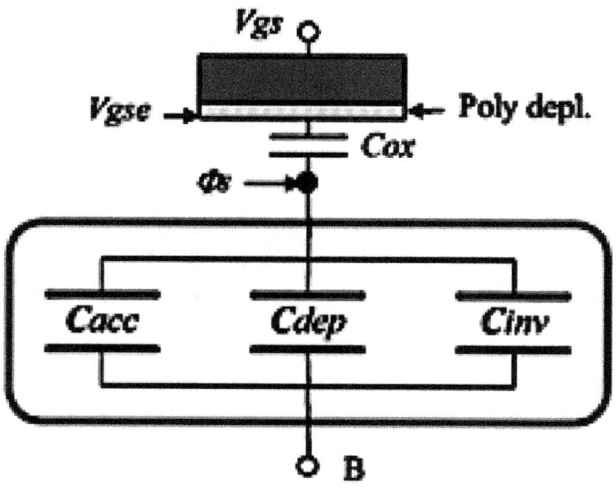

Short Circuit Power of a Transition: P_{sc}

During transition the NMOS and PMOS switch are ON simultaneously for some time. During this period the power supply gets connected to the ground via PMOS and NMOS network. A short circuit current flow in the circuit. Consider the basic inverter circuit as shown in Figure7. When input voltage lies in the range of V_{DD}-V_{TP} to V_{TN}, a short circuit current flows as shown in Figure 8. The peak short circuit current is directly proportional to transistor size (or gain, β). Larger the size of the transistors larger would be the peak short circuit current. It decreases with increase in load capacitance as depicted in Figure 9.

The Short-Circuit Energy per Transition can be depicted as follows:

$$E_{scf} = \int_{tB}^{tE} V_{DD} i_{SC}\left(t\right) dt = \left(t_E - t_B\right) I_{scmax} \frac{V_{DD}}{2} \tag{26}$$

Figure 7. Basic inverter circuit

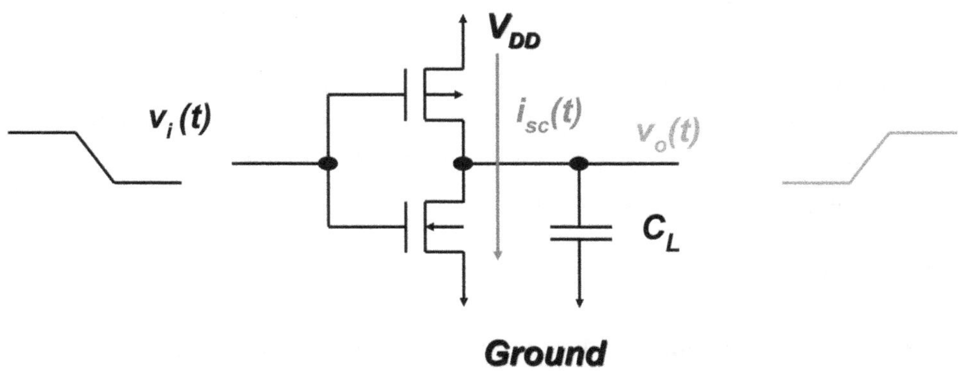

Figure 8. Short circuit current

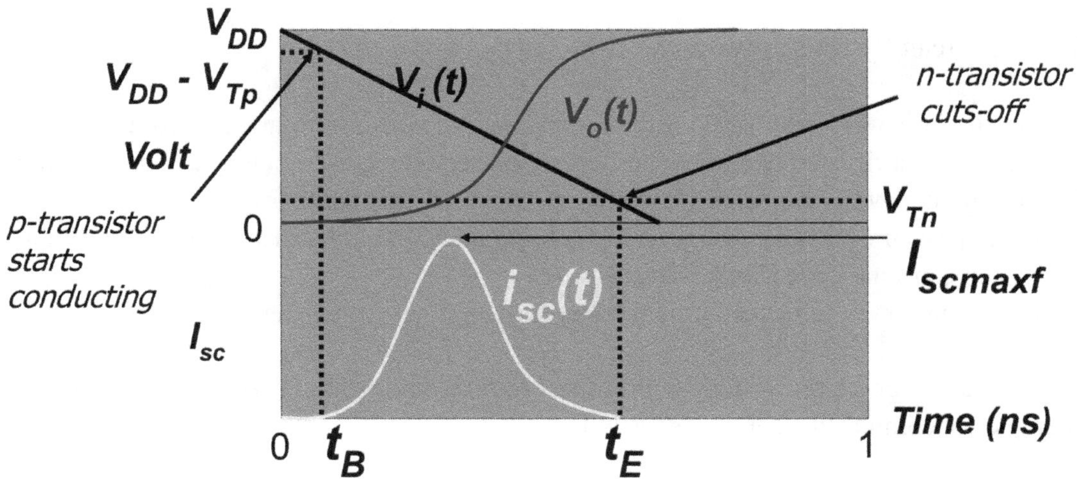

Figure 9. Peak short circuit current

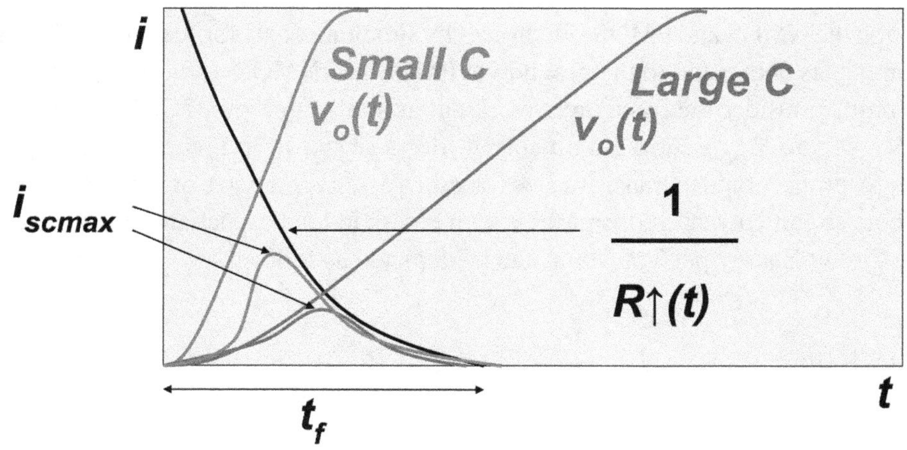

$$E_{scf} = t_f(V_{DD} - |V_{Tp}| - V_{Tn})I_{scmaxf}/2 \tag{27}$$

$$E_{scr} = t_r(V_{DD} - |V_{Tp}| - V_{Tn})I_{scmaxr}/2 \tag{28}$$

when $V_{DD} = |V_{Tp}| + V_{Tn}$ then $E_{scf} = E_{scr=} = 0$

Hence the Short-Circuit Energy, increases with rise and fall times of input signals and decreases for larger output load capacitances. Also it decreases and eventually becomes zero when V_{DD} is scaled down but the threshold voltages are not scaled down.

Short-Circuit Power Calculation

Assuming an equal rise and fall times as shown in Figure 10, the short circuit power can be depicted as Equation 29.

$$P_{SC} = \alpha f c_{kE} S_C \tag{29}$$

The short circuit P_{sc} can be summarized as:

- Short-circuit power is consumed by each transition (increases with input transition time).
- For given input rise and fall times short circuit power decreases as output capacitance increases.
- Short circuit power increases with increase of input rise and fall times
- Reduction requires that gate output transition should not be faster than the input transition (faster gates can consume more short-circuit power).
- Short circuit power is reduced if output rise and fall times are smaller than the input rise and fall times.
- Scaling down of supply voltage with respect to threshold voltages reduces short-circuit power; completely eliminated when $V_{DD} \leq |V_{tp}| + V_{tn}$.

Figure 10. Equal rise and fall times wrt inverter circuit

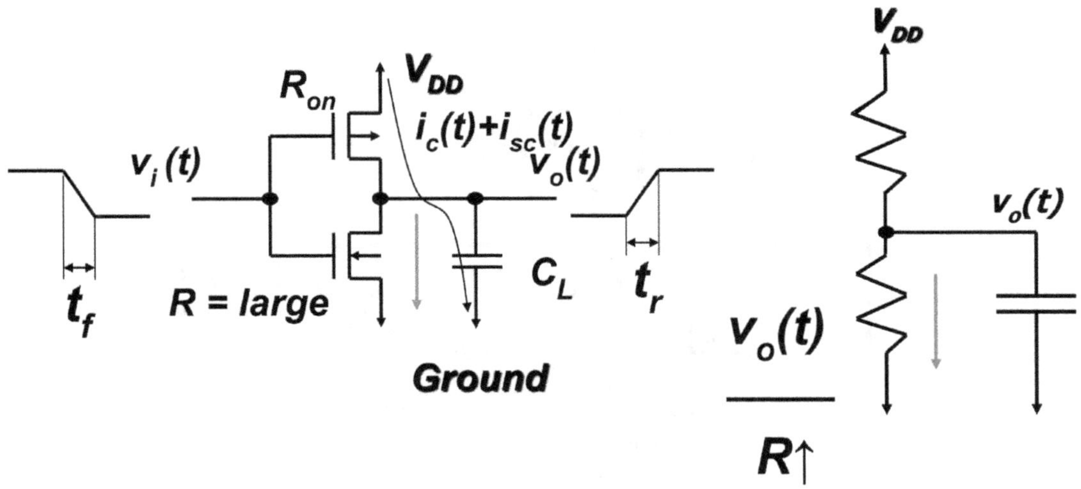

Leakage Power

The leakage current of a transistor is mainly the result of reverse-biased PN junction leakage, subthreshold leakage and gate leakage as illustrated in Figure 11. In submicron technology, the reverse-biased PN junction leakage is much smaller than subthreshold and gate leakage and hence can be ignored. The subthreshold leakage is the weak inversion current between source and drain of a MOS transistor when the gate voltage is less than the threshold voltage.

Leakage Current constitute following subcomponents

- Subthreshold conduction current, I_{sub}
- Reverse bias pn junction conduction current, I_D
- Gate induced drain leakage current, I_{GIDL} due to tunneling at the gate-drain overlap
- Drain source punchthrough current, I_{PT} due to short channel and high drain-source voltage
- Gate tunneling current, I_G through thin oxide; *may become significant with scaling*

Subthreshold Current

Subthreshold conduction current, I_{sub} can be estimated via relation shown in euayion 30,where μ0: carrier surface mobility, Cox: gate oxide capacitance per unit area, L: channel length, W: gate width, Vt = kT/q: thermal voltage, n: a technology parameter,VDS = drain to source voltage, η: a proportionality factor

$$I_{sub} = \mu_0 C_{ox} \left(W/L\right) V_t^2 e^{\{(V_{GS}-V_{TH}+\eta V_{DS})/nV_t\}} \tag{30}$$

Gate leakage is the oxide tunneling current due to the low oxide thicken and the high electric field which increases the possibility that carriers tunnel through the gate oxide. Tunneling current will become a factor and may even be comparable to subthreshold leakage when oxide thickness is less than 15-20Å.

Figure 11. Leakage current depiction

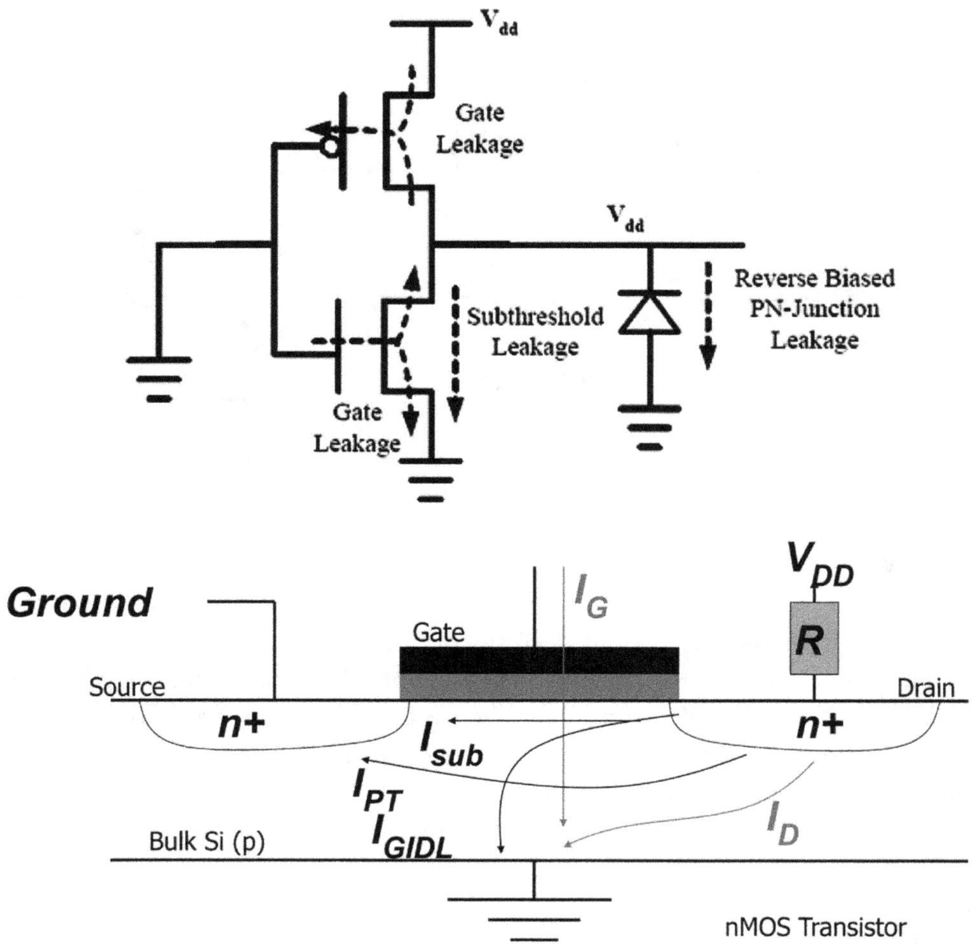

Unlike subthreshold leakage, which only exists in weakly turned-off transistors, gate leakage always exists no matter whether the transistor is turned on or turned off . Equation 31 gives the expression of the gate leakage. The gate leakage increase with scaling as T_{ox} decrease with scaling.

$$I_{gate} = W_{eff} L_{eff} A \left(\frac{V_{ox}}{T_{ox}}\right)^2 exp\left[\frac{-B\left(1-\left(1-\frac{V_{ox}}{T_{ox}}\right)^{\frac{3}{2}}\right)}{\frac{V_{ox}}{T_{ox}}}\right] \tag{31}$$

where V_{ox} is the potential drop across the thin oxide, Φ_{ox} is the barrier height for the tunneling particle (electron or hole), and T_{ox} is the oxide thickness. *A* and *B* are physical parameters given as illustrated in equation … Where m is effective mass of tunneling particle, q electronic charge, h is reduced Plank's constant.

$$A = \frac{q^3}{16\Pi^2 h_{ox}} and\, B = \frac{4\sqrt{2m_{ox}}^{\frac{3}{2}}}{3hq} \tag{32}$$

Leakage Power can be Summarized as:

- Leakage power as a fraction of the total power increases as clock frequency drops. Turning supply off in unused parts can save power.
- For a gate it is a small fraction of the total power; it can be significant for very large circuits.
- Scaling down features requires lowering the threshold voltage, which increases leakage power; roughly doubles with each shrinking.
- Multiple-threshold devices are used to reduce leakage power.

Alternatively for quick estimations the average power drawn from the source in time duration t_1 to t_2, can be find using Equation 33, where P(t) is the instantaneous power dissipated by the source. The instantaneous power can be calculated by multiplying the current flowing through the source and the voltage drop across the source at that time.

$$P(t_1, t_2) = \frac{1}{t_1 \bullet t_2} \int_{t1}^{t2} P(t)\, dt \tag{33}$$

The Circuit-Level Power Estimation is very difficult problem. Following should be kept in mind

- V_{DD}, fclk, CL are known
- Actually, the layout will determine the interconnect capacitances
- Need node-by-node accuracy
- Power dissipation is highly data-dependent
- Need to estimate switching activity accurately
- Simulation may take days to complete

Power Aware Design Techniques

The power aware design should start from the top OS, software, architecture, circuit level and transistor level. An inclusive approach provides effective solution. The optimization process is quite complex and multidimensional as depicted in Figure 12.

Power reduction techniques inherently involves following obvious techniques:

- Reduction in V_{dd}. Using high V_{dd} in critical path and low V_{dd} in non critical path
- Reducing switching activity via clock gating, input/state settings, proper logic synthesis
- Reducing physical capacitance via good layout an proper device sizing
- Using the minimum size transistor required for proper signal strength. Larger device would increase the operating speed but this increase the power dissipation.

Figure 12. Leakage current

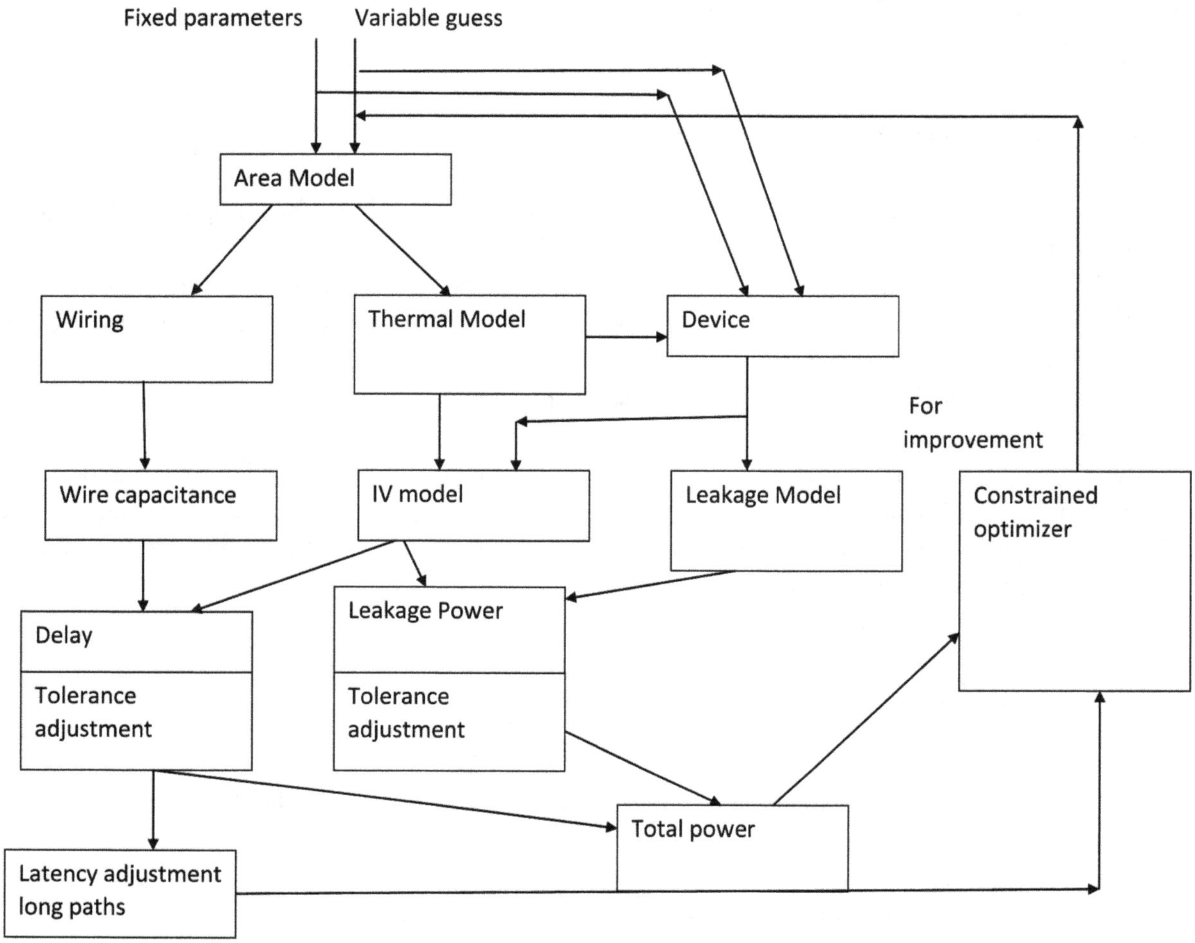

Other prevalent methodologies in use by circuit designers for power reduction are

- Parallelism
- Pipelining
- Pin ordering
- Data encoding
- Bus encoding
- Cache sub-banking
- Eliminating needless computation
- Combines circuit and architectural techniques
- clock gating
- Gate sizing
- Transistor sizing
- Dual-Vth approach
- Multi-Vth

- Dual *Vdd* technique
- Clustered Voltage Scaling (CVS)
- Extended clustered voltage scaling (ECVS)
- Glitch and hazard minimization design
- Adaptive body bias
- Transistor Stacking
- Dynamic power cutoff technique (DPCT)

Researcher reported a four parameter model for static power estimation model at architecture level as depicted by Equation 34. The parameters details are described in Table 3. It is also shown that the static power can be reduced by reducing these parameters.

$$P_{static} = V_{CC} N K_{design} I_{leak} \tag{34}$$

Static Power Estimator Model2

From the BSIM 3v3.2, the subthreshold drain current is depicted as Equation 35 V_{off} is a empirical parameter, V_t is physical parameter depending upon temperature, 'n' is multidimensional empirical constant.

$$I_{Dsub} = I_{s0}\left(1 - e^{\frac{-V_{dd}}{V_t}}\right) e^{\frac{V_{gs} - V_t - V_{eff}}{nV_t}} \tag{35}$$

$$I_{so} = I'_{so} W / L \tag{36}$$

The current I_{so} is geometry depended and can be given as in Equation 34. For 'off' state $V_{ds} = V_{CC}$ and $V_{GS}=0$. Therefore subthreshold current would be as shown in Equation 35, Where $k_{tech} = I'_{s0} e^{\frac{-V_{eff}}{nV_t}}$ and $S_T = 2.303nV_T$. S_t is subthreshold swing parameter.

Table 3 Summary of Static power model parameters

Parameter	Description	Scaling Behavior	Reducing
V_{CC}	Power supply voltage	Decreases by 30% per process generation	Multiple supply voltage, increase IPC
N	Number of transistor in design	Increase by 100% per process generation	Reduce functionality, optimized circuit implementation
K_{design}	Empirical parameter for average device	Appx constant	Use of efficient circuit style; reduce frequency
$I_{leakage}$	Technology parameter for device sub threshold leakage	Highly dependent on aggressiveness of V_t scaling	Partition device into frequency domain

$$I_{Dsub} = \left(\frac{W}{L}\right) I'_{s0} e^{\frac{-V_{eff}}{nV_t}} e^{\frac{-V_T}{nV_t}} = \left(\frac{W}{L}\right) k_{tech} e^{\frac{-V_T}{nV_t}} \quad (37)$$

$$I_{Dsub} = \left(\frac{W}{L}\right) k_{tech} 10^{\frac{-V_T}{S_t}} \quad (38)$$

For N transistors on same technology, the leakage is summation of subtrhreshold currents can be given as in Equation 37. Considering similar parameters for N and P MOSFETS, Equation 38 relates the different parameters.

$$I_{leakage} = \sum_{1}^{N} I_{Dsub} = NI'_{Dsub} = Nf\left[\left(\frac{W'}{L}\right) f_{off} k_{stack}\right] k_{tech} 10^{\frac{-V_T}{S_t}} = Nk_{design} k_{tech} 10^{\frac{-V_T}{S_t}} \quad (39)$$

$$I_{leakage} = N\left[f_N k_{designN} k_{techN} 10^{\frac{-V_{TN}}{S_{tN}}} + (1 - f_N) k_{designP} k_{techP} 10^{\frac{-V_{TP}}{S_{tP}}}\right] \quad (40)$$

Therefore the static power dissipated can be depicted as Equation 39, where k_{tech}, S_t, V_T are technology dependent parameters which can be combined into I_{leak} which is normalized leakage current.

$$P_{static} = V_{CC} Nk_{design} k_{tech} 10^{\frac{-V_T}{S_t}} \quad (41)$$

$$P_{static} = V_{CC} Nk_{design} I_{leak} \quad (42)$$

Manoj Sharma
IEEE, India

Section 1

Fundamental of Power Aware VLSI Design

Chapter 1
Low Power Design Techniques: Classical and beyond CMOS Era

Mohd Samar Ansari
Aligarh Muslim University, India

Shailendra Kumar Tripathi
Malaviya National Institute of Technology, India

ABSTRACT

Conventional CMOS based IC implementations are soon expected to attain a state of saturation mainly due to technological barriers and physical constraints. A number of nanoelectronic devices that may be viable alternatives to standard CMOS have been put forward, and these could pave way for even lower power circuits. Some worthy contenders are carbon nanotube field effect transistor, nanowire field effect transistor, tunnel field effect transistor, and the single electron transistor. In this chapter, device- and circuit-level techniques have been discussed for designing low power circuits in conventional CMOS. Moreover, various sources of power consumption have been reviewed. It is also attempted to review the major components of power i.e. dynamic, short-circuit, and leakage, followed by basics of device and circuit level techniques to reduce the power components. Further, pertinent features of emerging nano-electronic devices are discussed. These beyond-CMOS devices are expected to revolutionize the era of electronic systems with their low power & high performance characteristics.

1. INTRODUCTION

The major concerns of VLSI designers in 90's era were high speed, small area and low cost, but the issue of power somehow got less importance. However, since the past decade, the expansion of personal computing devices and wireless communication systems has forced designers to investigate low power techniques with high reliability. The portable and handheld products would exhibit lower battery life with difficulties in heat exhausting without the use of low-power design methodologies (Roy, 2009). Moreover, reliability is strongly affected by power consumption. In addition to this, high power dissipation implies high temperature operation, which could result in number of failures in the system. It could be said that, the lower the power dissipation of a product, the lesser electricity that is consumed and

DOI: 10.4018/978-1-5225-0190-9.ch001

the lower the impact on global environment. Moreover, utilizing low-power components in combination with low-power design techniques is desirable in the present time. Therefore, the requirements for low power consumption go on increasing drastically with the need of battery-powered multifunctional portable devices (Roy, 2009; Chandrakasan, 2012; Rabaey, 2012).

The motivations for reducing power consumption are application specific. Such as, for mobile phones, the objective is to maintain long battery lifetime and light weight with reasonable and low cost packaging. In the case of portable computers such as laptops, the goal is to decrease the power requirement of the electronics portion of the system to a point which is approximately half of the whole power consumption (Rabaey, 2009). Additionally, for high performance systems operated without battery such as workstations, the target of power minimization is to cut the system price while ensuring device reliability. These factors have motivated low power techniques for CMOS technology. Moreover, with technology scaling and increasing performance necessities, the power levels to the chip are rising with reduced supply voltages. This also puts a heavy demand on the battery technology which serves the actual circuit.

There are numerous techniques that have been introduced over the previous decade to tackle power reduction needs of the high performance circuits and systems. The fundamental low-power design methodologies, such as clock gating for decreasing dynamic power, or multiple voltage thresholds to reduce leakage current, are well-recognized and supported by existing tools (Horowitz, 2005; Thompson, 2006).

From the past several years, it is seen that the continued scaling in the device size of CMOS transistor leads to an increased risk of considerable power consumption, low yield, and reliability degradation in integrated circuits (Kuhn, 2012). The power consumption has turned into a critical issue in present time because of global warming. That leads to the requirement of low power circuits or even self-sustainable devices to meet the expectations of modern trends. In recent times, it is accepted that these limitations of the Si-MOSFET device require the semiconductor industry to investigate new materials and devices which are able to complement the CMOS transistor before silicon based technology reaches its limits towards the end of present decade, with the channel length of MOS approaching values under 10 nm (Skotnicki, 2005; Borkar, 2006; Marković, 2010). New ways and means to extend the saturating Moore's law are therefore needed.

Nanoelectronic devices such as carbon nanotube field effect transistor (Raychowdhury, 2007), tunnel field effect transistor (Seabaugh, 2010), single electron transistors, junctionless tunnel FETs, two dimensional devices, etc. have opened the way to the development of ultra low power circuits. The Carbon Nanotube Field Effect Transistor (CNFET) appears to be the most capable among various nanoelectronic devices from the point of their integration into future nanoelectronic transistors and circuits (Kim, 2009; Moaiyeri, 2011; Black, 2005). In this chapter, classical low-power design techniques for CMOS and futuristic design techniques for nano-electronics, such as hybrid (CMOS & CNFET) techniques, CNFET based circuit design techniques, Tunnel-FET based low-power circuit design, etc, are discussed. Additionally, a pertinent overview of recent nanoelectronic devices has also been included.

1.1 Kinds of Power Consumption in CMOS Circuits

Although the MOS technology is approaching its physical limit, yet it has significant contribution for electronic industries since last forty years. But presently in CMOS circuits, the main concern is high power dissipation due to increasing number of elements per chip area. The main contributions to the power consumption are short circuit, leakage and switching currents. The CMOS circuits have broadly two types of power consumption *i.e.* dynamic and static which can be further categorized as illustrated in Figure 1 for a standard CMOS inverter with all the leakage current marked on the diagram itself for clarity.

1.1.1 Short Circuit Power

Short circuit power is dissipated due to flow of short circuit current between the circuit's power supplies (typically V_{DD} and ground). In conventional Static CMOS designs, for stabilized input and output states, only one transistor (PMOS or NMOS) is ON and the other is turned off. The short circuit current flows in the transition state, when the input voltage changes from zero to V_{DD}. In this transition, there exists a short time interval where V_{in} is larger than V_{tn} but less than (V_{DD} - V_{tp}). In this situation both PMOS and NMOS transistors switch ON and short circuit current flows from V_{DD} to ground. For typical designs, short circuit power consumes less than 10% of total power (Chandrakasan, 2012). Figure 2 shows the graphical representation of different currents in a CMOS inverter. The short circuit power can be expressed as

$$P_{SC} = \frac{1}{12} k\tau f (V_{DD} - 2V_t)^3 \tag{1}$$

where, k is constant, τ is rise or fall time of signal, f is frequency of signal, V_{DD} is supply voltage and V_t is threshold voltage of transistor. The short circuit current flows in the following condition of input voltage, $V_{tn}<V_{in}<(V_{DD}-V_{tp})$.

1.1.2 Dynamic Power

The dynamic power is consumed due to switching currents. These currents are generated because of charging and discharging of node capacitances. Dynamic power can be expressed as

$$P = \alpha C_L f V_{DD}^2 / 2 \tag{2}$$

Figure 1. Types of power dissipation in CMOS circuits

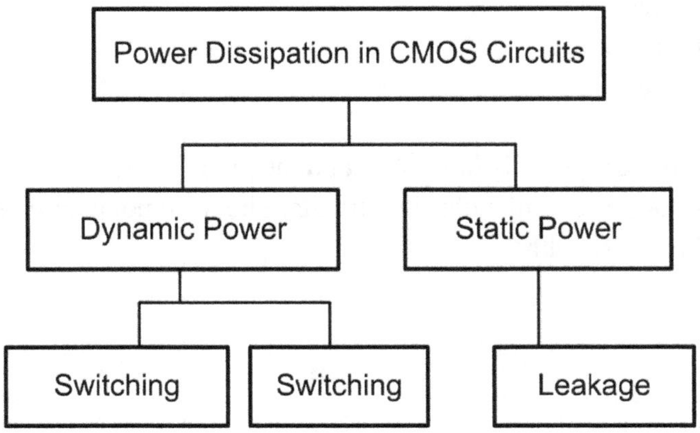

Figure 2. Various Components of Power Dissipation in a CMOS (Dynamic, Short circuit and Leakage Power)

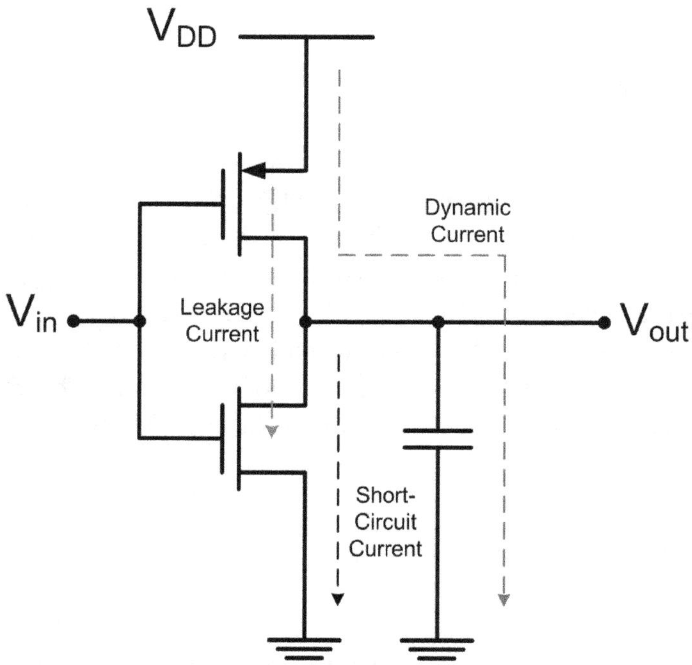

Here, f is operating frequency, V_{DD} is supply voltage, C_L is capacitance load and α is the switching activity constant. It is clear from Equation (2) that the dynamic power depends on following parameters: supply voltage, operating frequency, loads capacitance *etc.* (Roy, 2009; Chandrakasan, 2012). It is a significant portion of total power consumption. Therefore, dynamic power can, and should, be decreased by reducing switching currents, which forms one of the much used low power techniques.

1.1.3 Leakage Power

It is the unwanted power consumption by a transistor which cannot be removed completely by the present technological developments. The main factors responsible for leakage power are:

1. Reverse bias diode leakage
2. Gate oxide tunneling leakage
3. Sub-threshold leakage

Though the leakage current typically hovers around the picoampere (*pA*) range but it causes a severe power dissipation factor for present high density circuits. The most popular technique to reduce leakage current is to use multithreshold voltages.

1.2 Need of Low Power VLSI Circuits

The present era of microelectronics is shifting towards nanoelectronics. There are various factors which motivate low power methodologies. Some important factors are described below (Rabaey, 2009; Horowitz, 2005; Thompson, 2006).

1. **Issue of Scaling:** The current trend is continued miniaturization of transistor sizes. Each upgrade of integration technology, produces faster, smaller and increased number of transistors than its predecessor. Moreover, when these improved small transistors are placed on a chip, it further improves the operating frequency. The overall result is in the form of improved functional capability and increased power requirements. To sustain such ICs, one has to think about low power design methodologies.
2. **High Demand of Portable Devices:** Portable devices which operate on battery power are now in high demand. The penetration of consumer electronics equipment like laptops, mobiles, tablets etc. is increasing day by day. The most essential features of this equipment are smaller size, lighter weight and longer battery lifetimes. This trend forces the circuit designers to investigate and design low power techniques.
3. **Issue of High Operating Temperature:** The numbers of transistors in a chip are increasing drastically with time. This in turn implies that the component density also increases and leads to large power dissipation. Due to this operating temperature also increases which pushes up the packaging and cooling cost of these chips. This factor also raises concerns about the reliability issues of the devices.
4. **Environmental Issue:** Presently, due to high density ICs with increasing operating frequency and functionality, an enormous amount of heat is dissipated in the chip. This leads to a high impact on environment by the emission of CO_2 and other harmful gases. It is obvious that low power circuits generate low heat and need less cost of cooling which helps the environment conditions. These challenges can motivate the engineers to design ultra-low power and self-sustainable circuits.

1.3 Strategies and Design Flow for Low Power Methodologies

The low power design techniques need to be properly partitioned on the basis of the abstraction levels. Table 1 represents the various strategies that fall adequately on specific design level. For example, portioning is applied in the operation at system level, whereas regularity is more suited to apply at the software level. Similarly, pipelining and redundancy are more apt to be applied at the architectural level and strategy of transistor sizing could be efficiently fitted in circuit level and so on. Other strategies have been listed in Table 1.

The low power techniques would be less effective if these techniques are not placed in a proper design flow/sequence. The broad steps and their respective design flows have been categorized in Table 2. The process starts with definition and goes through verification, synthesis, physical implementation, checking and signoff. The corresponding design flow should be applied for better performance and time efficiency. For instance, it would not be appropriate to fine tune the Synthesis part of the design before the Verification part has been completely achieved. If such an out-of-sequence of optimization steps is attempted, the final design, in all probability would not be the best design.

Table 1. Strategies for low power techniques

Abstraction Level	Strategies
Operating System level	Portioning, Power down
Software level	Regularity, locality, concurrency
Architecture level	Pipelining, Redundancy, data encoding
Circuit/Logic level	Logic styles, transistor sizing and energy recovery
Technology level	Threshold reduction, multi-threshold devices

Table 2. Illustration of low power design flow

Steps	Design Flow
1. Definition	Define power domains and logical power intent
2. Verification	Simulation of power behavior (up/down) and multi-voltage static rule check
3. Synthesis	Top-down approach for multi-supply, multi-V_{th} synthesis
4. Physical Implementation	Power switch planning, clock gating
5. Checking	Clock tree synthesis and routing
6. Signoff	Low power formals and rule checks, power analysis

Low power design methodologies should be applied throughout the complete design process from system level to layout level abstraction, gradually refining the abstract specification. Further power optimization approaches at the higher level are more significant than lower levels because research results represent that higher level of abstraction has greater potential for power reduction as depicted in Table 3.

2. BASIC METHODOLOGIES OF LOW POWER CMOS DESIGN

There are various techniques which have been proposed to reduce power consumption in CMOS circuits. Some of them are reduction in supply voltage, frequency and capacitance, minimization of leakage currents, clock gating and multi-V_{DD}. These techniques deserve attention of design engineers due to their effectiveness, and are reviewed in depth in the current section.

2.1 Reduction in Switching Voltage

The component of dynamic power is a major factor of total power consumption. The expression of dynamic power reveals that it includes mainly three terms *i.e.* voltage, capacitance and frequency (2). From the expression, it can be seen that voltage has quadratic effect on dynamic power (2). So, by scaling supply voltage, power consumption can be dramatically reduced. Though, it also has some performance

Table 3. Power reduction at different levels

Abstraction Level	Power Reduction (%)
Behavior/Algorithm	90-99
Architecture	10-90
Logic	15
Circuit	30
Layout	20

trade-offs. The MOS transistors become slower at lower supply voltages, which results in decreased device performance (Roy, 2009; Chandrakasan, 2012; Rabaey, 2012; Rabaey, 2009). This is primarily because the threshold voltage of the transistors do not scale appropriately with reduction in allowable supply voltages. Moreover, noise immunity is also an issue at lower supply voltage swing.

2.2 Reduction in Capacitance

Generally, in digital circuits it is endeavored to reduce the parasitic capacitance in order to minimize power dissipation and enhancing performance. It has been investigated in the literature that nodes with high parasitic capacitance could not efficiently switch at higher frequencies (Chandrakasan, 2012). Such a parasitic optimization can be applied in different level of abstractions like physical design, circuit design, process technology *etc.*

2.3 Reduction in Switching Frequency

The operating frequency of communication systems is increasing day by day but there are some issues of reliability and increased power requirement. Thus, there is a method to obtain trade-off between power and frequency for the communication systems. It has been investigated that lower operating frequency can be applied to signals with large capacitance. Another way to decrease operating frequency is to eliminate logic switching, which is not so important for computation (Roy, 2009; Chandrakasan, 2012; Rabaey, 2012).

2.4 Reduction in Static and Leakage Currents

Though the portion of power generated by these currents is low but it is better to control the static and leakage currents too. Transistor sizing helps to reduce the leakage current with effective layout and appropriate circuit design techniques (Rabaey, 2012).

The leakage current is generated due to reverse bias junction and sub-threshold current. Figure 3(a) represents the current in reverse bias junction that makes a part in leakage power of the device. The sub-threshold current has been depicted in Figure 3(b) which is generated during sub-threshold conduction in a MOS transistor (Marković, 2010). Basically, leakage power exists in low frequency circuits. It can be reduced effectively in process and device level abstractions. In the process level, buried oxide layer and n/p-well processes are used to minimize the problem.

2.5 Implementation of Multi-Level Thresholds

As discussed earlier, reduction in supply voltage is a very common method to reduce power dissipation but it needs a simultaneous lowering of the threshold voltage of the transistor (V_t). This results in the exponential rise of sub-threshold leakage current. However, multi-threshold CMOS (MTCMOS) is a valuable circuit-level method that gives a high performance and low leakage design approach. A method of designing devices with multiple thresholds is to initiate the transistors at the standby mode to segregate the power supply (Rabaey, 2009; Horowitz, 2005).

Figure 3. Representation of (a) Reverse biased current (b) Sub-threshold current

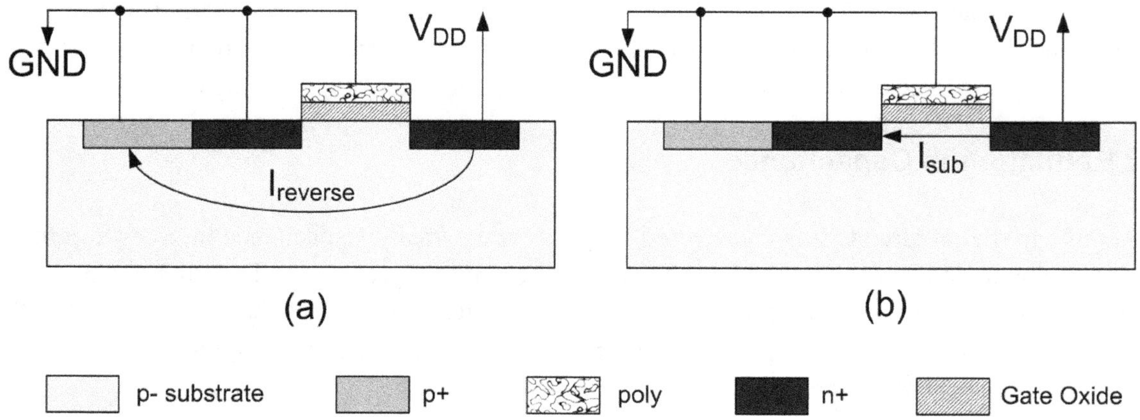

Figure 4 shows the general structure of a multi-threshold CMOS logic. The basic theory of the multi-level threshold method is to introduce low-V_t transistors to logic gates for which the switching speed is important. Additionally, the high-V_t transistors (sleep transistors) are used to successfully separate the logic gates in the standby state and decrease the leakage power dissipation. The complete design has various abstraction levels. So there are different approaches for low power design in the different level of abstraction as summarized in Table 1.

Figure 4. General structure of a multi-threshold logic

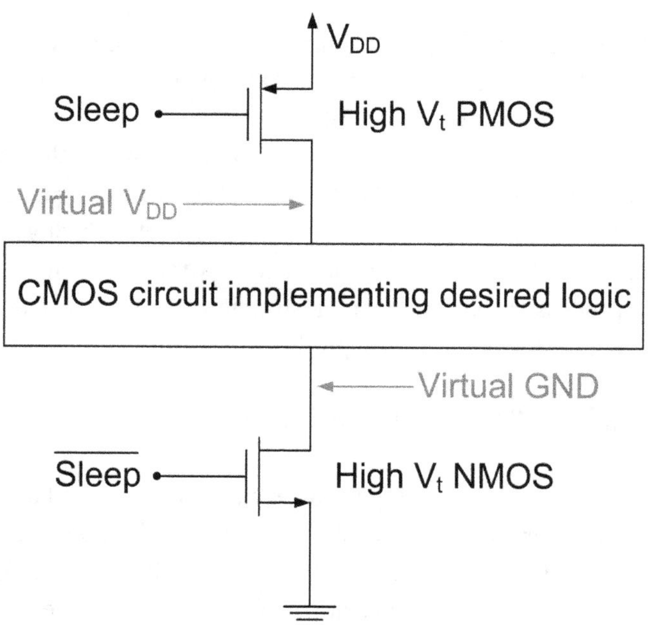

2.6 Clock Gating

Clock gating is an efficient dynamic RTL power reduction technique in digital synchronous systems. When there is no change in input data activity, then the clock signal may be gated to switch off using a clock enable/disable signal. Clock gating is done using a simple AND/OR gate which continuously compares the output and data input. The output is glitch free if enable signal is generated by a negative-edge triggered Flip Flop (FF) which ensures that the signal is changing after the falling edge of the clock signal (Roy, 2009; Chandrakasan, 2012). Similarly, while clock gating is done using an OR-gate, clock signal is generated by a positive-edge triggered FF in order to glitches being passed onto the FFs.

Clock enable signal is asserted if there is a change in data input activity. Setup time and data-to-output delay significantly increases due to the extra circuitry, but the overall energy consumption reduces due to less switching activity. Libraries are available which include specific clock gating cells that are automatically picked up by the synthesis tool (Rabaey, 2009; Horowitz, 2005). The combination of explicit clock gating cells and automatic insertion is being used for power minimization. The types of clock gating are listed below:

1. Global clock gating
2. Local clock gating
3. Asynchronous clock gating
4. Synchronous clock gating

2.7 Multi- V_{DD} Approach

The multiple V_{DD} approach gives a high-V_{DD} supply for high performance circuits and a low-V_{DD} supply for low performance circuits. For example, in a dual V_{DD} system, the lower voltage (Low-V_{DD}) is supplied to the circuit on non-critical paths, whereas the higher voltage (High-V_{DD}) is fed to the circuit components lying on critical paths. As the critical path of the design is unaffected, this technique conserves the performance of the circuit. Figure 5 depicts a pipeline stage of multi-V_{DD} system in which some of the paths have low-V_{DD} gates while some have high-V_{DD} gates (Roy, 2009; Chandrakasan, 2012; Rabaey, 2012; Rabaey, 2009; Horowitz, 2005). It is important to know that high-V_{DD} gates can drive low-V_{DD}, but not vice versa. For low to high-V_{DD} transition a level converter is used in the circuit.

It is important to review the low power techniques and their impact on different domains of the power and design. Here, three outcomes of low power techniques have been discussed over three domains such as dynamic power, static power and design impact. From Table 4, it can be observed that techniques like clock gating, multi-V_{DD}, and DVFS are much better for reduction in dynamic power. Similarly, low power techniques such as power gating and multi-V_{th} are more beneficial in static power reduction. Moreover, multi-V_{th}, power gating and DVFS have good impact on design.

2.8 CAD Tools for Low Power Techniques

The role of CAD tools is very important in the complete VLSI design cycle. One primary objective of CAD tools is to reduce the risk of failure at the time of porting the circuit to Silicon. These tools simulate the performance of a circuit before its actual fabrication. Moreover, CAD tools investigate various design alternatives and characterize them in terms of cost, performance with check against all possible

Figure 5. Representation of a Multi-V_{DD} system

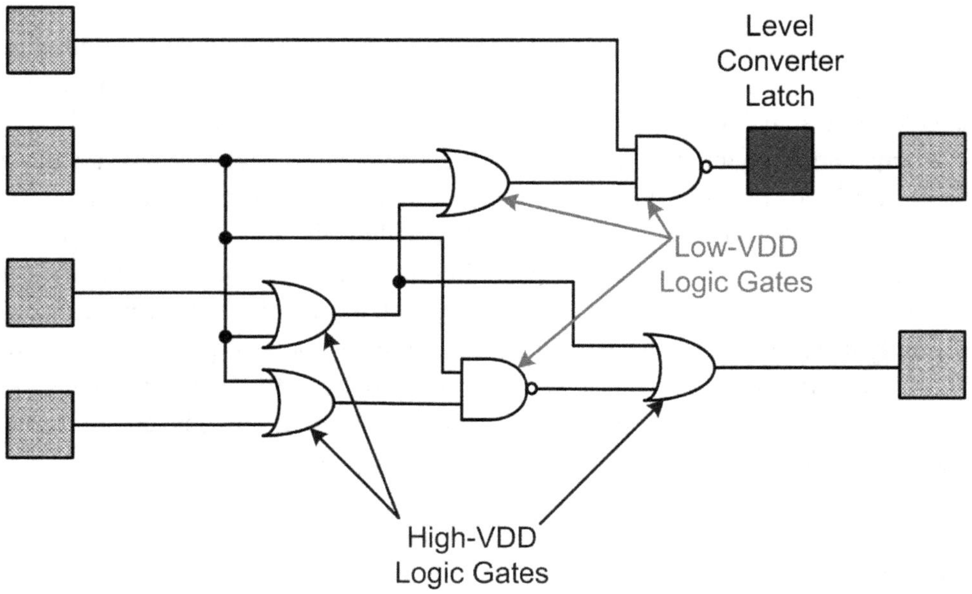

Table 4. Impact of different low power techniques on various domains

Technique	Benefit in Dynamic Power	Benefit in Static Power	Design Impact
Clock Gating	Large	Small	Small
Multi-V_{DD}	Large	Small	Little
Muti-V_{th}	Small	Large	Medium
Power Gating	Small	Very Large	Medium
Dynamic Voltage & Frequency Scaling (DVFS)	Large	Small	Medium

known fabrication process violations before final data is given to foundry for manufacturing. Low power techniques require various CAD support for power optimization of a specific design (Rabaey, 2012). However, a number of CAD tools are offered for this task from various leaders in market like Synopsys, Cadence, Sequence Design, Magma and Atrenta. All of them have developed and provide low power CAD tools but of course, their functionality may have some slight variations according to need of customers (Huijsing, 2013).

Synopsis proposed a CAD tool named 'Eclipse' which has a comprehensive approach, power-aware tools for various levels of design abstraction and hierarchy. It starts from architectural and system level analysis to verification. Then RTL synthesis tests the physical implementation and then finally Synopsis is invoked. It is not really a single tool, but more appropriately a suite of tools starting from design intent to sign-off. Similarly, Atrenta presented 'Spyglass power' CAD tool for power optimization at RTL level of design abstraction. Cadence system has introduced various CAD tool for power optimization like 'Allegro' (Sigrity power aware), 'Virtuso power system' and 'Design compiler' etc.

The CAD tools boost power savings during implementation of the design. Further, there are four level of abstractions such as logic synthesis level, system level, physical design level and circuit level. The power optimization can be applied through CAD tools over these levels of abstractions.

1. **System Level:** In this level, the hardware modules which are inactive could be turned off automatically in order to save power. A system designer should provide minimum input voltage to modules and place them with level converters. Thus use of this technique will cycle back the energy to save power.
2. **Logic Synthesis:** It a very important level of abstraction in concern with power minimization. The logic synthesis fits between register transfer level and net-list of gate specification. Various techniques are employed for the transformation and optimization of RTL description depending upon input target implementation, delay model and objective functions. Subsequent to the system level, architectural realization and technological choices are made.
3. **Physical Design Level:** The physical design level lies between net-list gate specification and layout, and depends on target design technology, packaging technology and objective functions. Different power optimization technologies are applied for partition, place, resize and route gate. Moreover, under zero-delay model switching action of the gate is invariable during layout optimization. Therefore, to reduce the power it is prudent to reduce the load on high switching activity gate by suitable net-list portioning, gate placement, and/or gate & wire resizing.
4. **Circuit Level Design:** This design utilizes adiabatic switching rule in which speed is traded for low power. Some other techniques are based on partial transfer of energy stored on a capacitance to a charge sharing capacitance, may also be used. However, DC/DC level converters and energy efficient level converters are required for circuit level design power reduction.

3. LOW POWER NANOELECTRONIC DEVICES

The continued scaling of MOSFETs raises several critical challenges in the performance of a transistor. Moreover, with technology approaching to sub-20 nm regime, performance issues like current tunneling effects (gate and source - drain tunneling), short channel effects and large power & heat dissipation begin to crop up in the MOS transistor. There are large numbers of nanoelectronic devices that may probably replace present CMOS based devices and could open a way to low power nanoelectronic circuits. Some of the worthy contenders include carbon nanotube transistors, nanowire-FETs, tunnel-FET, single electron transistors (SET) *etc*.

3.1 Emergence of Nanoelectronics

The physical constraints of device scaling dictate a shift in the elementary building blocks of IC technology. However, in the midst of the present research and development of nanotechnology, emerging nano-electronics can provide an answer to decrease fabrication cost and permit further device size scaling. The essential elements of present integrated systems are silicon based transistors and copper/aluminum wires (Borkar, 2006; Marković, 2010; Raychowdhury 2007). These wires may be replaced by carbon nanotubes (CNTs) and nanowires (NWs), and the transistors can be replaced by carbon nanotube transistors, thereby resulting in Nanowire-FETs which are suitable to nanoelectronics. Other emerging nanoelectronic devices are tunnel field effect transistor (TFET), single electron transistor (SET), quantum cellular automata (QCA) etc (Kim, 2012; Moaiyeri, 2011; Black, 2005).

Additionally, the CNT is used as a transistor component in the carbon nanotube field effect transistors (CNFET). It looks like MOS transistor with the channel material (silicon) has been replaced by a carbon nanotube. The carbon nanotube-FET shows potential characteristics to be exploited in futuristic low power and high performance circuit design (Seabaugh, 2010; Kim, 2009; Moaiyeri, 2011).

Active devices, such as field effect transistor (FET), can be implanted with nanowire by changing the doping at the time of fabrication. In addition to this, a nanowire can be applied as the channel material in the conventional field effect transistors that can be related to the CNFET. The benefits of Nanowire-FETs over MOS are quite analogous to CNFET. However, Nanowire-FET has the capability to work as high speed and high performance device (Black, 2005).

Moreover, tunnel field effect transistor (TFET) is very capable successor of MOSFET because of its promise of sub-60mV/decade sub-threshold swing (Khatami, 2009). This reduced swing is a significant need for ultra-low voltage, ultra-low power next generation integrated circuits.

3.2 Challenges of Nanoelectronic Devices

These nanoelectronic devices exhibit superior performance than conventional CMOS. The existing CMOS design rules and CAD tool cannot be applied for the design of emerging nanoelectronic devices. There may be several reasons, of which few are listed below:

1. Some nanoelectronic devices may require entirely different & novel logic forms than Boolean laws.
2. The defect rate of emerging devices could be higher than conventional CMOS.
3. These devices will require sophisticated CAD tools beyond the basic CMOS physics.
4. Fabrication issue is also a major concern for futuristic nanoelectronic devices.
5. Investigation of the appropriate computational models for nanoelectronics.
6. Partial use of current design methodologies for emerging devices.
7. The efficient exploitation of advantages of nanoelectronic devices for various abstraction levels.

3.3 Basics of Carbon Nanotubes

Carbon Nanotubes (CNTs) are cylindrical structure of Graphene with outstanding electrical, thermal, and mechanical properties. They have commonly two forms:

1. Single-walled carbon nanotubes (SWCNTs) and
2. Multi-walled carbon nanotubes (MWCNTs) as shown in Figure 6(a).

The SWCNTs are tubes of Graphene rolled in cylindrical shape with diameter of 1-2 nm and length a few microns (0.2 - 5μm). The MWCNTs appear as coaxial tube with layers of single walled CNTs (Chen, 2004; Javey, 2003). The diameter of MWCNTs range from 2-25 nm with inter layer spacing (pitch) about 0.36 nm. The performance of SWCNTs is better to MWCNTs, yet they are costlier while MWCNTs are easy to fabricate in large capacity.

Figure 6. (a) Single & Multi-walled Structure of CNTs (b) Different Types of Carbon nanotubes

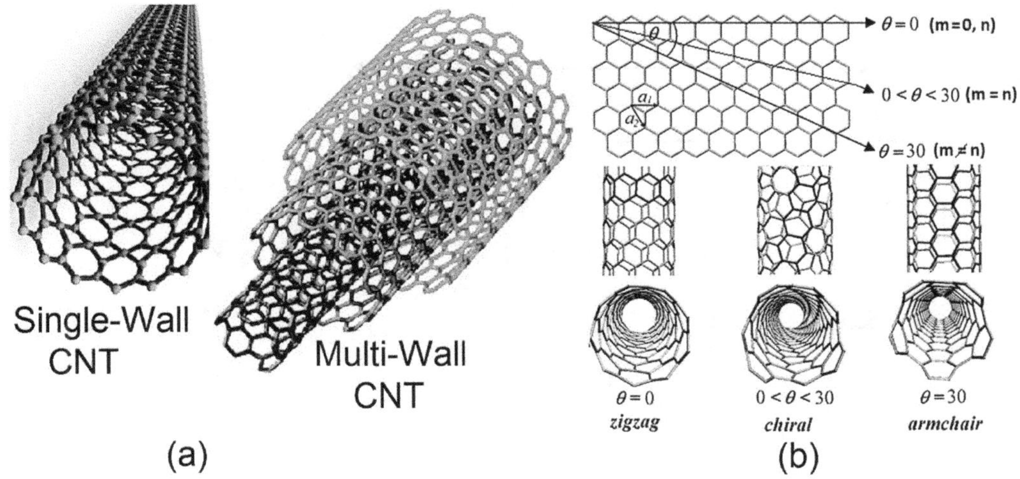

The CNTs have three types of structures, based on chiral vector (C) and chiral angel (θ). A carbon nanotube is identified as zigzag, if m = n and $\theta = 0^0$. If chiral vector $m \neq n$ and chiral angel lies between 0^0 and 30^0, CNT is said to be chiral type and if m = 0 and $\theta = 30^0$ called armchair CNTs as illustrated in Figure 6(b).

3.4 Carbon Nanotube Field Effect Transistor (CNFET)

Japanese scientist Sumio Iijima discovered the Carbon Nanotube in 1991 (Iijima, 1991). CNTs are cylindrical Graphene's nanostructures with excellent electrical, thermal, and mechanical characteristics. A CNT can be semiconducting or metallic as per the arrangement of carbon atoms in the CNT's twist. Moreover, a CNT is said to be metallic if little amount of energy is needed to excite an electron into an unfilled excited state. Alternatively, if a finite energy gap exists between the occupied and vacant states, the CNT is known as semi-conducting. The CNTFET have several distinctive properties over MOS such as higher mobility in semiconductor tube, higher drive current and easy incorporation of high-K dielectrics. Based on SWNT, carbon nanotube-FET demonstrates similar functionality to their silicon MOSFET counterpart. These circuits have benefit of low leakage current and good switching characteristics. Moreover, for fabricating a CNFET, an oxide film is placed in between a conductive gate and the SWNTs that act as the channel of the nano-device. Additionally, metal contacts are made on either end of the SWNTs to act as the source and drain. The CNT based transistors are also compatible with present MOS technology fabrication processes.

Figure 7 shows the top view of carbon nanotube field effect transistor. The main design parameters of carbon nanotube-FET, are number of CNTs in a transistor (N), inter-CNT space (S) and diameter of carbon nanotube (D_{CNT}). The gate length and width are represented by L_{gate} and W_{gate} respectively. The CNT's diameter (D_{CNT}) and threshold voltage (V_{th}), the width of CNFET-based transistor (W), number of CNTs in channel (N), inter CNT spacing (S) and energy gap (Σ_g) are related by Equations (3) through (6).

Figure 7. Top View of Carbon Nanotube Field Effect Transistor

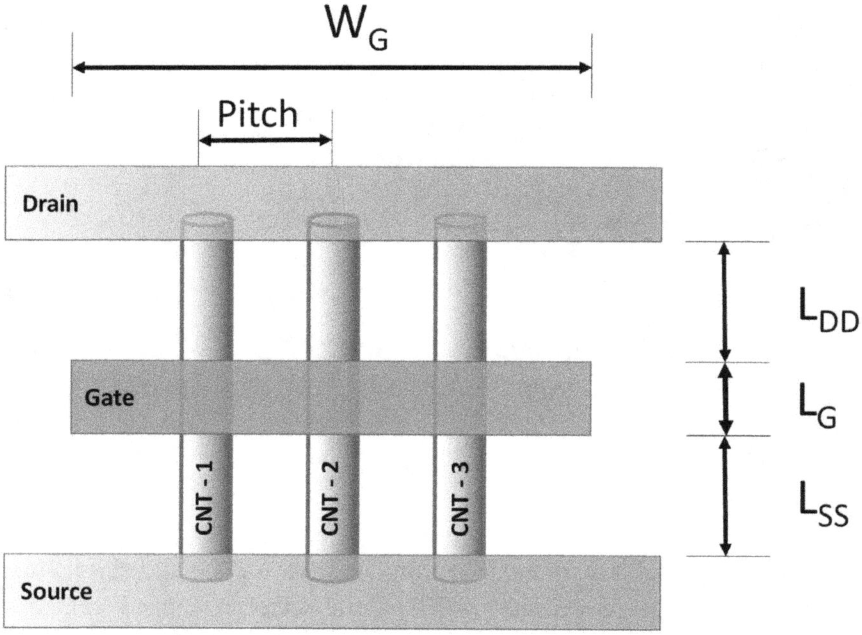

$$D_{CNT} = a\frac{\sqrt{n^2 + m^2 + nm}}{\Pi} \tag{3}$$

$$V_{th} = \frac{aV_{\Pi}}{qD_{CNT}\sqrt{3}} \tag{4}$$

$$W = (N - 1)*S + D_{CNT} \tag{5}$$

$$\Sigma g = \frac{0.84eV}{D_{CNT}} \tag{6}$$

Here, terms m and n are the indices of chiral vector of Graphene lattice and a = 2.49A° (lattice constant). Also, q is electronic charge and V = 3.033 eV, is the Carbon-Carbon bond energy. It is desirable to know about the relation between drain current and terminal voltages of a transistor. The drain current can be expressed in term of charge mobile densities (ξ_S & ξ_D) as given in (7).

$$I_D = \frac{4qkT}{h}[\ln(1 + \exp(\xi_S))\ln(1 + \exp(\xi_D))] \tag{7}$$

The charge mobile densities (ξ_S & ξ_D) which are illustrated in (8) and (9). Moreover, the charge mobile densities of CNFET are closely related to surface potential (ϕ_S) as given in (10).

$$\xi_S = \frac{(\phi_S - \delta_1)}{V_T} \tag{8}$$

$$\xi_D = \frac{(\phi_S - \delta_1 - V_{DS})}{V_T} \tag{9}$$

$$\phi_S = \frac{V_{GS} - qN_{mobile}}{C_{ox}} \tag{10}$$

Also, (11) shows the relation among diameter of nanotube (d), oxide thickness (t_{OX}) and oxide capacitance (C_{OX}) of CNFET with some constants. The relation of thermal voltage is given in (12).

$$C_{ox} = \frac{2\Pi\varepsilon_0\varepsilon_{ox}}{\ln(2t_{ox} / d) + 1} \tag{11}$$

$$V_T = \frac{kT}{q} \tag{12}$$

Here, δ_1 is the equilibrium sub-band minima of first sub-band (EG/2). The constant terms q, k, h and T represent charge of electron, Boltzmann's constant, Planck's constant and absolute temperature receptively, while V_{GS} and V_{DS} are the gate-source and drain-source voltages of the CNFET model.

3.5 CNFET Based Low Power Circuits

Carbon Nanotube-based FET (CNTFET) is similar to a field-effect transistor, which uses one or more CNTs as medium of carrier transport in the place of silicon channel. Theoretically, carbon nanotubes are also capable to conduct heat almost as diamond or sapphire and due to their miniaturized size; the CNTFET should switch consistently using much less power than a silicon based MOS device. There are several circuits which have been proposed using CNFET; some of them are discussed in present section.

3.5.1 CNFET Based Current Conveyor

Current conveyors (CCs) are gaining a lot of research attraction for their advantages like wider bandwidth, larger dynamic range, simpler circuit *etc*. It is a very versatile building block and can be used to realize voltage-mode as well as current-mode circuits. Here a CNFET based inverting current conveyor has been discussed. Inverting current conveyor (ICC-II) is also a variant of current conveyor family (Tripathi et al., 2014). The symbol of inverting current conveyor and its node relations are illustrated in Figure 8.

The CNFET-based implementation of the ICC-II is introduced, and is presented (Tripathi et al., 2014) in Figure 9. The performance of the circuit of Figure 9 was tested through HSPICE simulation with 32 nm CNFET model parameters. The power supplies were kept at ± 0.9V. The AC, DC and transient characteristics of node voltages and currents have been verified.

Here, power optimization has been considered on the basis of number of CNTs in a transistor given in Figure 10. It has been investigated that as the number of CNTs are increased, the power dissipation also increases. Moreover, by using six CNTs in this design, the average dissipated power has been estimated 53.66 μW (HSPICE Simulation). This optimization provides better current and voltage bandwidths and work efficiently up to 5 GHz. The change in number of CNTs and estimated power dissipation as calculated by HSPICE has been given in Figure 10. The circuit has further been used to design an active filter which can be used for ZigBee applications (Tripathi et al., 2014).

Figure 8. (a) Circuit symbol of ICC-II (b) Representation of port relations

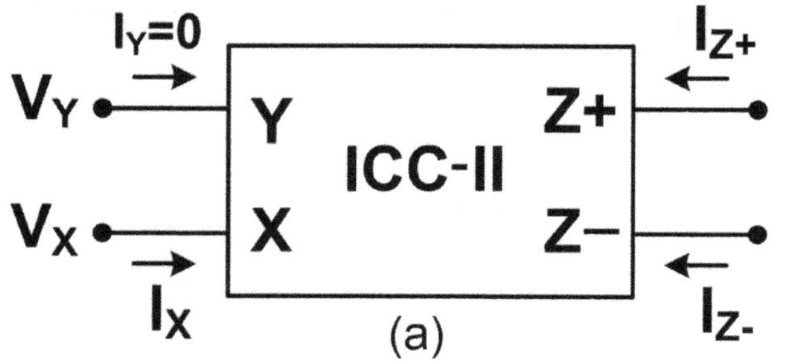

$$\begin{bmatrix} I_Y \\ V_X \\ I_{Z-} \\ I_{Z+} \end{bmatrix} = \begin{bmatrix} 0 & 0 & 0 \\ -1 & 0 & 0 \\ 0 & -1 & 0 \\ 0 & 1 & 0 \end{bmatrix} \begin{bmatrix} V_Y \\ I_X \\ V_Z \end{bmatrix}$$

(b)

Figure 9. Transistor level realization of ICC-II using CNFET

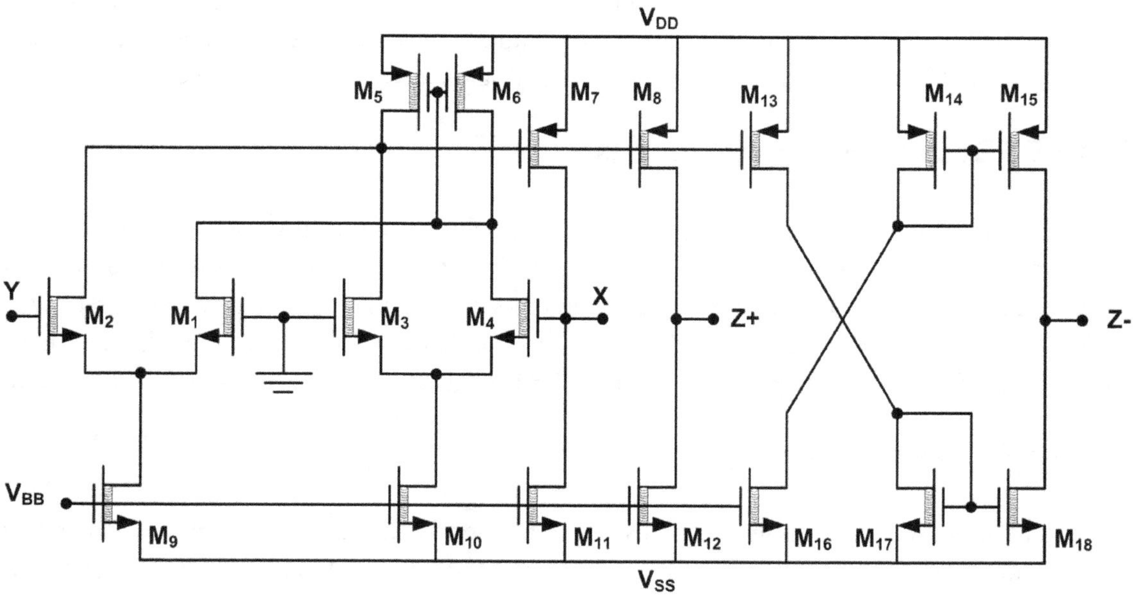

Figure 10. Average power dissipation of CNFET based current conveyor

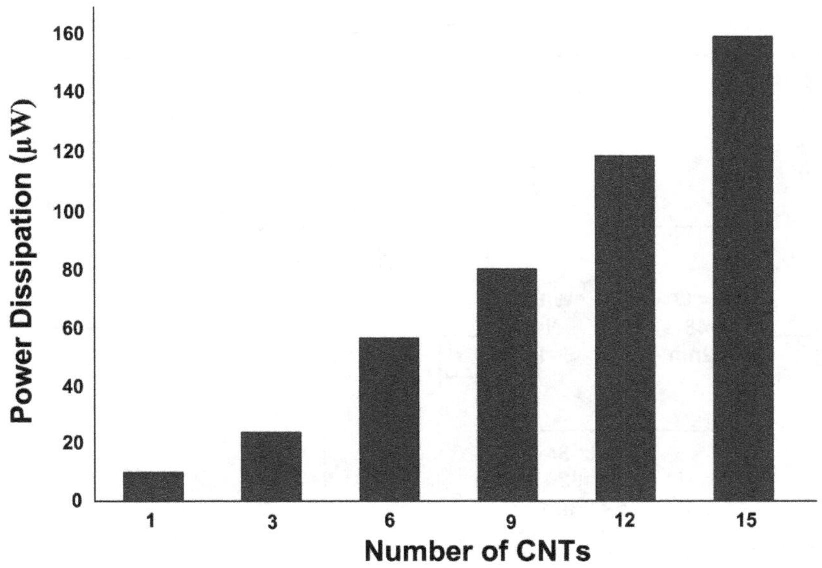

Some other variants of CNFET based current conveyor have also been discussed in literature (Imran *et al.* 2012; Sharma J. et al. 2014). Imran *et al.* (2012) proposed a CNFET based second generation current conveyor with optimization of bandwidth, power, gain *etc.* A current controlled current conveyor based on CNFET has been presented (Sharma J. et al., 2014) with application to a resistor-less filter. The design is well suited to low power high performance current mode circuits.

3.5.2 Optimized Inverting Amplifier using CNFET Op-Amp

Due to continuing miniaturization in the size of MOS transistor, various challenges came into picture. As technology approached the nano-meter range, device performance issues such as short channel effects, current tunneling effects and increased heat & power dissipation in MOS transistors became dominant. Designers and industries are investigating for new materials and devices to overcome the challenges of current MOS technology.

A low power operational amplifier has been proposed (Usmani et. al., 2010) which uses carbon nanotube field effect transistor as active elements. Figure 11 shows a two stage Op-Amp with Miller compensation. This design is further extended to an inverting amplifier. For a better understanding of the circuit, the design has been tested for pure CNFET and hybrid (CMOS & CNFET) inverting amplifiers. The results are quite satisfactory with hybrid design exhibiting better frequency response and CNFET amplifier returning good amplification characteristics. The average power dissipation has been also tested with CNFET and hybrid designs. Figure 12 shows the average power dissipation with the distinct design parameters (No. of CNT, Diameter of CNT and Pitch) of CNTs for different technology variants. In some cases [Figure 12(a)] hybrid designs are superior to pure CNFET design from power dissipation viewpoint.

Figure 11. CNFET based two stage loaded CMOS op-amp with Miller compensation

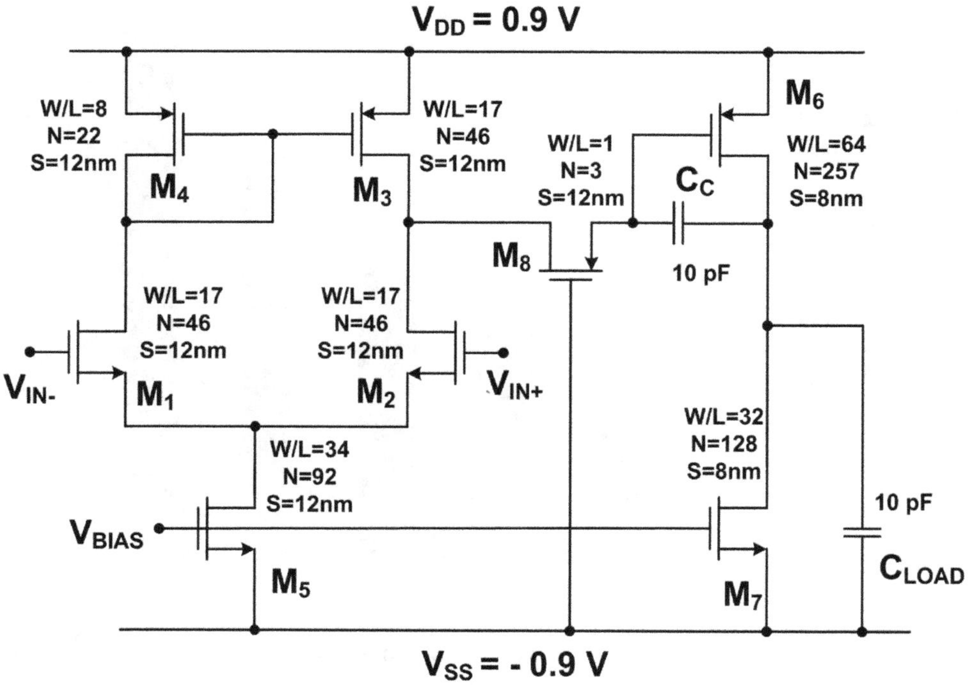

Figure 12. Representation of average power dissipation (a) Based on number of CNT (b) Based on CNT diameter (c) Based on pitch of CNT

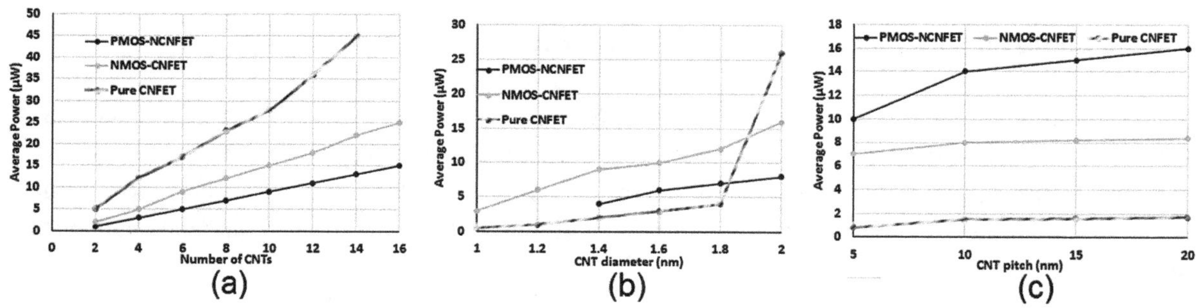

3.5.3 CNFET based Low Power OTA

The Operational Transconductance Amplifier (OTA) is an analog building block which accepts differential voltage inputs and produces an output current proportional to the input potential difference. It is widely used in integrated amplifier and filter applications. There are also some specific discrete applications where the device can be used such as low power biomedical applications. With the use of carbon nanotubes in the transistors, OTA circuit has been also realized. A purely CNFET based operational transconductance amplifier is proposed by Tanísia Possani *et al.* (Possani, 2012). Figure 13 presents a two stage Miller OTA whose performance is better than CMOS.

Figure 13. Schematics of a two-stage Miller operational transcondutance amplifier

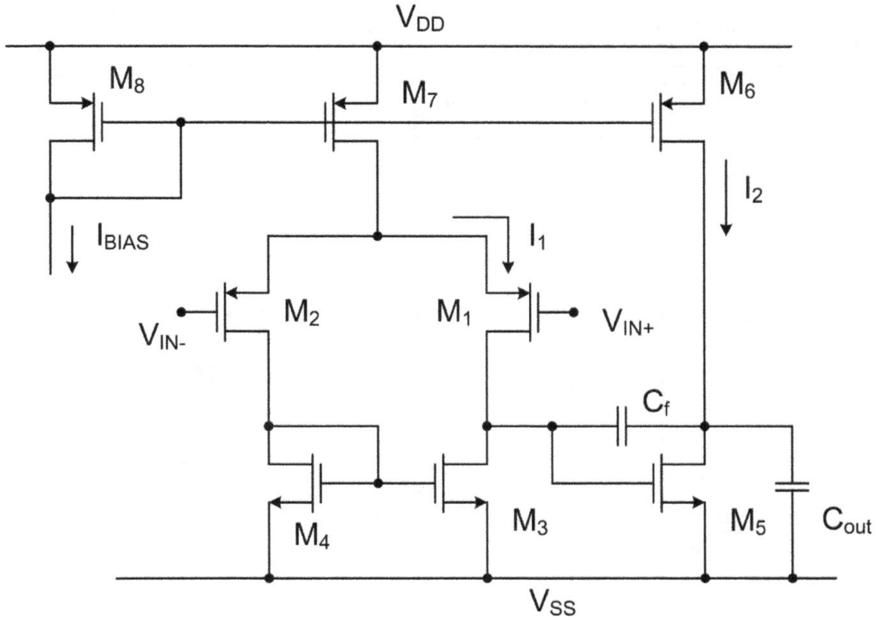

Here, two versions of the Miller's OTA have been implemented *i.e.* CNTFET and CMOS. Both designs were realized with UCAF (simulation-based optimization engine) tool with same specifications, while predictive models for CMOS and CNTFET transistors were used for the 32nm technology. Synopsys HSPICE is used as an electrical simulator. The comparison of performance has been illustrated in Table 5. Investigations show that CNFET based design consumes about half the power as compared to CMOS.

3.6 Analysis of Tunnel Field Effect Transistor (TFET)

TFET is another promising active device due to not only its strong similarity with the MOSFET configuration but also because it allows significant use of the existing MOSFET fabrication expertise in TFETs based VLSI circuits. Because of their potential for sub-60 mV/decade sub-threshold swing, TFETs are suitable for ultra-low power, ultra-low voltage and high speed operation of next generation VLSI circuits (Mukundrajan, 2012). A simplified structure of tunnel-FET shown in Figure 14(a), it consists of highly doped *p* and *n* regions and undoped middle region. Moreover, the characteristic between drain current and gate voltage has been illustrated in Figure 15, from where it is clear that TFET has sub-threshold slope less than 60 mV.

Table 5. Performance comparison of CMOS and CNFET based OTAs

Parameters	CMOS Design	CNFET Design
Slew Rate	1 V/µS	2.17 V/µS
BW	1.42 MHz	5.63 MHz
Gain	60 dB	51.63 dB
Power Dissipation	43.61 µW	26.31 µW

Figure 14. (a) Schematics of tunnel field effect transistor (b) Equivalent circuit

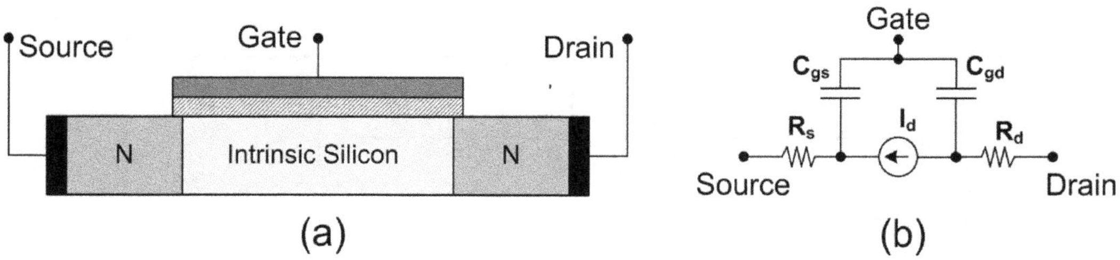

Figure 15. $I_D - V_G$ Curve of Tunnel FET with sub-threshold slope less than 60 mv

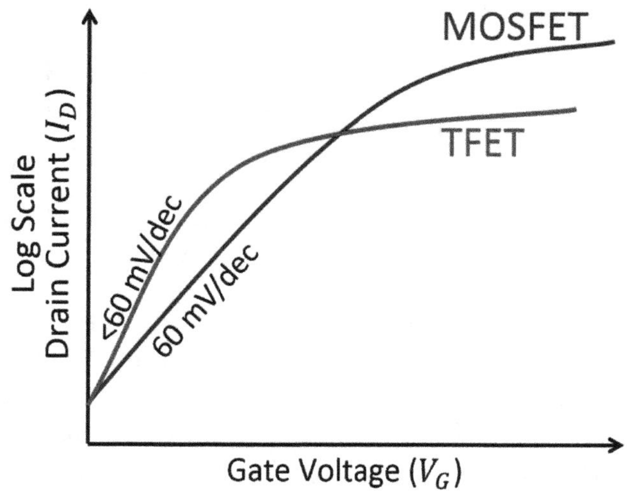

Since 1980s transistor action has been investigated in gated diode structures. These structures consist of a heavily doped p-region and a heavily doped n-region with an intrinsic or undoped region in between and are called p-i-n diodes. The fundamental characteristic of these gated *p-i-n* (diode) transistors has been observed to be Zener tunneling, (Pal et al., 2011; Boucart et al., 2007) called as Band-to-Band tunneling (BTBT) or inter-band tunneling. Due to this, the device became popular as Tunnel Field-effect transistor (TFET). Like the MOSFET, TFET has three terminals: source, gate and drain (but not bulk/body/substrate).

Distinct from conventional MOSFETs, where an inverted region (channel) is created to allow conduction, TFETs operate by creating sufficient amount of potential between the source and channel, so that charge carriers can tunnel from the conduction band of one region to the valence band of the other.

Compared to MOSFET, TFET has several advantages as follows

1. Suitable for low power applications due to lower leakage current
2. Better immunity to short channel effects
3. Sub-threshold (SS) swing not limited to 60mV/decade
4. Enhanced operating speed due to tunnelling
5. Smaller threshold voltage roll-off
6. Higher on/off current ratio

3.6.1 Principle of Operation

In equilibrium state, the built-in potentials of the *p-i* and *n-i* junctions result in a staircase-like band-profile. If a small V_{DS} is applied to the equilibrium state, electron and hole currents are blocked by the built-in potential barriers. This is the off-state of the TFET (Mukundrajan, 2012; Pal et al., 2011; Boucart et al., 2007). Now, if a negative gate-bias is applied, the bands in the channel move up. For negative V_{DS}, charge carriers can tunnel through the band gap at the source/channel junction as soon as the valence band in the channel is lifted above the conduction band in the source. This operating mode is the p-channel on-state of the TFET because holes accumulate in the channel. The band diagrams of on/off states are illustrated in Figure 16.

3.6.2 Equivalent Circuit and Estimation of Drain Current in a TFET

The equivalent circuit of TFET [given in Figure 14(b)] has a voltage controlled current source I_d to model the drain current, capacitors C_{gs} and C_{ds}, and source and drain series resistors R_S and R_d. Further, the asymmetric source/drain junction consequences in asymmetric character both as a function of drain-source bias (V_{ds}) and gate-source bias (V_{gs}). In the negative drain-source bias polarity, $V_{ds} < 0$ with positive V_{gs}, band-to-band tunneling current initially increases to a current maxima, followed by an area of negative drain resistance (NDR) and further by an exponentially-increasing diffusion current. The total drain current can be modeled by summing three currents, as illustrated by the given Equation (13).

$$\mathbf{I_d = I_{dt} + I_{da} + I_{de}} \tag{13}$$

Here,

I_{dt}: drain-source tunneling current
I_{da}: Ambipolar current
I_{de}: Esaki drain-source tunneling current and diode current

Figure 16. Simplified energy band diagram for a typical TFET (a) on-state (b) off-state

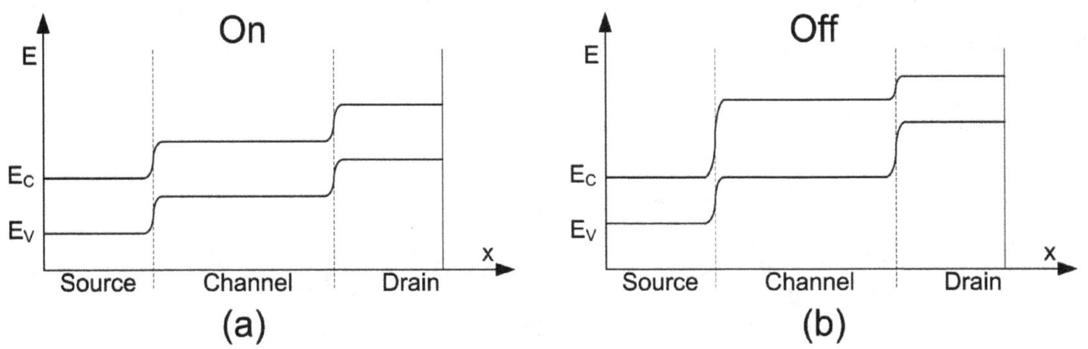

The ambipolar current (I_{da}) is added to the model by copying the current for $V_{gs} > 0$ to $V_{gs} < 0$ and multiplying the current by a scaling factor of S. Moreover, Esaki drain source tunnelling current and diode current established when the drain-source bias, V_{ds}, is negative, the source-channel junction is forward biased and works like a forward-biased tunnel diode.

3.7 Overview of Other Nanoelectronic Devices

In last section CNFET and Tunnel-FET have been discussed as futuristic non-CMOS low power high performance devices. Some circuit applications based on CNFET have also been discussed. Among the recent nanoelectronic devices, few emerging devices with features of miniature size, high speed and ultra-low power have been discussed in current section such as nanowire-FET, single electron transistor (SET) and quantum dot cellular automata (QCA).

3.7.1 Nanowire Field Effect Transistor (NWFET)

Generally, the nanowire field effect transistor (NWFET) is similar to carbon nanotube field effect transistor except silicon nanowires are used in place of CNTs to form the conductive channel below the gate (Cui. et al., 2003). The NWFET has gained attention in nanoelectronic research because of several important factors such as high yield reproducibility, low power requirement, cost effective fabrication, higher drive current (attributed to reduced scattering) and ability to create radial and axial nanowire structures (Gnani et al., 2011; Colinge et al. 2010). Figure 17 shows a simpler schematic of nanowire-FET, where unlike the conventional MOSFET channel material is replaced by nanowires.

Beside their various advantages NWFETs have some challenges too. The alteration in nanowire dimension can cause scattering and degrade the transport of charge carrier with perturbation in carrier potential (Colinge et al., 2010). Moreover, change in diameter of nanowire could change the threshold voltage of FET. In addition to this, compact models are required to understand physics related to operation of NWFET while considering ballistic transport.

3.7.2 Single Electron Transistor (SET)

Single-electron transistor (SET) is also a promising candidate of present research area of nanoelectronics that can offer lower power consumption and higher operating speed than MOSFET (Shin et al., 2010; Khondaker et al., 2010). The SET has been introduced as a new type of switching device that uses tun-

Figure 17. Schematic of Nanowire Field effect transistor

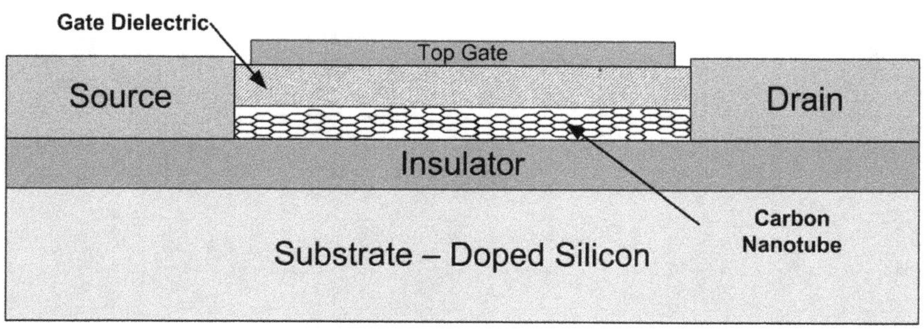

neling of electron in a controlled manner to amplify current (Roche et al., 2012). Present research in SET explores novel ideas which can transform the current memory and data storage technologies.

Considering Figure 18, a single electron transistor (SET) consists of an island coupled by two tunneling junctions with a drain and a source terminal, and a capacitor forming at the gate terminal. When no bias is applied on any electrode, electrons in the device do not possess adequate energy to undergo tunneling through the junctions.

Unlike a MOSFET which turns ON as electrons are added to a semiconductor material (Channel) and turns off when they are detached (Khondaker et al., 2010; Roche et al., 2012), SET is a new kind of transistor, in which the electrons are restricted within a tiny volume and move by tunneling. Thus, it is type of transistor that turns ON and off, each time one mobile electron has been added to it (Shin et al., 2010). The single electron transistor has very remarkable application like charge sensor, infrared detector, ultrasensitive microwave detector, sensitive electrometer *etc.*

3.7.3 Quantum-Dot Cellular Automata (QCA)

Basically, quantum dot are nanoelectronic particles, which can be used in ultra-low power circuit applications. The quantum dots are devices that are either empty or contain an electron. An array of these dots, called Quantum-dot Cellular Automata, is used for implementing logic functions (Vankamamidi, 2008). Figure 19 illustrates a general structure of four dot QCA with representation of logic '0' and logic '1' states. A four-dot QCA cell has exactly two free electrons moving among various quantum dots by tunneling. However, the electrons will only move between dots within the cell boundaries but not across cells. Therefore, QCAs give a form of information communication not by flowing electric current only but transmitting the state of the device too *i.e.* logic function (Navi et al., 2010). The QCA based low power full adder and memory cell have been discussed in literature (Navi et al., 2010; Kong et al., 2010).

4. CONCLUSION

Presently, the feature size of transistor is approaching the fundamental limit set by physical constraints. Continuous CMOS scaling is not the only solution to achieve high performance system in terms of re-

Figure 18. General structure of single electron transistor

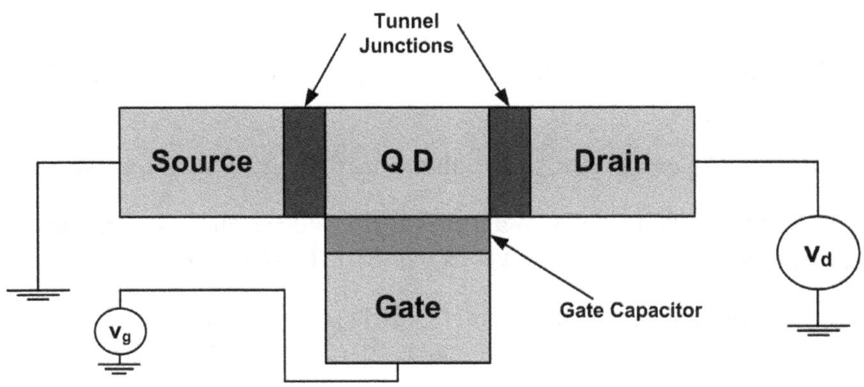

Figure 19. Representation of (a) QCA (b) logic state '0' (c) logic state '1'

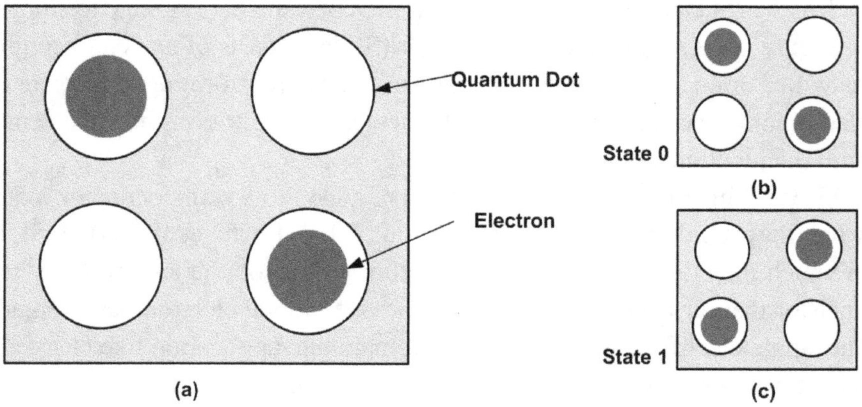

duced area, low power consumption, higher frequency, improved logic swing, better noise immunity and high gain. Various optimization techniques are discussed at different abstraction level, namely device level, circuit level *etc*. Moreover, various sources of power consumptions have also been discussed and it is attempted to highlight the major components of power dissipation in any CMOS circuit. Further, the features of emerging nanoelectronic devices such as carbon nanotube-FET, nanowire-FET, tunnel-FET *etc*. have been also presented. These emerging beyond-CMOS devices will change the perspective of present electronic circuits and systems with their low power & high performance characteristics.

REFERENCES

Black, C. T. (2005). Self-aligned self assembly of multi-nanowire silicon field effect transistors. *Applied Physics Letters, 87*(16).

Borkar, S. (2006). Electronics beyond nano-scale CMOS. In *Proceedings of the 43rd annual Design Automation Conference*. ACM.

Boucart, K., & Adrian, M. I. (2007). Double-gate tunnel FET with high-κ gate dielectric. *Electron Devices, IEEE Transactions on, 54*(7), 1725–1733. doi:10.1109/TED.2007.899389

Chandrakasan, A. P., & Robert, W. B. (2012). *Low power digital CMOS design*. Springer Science & Business Media.

Chen, Z. (2004). Nanotubes for nanoelectronics. Encyclopedia of Nanoscience and Nanotechnology, 7, 919-942.

Colinge, J-P., et al. (2010). Nanowire transistors without junctions. *Nature Nanotechnology, 5*(3), 225-229.

Cui, Y., Zhong, Z., Wang, D., Wang, W. U., & Lieber, C. M. (2003). High performance silicon nanowire field effect transistors. *Nano Letters, 3*(2), 149–152. doi:10.1021/nl0258751

Gnani, E., Gnudi, A., Reggiani, S., & Baccarani, G. (2011). Theory of the junctionless nanowire FET. *Electron Devices, IEEE Transactions on, 58*(9), 2903–2910. doi:10.1109/TED.2011.2159608

Horowitz, M. (2005). *Scaling, power, and the future of CMOS*. Electron Devices Meeting. IEDM Technical Digest. IEEE. doi:10.1109/IEDM.2005.1609253

Huijsing, J. (2013). *Analog circuit design: low-noise, low-power, low-voltage; mixed-mode design with CAD tools; voltage, current and time references*. Springer Science & Business Media.

Iijima, S. (1991). Helical microtubules of graphitic carbon. *Nature, 354*(6348), 56-58.

Imran, A., Hasan, M., Islam, A., & Abbasi, S. A. (2012). Optimized design of a 32-nm CNFET-based low-power ultra wideband CCII. *Nanotechnology, IEEE Transactions on, 11*(6), 1100–1109. doi:10.1109/TNANO.2012.2212248

Javey, A., et al. (2003). Ballistic carbon nanotube field-effect transistors. *Nature, 424*(6949), 654-657.

Khatami, Y., & Kaustav, B. (2009). Steep subthreshold slope n-and p-type tunnel-FET devices for low-power and energy-efficient digital circuits. *Electron Devices, IEEE Transactions on, 56*(11), 2752–2761. doi:10.1109/TED.2009.2030831

Khondaker, S. I., Kang, L., & Zhen, Y. (2010). The fabrication of single-electron transistors using dielectrophoretic trapping of individual gold nanoparticles. *Nanotechnology, 21*(9).

Kim, Y. B. (2010). Challenges for nanoscale MOSFETs and emerging nanoelectronics. *Transactions on Electrical and Electronic Materials, 11*(3), 93–105. doi:10.4313/TEEM.2010.11.3.093

Kim, Y. B., Yong-B, K., & Fabrizio, L. (2009). A novel design methodology to optimize the speed and power of the CNTFET circuits. *Circuits and Systems, 2009. MWSCAS'09. 52nd IEEE International Midwest Symposium on*. IEEE. doi:10.1109/MWSCAS.2009.5235967

Kong, K., Yun, S., & Ruqian, L. (2010). An optimized majority logic synthesis methodology for quantum-dot cellular automata. *Nanotechnology, IEEE Transactions on, 9*(2), 170–183. doi:10.1109/TNANO.2009.2028609

Kuhn, K. J. (2012). Considerations for ultimate CMOS scaling. *IEEE Transactions on Electron Devices, 59*(7), 1813–1828. doi:10.1109/TED.2012.2193129

Marković, D., Wang, C. C., Alarcon, L. P., Tsung-Te Liu, , & Rabaey, J. M. (2010). Ultralow-power design in near-threshold region. *Proceedings of the IEEE, 98*(2), 237–252. doi:10.1109/JPROC.2009.2035453

Moaiyeri, M. H., Akbar, D., & Keivan, N. (2011). Design of energy-efficient and robust ternary circuits for nanotechnology. *IET Circuits, Devices & Systems, 5*(4), 285–296. doi:10.1049/iet-cds.2010.0340

Mukundrajan, R. (2012). Ultra low power circuit design using Tunnel FETs. *VLSI (ISVLSI), 2012 IEEE Computer Society Annual Symposium on*. IEEE. doi:10.1109/ISVLSI.2012.70

Navi, K., Farazkish, R., Sayedsalehi, S., & Rahimi Azghadi, M. (2010). A new quantum-dot cellular automata full-adder. *Microelectronics Journal, 41*(12), 820–826. doi:10.1016/j.mejo.2010.07.003

Pal, A., Sachid, A. B., Gossner, H., & Rao, V. R. (2011). Insights into the design and optimization of tunnel-FET devices and circuits. *Electron Devices, IEEE Transactions on, 58*(4), 1045–1053. doi:10.1109/TED.2011.2109002

Possani, T., Lucas, C. S., & Alessandro, G. (2012). *Automatic Design of Micropower Carbon Nanotube Operational Transconductance Amplifiers*. SIM - South Symposium on Microelectronics.

Rabaey, J. (2009). *Low power design essentials*. Springer Science & Business Media. doi:10.1007/978-0-387-71713-5

Rabaey, J. M., & Massoud, P. E. (2012). *Low power design methodologies* (Vol. 336). Springer Science & Business Media.

Raychowdhury, A., & Kaushik, R. (2007). Carbon nanotube electronics: design of high-performance and low-power digital circuits. *Circuits and Systems I: Regular Papers, IEEE Transactions on, 54*(11), 2391-2401. doi:10.1109/TCSI.2007.907799

Roche, B., et al. (2012). A tunable, dual mode field-effect or single electron transistor. *Applied Physics Letters, 100*(3).

Roy, K., & Sharat, C. P. (2009). *Low-power CMOS VLSI circuit design*. John Wiley & Sons.

Seabaugh, A. C., & Qin, Z. (2010). Low-voltage tunnel transistors for beyond CMOS logic. *Proceedings of the IEEE, 98*(12), 2095–2110. doi:10.1109/JPROC.2010.2070470

Sharma, J., Ansari, M. S., & Sharma, J. (2014). Current-Mode Electronically Tunable Resistor-less Universal Filter in ±0.5V 32 nm CNFET. *Devices, Circuits and Communications (ICDCCom), 2014 International Conference on*. IEEE.

Shin, S. J., et al. (2010). Si-based ultrasmall multiswitching single-electron transistor operating at room-temperature. *Applied Physics Letters, 97*(10).

Skotnicki, T., Hutchby, J. A., King, T., Wong, H.-S. P., & Boeuf, F. (2005). The end of CMOS scaling: Toward the introduction of new materials and structural changes to improve MOSFET performance. *Circuits and Devices Magazine, IEEE, 21*(1), 16–26. doi:10.1109/MCD.2005.1388765

Thompson, S. E., & Srivatsan P. (2006). Moore's law: the future of Si microelectronics. *Materials Today, 9*(6), 20-25.

Tripathi, S. K., & Ansari, M. S. (2014). Voltage-mode universal filter for ZigBee using ±0.9 V 32nm CNFET ICC-II. (2014). *Confluence The Next Generation Information Technology Summit (Confluence), 5th International Conference*. IEEE.

Usmani, F. A., & Mohammad, H. (2009). *Design and parametric analysis of 32nm op-amp in cmos and cnfet technologies for optimum performance. In Micro-Nanoelectronics, Technology and Applications. EAMTA 2009*. IEEE.

Usmani, F. A., & Mohammad, H. (2010). Carbon nanotube field effect transistors for high performance analog applications: An optimum design approach. *Microelectronics Journal, 41*(7), 395–402. doi:10.1016/j.mejo.2010.04.011

Vankamamidi, V., Marco, O., & Fabrizio, L. (2008). A serial memory by quantum-dot cellular automata (QCA). *Computers, IEEE Transactions on, 57*(5), 606–618. doi:10.1109/TC.2007.70831

Chapter 2
Low Power Strategies for beyond Moore's Law Era:
Low Power Device Technologies and Materials

B. Shivalal Patro
Kalinga Institute of Industrial Technology University, India

Vandana B.
Kalinga Institute of Industrial Technology University, India

ABSTRACT

Semiconductor industries are facing a lot of problems in designing the chips consist of transistors with less than 10nm technology. Moore's law which predicted the scaling down of semiconductor devices has forced the researchers to look upon the devices in another aspect. So, various architectures and materials are invented to increase the reliability, speed and most importantly low power operation without increasing the size of devices. The on-set of nanotechnology and nano-science leads to unconventional 3D structure devices to and 0-dimensional structures. This chapter gives a general overview of the various technologies; materials and architectures researchers are concentrating to continue the technology beyond Moore's law with low power consumption.

1. INTRODUCTION

Moore's law: It is the observation by Gordon Moore that the number of transistors in a square inch area for the integrated circuits had doubled every year. But in subsequent years, this enhancement of density has slowed down i.e. doubling in two years but still it will be continue living for another some years.

Traditionally, the MOSFETs are designed in planar structure form to design the integrated circuits ICs. With the down scaling of transistor sizes beyond 45nm technology, introduction of short channel effects come into consideration. This effect hence introduces leakage current which increases the power

DOI: 10.4018/978-1-5225-0190-9.ch002

consumption even at the zero operation state. Also passive power dissipation and non uniformity in device structures and doping concentrations are faced by the fabrication members. These problems are somewhat solved in considerable amount by introduction of multi-gate structure which is explained in below section.

Background

Researchers are also going for various architectures which are something different from present designs of transistors. These new architectures are found to be very good in achieving the most important specifications especially low power implementation of devices in various circuits. Also, silicon on insulator based technologies; materials having high k-dielectric are also source of achieving some of the design constraints like low scaled devices and low power.

With the more demanding of low power as well as compact size devices in market makes the researchers to think more out of box ideas in order to compensate the demands. These ideas strengthen Moore's law to get extended for further some decades as the size of transistors has to be minimized beyond the optimized level taking silicon as major ingredient of semiconductor. So, different materials exhibiting nearly semiconductor properties especially lower band-gap property have been the source of replacement for the silicon based gates of transistors. Not only that to miniaturize the device and hence the integrated circuits, researchers opted for reducing dimensionality of the devices which can helpful for miniaturizing them.

In this chapter, there is a general overview of various types of modified devices in structures, materials and miniaturizing the dimensions of the semiconductor materials.

2. LIST OF LOW POWER DEVICES

2.1 Floating Gate MOSFET

It is a type of Metal Oxide Semiconductor Field Effect Transistor where gate is electrically isolated. A number of secondary gates are deposited over the isolated floating gate. This isolated gate is surrounded by highly resistive material and hence charge stored inside it can be retained for a longer period of time. As the construction suggests it can be a building block of various flash memories like EPROM, EEPROM. Also it has other applications in designing digital potentiometers, neural network constitutional elements and also digital to analog converters (DACs) as mentioned by Holer et al. (1989). When implemented various computational circuits instead of traditional MOSFETs, the average power was found to be less as it reduces the leakage currents especially during the OFF state as it stores charge.

This type of transistor was first made by Kahng and Sze in1967 to store data. But it has some drawbacks of low input transconductance and low output resistance.

Implementation of FGMOS allows threshold voltage (V_{th}) controllability without reducing the feature size, thus operates at power supply voltage levels which are well below the intended operational limit was shown by Thomsen & Brooke (1991). A typical multi-input floating gate transistor is shown in Figure 1. It is a conventional MOSFET in which the gate is capacitively coupled to the input using another polysilicon layer. FGMOS device provides impressive features relevant to low-voltage, low-power context.

Figure 1. Equivalent schematic of N-input n-channel FGMOS

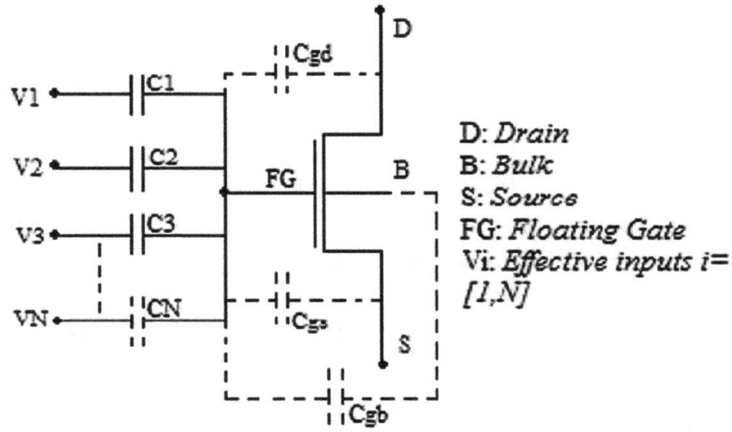

It shows flexibility in implementing both linear as well as non-linear functions. Independent control of threshold voltage (V_{th}) accounts for controllability of the device. Also, since it is a multi-input device, FGMOS allows addition of extra inputs as per designer's requirement and hence it helps to tuning the circuit. The simulated output characteristics and power dissipation on comparative basis is shown in Figure 2 and Figure 3 respectively.

In general it can be implicit that implementing FGMOS in various logical and low power requirement circuits in place of traditional MOSFET can be a very good replacement. The main advantage of charge storing or controllability factor helps in multi-area of applications. In case of charge storing,

Figure 2. DC response of N-type FGMOS and conventional NMOS

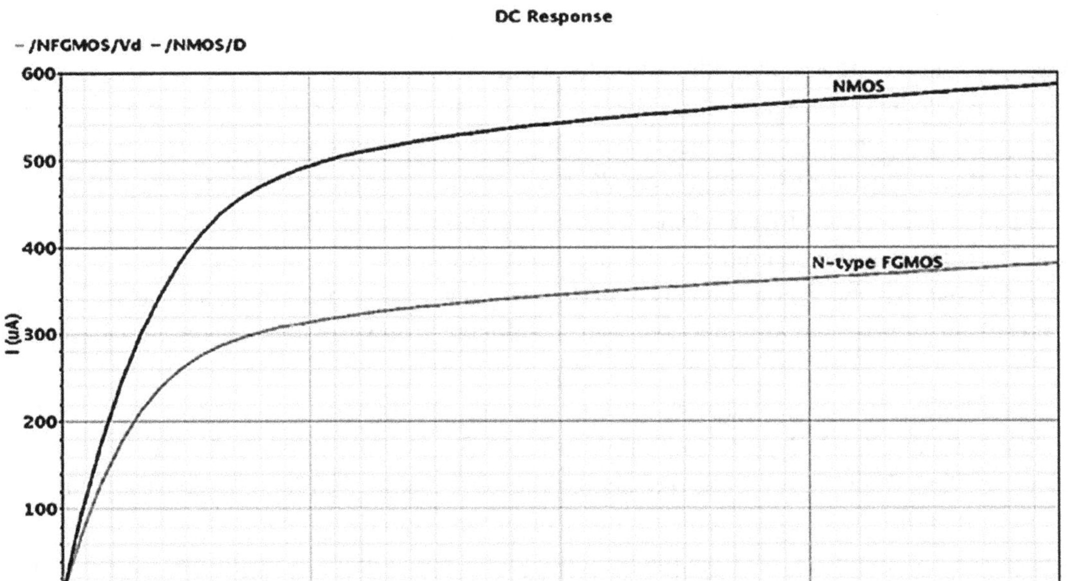

Figure 3. Power dissipation of N-type FGMOS and conventional NMOS

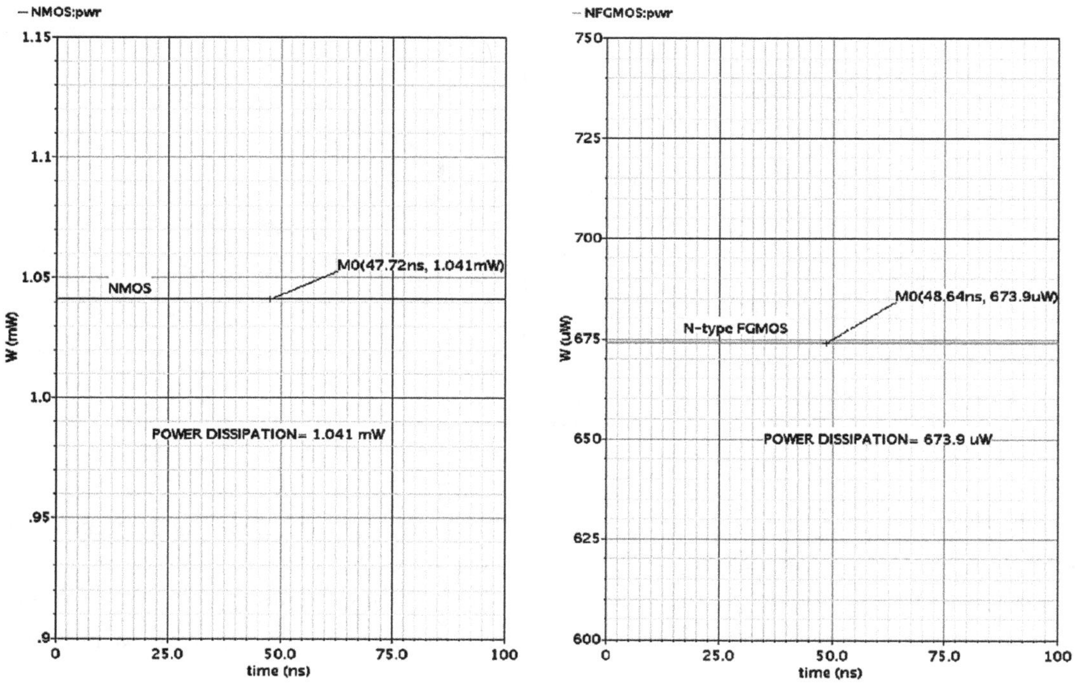

these can be used in memory which are non-volatile in nature as well as computational elements like neural networks. In the area of capacitively coupled, these can be used in logical applications like adders, multipliers, converters etc.

2.2 Multi-Gate Mosfet

Multiple Gate Metal Oxide Semiconductor Field Effect Transistor as the name suggests incorporation of more than one gate in a single transistor. These multiple gates can be controlled by single gate electrode or by independently. These independent multi gated devices are also termed as Multiple Independent Gate Field Effect Transistor (MIGFET) is well shown by Subramanian (2010). Figure 4 shows some of the typical structures of multi-gate type of MOSFETs. Introducing these types of structures in place of traditional transistors is the continuation beyond Moore's law era.

Figure 5 shows planar and non planar type of double gates. Double gate transistor was well analyzed by Wong et.al. (1997) where source to drain channel is sandwiched between two gates stacked independently.

1. **Trigate FET**: The name itself suggests the containing of three gates. This transistor was first used by the company Intel Corporation and further AMD also join into the race for this type of non planar device. They reported that enhancing the surface area for electrons to travel and confining them in that area has helped reduction of leakage into greater extent and hence consumption of less power. Also, during on state the current flow is much more as against anticipated. This new architecture has also the capability of high switching speed. Gate all around FET: As the name suggests the gate is surrounded to the channel on all sides. Various structures of surrounding gate architectures are shown in Figure 4 were implemented by Singh et.al. (2006).

Figure 4. Different types of multi gate structure of MOSFETs

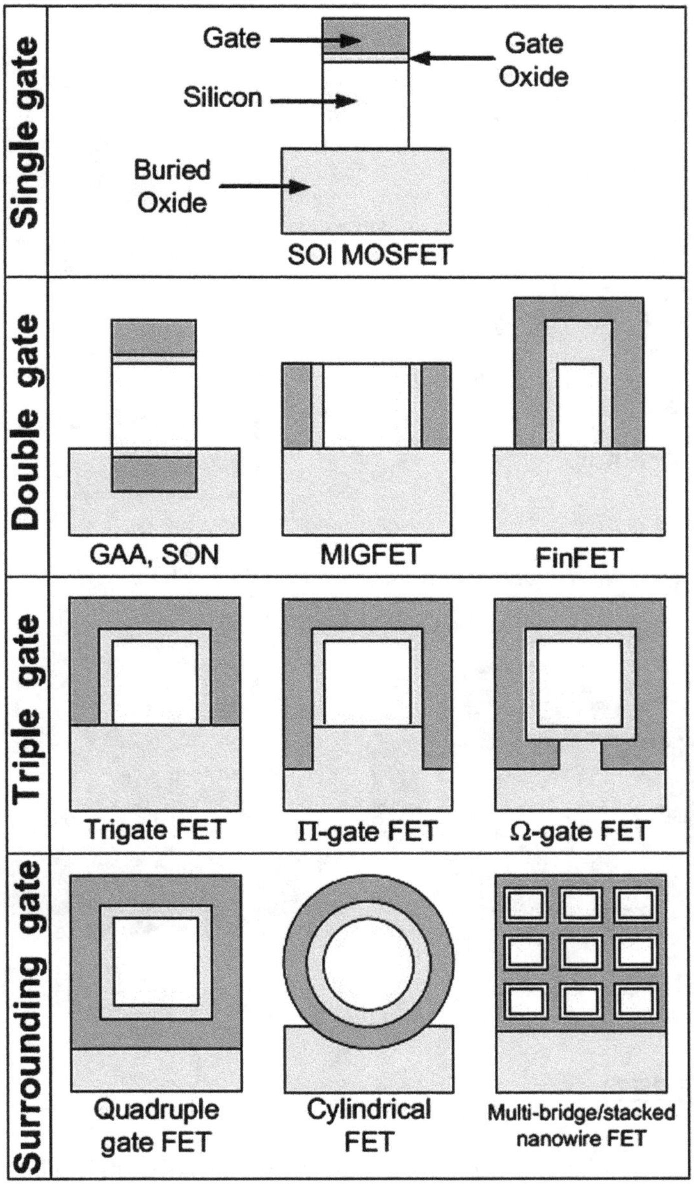

2. **FinFET:** A non-planar multi gate transistor built on silicon on insulator (SOI) substrate. This name was first coined by researchers of Berkeley from University of California. The term "fin" comes from the body of the device made up of thin silicon wrapped to the conducting channel. Figure 6 shows a general structure of a FinFET in comparison to general planar MOSFET. The thickness of this fin determines the effective channel length of the device. The wrapped gate reduces the short channel effects and hence leakage current as it provides a better electrical control over the channel.

Figure 5. Double gate MOSFET

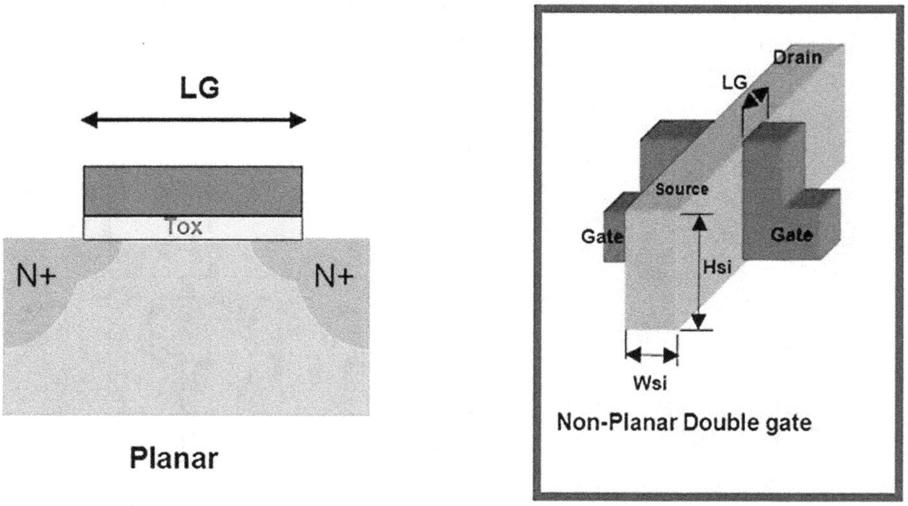

Figure 6. A comparative structure between planar FET and FinFET

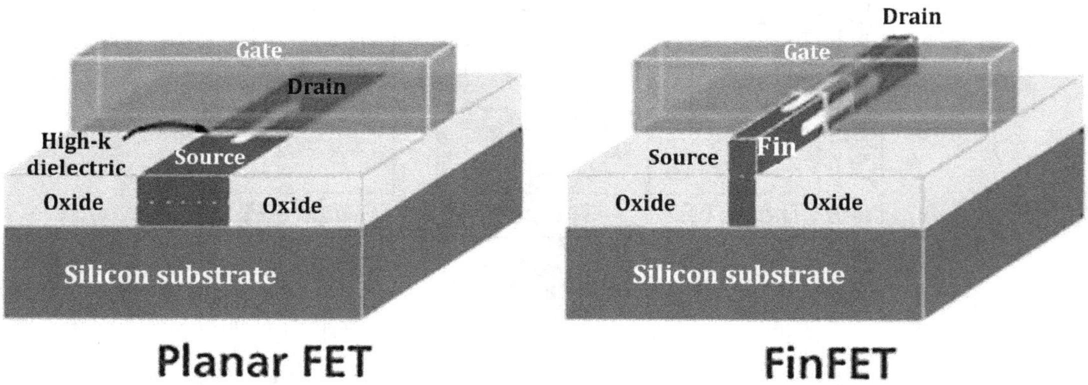

3. TUNNEL FET (TFET)

It is a same type of transistor like MOSFET but the operation differs from it. The main switching mechanism is done by the process of quantum tunneling instead of thermionic emission of charge carriers over a barrier in traditional MOSFET. The researchers in IBM carried out the first design and implementation and found out the low subthreshold swing are possible in this type of transistors. When implemented in various logic circuits, the circuits can be operated at very low voltage and hence low power is dissipated by the circuits. Introduction of these transistors have helped the researchers to work in high speed and downscaling of operating voltage as well as threshold voltage was possible and well analyzed by Seabaugh & Zhang (2010). Figure 7 shows a general architecture of TFET.

The vital structure is similar to MOSFET but the source and drain terminals are doped oppositely i.e. one is of p-type dopant and other is of n-type dopant. The middle portion is made up of intrinsic type material which is obviously controlled by the gate terminal. In simple manner, it can be told that the structure is fabricated in P-I-N (p-type, intrinsic type, n-type) pattern.

Figure 7. A typical tunnel FET structure

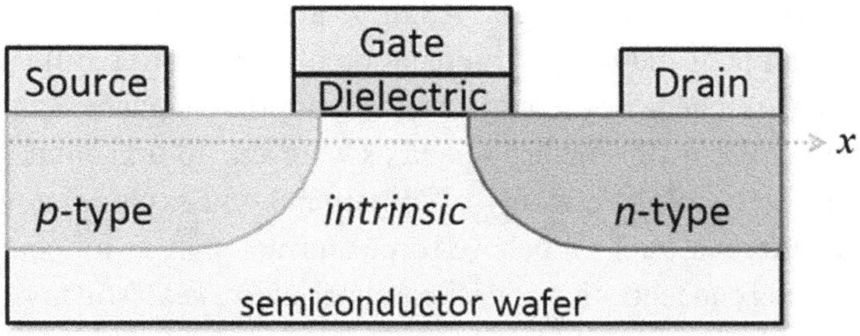

On proper biasing the gate, the electrons gate accumulated at the intrinsic region and further if sufficient biasing is present then the band to band tunneling occurs and hence channel is created. Similarly by reducing the gate biasing voltage the misalignment of band stops the conduction of electrons as shown by Teherani et.al(2013). But still researchers are finding difficult to implement practically as there is a large amount of gate leakage was found and also the steep slope of subthreshold at drive currents. Probably going for the ultra sharp doping concentration may be able to eradicate the above tribulations.

4. HIGH ELECTRON MOBILITY TRANSISTOR (HEMT)

Unlike MOSFETs, this type of transistor is having junction between two different types of materials which exhibits different band-gaps. Due to this hetero-structure nature of the transistor also named as Hetero-structure Field Effect Transistor (HFET). Most commonly used materials are GaAs (Gallium Arsenide) and AlGaAs (Aluminum Gallium Arsenide).To increase the range of operation of the transistor devices Indium is preferred most of the time. But GaN (Gallium Nitride) based devices have shown results especially in the area of low power based devices.

The main working principle depends on the factor that one material is having rich source of free electrons and other having depleted electrons like doping p-type or n-type dopant to semiconductors. Instead of using semiconductors directly, compound semiconductors (e.g GaAs etc.) has become a key role in increasing the operation range especially in millimeter and high speed of operation without changing the other important parameters especially the size and power consumption of the devices.

These are mainly used in optical devices as they are operating in GHz to THz ranges and requires very high gain and low noise operations. These devices have given boost to Moore's law and hence 4G and 5G communication is possible.

5. GRAPHENE

Graphene like graphite is an allotrope of carbon. In atomic scale it is visualized like a two dimensional hexagonal lattice of indefinitely large aromatic flat polycyclic hydrocarbon. Practically, it can be visualized like a honeycomb. Its high durability, strength and almost transparent material make it an eye catching of various researchers and industrialist to focus on its productivity. Not only this it has very high rate of conductivity and can be made to use as a switch can be a good replacement of transistors. Also it has good sensing capabilities for gas which can be used as gas sensors in various industries dealing with highly poisonous gas to determine any leakages. All these properties make it very essentially important to produce in large scale but the isolation and preparation of pure form is still very challenging task.

Figure 8 shows a typical graphene hexagonal structure pattern. Graphene was first coined by scientist Hannspeter Boehm in 1962. From 18th century scientists were able to find out the amazing properties of graphene but after the nanotechnology era only its importance and exploration were further developed into larger scale. Unlike other materials, it behave some unusual manner for conductivity, relation with temperature in various doping concentration of various materials makes it still mysterious material. Hence even after studying after so long period and its unstable behavior, it is estimated to be the backbone of quantum computing and main ingredient to design the quantum computers.

Graphene forms sp^2 orbital hybridization which is combination of σ bonds and π bonds which is very much stronger than even the steel. Unlike semiconductor, its conduction band and valence band are very nearer (can range from 0 to 0.25 eV under controlled conditions) to each other which help it to exhibit high conductivity (upto $5.3kW.m^{-1}.K^{-1}$) when compared with general graphite ($2kW.m^{-1}.K^{-1}$) at room temperature Geim in 2009. Its resistivity is around 10^{-6} Ω cm which is even less than silver at room temperature. So even slightest contamination with moisture (water) or oxygen possess such high chemical reactivity makes it difficult to produce in larger scale. So, researchers are finding suitable coating to protect it from reaction and hence its electrical behaviors are studied in vacuum by Akturk & Goldsman, in 2008. Arranging on a certain pattern can make it to behave like quantum dot which is further described in section 7.

It possesses very high optical conductivity due to its high saturation absorption under strong excitation beyond threshold level. If practically realizable, it can make the lasers used in optical fibers to operate at terahertz range and can increase the speed of communicating beyond thinking power of human-beings.

Figure 8. A graphene structure

6. OTHER 2-D STRUCTURES

A. Silicene

Similar to the graphene, it is also a two dimensional structure of allotrope of silicon shown by Cahangirov in 2009, with a hexagonal honeycomb structure. Figure 9 shows a two dimensional silicene structure showing the absence of flatness. Its existence was estimated long back but in 2010, its details are studied by the researchers. It is highly unstable than graphene. So, assembling and preparing devices using silicene and studying the behavior and properties is a herculean task for the researchers. Unlike graphene, its hexagonal structure is not completely flat but buckled hexagonal shape which helps to tune the band-gap with the help of external electric field. The bonding forces are of sp^3 type of hybridization which makes it highly chemically active.

Reducing power dissipation in traditional metal oxide field effect transistors (MOSFETs) in present generation has become a hell of a task for the researchers. Various techniques of power reduction have come into a level of saturation and cannot be reduced further. So, using this type of tunable band-gap material, the transistors like Tunnel field effect transistors (TFETs) can be good replacement for present type of transistors. A detail discussion about TFETs is already done in section 3.

As silicene is highly chemically active, its hydrogenation i.e. converting silicene to silicane can be fruitful for storing hydrogen. The best substrate to grow silicene is silver and its isotopes but researchers are trying other materials for more stabilized and pure form especially deprived of oxygen to know the features in lot more details.

B. Phosphorene

As similar to graphene, it is also one of the allotrope of Phosphorus (P) exhibiting two dimensional structures. Unlike graphene, it has very low band gap which helps the researchers to implement it as a semiconductor. The simplest or single layer exhibits a band gap of 1.75eV at room temperature and the band gap can be tuned by adding more layers.

Figure 9. A two dimensional silicene structure

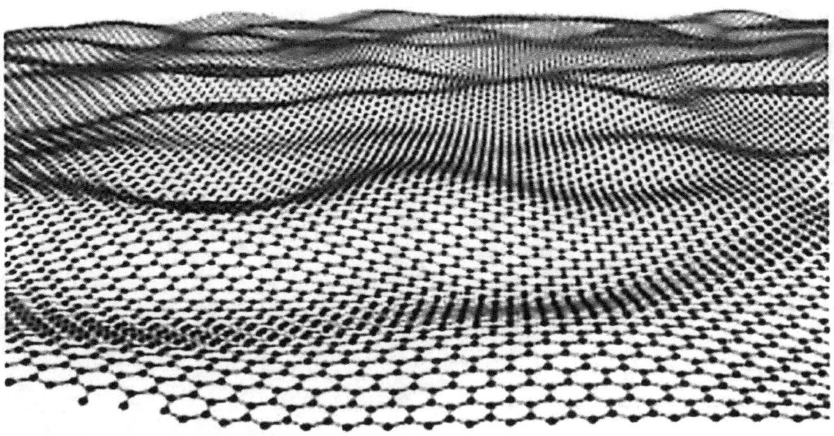

C. Germanene

As similar to silicene and graphene it is also a two dimensional structure. But the researchers are stating that it has some unusual properties unlike graphene and silicene and can be a useful replacement for the germanium made devices. It can also be a highly rated substitute for other type of semimetals. It has also the added advantage of exhibiting small size.

D. Stanene

It is an allotrope of Stannum (tin). At room temperature, it exhibits the property of insulator unlike phosphorene. But scientists have found out that addition of fluorine atoms into the lattice can make it to behave as a semimetal. The integrated circuits (ICs) made from it are estimated to perform at a very high speed and more energy efficient levels.

E. Transition Metal Dichalcogenide Monolayers

These are two dimensional semiconductors made of type MX_2 form from M as transition metal (Molybdenum (Mo), Tungsten (W) etc.) atom X as chalcogen atoms (Sulphur (S), Selenium (Se) etc.). Figure 10 shows the internal architecture of Molybdenum sulphide (MoS_2).

The best example is the graphene where different extraordinary physical properties come into sight of the researchers when the crystal is thinned down to atomic level i.e. two dimensional structures. Although most of the properties of these monolayers are different from graphene yet the most important property required is semiconductor with direct band gap has put the researchers into fascinating state to study in details about these materials.

Figure 10. Different sections of a field effect transistor based on a monolayer of MoS_2

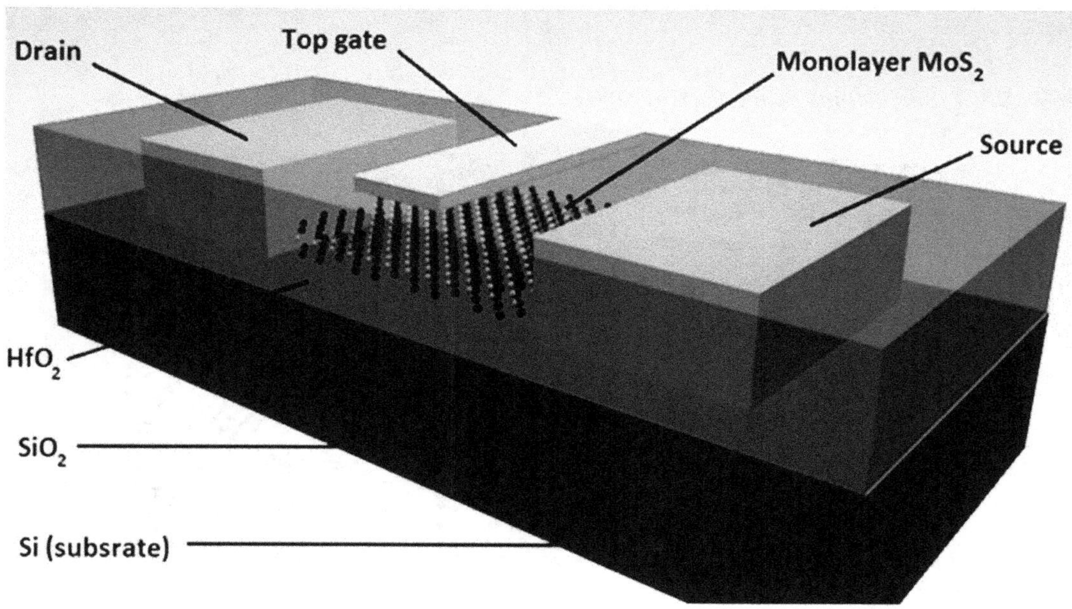

These are the very good materials as a replacement against traditional silicon based transistors as they are thinner, faster conductivity and also the most important factor is low power. Also the band gap can be tuned and hence a variety of specifications of transistors can be found out. This is the main advantage over graphene as the later possess almost zero band-gap which makes it difficult to do the switching property of transistors.

F. Quantum Wire

It is like a wire which conducts electricity besides possessing quantum effects which can influence various properties like conduction of electricity. The electrons are quantum confined to make it to conduct in only one direction which can be possible by diminishing the size of the wire sufficiently small. Figure 11 shows the quantum confinement of electrons in quantum wire and quantum well or reservoir.

Conductivity can be represented mathematically as

$$\sigma = \frac{2nq^2}{h} \tag{1}$$

where q=charge of the carrier, h= Planck's constant, n= positive integer

While dealing with conduction in quantum wires, the carriers are mostly considered as electrons. If the wire size is not small, then the conductivity relation is not valid and can be replaced by general formula of electrical as

$$\sigma = \frac{l}{R * A} \tag{2}$$

where l is the length of the semiconductor, R is the resistance, A is the cross sectional area of the wire.

The visibility of quantum property come into picture when the cross sectional area of the material is too small comparable to the size of atoms. So it can be inferred as the quantization is inversely proportional to the size of the wire.

Figure 11. Quantum wire (a) Quantum confinement (b) Charge distribution

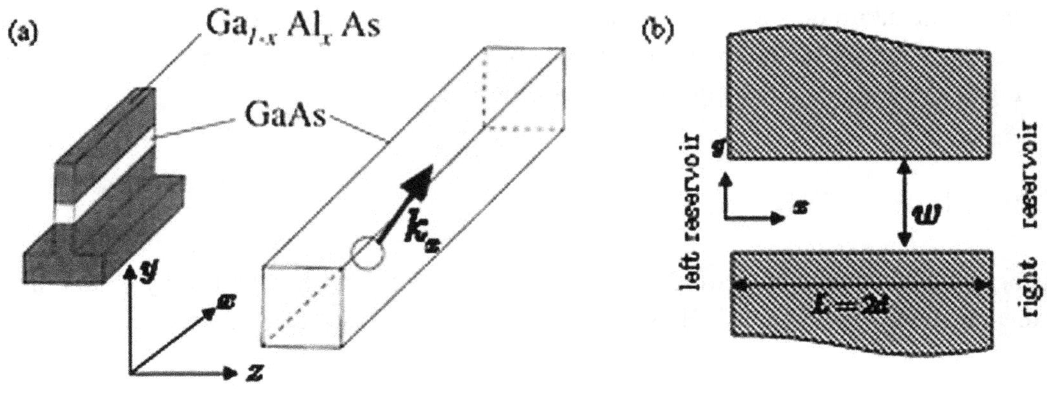

One of the widely proposed quantum wires is carbon nanotube which is sufficiently small and has no internal scattering to make the flow of carriers/electrons in a single direction. Its quantum conductance relation is

$$\sigma = \frac{4q^2}{h} \tag{3}$$

Here n=2 as carbon nanotube have two spatial channels.

7. QUANTUM DOT

It is a nano-crystal made from semiconductor materials which exhibits quantum mechanical properties. Sometimes it is also called as single electron transistor. It is very small enough yet confined in all three spatial dimensions. The term quantum dot was first coined by Mark Reed but it was first discovered by Alexey Ekimov in 1981 in a glass matrix and further by Louis Brus in 1985 in colloidal solutions.

It has vast area applications in medical imaging, quantum computing, laser communication, diodes, transistors, etc. The first commercially product made from it was launched by SONY in one of its series of flat panel television XBR X900A in 2013. It is a very promising contender for revolutionizing the future computing and communication areas along with medical applications. Figure 12 shows the color changing pattern by the quantum dots for various sizes.

The working principle is mainly based on the frequency of light emitted by it which is indirectly proportional to the size of the quantum dot. In simple words, to change the color of light from green to red one has to increase the size of the quantum dot. So, by controlling the size as well as the controlling the conductive properties and hence the emission of quantum dots can help us to tune different range of frequencies.

While dealing with quantum dots main technical term mainly linking with it is quantum confinement. Previously, scientist found out that electron behaves both particles as well as wave pattern. When the particle size is less than 10nm, then the behavior of the particle is found to be free yet confining its dimension as large as comparable to wavelength of the particle. The band gap now depends upon completely size of the particle and can be tuned as per the requirements. These energy levels possessed by the electrons and holes can be modeled usually the particle in a box model. Using the length of the box the electronic properties as well as the optical properties can be tuned easily. There are various techniques to generate the quantum dots and tuning the band gaps in various labs by the researchers. Still it is a very challenging task for the industries to produce practically in large numbers in a controlled manner considering the cost of production during the same period of time.

8. MEMRISTOR: INTRODUCTION

Memristor is combination of two words *memory* and *resistor*. It was first coined by one of the famous circuit theorist Leon Chua in 1971. He stated that it is a passive electronic component consists of two

Figure 12. Quantum dots generating various colors

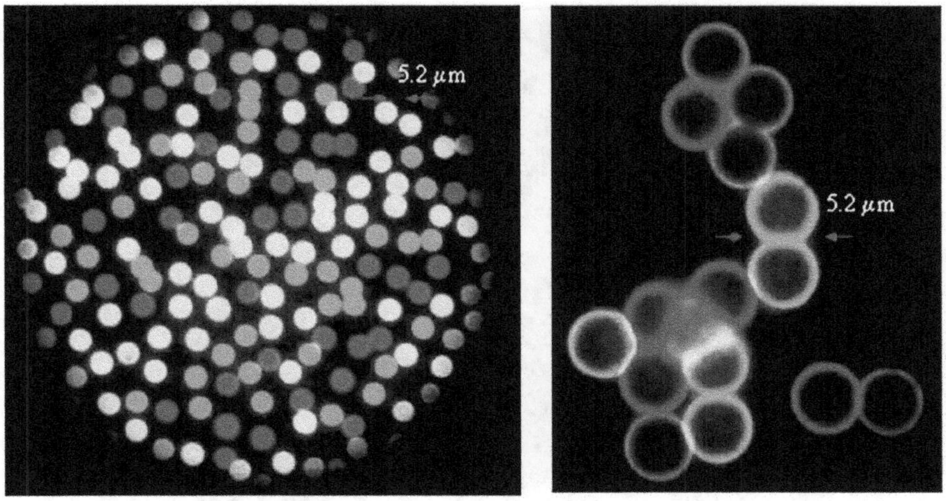

terminals relating to electric charge and magnetic flux linkage. The symbolic representation of memristor is given in Figure 13.

The main motivation behind going for this type of devices is to become the replacement device against the transistor. There are many problems faced while miniaturizing the transistors and increase of more and more flip flops in digital circuits which enhance the power consumption of the circuits. On the other hand it can act as a hardware security device. So, researchers are trying to formulate various models and try to practically implement in various circuits especially in digital domain.

On the basis of derived characteristics using mathematical relations, the working of memristor is basically the resistance is going to be altering or changing depending upon the current that had previously flowed through the device. That means as the memory is non volatile, so even after switching on the circuit after some time the resistor will be remembering the previously operated resistance.

Figure 14 shows oxygen depleted TiO_2 memristor built at HP labs. According to its behavior, its applications are intended in the area of computer logic, neural based machine learning, neuro-memristive computer and also in nano-electronic device memories. Various teams like HP (Hewlett-Packard) stated that these components will be commercially available by 2018 if everything goes well. If these devices come into market then they will become the replacements for the presently available memory devices like SRAM, DRAM, Flash memories.

Figure 13. Symbolic representation of memristor

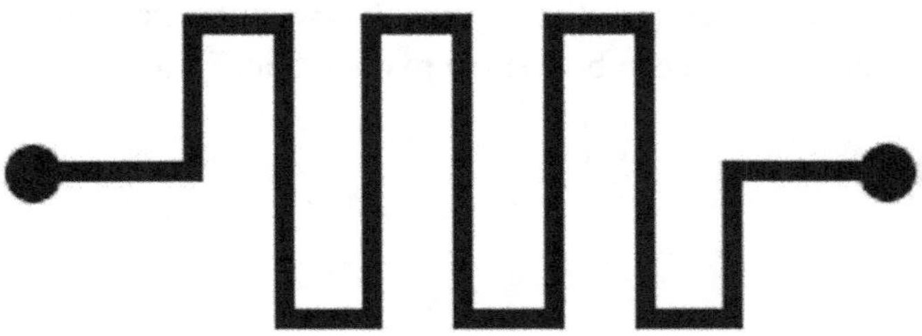

Figure 14. A memristive material made up of TiO_2

Mathematical Analysis

In accordance to Chua, he described the memristor as a passive two terminal element that maintains a functional relationship between the time integral of current (charge) and time integral of voltage (flux). The slope of this function is called as memristance M as similar to resistance.

Mathematically it can be written as

$$M(q) = \frac{d\varphi}{dq} \tag{4}$$

where φ is magnetic flux linkage, q is electric charge and M is memristance.

Substituting φ as time integral of voltage and q as time integral of current, the equation (1) is rewritten as

$$M(q(t)) = \frac{\frac{d\varphi}{dt}}{\frac{dq}{dt}} = \frac{V(t)}{I(t)} \tag{5}$$

It can be inferred from the equation (2) that even if I(t) =0 and hence V(t) = 0, M(q(t)) is constant.

Similarly, the power consumption can be written mathematically like resistance (I^2R) is

$$P(t) = V(t) \times I(t) = I(t)^2 M(q(t)) \tag{6}$$

The value of M(q(t)) can be constant, negative value, positive value. The positive value cannot be valid for longer duration as it will lead to superconductivity which is not practically realizable. The main important reason or advantage of memristor is the very high switching speed.

9. RESISTIVE RANDOM ACCESS MEMORY (RRAM)

It is type of non volatile memory which consists of the dynamic resistive material called memristor. It is a replacement to the present day flash memories but the cost of production and performance benefits are doubtful for this RRAM. But the main advantages carried by this type memory are very low voltage operation and hence power consumption. Presently many companies are trying to get the product using the help of material with high k gate dielectric like Hafnium dioxide (HfO_2). Correspondingly, Panasonic launched one transistor-one resistor based memory cell architecture made from tantalum dioxide. Crossbar introduced postage stamp size RRAM which can be able to store upto 1TB of data.

The biggest merit carried out by this type of memory devices is the very high switching speed even less than 10ns can be obtained. But the large scale is still a challenging task. One of the issues found was the sneak path problem where current interference occurs. To eradicate this problem, researchers proposed some solutions which may reduce this interference problem.

CONCLUSION

This chapter can be concluded stating that there is a huge scope of extending the law stated by Moore. With more demanding market of low power and miniaturized devices without any compromising the speed of operation has pushed the researchers to work differently. The more challenging aspect while miniaturization is the extra problems like short channel effects which made the researchers to go for quantum modeling of the devices to make them practically feasible.

Starting from FGMOS to multi-gate devices or FINFETs is all in fact device architecture modification both in 2-D and 3-D aspects. FGMOS found to be a very good device for low power digital circuits and memory devices and FINFETs proved to be worthy way of developing low size transistors. Further

Figure 15. Interconnecting metal layers in one of RRAM

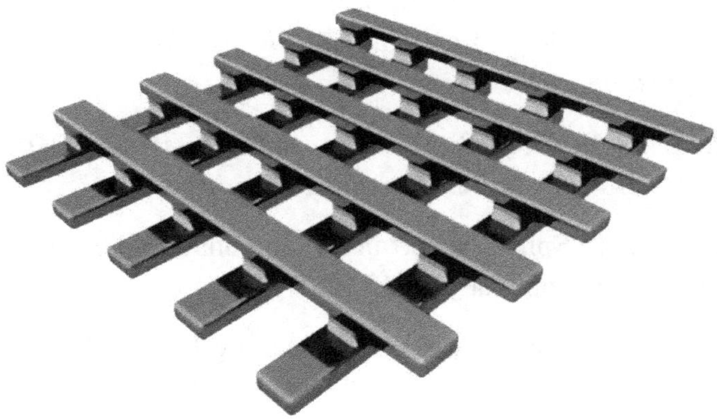

various other types of transistors are being proved and can be replacements against traditional transistors. TFETs and HMETs can be fruitful for extending the Moore's law to further some years.

Although these 3-D structured based devices can continue for some years but the results shown by the 2-D structures like graphene, silicene and 1-D, 0-D structures will completely change the technical aspect of the world. If correctly implemented, everything can be made automated and all dreams can make into truth. All depends upon the nanotechnologists and other associated researchers to make it realizable. The outcomes are found in structures like quantum-wire and quantum dots.

Similarly there are some other replacement devices which atleast can work side by side with the silicon based transistors. Such examples are found in memristors and RRAMs. All these devices can play a critical role in taking the technology into next level and increase the period of existence of Moore's law.

REFERENCES

Akturk, A., & Goldsman, N. (2008). Electron transport and full-band electron–phonon interactions in graphene. *Journal of Applied Physics*, *103*(5), 053702. doi:10.1063/1.2890147

Alamo, J. A. (2010). Is nanometer-scale III–V CMOS cool enough to rejuvenate Moore's Law? *Compound Semiconductors*, *5*(16), 19–22.

Alper, C., Lattanzio, L., Michielis, L. D., Palestri, P., Selmi, L., & Ionescu, A. M. (2013). Quantum mechanical study of the germanium electron–hole bilayer tunnel FET. *IEEE Transactions on Electron Devices*, *9*(60), 2754–2760. doi:10.1109/TED.2013.2274198

Anand, A., Das, A., Mandal, S.K., & Patro, B.S. (2013). FGMOS based low-voltage low-power high output impedance regulated cascode current mirror. *International Journal of VLSI Design & Communication Systems, 4*(2).

Ascoli, A., Corinto, F., Senger, V., & Tetzlaff, R. (2013). Memristor model comparison. *IEEE Circuits and Systems Magazine*, *13*(2), 89–105. doi:10.1109/MCAS.2013.2256272

Aufray, B., Kara, A., Vizzini, S. B., Oughaddou, H., Léandri, C., Ealet, B., & Le Lay, G. (2010). Graphene-like silicon nanoribbons on Ag(110): A possible formation of silicene. *Applied Physics Letters*, *96*(18), 183102. doi:10.1063/1.3419932

Basker, V., Standaert, T., Kawasaki, H., Yeh, C., Maitra, K., Yamashita, T., & Kumar, A. et al. (2010). A 0.063 μm2 FinFET SRAM cell demonstration with conventional lithography using a novel integration scheme with aggressively scaled fin and gate pitch*Proceedings International Symposium VLSI Technology, Systems Application*.

Bhoj, A. N., Joshi, R. V., & Jha, N. K. (2013). *Efficient methodologies for 3D-TCAD modeling of emerging devices and circuits. IEEE Transanctions Computer-Aided Design*.

Bishnoi, B., & Ghosh, B. (2014). Spin transport in buckled bilayer silicene. *Computational Materials Science*, *85*, 16–19. doi:10.1016/j.commatsci.2013.12.041

Bullis, K. (2013). Quantum Dots Produce More Colorful Sony TVs. *MIT Technology Review*.

Cahangirov, S., Topsakal, M., Aktürk, E., Şahin, H., & Ciraci, S. (2009). Two- and One-Dimensional Honeycomb Structures of Silicon and Germanium. *Physical Review Letters, 102*(23), 236804. doi:10.1103/PhysRevLett.102.236804 PMID:19658958

Cahay, M. (2012). Quantum Confinement VI: Nanostructured Materials and Devices. *Proceedings of the International Symposium*. The Electrochemical Society.

Carlson, A., Guo, Z., Balasubramanian, S., Zlatanovici, R., Liu, T.-J. K., & Nikolic, B. (2011). Modeling of width-quantization-induced variations in logic FinFETs for 22 nm and beyond.*Proceedings International Symposium VLSI Technology, Systems Applications*.

Castellanos-Gomez, A., Vicarelli, L., Prada, E., Island, J., Narasimha, A. K. L., Blanter, S. I., … van der Zant, H. S. J. (2014). Isolation and characterization of few-layer black phosphorus. *2D Materials, 1*.

Chua, L. O. (1971). Memristor—The Missing Circuit Element. *IEEE Transactions on Circuit Theory, CT-18*(5), 507–519. doi:10.1109/TCT.1971.1083337

DasGupta, S., McMorrow, D., Reed, R. A., Schrimpf, R. D., Boos, J. B., & Ramachandran, V. (2010). Process and contamination effects on the single-event response of AlSb/InAs HEMTs. *IEEE Transactions on Nuclear Science, 57*, 3262–3266.

Davila, M. E., Xian, L., Cahangirov, S., Rubio, A., & Le Lay, G. (2014). Germanene: A novel two-dimensional germanium allotrope akin to graphene and silicene. *New Journal of Physics, 16*(9), 095002. doi:10.1088/1367-2630/16/9/095002

Denis, P. A., & Iribarne, F. (2013). Comparative Study of Defect Reactivity in Graphene. *The Journal of Physical Chemistry C, 117*(37), 19048–19055. doi:10.1021/jp4061945

Dewey, G., et al. (2011). Fabrication, characterization, and physics of III–V heterojunction tunneling field effect transistors (H-TFET) for steep sub-threshold swing. *Proceedings IEEE IEDM*.

Dresselhaus, M. S., Dresselhaus, G., & Avouris, P. (2001). *Carbon nanotubes: synthesis, structure, properties, and applications*. Springer. doi:10.1007/3-540-39947-X

Fei, W., & Yu, H. (2011). Design exploration of hybrid cmos and memristor circuit by new modified nodal analysis. *IEEE Transactions on Very Large Scale Integration (VLSI) Systems, 99*, 1–15.

Geim, A. (2009). Graphene: Status and Prospects. *Science, 324*(5934), 1530–1534. doi:10.1126/science.1158877 PMID:19541989

Guerra, D., Akis, R., Marino, F. A., Ferry, D. K., Goodnick, S. M., & Saraniti, M. (2010). Aspect ratio impact on RF and DC performance of state-of the- art short-channel GaN and InGaAs HEMTs. *IEEE Electron Device Letters, 11*(31), 1217–1219.

Gusynin, V. P., & Sharapov, S. G. (2005). Unconventional Integer Quantum Hall Effect in Graphene. *Physical Review Letters, 95*(14), 146801. doi:10.1103/PhysRevLett.95.146801 PMID:16241680

Haug, H., & Koch, S. W. (2012). *Quantum Theory of the Optical and Electronic Properties of Semiconductors*. World Scientific.

Ho, Y., Huang, G., & Li, P. (2011). Dynamical properties and design analysis for nonvolatile memristor memories. *IEEE Transactions on Circuits and Systems. I, Regular Papers*, *58*(4), 724–736. doi:10.1109/TCSI.2010.2078710

Holler, M., Tam, S., Castro, H., & Benson, R. (1989). An electrically trainable artificial neural network with 10240 'floating gate' synapses.*Proceeding of the International Joint Conference on Neural Networks*, (vol. 2, pp. 191-196). doi:10.1109/IJCNN.1989.118698

Jo, S., Chang, T., Ebong, I., Bhadviya, B. B., Mazumder, P., & Lu, W. (2010). Nanoscale memristor device as synapse in neuromorphic systems. *Nano Letters*, *10*(4), 1297–1301. doi:10.1021/nl904092h PMID:20192230

Jose, D., & Datta, A. (2014). Structures and Chemical Properties of Silicene: Unlike Graphene. *Accounts of Chemical Research*, *47*(2), 593–602. doi:10.1021/ar400180e PMID:24215179

Kahng, D., & Sze, S. M. (1967). A floating-gate and its application to memory devices. *The Bell System Technical Journal*, *4*(46), 1288–1295. doi:10.1002/j.1538-7305.1967.tb01738.x

Kim, D., & Alamo, J. A. (2010). Scalability of sub-100 nm InAs HEMTs on InP substrate for future logic applications. *IEEE Transactions on Electron Devices*, *7*(57), 1504–1511. doi:10.1109/TED.2010.2049075

Kim, K.-H., Gaba, S., Wheeler, D., Cruz-Albrecht, J. M., Hussain, T., Srinivasa, N., & Lu, W. (2012). A functional hybrid memristor crossbar-array/CMOS system for data storage and neuromorphic applications. *Nano Letters*, *12*(1), 389–395. doi:10.1021/nl203687n PMID:22141918

Lattanzio, L., Michielis, L. D., & Ionescu, A. M. (2012). Complementary germanium electron–hole bilayer tunnel FET for sub-0.5-V operation. *IEEE Electron Device Letters*, *2*(33), 167–169. doi:10.1109/LED.2011.2175898

Lattanzio, L., Michielis, L. D., & Ionescu, A. M. (2012). The electron–hole bilayer tunnel-FET. *Solid-State Electronics*, *74*, 85–90. doi:10.1016/j.sse.2012.04.016

Lay, G. L. (2015). 2D materials: Silicene transistors. *Nature Nanotechnology*, *10*(3), 202–203. doi:10.1038/nnano.2015.10 PMID:25643254

Leatherdale, C. A., Woo, W.-K., Mikulec, F. V., & Bawendi, M. G. (2002). On the Absorption Cross Section of CdSe Nanocrystal Quantum Dots. *The Journal of Physical Chemistry B*, *106*(31), 7619–7622. doi:10.1021/jp025698c

Lee, D., Seong, D. J., Jo, I., Xiang, F., Dong, R., Oh, S., & Hwang, H. (2007). Resistance switching of copper doped MoO(sub x) films for nonvolatile memory applications. *Applied Physics Letters*, *90*(12), 122104. doi:10.1063/1.2715002

Lee, S. W., Mao, C., Flynn, C. E., & Belcher, A. M. (2002). Ordering of quantum dots using genetically engineered viruses. *Science*, *296*(5569), 892–895. doi:10.1126/science.1068054 PMID:11988570

Lin, C. H., Chang, J., Guillorn, M., Bryant, A., Oldiges, P., & Haensch, W. (2010). Non-planar device architecture for 15 nm node: FinFET or trigate?. In *Proceedings of International SOI Conference*.

Liu, H., Neal, A. T., Zhu, Z., Luo, Z., Xu, X., Tománek, D., & Ye, P. D. (2014). Phosphorene: An Unexplored 2D Semiconductor with a High Hole Mobility. *ACS Nano*, *8*(4), 4033–4041. doi:10.1021/nn501226z PMID:24655084

Maitra, K., Khakifirooz, A., Kulkarni, P., Basker, V. S., Faltermeier, J., Jagannathan, H., & Bu, H. et al. (2011). Aggressively scaled strained-silicon-on-insulator undoped-body high-κ/metal-gate nFinFETs for high-performance logic applications. *IEEE Electron Device Letters*, *6*(32), 713–715. doi:10.1109/LED.2011.2126556

Mak, K. F., Lee, C., Hone, J., Shan, J., & Heinz, T. F. (2010). Atomically Thin MoS_2: A New Direct-Gap Semiconductor. *Physical Review Letters*, *105*(13), 136805. doi:10.1103/PhysRevLett.105.136805 PMID:21230799

Mayer, F. (2011). Demonstration of MOSFET-like ON-current performance in arsenide/antimonide tunnel FETs with staggered hetero-junctions for 300 mV logic applications. In Proceedings IEEE IEDM.

Mehonic, A., Cueff, S. B., Wojdak, M., Hudziak, S., Jambois, O., Labbé, C., & Kenyon, A. J. et al. (2012). Resistive switching in silicon suboxide films. *Journal of Applied Physics*, *111*(7), 074507. doi:10.1063/1.3701581

Meuffels, P., & Soni, R. (2012). *Fundamental Issues and Problems in the Realization of Memristors*. Academic Press.

Ni, Z., Liu, Q., Tang, K., Zheng, J., Zhou, J., Qin, R., & Lu, J. et al. (2011). Tunable bandgap in silicene and germanene. *Nano Letters*, *12*(1), 113–118. doi:10.1021/nl203065e PMID:22050667

Ni, Z., Qihang, L., Tang, K., Zheng, J., Zhou, J., Qin, R., & Yu, D. et al. (2012). Tunable Bandgap in Silicene and Germanene. *Nano Letters*, *12*(1), 113–118. doi:10.1021/nl203065e PMID:22050667

Ni, Z., Zhong, H., Jiang, X., Quhe, R., Luo, G., Wang, Y., & Lu, J. et al. (2014). Tunable band gap and doping type in silicene by surface adsorption: Towards tunneling transistors. *Nanoscale*, *6*(13), 7609–7618. doi:10.1039/C4NR00028E PMID:24896227

Norris, D. J. (1995). *Measurement and Assignment of the Size-Dependent Optical Spectrum in Cadmium Selenide (CdSe) Quantum Dots*. (PhD thesis). MIT, Cambridge, MA.

Oktyabrsky, S., & Ye, P. D. (2010). *Fundamentals of III-V Semiconductor MOSFETs*. New York: Springer. doi:10.1007/978-1-4419-1547-4

Park, J. Y., Kim, S.-H., Hong, S.-M., & Kim, K. R. (2015). Physical analysis and design of resonant plasma-wave transistors for terahertz emitters. *IEEE Transactions THz Science Technology*, *5*, 244-250.

Prati, E., De Michielis, M., Belli, M., Cocco, S., Fanciulli, M., Kotekar-Patil, D., & Sanquer, M. et al. (2012). Few electron limit of n-type metal oxide semiconductor single electron transistors. *Nanotechnology*, *23*(21), 215204. doi:10.1088/0957-4484/23/21/215204 PMID:22552118

Rajendran, J., Rose, G., Karri, R., & Potkonjak, M. (2012). Nano-PPUF: A memristor-based security primitive. In *IEEE Computer Society Annual Symposium on VLSI (ISVLSI)*, (pp. 84–87). IEEE.

Reed, M. A., Randall, J. N., Aggarwal, R. J., Matyi, R. J., Moore, T. M., & Wetsel, A. E. (1988). Observation of discrete electronic states in a zero-dimensional semiconductor nanostructure" (PDF). *Physical Review Letters*, *60*(6), 535–537. doi:10.1103/PhysRevLett.60.535 PMID:10038575

Risch, L. (2005). Pushing CMOS Beyond the Roadmap.*Proceedings of ESSCIRC*.

Rose, G., McDonald, N., Yan, L.-K., Wysocki, B., & Xu, K. (2013). Foundations of memristor based PUF architectures. In *IEEE/ACM International Symposium on Nanoscale Architectures (NANOARCH)*, (pp. 52– 57). doi:10.1109/NanoArch.2013.6623044

Schedin, F., Geim, A. K., Morozov, S. V., Hill, E. W., Blake, P., Katsnelson, M. I., & Novoselov, K. S. (2007). Detection of individual gas molecules adsorbed on graphene. *Nature Materials*, *6*(9), 652–655. doi:10.1038/nmat1967 PMID:17660825

Seabaugh, A. C., & Zhang, Q. (2010). Low-Voltage Tunnel Transistors for Beyond CMOS Logic. *Proceedings of the IEEE*, *98*(12), 2095–2110. doi:10.1109/JPROC.2010.2070470

Shang, Y., Fei, W., & Yu, H. (2012). Analysis and modeling of internal state variables for dynamic effects of nonvolatile memory devices. *Circuits and Systems I: Regular Papers, IEEE Transactions on*, *99*. doi:10.1109/TCSI.2011.2180441

Singh, N., Agarwal, A., Bera, L. K., Liow, T. Y., Yang, R., Rustagi, S. C., & Kwong, D.-L. et al. (2006). High-Performance fully depleted Silicon Nanowire Gate-All-Around CMOS devices. *IEEE Electron Device Letters*, *27*(5), 383–386. doi:10.1109/LED.2006.873381

Splendiani, A., Sun, L., Zhang, Y., Li, T., Kim, J., Chim, C. Y., & Wang, F. et al. (2010). Emerging Photoluminescence in Monolayer MoS_2. *Nano Letters*, *10*(4), 1271–1275. doi:10.1021/nl903868w PMID:20229981

Subramanian, V. (2010). Multiple gate field-effect transistors for future CMOS technologies. *IETE Technical Review*, *27*(6), 446–454. doi:10.4103/0256-4602.72582

Tao, L., Cinquanta, E., Chiappe, D., Grazianetti, C., Fanciulli, M., Dubey, M., & Akinwande, D. et al. (2015). Silicene field-effect transistors operating at room temperature. *Nature Nanotechnology*, *10*(3), 227–231. doi:10.1038/nnano.2014.325 PMID:25643256

Teherani, J. T., Agarwal, S., Yablonovitch, E., Hoyt, J. L., & Antoniadis, D. A. (2013). Impact of Quantization Energy and Gate Leakage in Bilayer Tunneling Transistors. *IEEE Electron Device Letters*, *34*(2), 298–300. doi:10.1109/LED.2012.2229458

Teherani, T. J., Agarwal, S., Yablonovitch, E., Hoyt, J. L., & Antoniadis, D. A. (2013). Impact of quantization energy and gate leakage in bilayer tunneling transistors. *IEEE Electron Device Letters*, *2*(34), 298–300. doi:10.1109/LED.2012.2229458

Thomsen, A., & Brooke, M. A. (1991). A floating-gate MOSFET with tunneling injector fabricated using a standard double-polysilicon CMOS process. *IEEE Electron Device Letters*, *12*(12), 111–113. doi:10.1109/55.75728

Wallace, P.R. (1947). The Band Theory of Graphite. *Physical Review*, *71*(9), 622–634.

Wan, J., Royer, C. L., Zaslavsky, A., & Cristoloveanu, S. (2011). Tunneling FETs on SOI: Suppression of ambipolar leakage, low-frequency noise behavior, and modeling. *Solid-State Electronics*, *65-66*, 226–233. doi:10.1016/j.sse.2011.06.012

Wong, H.-S., Chan, K., & Taur, Y. (1997). Self-Aligned (Top and Bottom) Double-Gate MOSFET with a 25 nm Thick Silicon Channel. IEDM, 427.

Xu, Y., Yan, B., Zhang, H. J., Wang, J., Xu, G., Tang, P., & Zhang, S. C. et al. (2013). Large-Gap Quantum Spin Hall Insulators in Tin Films. *Physical Review Letters*, *111*(13), 136804. doi:10.1103/PhysRevLett.111.136804 PMID:24116803

Yang, X., Maitra, K., Yeh, C., Zeitzoff, P., Raymond, M., Kulkarni, P., & Miller, R. J. et al. (2011) Analysis of parasitic resistance in double gate FinFETs with different fin length. In *Proceedings of International SOI Conference*. doi:10.1109/SOI.2011.6081799

Ye, X. S., Zhi-G., S., Hongbo, Z., Lei, Y., & Cang-L., W. (2014). Intrinsic carrier mobility of germanene is larger than graphene's: First-principle calculations. *RSC Advances*, *4*(41), 21216–21220. doi:10.1039/C4RA01802H

Chapter 3
Challenges and Limitations of Low Power Techniques:
Low Power Methodologies in Analog and Digital Circuits

Vandana B.
KIIT University, India

Patro B. S.
KIIT University, India

ABSTRACT

In contemporary world the technology has kept its vast identity in developing ultra NANO devices to give up the compact device utilities, in VLSI, Metal Oxide Semiconductor device plays an key role in power dissipation product, in terms of MOS theory characteristics it is predefined that a MOS transistor can conduct easily with low voltage which gives low power but in DSM technology there is a likelihood to achieve ultra low power, so this can be achieved due to the rapid shrinking of gate length, here the chapter deals with challenges and limitations of low power techniques. The predominant way to generate low power is to start with the fundamental principles that are defined in the existing technologies that it gives low power with less leakage current. Apart from this parameter consideration is also required to achieve this. The successful and the major parameter in generating low power is that the shrinking of supply voltage. To go through this, upcoming sections gives the brief idea about the different techniques that are utilized to generate low power with less leakage.

INTRODUCTION

VLSI is the domain, dealing with MOS transistors, according to the principle given by scientist Moore's the count of the transistors on printed circuit board has doubled for every 18 months as this got influenced the engineers to develop Compaq board. On the other hand circuit logic and managing power was the main focused to enhance the technology. So there after change in the technology has forced the device

DOI: 10.4018/978-1-5225-0190-9.ch003

developers to manufacture and operate with ultra low power. Here the chapter looks forward with why low power, techniques that achieve low power, limitations and challenges.

Before dealing with low power techniques, let us know why power has become the major parameter in VLSI, as it is the collaboration of MOSFET devices, due to its characteristics properties it has the feather of conducting with less power, taking this in to account design engineers has contributed the work with several techniques to achieve ultra low power. Due to the rapid growth in technology, power become the main concern parameter for every operating system because now a days we are surviving with smart technology before this came into existence few decades ago power dissipation was the major issue rather than the power consumption because as now we have a high end technology resources which are further developed for upcoming technologies.

BACKGROUND

This chapter deals with the brief description of the challenges facing by the present semiconductor industries and researchers. There are many techniques implemented to improve analog and digital circuits which can enhance the reduction of power consumption as well as leakage current. Another crucial driving factor is that excessive power consumption is becoming the limiting factor in integrating more transistors on a single chip or on a multiple-chip module. Unless power consumption is dramatically reduced, the resulting heat will limit the feasible packing and performance of VLSI circuits and systems.

WHY LOW POWER?

Digital integrated circuits are the motivating source in VLSI for innovating performance of various applications like computing and other digitized simulation, related to technology and science. The insist for digital integrated circuits will persist to enhance in near future, due to its some of the salient features like low power, reliable performance and improvements in the processing technology.

According to Moore's law, the prediction in development of integrated circuits has various interpretations such as the density of devices gets incremented to ten folds for every seven years. So the need of low power got arises for such progression forces of integrated circuits. The Intel 4004 microprocessor, developed in 1971, had 2300 transistors, dissipated about 1 watts of power and clocked at 1 MHz. Then the Pentium in 2001, with 42 million transistors, dissipating around 65 watts of power and clocked at 2.40 GHz.

While the transistor count got doubled as years go by, therefore the power dissipation increases exponentially, because of ever-shrinking size of integrated circuits, such power incrimination introduces a reliability concern such as electro migration and hot carrier induced device degradation, resulting in the loss of performance. The most convenient way to diminish the power consumption of digital circuits is finite reduction in supply voltage, in fact the average power consumption of CMOS circuits is proportional to the square of the supply voltage, and on other hand electric field should also be maintained at acceptable level. To ensure performance loss can be conquer for standard CMOS technologies by introducing more parallelism is shown and authors discusses about the research done in Veendrick (1984), Rabaey (2002), Horowitz, et.al (1994), and/or to modify the process and optimize it for low voltage operation is given in Horowitz, et.al. (1994), Sakurai (1990), Chandrakasan (1995).

The rules followed for analog circuits are moderately unlike to those of applied digital circuits. In order to elucidate these changes, the elementary limits in reducing of power is given in Blalock (1998) here the author reported that shrinking down the supply voltage will regrettably not reduce the power consumption of analog circuit, but can consume power in a circuit at a definite temperature with adjusted or related with the required signal-to-noise ratio (SNR) and the operating frequency (or the required bandwidth). Along with this first-order analysis is also obtain for an absolute minimum power consumption required to process analog signals, is almost independent of the supply voltage. Along with this there are some of the basic limits and the practical limits which give additional obstacles for the power reduction is also described. This trend allows low-power to emphasize some aspect of MOS modeling for low-voltage and low-current analog circuit design and simulation. Particularly, there is a need to have the clear concepts of MOS transistor operating at very low current and to properly model the operation of the device in the weak and moderate inversion regions, which have been ignored by most designers for years, has become a must. The availability of a good MOS transistor model has thus become a real issue for the efficient design and simulation of high performance analog and digital integrated circuits.

SOURCES OF LOW POWER

In 1990's the growth of electronic market was mainly concern and drew attention on low power design approach An overview describing change in power dissipation of CMOS circuits changed since 1980 has been given by Kuroda and Sakurai (1995). Commercial digital signal processor (DSP) and microprocessor unit (MPU) were used as examples. Power dissipation of commercial CMOS circuits got increased due to increased clock frequency. For example, in 1982, power dissipation of a 5 V DSP chip was 0.1 W and increased to 10 W in 1991. Power dissipation peaked in 1990-1991. After 1991, power dissipation of commercial CMOS chips have decreased since due to increasing demand of large number of portable systems such as notebook computers and cellular phones where very low to ultra-low power dissipation is a primary requirement.

In 1996, Montanaro et al. (1996) discussed typical power saving techniques related to a low power CMOS microprocessor design. They used a DEC Alpha 20164 CMOS microprocessor as a benchmark and reported power dissipation of 26 W at 200 MHz and 3.45 V supply voltage. Among several reported methods, method-1 reduces functions of a microprocessor (e.g. floating point function is removed for some special target applications). As a result, power dissipation is reduced by a factor of 3. Method-2 uses a processing scaling technique, which reduces node capacitances, and thus the power consumption. The power is reduced by a factor of 2 from 0.75 μm CMOS technology to 0.35 μm CMOS technology. In method-3, the capacitive load in the clock is reduced. The power is reduced by a factor of 3 after the latch circuits are redesigned. Method-4 reduces the clock rate from 200 MHz to 160 MHz, and the power dissipation is reduced by a factor of 1.25. Method-5 reduces the power supply voltage, VDD from 3.45 V to 1.5 V. The power dissipation is reduced by a factor of 5.3. From the above methods, the voltage scaling technique is recognized as a straightforward and also an aggressive technique due to dependence of power dissipation on the square of the supply voltage value. The power dissipation, P_{total} in digital CMOS circuits is given in the following relation

$$P_{total} = P_t\left(C_L \times V \times V_{DD}\right) \times f_{clk} + P_t\left(I_{sc} \times V_{DD} \times t_{switch}\right) \times f_{clk} + I_{leakage} . V_{DD} \tag{1}$$

From the Equation (1) the Ptotal is represented as total power dissipation and is given as summation of dynamic power, short circuit power and the static power. The first term of equation represent the switching of dynamic power dissipation, where CL is the equivalent load capacitance and *f*clk is the clock frequency, V is the voltage swing and pt is the probability that a power dissipation transition occurs. The probability pt is also called the switching activity factor and is normally taken as unity. The voltage swing, V is same as voltage supply, VDD except in pass-transistor based implementations where voltage swing is reduced by the threshold voltage Veendrick (1984). The second term in Equation (1) is denoted as a short-circuit power dissipation which is due to moderately switching-on of both transistors (n-MOS and p-MOS transistors) in CMOS for a switching time, t_{switch} and characterized by the short circuit current, I_{SC}. The third term in Equation (1) is due to sub-threshold currents in MOSFETs and substrate injection effects. $I_{leakage}$ is the leakage current. The power-delay product or the energy per transition (P_{total}/f_{clk}) can be obtained from Equation (1).

From Equation (1) short-circuit power dissipation part can be neglected in comparison to the switching power, which reduces to the following equation and is normally used in CMOS design.

$$P_{total} = P_{dynamic} + P_{static} = C_L \times V_{DD}^2 \times f_{clk} + I_{leakage} \times V_{DD} \qquad (2)$$

From Equation (2) neglecting the switching power gives the total power that gives dynamic power dissipation and static power dissipation. A brief idea about dynamic power dissipation is discussed above and can be scaled directly depending on the supply voltage, and one of the important merit to achieve low power dissipation is the direct way of scaling the Vdd such that can obtain low power dissipation

In order to operate the device with low power the characteristics property is that the usage of low voltage, and the voltage drop occur across gate and source, thus acquire low threshold voltage require to operate at low power applied voltage, which employ low threshold fabrication process. A new innovations and new design circuit techniques are ahead to achieve low threshold voltage techniques, which will the increase the fabrication cost and at the same level should maintain moderate thermal budget.

Power Consumption Components

Before discussing the issues of the power consumption component, how power is correlated with other parameters like frequency is also discussed frequency which is directly proportional to the power. This can be applicable with low or high frequencies, if the issue is with low frequency then the applied voltage is low and generated power is also which is not of that concern. But the systems are dealing with high frequencies it enforces a strict limit on the power consumption and will have minimum power consumption on boards. So here in this section it is given that theoretical calculations of the power consumption is solved with the parameters like determine power-supply sizing, current requirements, cooling/heat sink requirements, and criteria for device selection and can also establish the maximum reliable operating frequency, that are clearly visualized with the help of a CMOS inverter

Two components that is very essential in determining the power consumption in a CMOS circuit:

- Static power consumption
- Dynamic power consumption

Static power consumption is said to be very low power consumption in MOS circuit, which results in leakage currents. This can be achieved when all inputs are having with same applied logic levels and the circuit is not in conduction states. But, when switching at a high frequency, dynamic power consumption can contribute significantly to overall power consumption. Depending on the Charging and discharging a capacitive load further increases this dynamic power consumption.

Static Power Consumption

Normally, the circuits developed with low-voltage devices have a CMOS inverter as the basic element across input and output stage. In Figure 1, the CMOS implementation is shown. To understand the idea of static power consumption refers to the CMOS inverter modes shown in Figure 2.

From the above Figure 2, in CMOS circuit the drive device is PMOS and the load device is NMOS, here MOS transistor acts as a switch, as like if given input is logic 0, the n-MOS is OFF, and the p-MOS is ON (Case 1), and the brief discussion of static power is given in Chandrakasan, A. P.Et.Al (1999), Veendrick, H. J. M. (1984), Rabaey, J. M., & Pedram, M. (2002), Horowitz, M.Et.Al (1994), Sakurai, T. (1990) And Chandrakasan, A. P. (1995) the output voltage directly draw current from the source and current flow from source to drain and the output is logic 1. Similarly, when the input is at logic 1, the associated n-MOS device is biased ON and the p-MOS device is OFF. The output voltage is *GND*, or logic 0. Here we can observe that always one of the transistors always OFF when the gate is in either of these logic states. Since no current flows into the gate terminal, and there is no dc current path from V_{CC} to *GND*, the resultant quiescent (steady-state) current is zero; hence, static power consumption (*Pq*) is zero.

However, there is a small amount of static power consumption due to reverse-bias leakage between the junctions of diffused N-regions of source/drain and the substrate. Leakage current is developed inside a device due to the sub-threshold condition ie there is a formation of reverse biased diode across the source/drain regions and the substrate due to this reason there is a possibility of leakage current in the device usually in the range of 10-12 to 10-15 can be explained with a simple model that describes the parasitic diodes of a CMOS inverter. Mostly it can be neglected even. The source drain diffusion and

Figure 1. CMOS inverter mode for static power consumption

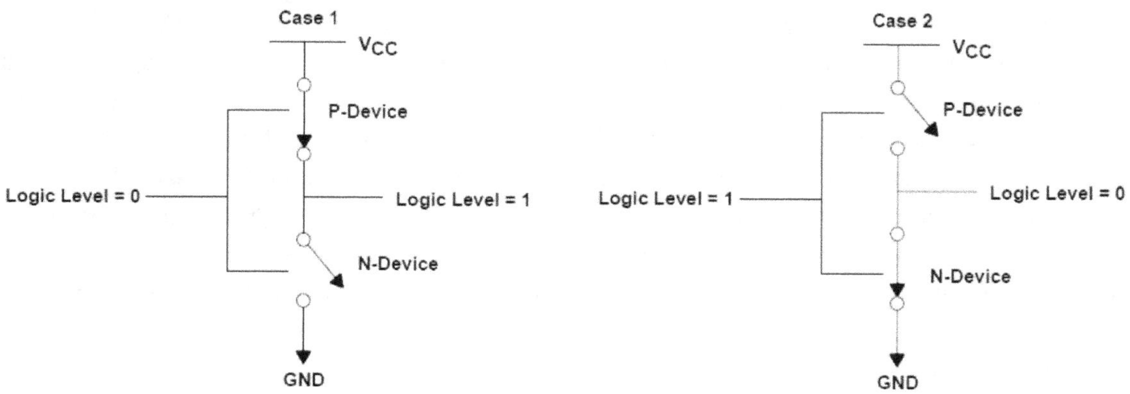

Figure 2. Dynamic Power Analyses

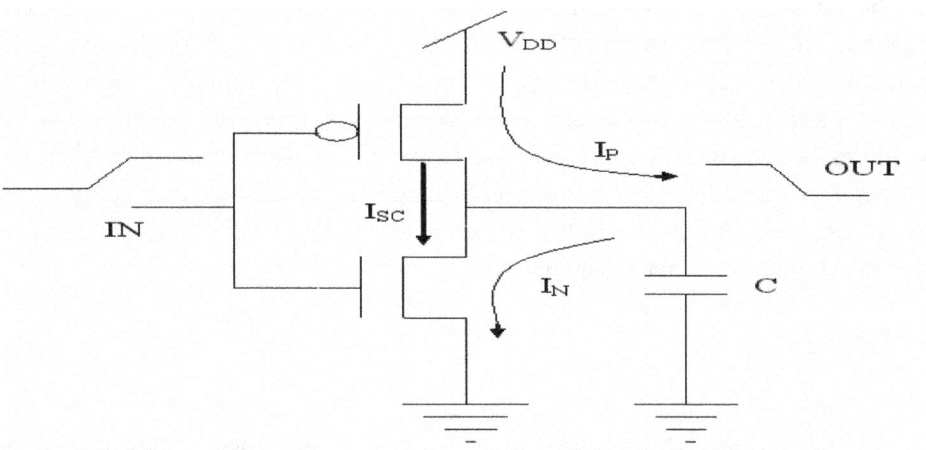

N-well diffusion form parasitic diodes. The parasitic diodes are between the N-well and substrate. These parasitic diodes are reverse biased; only their leakage currents contribute to static power consumption. The leakage current (I_{lkg}) of the diode is described by the following equation:

$$I_{lkg} = i_s \left(e^{\frac{qv}{kT} - 1} \right) \tag{2.1}$$

Static power consumption is the product of the device sb-threshold leakage current and the given supply voltage. Total static power consumption, P_S, can be obtained as shown in Equation (2.2).

$$P_s = I_c \times V_s \tag{2.2}$$

Most of the CMOS data sheets specifies ICC maximum in the range of 10mA to 40mA range. The source leakage current I_{CC} (current into a device), along with the supply voltage, causes static power consumption in the CMOS devices. This static power consumption is defined as quiescent, or PS, and can be calculated by Equation (2.3). The other source for static current $I_{CC,}$ which is because when input current will not drive completely to the rail, which gives transistor for partial switch OFF.

$$P_s = V_{cc} \times I_{cc} \tag{2.3}$$

Dynamic Power Consumption

The dynamic power consumption is defined and calculated by adding of transient power consumption (P_T), and capacitive-load power consumption (P_L) well shown by Kang & Leblebici (2003).

1. Transient Power Consumption

Difference in the current that occur when the switching takes place from OFF to ON condition or vice versa, this change in the current leads to charge the internal nodes plus through current. The frequency at which device is switching plus the rise and fall time of the input signal as well as the internal node of the device has a different effect of the current spike. For fast transition rate, through current is quiet negligible with that of the switching current. Due to the above effect dynamic current is governed by the internal capacitance and also the charge and discharge rate of the load capacitance

Transient power consumption can be calculated using equation (2. 4). N_{SW} in equation (2. 4) is 1 probability of switching power factor can be unity.

$$P_T = C_{pd} \times V_{cc}^2 \times f_i \times N_{sw} \tag{2.4}$$

In dynamic power consumption, the movement of the charge in parasitic capacitance consumes most of the power; hence dynamic power is the dominant factor in the CMOS circuit analysis. Assuming that if CMOS circuit consisting of several gates can be viewed as capacitor, that has high value, which charges and discharges between the power rails. Therefore the power dissipation capacitance is often specified as a measure of the equivalent capacitances and is use to approximate dynamic power dissipation. C_{pd} is defined as the internal equivalent capacitance of a device calculated by measuring operating current without load capacitance. Depending on the output switching capability, C_{pd} can be measured with no output switching (output disabled) or with any of the outputs switching (output enabled).

2. Capacitive-Load Power Consumption

Suppose if a circuit is operating with the external power supply then the power is consumed in charging the load capacitance and also depend on the switching frequency, following Equation (2.5) is used when all outputs have the same load and also switching at the same frequency

$$P_L = C_L \times V_{cc}^2 \times f_a \times N_{sw} \tag{2.5}$$

In the case of different loads and different output frequencies at all outputs, Equation (2.6) is used to calculate capacitive-load power consumption.

$$P_L = \sum_{i=1}^{n} C_{Ln} \times f_{on} \times V_{cc}^2 \tag{2.6}$$

Therefore, dynamic power consumption (P_D) is the sum of these two power consumptions and can be expressed as shown in Equation (2.7), Equation (2.8) (single-bit switching), and (multiple-bit switching with variable load and variable output frequencies).

$$P_D = P_T + P_L \tag{2.7}$$

$$P_D = (C_{pd} \times f_I \times V_{cc}^2) + (C_L \times f_O \times V_{cc}^2) \quad (2.8)$$

$$P_D = [(C_{pd} \times f_I \times N_{sw}) + \sum (C_L \times f_O)] \times V_{cc}^2 \quad (2.9)$$

Total power consumption is the sum of static and dynamic power consumption.

$$P_{total} = P_{static} + P_{dynamic} \quad (2.10)$$

Energy and Power Calculation

It is important to distinguish between energy and power. The power that consumed by a circuit is that the energy consumed per unit time. Equation (2.12) gives the power analysis of the CMOS

$$E = \int_0^{T_{op}} p(t)dt \quad (2.11)$$

$$P = CV_{DD}V_{sf} \quad (2.12)$$

where, C is the capacitance being recharged during a transition. V_{DD} is the supply voltage, V_s is the voltage swing of the signal, and f is the clock frequency. If it is assumed that an operation requires n clock cycles, T_{op} can be expressed as n / f. Hence, Equation (2.11) can be rewritten as

$$E = \eta VD_{DV}S_S \quad (2.13)$$

It is to be noted that energy obtained per calculation is independent to that of the clock frequency, as lowering the frequency will lower the power consumption where no change in the energy and perform for given operation was well analyzed by Weste (2008). As energy consumption determines the battery life, and it is very important to reduce the energy rather than reducing the power. However, to note that power is the major cause for the heat dissipation.

Propagation Delay

Several observations can be made from the analysis: PMOS was widened to match resistance of NMOS by 3 - 3.5. This was done to provide symmetrical H-to-L and L-to-H propagation delays. This also increases three times the PMOS gate and diffusion capacitances. Another source to speed-up the CMOS is by reducing the PMOS width (at the expense of symmetry and noise margins)! Widening PMOS reduces tpLH by increasing the charging current, but it also degrades the tpHL by causing a larger parasitic capacitance. This implies that there is an optimal ratio that balances the two contradictory effects.

Consider two identically sized CMOS inverters. The load cap of the first gate is approximated by:

$$C_L = (C_d p1 + C_d n1) + (C_g p2 + Cgn) + C_w \tag{2.14}$$

Now assume PMOS devices are made b times larger than NMOS

$$C_d p1 = \beta C_d n1 \ \& \ C_g p1 = \beta C_g \tag{2.15}$$

$$C_L = (1+\beta)(C_d n1 + C_g n2) + C_w \tag{2.16}$$

$$t_p = t_{pHL} + \frac{t_{pLH}}{2} = 0.69 C_l (\mathrm{R}_{eqn} + \mathrm{R}_{eqp} / \beta)$$

$$t_p = (0.345(1+\beta)(C_d n1 + C_g n2) + C_w) \mathrm{R}_{eqn} (1 + \gamma / \beta) \tag{2.17}$$

The calculated data is independent to the propagation delay of CMOS circuit, and the calculation of the delay was reported by Arreguit (1989).

$$t_{pd\,inv} = KC_{load} V_{DD} / I_o \exp\left[\frac{\frac{V_{DD} - V_{th}}{\eta Kt}}{q}\right] \tag{2.18}$$

System designer's perception, the performance of a digital circuit expresses the computational load that the circuit can manage. To know its performance it is often expressed by the duration of the clock period (clock cycle time) or its clock frequency. The minimum value of the clock period for a given technology and design is set by a number of factors such as the time it takes for the signals to propagate through the logic, the time it takes to get the data in and out of the registers, and the uncertainty of the clock arrival times. The propagation delay tp of a gate defines how quickly it responds to a change at its input(s). It expresses the delay experienced by a signal when passing through a gate. It is measured between the 50% transition points of the input and output waveforms, as shown in Figure 3 for an inverting gate.2 because a gate displays different response times for rising or falling input waveforms, two definitions of the propagation delay are necessary. Equation (2.19) defines the propagation delay the *tpLH* defines the response time of the gate for a *low to high* (or positive) output transition, while *tpHL* refers to a *high to low* (or negative) transition. The propagation delay *tp* is defined as the average of the two.

$$t_p = t_{pLH} + \frac{t_{pHL}}{2} \tag{2.19}$$

Figure 3. Propagation delays and rise and fall time

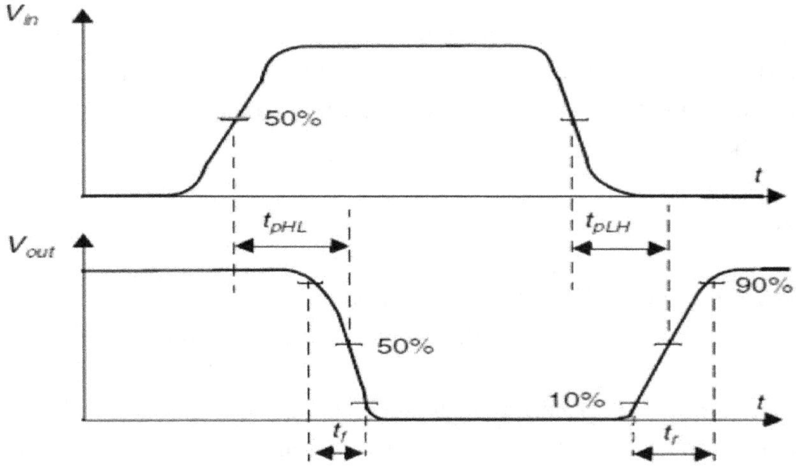

Power Delay Product

Figure 4 shows another important advantage of sub-threshold circuits: low power-delay product (PDP). The PDP for V_{dd}=0.5V is about 40 times smaller than the PDP for V_{dd}=3V. The lower PDP is because the reduction in power consumption outweighs the increase in delay. The PDP is the amount of energy per switching. Thus, having a lower PDP means the sub-threshold circuit consumes less energy than the normal strong inversion circuit when both circuits operate with the same amount of switching activities. This reduction is expected as the energy per switching activity (PDP) is proportional to CV_{dd}^2.

Figure 4. Power Delay Product for Various Power Supplies

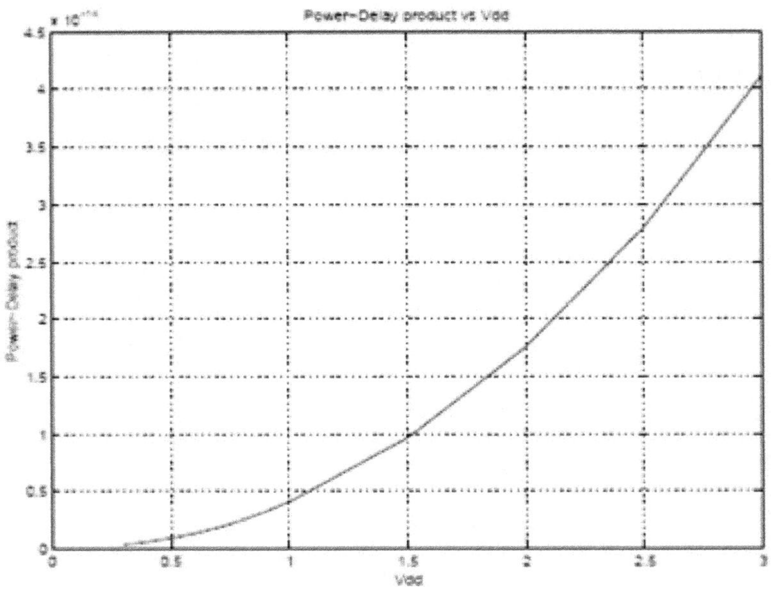

CMOS LOW POWER ANALOG TECHNIQUES

The section presents the low power analog techniques which give brief description about the parameters that involve in limiting the low power. By limiting the technology some disadvantages may arise in both analog and digital low power technique but on the other hand advantages plays a major role in limiting the power from the above Figure 5.

Practical Limits

The limits discussed are not dependent on technology or on the power supply voltage. However, there are number of obstacles or technological limitations which are then limited with practical circuitry approach:

1. Capacitors increase the power necessary to achieve a given bandwidth. They are only acceptable if their presence reduces the noise power by the same amount (by reducing the noise bandwidth). Therefore, parasitic capacitors very often increase power consumption.
2. The power depleted in bias circuitry is wasted and should in principle be minimized. However, inadequate bias schemes may increase the noise and therefore require a proportional increase in power. For example, a bias current is noisier if it is obtained by multiplying a smaller current.
3. According to Yeo, Rofail & Goh (2002), power is increased if the signal at any node corresponding to a functional pole (pole within the bandwidth, or state variable) has peak-to-peak voltage amplitude smaller than the supply voltage V_B. Thus, care must be taken to amplify the signal as early as possible to its maximum possible voltage value, and to maintain this level all along the processing path. Using current-mode circuits with limited voltage swings is therefore not a good approach to reduce power, as long as the energy is supplied by a voltage source. It only becomes attractive if voltage companding techniques can be used.
4. When capacitive loads are imposed (for example by parasitic capacitors), the current I necessary to obtain a given bandwidth is inversely proportional to the transconductance-to-current ratio of

Figure 5. Minimum powers for analog and digital circuits

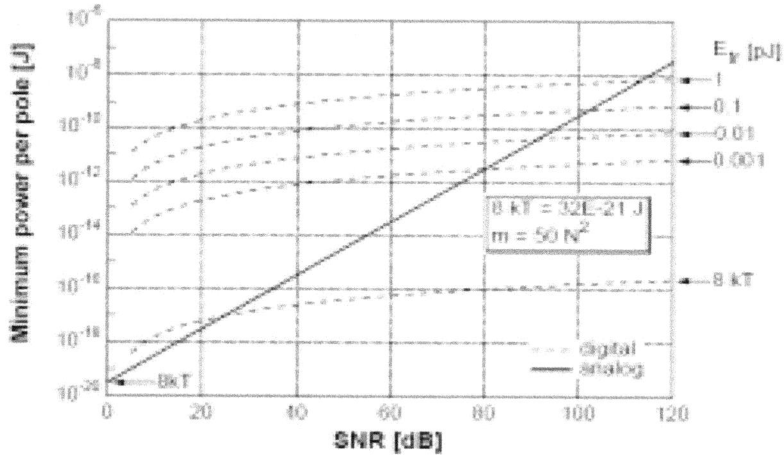

the active device. The small value of inherent to MOS transistors operated in strong inversion may therefore cause an increase in power consumption.

5. The need of precision usually leads to the use of larger dimensions for active and passive components, with a resulting increase in parasitic capacitors and power.
6. Switched capacitors should be clocked at a higher frequency and is twice the signal frequency. In some applications we can find the power consumption which reduces the effect of various analog design ranging from device to system.

Other Obstacles to Low-Power

In addition to the fundamental and practical limitations discussed previously, there are also historical or even psychological barriers to the efficient design of LP analog circuits. The most important can be listed as:

1. Analog blocks must often be taken from existing libraries with bias currents at the milli-ampere level and with architectures that are not compatible with low-voltage or low-current.
2. The use of very low bias currents is often discarded due to a lack of adequate transistor models and correct characterization of transistors parameters as well as worst case leakage currents. Another obstacle is the fear of breaking the psychological microampere barrier.
3. The requirements on PSRR are often exaggerated and mistaken for insensitivity to noise generated on chip.

To obtain low power techniques the Chen et al. (1996) proposed the back gate forward body-bias method to lower threshold voltage of a MOSFET in standard CMOS technology. Based on this method, inverter with power supply voltage < 1 V was designed. Another low power design using the forward body-bias is the dynamic threshold MOSFET (DTMOS) Assaderaghi, et.al (1994) technique where the body is tied to the gate. This method forces a forward body-bias when the transistor is on and a reverse body bias when the transistor is off. The DTMOS mode was demonstrated for a 0.6 V low power CMOS circuits. The forward body-bias method was used for the analog circuit design by Blalock et al. [1998]. A 1-V op-amp design in a standard digital CMOS technology was proposed.

Forward Body-Bias Method

Traditionally MOSFET is a four terminal device, but the utilization is only with three terminals i.e. source, drain, gate and body terminals from the Figure 6 (a) source and substrate terminals are shorted where there is no body effect means zero body bias threshold device. And on the other hand Figure 6 (a) replicates that source and body terminals are forward or reverse bias. Under forward bias the threshold voltage of MOSFET is reduced and in reverse bias threshold voltage increases. Due to this, the relation between threshold voltage and the body bias implies that MOSFET utilizes its forth terminal as body bias effect.

The back gate forward body-bias method shown in Figure. 6 (b) is compatible with CMOS processes. The threshold voltage of MOS transistors can be reduced electrically without any technology modification (Figure 7).

Figure 6. (a). shows a standard CMOS inverter with shorted body-source transistors. Figure 6 (b). shows a CMOS inverter under forward body-bias

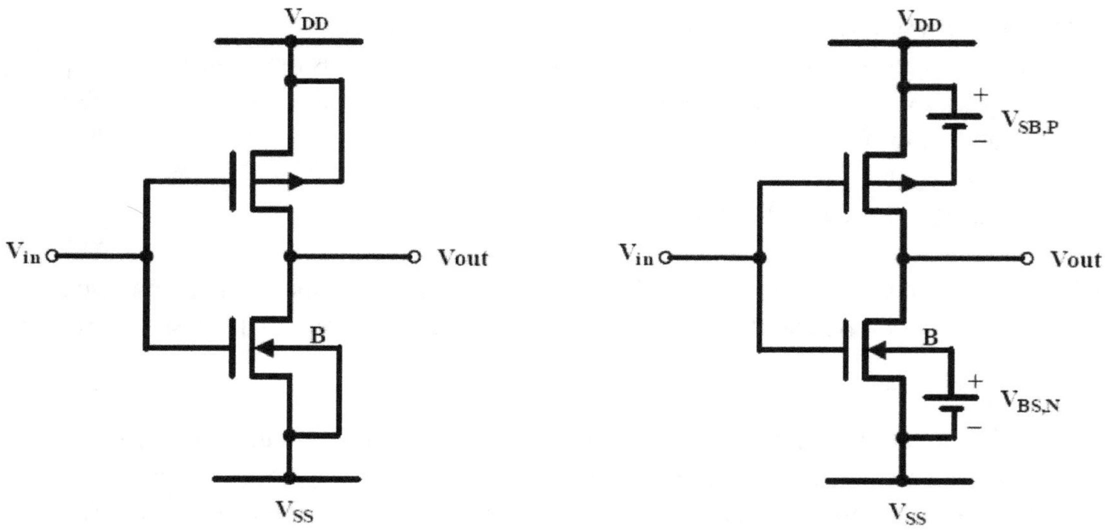

(a) Body-source shorted (b) Forward-body bias

Figure 7. Circuit diagram of a bulk-driven CMOS amplifier

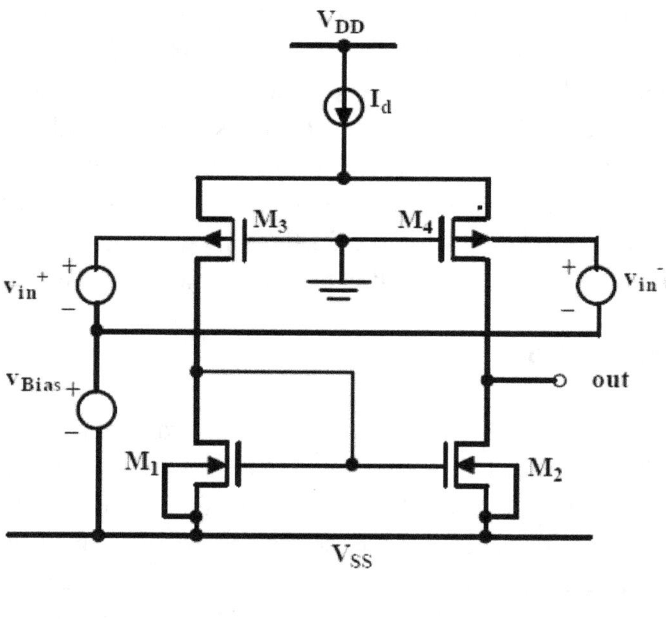

In the papers of Assaderaghi, et.al (1994), Blalock (1998), MOS transistor has been used as a four terminal device with drive-in voltage applied to the bulk or the substrate as shown in Figure 8. In Figure 8, a 1-V op-amp was designed using the bulk-drive technique, which provided an important solution to the threshold voltage limitation. A drawback of the bulk-driven technique is that it reduces the input

Figure 8. Cross-section of an idealized n-channel MOS transistor and the corresponding symbol. All voltages are referred to the local p-type substrate (21-25)

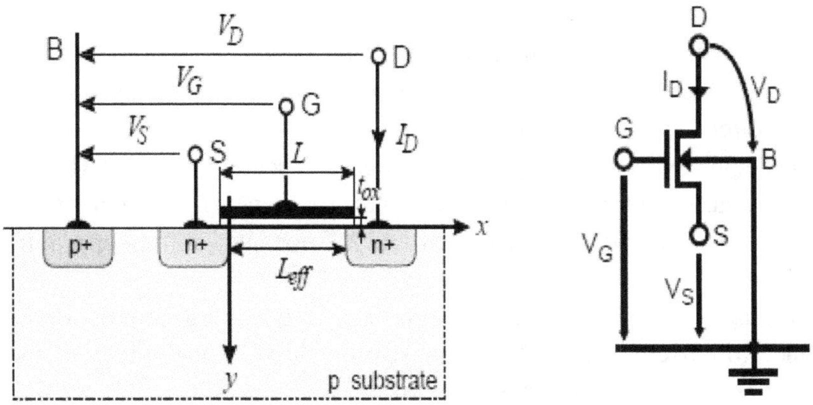

impedance of an amplifier. Recently, the back-gate forward bias technique has been used extensively to design low power digital and analog circuits Kuroda, et.al (1995), Montanaro, et.al. (1996), ITRS (1997), Chatelain, et.al (1979) and Chen, et.al (1996).

Implications of Supply Voltage Reduction

Unlike digital circuits, where the dynamic power decreases with the square of the supply voltage, according to Rabaey, et.al (2002), reducing the supply voltage of analog circuits while preserving the same bandwidth and SNR, has no fundamental effect on their minimum power consumption. However, this absolute limit was obtained by neglecting the possible limitation of bandwidth B due to the limited transconductance g_m of the active device. The maximum value of B is proportional to voltage replacing the capacitor value C by and expressing the product of the SNR times the bandwidth yields:

$$SNR \times B = V_{pp}^2 \times g_m / 8kT \tag{3.1}$$

In most cases, scaling the supply voltage V_B by a factor K requires a proportional reduction of the signal swing V_{PP}. Maintaining the bandwidth and the SNR is therefore only possible if the transconductance gm is increased by a factor K^2. If the active device is a bipolar transistor (or a MOS transistor biased in weak inversion), its transconductance can only be increased by increasing the bias current I by the same factor K^2; power is therefore increased by K. The situation is different if the active device is a MOS transistor biased in strong inversion. Its transconductance can be shown to be proportional to, where V_p is the pinch-off or saturation voltage of the device. Since this saturation voltage has to be reduced proportionally with V_B, increasing g_m by K^2 only requires an increase of current I by a factor K and hence the power remains unchanged.

However, the maximum frequency of operation may be affected by the value of the supply voltage. For a MOS transistor in strong inversion, the frequency f_{max} for which the current gain falls to unity is approximately given by:

$$f_{\max} = \frac{\mu V_p}{L^2} \tag{3.2}$$

Therefore, if the process is fixed (channel length L constant) a reduction of V_B and V_P by a factor K causes a proportional reduction of f_{max}. However, there is no fundamental reason to reduce the supply voltage of an analog circuit in a given process.

To maintain electrical field constant V_B is unavoidable at scaling process. Therefore both V_P and L are then scaled by the same factor K and the maximum frequency f_{max} is increased by K. hence for a bipolar transistor, V_P in Enz, (1995) is replaced by and f_{max} does not, in first approximation, depend on the supply voltage V_B.

Low-voltage limitations are not restricted to power or frequency problems. Reducing V_P increases the transconductance-to-current ratio of MOS transistors which in turn increases the noise content of current sources and drastically degrades their precision.

Conductance in analog switches is difficult to ensure when the supply voltage falls below approximately the sum of the p- and n-channel transistor threshold voltages. For a given value of time constant, charge injection in a switch does not depend on V_B in absolute value, but it increases in relative value if V_B and V_{pp} are decreased. The same is true for any constant voltage overhead such as the base-emitter voltage in bipolar transistors or the threshold voltage in MOS transistors.

Drain Current, Pinch-Off Voltage and Modes of Operation

Figure 8 shows the cross-section and the corresponding symbol of an idealized n-channel MOS transistor. The intrinsic geometric and operating symmetry of the device with respect to the source and the drain can be preserved in the model by referring the source voltage V_S, the gate voltage V_G and the drain voltage V_D to the local substrate.

This is clearly not the convention adopted for SPICE models for which all potentials are referred to the source electrode. The drain current is defined positive entering the drain electrode which is the most positive electrode between source and drain.

A general expression for the drain current that includes both the drift and the diffusion transport mechanisms is shown in Equation 3.3

$$I_D = W\mu_n(-Q'_{inv})\frac{dVch}{dx} \tag{3.3}$$

From the above equation it is assumed that the mobility *is* constant along the *y* axis and that the carrier velocity is not saturated. According to Enz, (1996), the drain current is proportional to the gradient of the channel voltage (or the longitudinal electrical field *Ex*) and to the *mobile inversion charge density.* The *channel voltage* V_{ch}, which depends on the position along the channel, is defined as the difference between the quasi-Fermi potential of the mobile carriers forming the channel and the Fermi potential *F.* This channel potential represents the disequilibrium in electron distribution produced by the source and the drain voltages.

As shown in Figure 9, for a given gate voltage larger than the threshold voltage V_{TO}, the inversion charge is almost a linear function of the channel voltage (23, 26). The channel voltage for which becomes

Figure 9. Inversion charge versus the channel potential

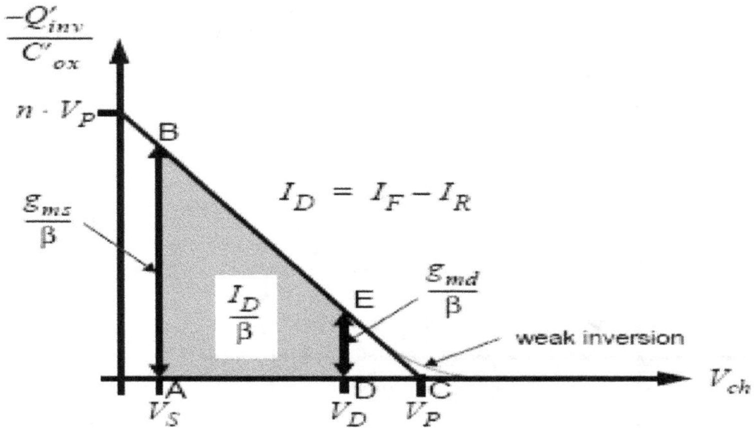

zero is defined as the *pinch-off voltage* V_P and corresponds in Figure 10 to the intersection of with the *x*-axis (point C). V_P represents the voltage that should be applied to the equi-potential channel (source and drain connected together) to cancel the effect of the gate voltage. It can be interpreted as the effect of the gate voltage referred to the channel and is thus directly related to the gate voltage by (Figure 11 and Table 1):

$$V_P = V_G - V_{TO} - \gamma \left[\sqrt{V_G - V_{TO} + \left(\sqrt{\Psi_o} + \frac{\gamma}{2} \right)^2} - \left(\sqrt{\Psi_o} + \frac{\gamma}{2} \right) \right] \quad (3.4)$$

$$V_G = V_{TO} + V_P + \gamma \left[\sqrt{\Psi_O + V_P} - \sqrt{\Psi_O} \right] \quad (3.5)$$

Figure 10. Pinch-off voltage and slope factor versus gate voltage

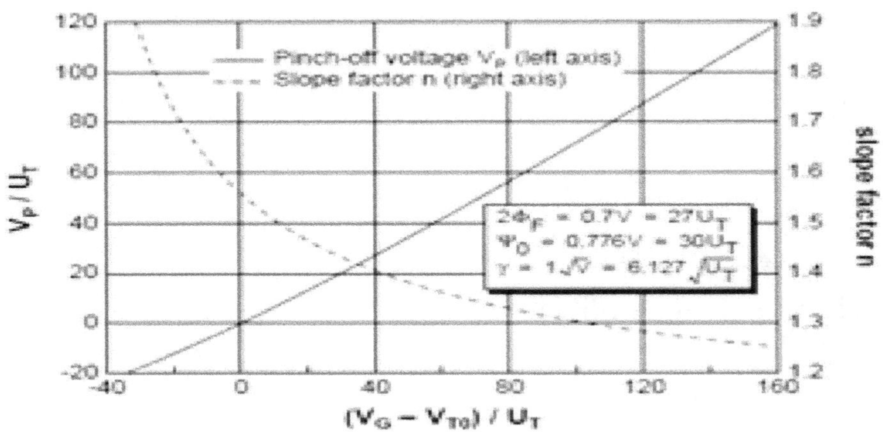

Figure 11. Modes of operation of the transistor

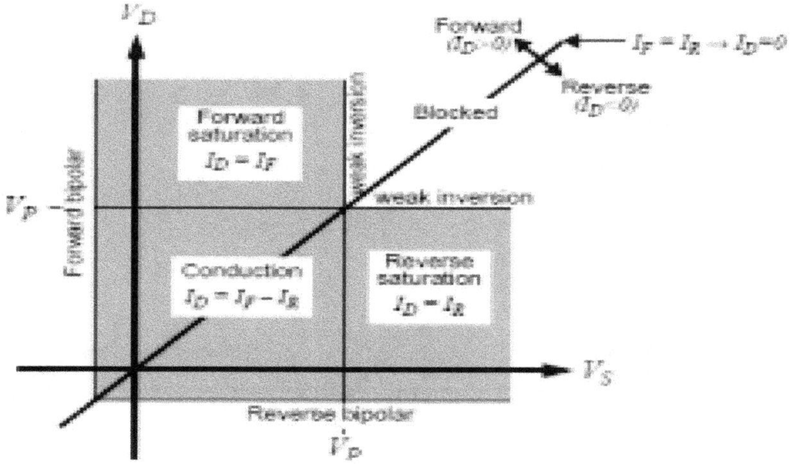

Table 1. Drain current in strong inversion

Mode	Drain Current Expression	Condition
Conduction	$\eta \times \beta \times \left[V_P - \frac{V_S + V_D}{2}\right] \times (V_D - V_S) \cong \beta \times \left[V_G - V_{TO} - \frac{\eta}{2}.(V_S + V_D)\right] \times (V_D - V_S)$	$V_S \leq V_P$ $V_D \leq V_P$
Forward Saturation	$\frac{\eta \times \beta}{2} \times (V_P - V_D)^2 \cong \frac{\beta}{2n} \times (V_G - V_{TO} - \eta \times V_D)^2$	$V_S \leq V_P$ $V_D > V_P$
Reverse Saturation	$\frac{-\eta \times \beta}{2} \times (V_P - V_D)^2 \cong \frac{-\beta}{2\eta} \times (V_G - V_{TO} - \eta \times V_D)^2$	$V_S > V_P$ $V_D \leq V_P$
Blocked	$I_F = I_R \Rightarrow I_D = 0$	$V_S = V_D$

$$\gamma = \frac{\sqrt{2q\varepsilon_S N_{sub}}}{C_{ox}} \tag{3.6}$$

$$n = \frac{dV_G}{dV_p} = 1 + \frac{\gamma}{2\sqrt{\Psi_O + V_P}} \tag{3.7}$$

$$\frac{1}{n} = \frac{dV_P}{dV_G} = 1 - \frac{\gamma}{2\sqrt{V_G - V_{TO} + (\frac{\gamma}{2} + \sqrt{\Psi_O})^2}} \tag{3.8}$$

$$V_P = \frac{V_G - V_{To}}{n(V_G)} \tag{3.9}$$

$$I_D = \beta \int_{V_S}^{V_D} \left[\frac{-Q_{inv}(V_{ch})}{C_{ox}} \right] dV_{ch} \tag{3.10}$$

where

$$\beta = \mu_n C_{ox} \frac{W}{L} \tag{3.11}$$

$$I_D = \beta \int_{V_S}^{\infty} \left[\frac{-Q_{inv}(V_{ch})}{C_{ox}} \right] dV_{ch} - \beta \int_{V_D}^{\infty} \left[\frac{-Q_{inv}(V_{ch})}{C_{ox}} \right] dV_{ch} \tag{3.12}$$

$$I_{F(R)} = I_S \exp\left[\frac{V_P - V_{S(D)}}{U_T} \right] \tag{3.13}$$

$$I_S = 2n\beta U_T^2 \tag{3.14}$$

At given temperature, depends essentially on the aspect ratio of the device and on the mobility. As shown in Figure 11, I_S corresponds in fact to the intersection of the weak and strong inversion asymptotes of the normalized trans-conductance versus drain current plot. *IS* can therefore also be used as a delimiter between weak and strong inversion. If drain current reaches to forward saturation, i.e. if drain current larger than I_S forces the device to operate in strong inversion, while the drain current smaller than I_S then transistor operates in weak inversion. The specific current is thus a convenient design parameter that helps sizing the transistor according to bias current and a given mode of operation.

It is convenient to explicit the current in terms of the gate voltage by using the approximation (Table 2):

$$I_D = I_F \left\{ 1 - \frac{I_R}{I_F} \right\} = I_S exp\left[\frac{V_P - V_s}{U_T} \right] \left\{ 1 - exp\left[-\frac{V_D - V_S}{U_T} \right] \right\} \tag{3.15}$$

Table 2. Drain current in weak inversion

Mode	Drain Current Expression	Conditions
Conduction	$I_s \times e^{\frac{V_p}{U_T}}[e^{\frac{V_s}{U_T}} - e^{\frac{V_D}{U_T}}] = I_{D0} \times e^{\frac{V_G - n\times V_s}{n\times U_T}}[1 - e^{-\frac{V_D - V_S}{U_T}}]$	$V_S > V_P$, $V_D > V_P$, $I_{D0} \equiv I_S \times e^{\left[\frac{-V_{T0}}{n.U_T}\right]}$
Forward saturation	$I_F = I_S \times e^{\left[\frac{V_P - V_S}{U_T}\right]} \simeq I_{D0} \times e^{\left[\frac{V_G - n\times V_S}{n\times U_T}\right]}$	$V_S > V_P$, $V_D > V_P$, $V_D - V_S \gg U_T$
Reverse saturation	$-I_R = -I_S \times e^{\frac{V_P - V_D}{U_T}} \cong -I_{D0} \times e^{\frac{V_G - n\times V_D}{n\times U_T}}$	$V_S > V_P$, $V_D > V_P$, $V_D - V_S \gg U_T$
Blocked	$I_F = I_R \Rightarrow I_D = 0$	$V_S \gg V_P$, $V_D \gg V_P$, $V_S = V_D$

The Small Signal Analysis

The small-signal trans-conductance of a MOS transistor biased in conduction is defined using the voltage referred to the substrate:

$$g_{mg} = \frac{\partial I_D}{\partial V_G}\bigg|_{V_S V_D}, \qquad g_{ms} = \frac{\partial I_D}{\partial V_S}\bigg|_{V_G V_D} \quad g_{md} = \frac{\partial I_D}{\partial V_D}\bigg|_{V_G V_S} \tag{3.16}$$

Where g_{mg}, g_{ms} and g_{md} are respectively the *gate, source* and *drain trans-conductance* of the model with voltages referred to the substrate. It is important to note that since the variation of the pinch-off voltage is n times smaller than the corresponding gate voltage variation and the forward current only depends on the voltage difference, the gate trans-conductance in saturation (i.e. for) is n times smaller than the source trans-conductance:

$$g_{mg} = g_{ms}/n \tag{3.17}$$

This relation is valid in saturation from weak to strong inversion.

The value of the trans-conductance's in strong and in weak inversion can be calculated respectively by Arreguit, X. (1989) & Vittoz, E. A. (1985, June). They are summarized in Table 3. Source and gate trans-conductance's in saturation are proportional to the drain current when the transistor is biased in weak inversion and proportional to the square root of the drain current in strong inversion.

In forward saturation, the drain trans-conductance becomes negligible compared to the output conductance g_{ds} due to the channel length modulation. It is important to note that although the gate trans-conductance g_m and output conductance *gds* of the model having the voltages referred to the source are respectively equal to g_{mg} and g_{ds} of the model with voltages referred to the substrate, the trans-conductance from the bulk is related to g_{mg}, g_{ms} and g_{ds} according to Enz, (1996):

Table 3. Trans-conductance's in weak and in strong inversion

	Strong Inversion		Weak Inversion
	Conduction	**Forward Saturation**	
g_{mg}	$\beta.(V_D - V_S)$	$\beta.(V_P - V_S) = \sqrt{\frac{2.\beta.I_D}{n}} = \frac{2.I_D}{n.(V_P - V_S)} = \frac{2.I_D}{V_G - V_{T0} - n.V_S}$	$\frac{I_D}{n.U_T}$
g_{ms}	$n.\beta.(V_D - V_S) = \sqrt{2.n.I_F} = \frac{2.n.I_F}{V_G - V_{T0} - n.V_S}$		$\frac{I_F}{U_T}$
g_{md}	$n.\beta.(V_P - V_D) = \sqrt{2.n.\beta.I_R} = \frac{2.n.I_R}{V_G - V_{T0} - n.V_D}$	0	$\frac{I_R}{U_T}$

$$g_{mb} = g_{ms} - g_{mg} - g_{ds} = g_{ms} - g_{mg} = (n-1)\,g_{mg} = \frac{n-1}{n}\,g_{ms} \tag{3.18}$$

Figure 12 shows the gate and source trans-conductance in saturation normalized to their value in weak inversion versus the inversion coefficient. As mentioned earlier, the strong inversion a symptote crosses the weak inversion asymptote (=1) at a saturation current equal to the specific current. At this particular point the error made on the trans-conductance estimation when using the weak inversion expression of Table 2 and 3 reaches its maximum which is about 40%. This is clearly unacceptable for doing analog design. Furthermore, transistors are very often biased in this moderate inversion region since it is a good compromise between current efficiency and gate area. A better estimation of the trans-conductance in the moderate inversion region is obtained by differentiating the large-signal drain current interpolation given by Vittoz & Neyroud (1979):

Figure 12. Trans-conductance interpolation functions

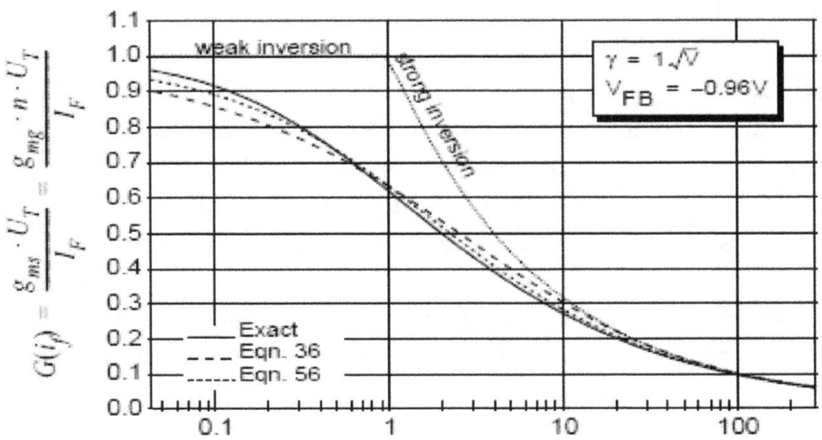

This interpolation function and the result obtained from a numerical computation are plotted in Figure 12. The simplest approximation given by Arora, N. D. (1993) tends to overestimate the result in strong inversion, while it is slightly pessimistic in weak inversion. Therefore, the maximum approximation error is now reduced from 40% to typically 5%, which is sufficient for manual calculations; Of course for computer simulation more accurate small-signal interpolation function is required. Although simple small-signal interpolation functions can be found, they have to be integrated to obtain the appropriate large-signal interpolation function used for the current. Additionally, in some circuit simulators the derivatives are calculated numerically instead of using an analytic expression. In such a case, the use of a large-signal interpolation function obtained by integration of the small signal function ensures not only an accurate estimation of the operating point but also of the corresponding trans-conductances.

In above sections the description shows how to reduce power, by parameters utilizing and showing either in reducing the channel length decreasing the trans-conductance, overcoming the pinch off voltage, reducing the threshold voltage and knowing the forward and reverse bias etc. we can also achieve low power by designing techniques like adiabatic, GDI, PTL, ECRL etc .

LEAKAGE CURRENT TECHNIQUES

This section introduces leakage current techniques, let us know how the leakage current are correlated with low power, as this chapter deals with low power techniques leakage currents is the parameter that play key role in power dissipation. From the above concept it is clear that MOS transistor will not conduct in cut off or sub-threshold condition, theoretically it is correct but if practical conditions are considered a small amount of current flow through the device. But if the current in the range of 10^{-15} or 10^{-18} then it can be negligible but if is in range of pico or nano then there is a possibility of leakage currents. Usually leakage currents can be observed in short channel effects. Generally considering the leakage current concept can achieve ultra low power; this can be obtained by utilizing leakage current techniques. Such as MTCMOS, VTCMOS, stack gate etc which is then reported by K Roy (2003).

SUMMARY

The chapter summarise with the brief idea about the low power in CMOS VLSI design and the processes to achieve and the possibility of various low power techniques in both analog and the digital circuits. The applications of MOSFET is in digital design is, it act as a switching circuit. And in analog design it act as an amplifier circuit. These are the major considered application in CMOS VLSI design. The chapter also shown the different low power techniques and challenges to achieve low power techniques are also proposed. A more detailed analysis include minimum power consumption of analog circuits is proportional to the ratio between the supply voltage and the signal peak-to-peak amplitude. Power efficient analog circuits should thus be designed to maximize the voltage swing and therefore should handle rail-to-rail signal voltages.

Many of the problems and solutions encountered in CMOS low-power analog circuit design are directly related to the properties of the MOS transistor itself, which must therefore be properly understood and modeled down to very low currents.

Suppressing leakage currents, an adaptive body biasing generator is used in Iddq testing in deep submicron technology, which is an important issue in current CMOS technique. Another important technique to improve power in microprocessors and multi-core processors an adaptive DC/DC converter are used. There are a different leakage currents technique which gives low power, among them MTCMOS, stack NMOS techniques is the effective one which gives appreciable low leakage and low power.

REFERENCES

Arora, N. D. (1993). MOSFETs models for VLSI Circuit Simulation, Theory and Practice. In Computational Microelectronics. Springer-Verlag.

Arreguit, X. (1989). *Compatible lateral bipolar transistors in CMOS technology: model and applications* (PhD Thesis N 817). EPFL, Lausanne.

Assaderaghi, F., Sinitsky, D., Parke, S., Boker, J., Ko, P. K., & Hu, C. (1994). A dynamic threshold voltage MOSFET (DTMOS) for very-low voltage operation. *IEDM Tech. Dig.*, 809-812.

Blalock, B. J., Allen, V., & Rincon-Mora, G. A. (1998, July). Designing 1-V op-amps using standard digital CMOS technology. *IEEE Trns. Circuits and System-II. Analog and Digital Signal Processing, 45*(7), 769–780.

Bult, K., & Geelen, J. G. M. (1992, December). An inherently linear and compact MOST-only current division technique. *IEEE Journal of Solid-State Circuits, 27*(12), 1730–1735. doi:10.1109/4.173099

Chandrakasan, A. P., & Brodersen, R. W. (1995). Low-power CMOS digital design. Kluwer Academic.

Chandrakasan, A. P., Sheng, S., & Brodersen, R. W. (1999). Low Power CMOS Digital Design. *IEEE Journal of Solid-State Circuits, 27*(4), 473–484.

Chatelain, J. D. (1979). *Dispositifs à semiconductours. In Traité d'électricité* (Vol. 7). Presses Polytechniques Romandes.

Chen, M. J., Ho, J. S., Huang, T. H., Yang, C. H., Jou, Y. N., & Wu, T. (1996, June). Back-gate forward bias method for low-voltage CMOS Digital Circuits. *IEEE Transactions on Electron Devices, 43*(6), 904–910. doi:10.1109/16.502122

Enz, C. C. (1989). *High Precision CMOS Micro-power Amplifiers* (PhD Thesis N 802). EPFL, Lausanne.

Enz, C. C., Krummenacher, & Vittoz, E. A. (1995, July). An Analytical MOS Transistor Model Valid in All Regions of Operation and Dedicated to Low-Voltage and Low- Current Applications. *Analog Integrated Circuits and Signal Processing Journal, 8*, 83-114.

Enz, C. C. (1996). The EKV model: a MOST model dedicated to low-current and lowvoltage analog circuit design and simulation in Low-power HF microelectronics. In IEEE Circuits. and Systems. IEEE Publications.

Horowitz, M., Indennaur, T., & Gonzalez, R. (1994, October). Low Power Digital Design. *Technical Digest IEEE Symposium Low Power Electronics.*

Kang, S., & Leblebici, Y. (2003). CMOS Digital Integrated Circuits - Analysis and Design. McGraw-Hill.

Kao, J. T., & Chandrakasan, A. P. (2000, July). Dual-threshold voltage techniques for low-power digital circuits. *IEEE Journal of Solid-State Circuits*, *35*(7), 1009–1018. doi:10.1109/4.848210

Kuroda, T., & Sakurai, T. (1995, April). Overview of low-power ULSI circuit techniques. *Inst. Of Electronics, Information & Communication Engineers Tran. Electron*, *78*(4), 334–343.

Montanaro, J., Witek, R. T., Anne, K., Black, A. J., Cooper, E. M., Dobberpuhl, D. W., & Thierauf, S. C. et al. (1996, November). A 160 MHz, 32-b 0.5-W CMOS RISC microprocessor. *IEEE Journal of Solid-State Circuits*, *31*(11), 1703–1714.

Pao, H. C., & Sah, C. T. (1996). Effects of diffusion current on the characteristics of metal oxide (insulator)-semiconductor transistors. *Solid-State Electronics*, *10*, 927–937.

Rabaey, J. M., & Pedram, M. (2002). Low Power Design Methodologies. Kluwer Academic Publishers.

Sakurai, T., & Newton, A. R. (1990, October). Alpha-Power Law MOSET Model and its Applications to CMOS Inverter Delay and other Formulas. *IEEE JSSC, 25*(2), 584-594.

Tomishima, S., Tsuji, T., Kawasaki, T., Ishikawa, M., Inokuchi, T., Kato, H., & Takahashi, K. et al. (2001, November). A 1.0-V 230-MHz column access embedded DRAM for portable MPEG applications. *IEEE Journal of Solid-State Circuits*, *36*(11), 1728–1737.

Tsividis, Y., & Suyama, K. (1993, May). MOSFET Modeling for analog circuit CAD: problems and prospects.*Proc. IEEE Custom Integrated Circuits Conf.* doi:10.1109/CICC.1993.590710

Tsividis, Y. P. (1987). *Operation and modelling of the MOS Transistor*. Mc Graw-Hill.

Veendrick, H. J. M. (1984, August). Short-circuit Dissipation of Static CMOS Circuitry and its Impact on the Design of Buffer Circuits. *IEEE JSSC*, 468-473.

Vittoz, E. A. (1985, June). The Design of High-Performance Analog Circuits on Digital CMOS Chips. *IEEE Journal of Solid-State Circuits*, *20*(3), 657–665. doi:10.1109/JSSC.1985.1052365

Vittoz, E. A., & Neyroud, O. (1979, June). A low-voltage CMOS bandgap reference. *IEEE Journal of Solid-State Circuits*, *14*(3), 573–577. doi:10.1109/JSSC.1979.1051218

Wei, G. Y., & Horowitz, M. (1999, April). A fully digital, energy-efficient, adaptive power supply regulator. *IEEE Journal of Solid-State Circuits*, *34*(4), 520–528. doi:10.1109/4.753685

Weste, N. H. E., Harris, D., & Banerjee, A. (2008). *CMOS VLSI Design: A circuits and system perspective* (3rd ed.). Pearson education.

Yeo, K. S., Rofail, S. S., & Goh, W. L. (2002). *CMOS/BiCMOS ULSI-Low Voltage, Low Power*. Prentice Hall PTR.

Chapter 4
Leakage Minimization in CMOS VLSI Circuits:
A Brief Review

Saurabh Chaudhury
NIT Silchar, India

Rohit Lorenzo
NIT Silchar, India

ABSTRACT

Ever increasing demand for portable and battery-operated systems has lead to aggressive scaling. While technology scaling facilitates faster and high performance devices, at the same time it causes excessive power dissipation. Leakage power dissipation is now a dominating component of total power consumption in such portable devices. So there is a tremendous need to limit the power dissipation in high density chips which has initiated many innovative techniques to develop in the design of low power circuits and systems. Today's nano-scaled VLSI chips have ultra-thin gate oxide, very low threshold voltage and having short channels. As such leakage power dissipation has emerged as the most challenging issue in VLSI circuit and systems. This Chapter review and compare the state of the art circuit techniques for leakage minimization. It also conceptually classifies the different techniques of leakage minimization. Moreover, a detailed comparison based on trading-offs with other design parameters is also given along with leakage minimization.

1. INTRODUCTION

Modern digital integrated circuits (ICs) are characterized by high performances and higher packing densities. This is because of aggressive device scaling that is taking place in every generation of technology, which may continue for a few more years till the device physics continue to hold true. Higher integration is economically attractive because it leads to a reduction in number of components per system, thereby reducing the cost of interconnection and packaging. Moreover, it enhances the reliability of the overall

DOI: 10.4018/978-1-5225-0190-9.ch004

system. Higher integration makes a circuit faster because of the reduction in parasitic effects in devices and interconnects. With the increasing growth of portable, battery-operated digital equipments, like, cell phone, laptop, personal digital assistant, multimedia products etc., power consumption has become one of the major design issues today, besides area, performance, cost and yield. On the other hand, life of battery is also limited. In spite of remarkable developments in battery technology, such as, high capacity rechargeable Lithium-ion cells, a revolutionary change in battery technology in near future also seems feeble. Under these circumstances, the only possibility to reduce power consumption is by adopting some innovative low power circuit design strategies. Moreover, unnecessary power dissipation, when the circuit or its constituent parts are inactive, has to be limited by dynamic power management schemes. When a circuit is optimized in terms of power, there are a lot of advantages. It helps to develop and use less expensive packaging and cooling, since core is less heated-up. This is essentially very important from the point of view of overall cost of the system and hence low power strategy is even applied now-a-days to desktop or workstation computers to reduce packaging and cooling cost.

To achieve low power, more emphasis is given in each level of the design cycle, starting from system or architectural level, down to the physical level. Economics of production, on the other hand, not only depends on volume of production but also on design styles, intended applications, time-to-market and quality of circuit design and fabrication. In reality, very few circuits enjoy large volume of sales and a long life, e.g. some general purpose processors, but unfortunately technology advancements make them obsolete in a short time. Application Specific Integrated Circuits (ASICs), on the other hand, are designed to perform some limited set of tasks, but they capture a large portion of the market share. Computer Aided Design (CAD) techniques play a vital role in circuit optimization for some specific quality and in the reduction of design time. As the level of integration is continually increasing (of the order of millions of devices in a single chip), the task of carrying out a complete design without errors is becoming increasingly difficult for human designers. The technological advancements in the field of electronics have resulted in more and more cores and processors that are being put on smaller System on Chip (SoC) (Kaur, 2013). These high performance systems are mostly power hungry. CAD tools have been used since the inception of integrated circuits. CAD techniques achieved a fairly good level of maturity in many areas by today and will continue to support designing increasingly larger circuits with its enhanced capabilities. In order to reduce power consumption, supply voltage is continually scaled down, as the dynamic or the switching power is directly proportional to square of the supply voltage. Scaling of power supply demands for scaling the threshold voltage also, to avoid any performance degradation. Leakage power is of no concern in long channel devices. Reduction in threshold voltage gives rise to more and more leakage power consumption as subthreshold leakage is exponentially dependent on threshold voltage. Subthreshold leakage dominates over all other leakages above 100 nm devices. For technologies below 90 nm, oxide leakage and band-to-band tunnelling leakage predominates. Leakage power consumption has become comparable to dynamic power consumption at recent technology level and even higher than dynamic power in many applications and hence need to be emphasized early in the design cycle over and above dynamic power. The increasing trend of power dissipation according to international technology roadmap for semiconductor (ITRS) (http//www.itrs.net) is depicted in Figure 1.

The chapter is organized as follows. Section 2 describes different types of leakage currents. Section 3 elucidates the circuit level leakage minimization technique briefly and classifies them based on their strategies. Finally section 4 concludes the Chapter.

Figure 1. ITRS power dissipation trend

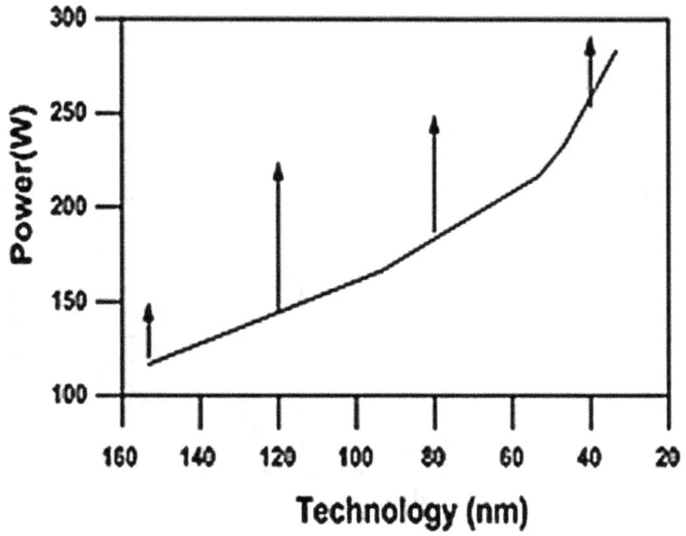

2. SOURCES OF POWER DISSIPATION

Power dissipation in digital circuits comprise of three main components. These are *static power dissipation* (P_{static}), which is mainly due to leakage current during standby mode, *dynamic power dissipation* ($P_{dynamic}$) is due to charging and discharging of output node capacitance during active mode and *short circuit power dissipation* ($P_{short\text{-}circuit}$) is due to the existence of any conducting path between power supply and ground. Thus the total power dissipation can be expressed as

$$P_{total} = P_{dynamic} + P_{static} + P_{short-circuit} \quad (1)$$

Amongst all, dynamic power dissipation is the dominant contributor when the technology is above 0.18 μm. As the CMOS technology is scaled down, the feature size and threshold voltage has shrunk drastically. This has led to exponential increase in leakage power (Borkar, 1999) as illustrated in Figure 2. Leakage power has emerged as the dominant contributor of total power consumption today and so the chapter is devoted on to it.

Figure 3 shows the mechanism of several sources of leakage current in MOSFET (metal oxide field effect transistor) devices. (I_1) is the reverse-biased junction leakage which flows when the p-n junction between the drain and bulk of the transistor is reverse biased (Pierret, 1996), sub threshold leakage current (I_2) due to feeble conduction below threshold voltage (Kahng, 2007; Kim, Choi, & Park, 2007), drain induced barrier lowering (I_3) occurs when depletion region of the drain interacts with the source (Roy, & Prasad, 1990), Gate induced drain leakage (I_4) arises in the high electric field under the gate-drain overlap which causes a depletion region (Brews, 1990). Punch through (Kang & Lebleblici, 2003) occurs when source and drain depletion region touch each other deep in the Channel (I_5). Narrow channel effect (I_6) occurs in MOS transistors which have channel width W of the same order of magnitude as the maximum depletion region thickness and are defined as narrow channel devices (Bohr, 1996). Gate oxide tunneling (Lee, Kwong, Blaauw, & Sylvester, 2003) arises due to the finite probability of an electron directly

Figure 2. Technology trend and power dissipation scenario

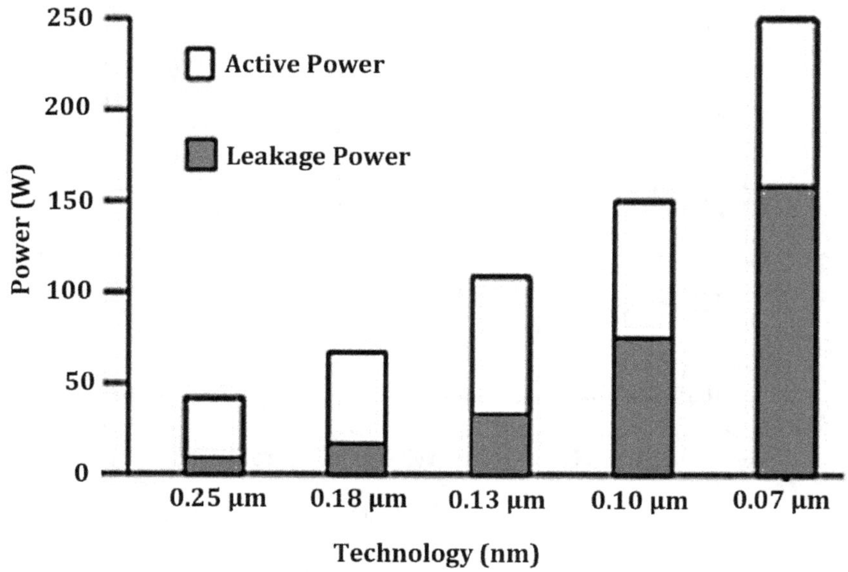

tunneling through the insulating sio_2 layer (I_7). Hot carrier injection (I_8) electron and holes that gain high kinetic energy by the electric field can be injected into the gate oxide (Kao, Narendra, & Chandrakasan, 2002). Now, if we look into the technological evolution from 90nm down to 11nm technology node (as shown in Table 1) as predicted by the international technology roadmap for semiconductor (ITRS'09), we see a trend in down scaling the important devices parameters of integrated circuits from 2003-2015. This trend in CMOS technology scaling will continue until it crosses the physical limits and device physics will not hold true, as long as there is a need to integrate more functionality in the same silicon area, reduced fabrication cost, and increased operating speed and less power dissipation. A demonstration on cost effective 32nm technology has been reported by Intel and IBM in 2008 (Chen et al., 2008).

3. CLASSIFICATION OF LEAKAGE CURRENTS

Here we present different components of leakage current, their behavior and expressions. The overall leakage in a device is the summation of three major leakage components--*band-band tunneling leakage* (BTBT), *subthreshold leakage* and *gate leakage*. Mathematically, this can be written as,

$$I_{overall} = I_{BTBT} + I_{SUB} + I_{gate} \quad (2)$$

3.1 Gate Direct Tunneling (Igate)

Gate direct tunneling current is the tunneling of electrons (holes) from the bulk silicon and source drain overlap region through the gate oxide potential barrier into the gate. The direct tunneling current density as in (Taur & Ning, 1998) can be expressed as,

Figure 3. Summary of leakage mechanism in MOSFET

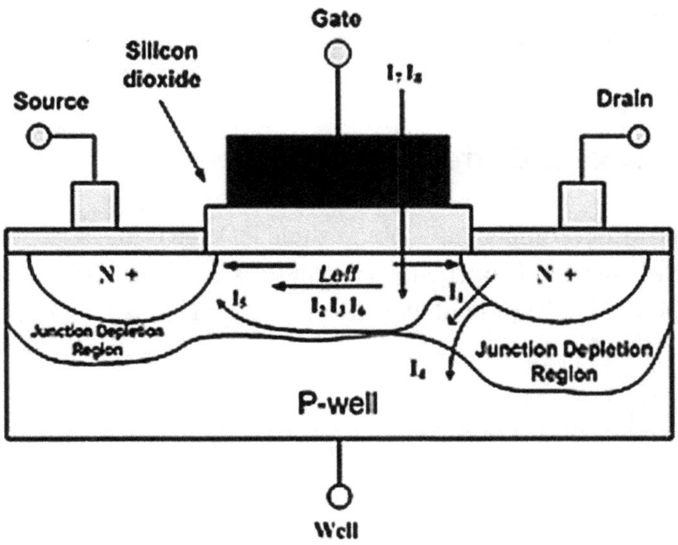

Table 1. Technological evolution and forecast up to 2015

Technology node (nm)	90	65	45	32	22	16	11
First Production	2003	2005	2007	2009	2011	2013	2015
effective gate length(nm)	50	35	30	25	18	12	9
gate material	poly	Poly	Metal	Metal	Metal	Dual	triple
Gate dielectric	Sio_2	SioN	HighK	HighK	HighK	HighK	HighK
Raw Mgates/mm^2	0.4	0.8	1.5	2.8	5.2	9	16
Memory Point(μ^2)	1.3	0.6	0.3	0.17	0.1	0.06	0.06

$$J_{DT} = A_g \left(\frac{V_{ox}}{T_{ox}}\right)^2 \exp\left[\frac{-B_g(1-(1-\frac{V_{ox}}{\varphi_{ox}})^{\frac{3}{2}})}{V_{ox}/T_{ox}}\right] \tag{3}$$

where,

$$A_g = \frac{Q^3}{16\pi^2\varphi_{ox}} \text{ and } B_g = \frac{4\sqrt{2m^*}\varphi_{ox}^{\frac{3}{2}}}{3\left(\frac{h}{2\pi}\right)\varphi_{ox}}$$

J_{DT} is the direct tunneling current density, V_{ox} is potential drop across oxide, φ_{ox} is the barrier height of tunneling electron, m^* is the effective mass of electron in conduction band of silicon and T_{ox} is the oxide thickness. The tunneling current increases exponentially with decrease in oxide thickness and increase in potential drop across oxide.

3.2 Subthreshold Leakage Current (ISUB)

Out of all other leakage current components, subthreshold leakage is the most dominant leakage component (Sakurai, 2004). Sub threshold conduction current between source and drain occurs when the gate voltage is below threshold (V_{th}). The sub threshold leakage current can be modeled as,

$$I_{sub} = I_0 e^{\left(V_{gs} - V_{th0} - \eta V_{ds} - \gamma \frac{V_{sb}}{nV_\theta}\right)} \left(1 - e^{\frac{-V_{ds}}{V_\theta}}\right) \tag{4}$$

Where, $I_0 = \mu C_{ox} \frac{W}{L} V^2 e^{1.8}$, $V_e = \frac{kT}{q} = 0.026v$ and $\eta = 1 + \frac{c_{dm}}{c_{ox}}$.

µ denotes carrier mobility, C_{ox} is the gate oxide capacitance per unit area, *W* and *L* denotes width and length of the channel in MOS transistor, *K* is the Boltzmann constant, *T* is the absolute temperature, q is the electrical charge of conduction, η denotes the subthreshold ideality factor, C_{dm} is the capacitance of the depletion layer and n is subthreshold swing coefficient.

Figure 4 shows how the subthreshold leakage current (I_{off}) increases exponentially as process technology scales down ("*IEEE devices and circuit magazine*", 2006). In Figure 5 sub threshold leakage current is plotted against the channel length at 65nm technology, where minimum channel length is 60nm.

Figure 4. Subthreshold leakage current trend with different technologies

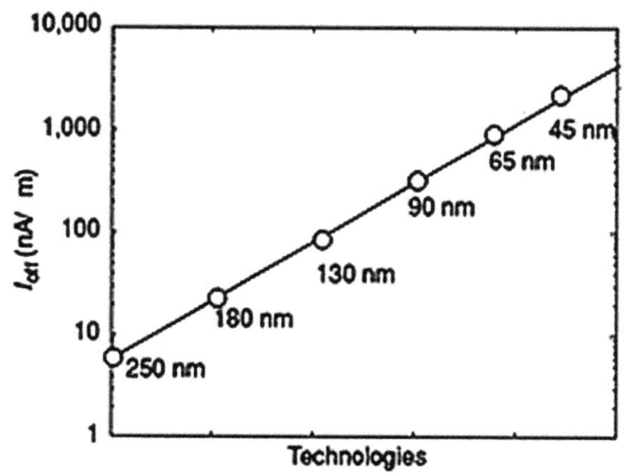

Figure 5. Subthreshold leakage current for NMOS /PMOS transistor

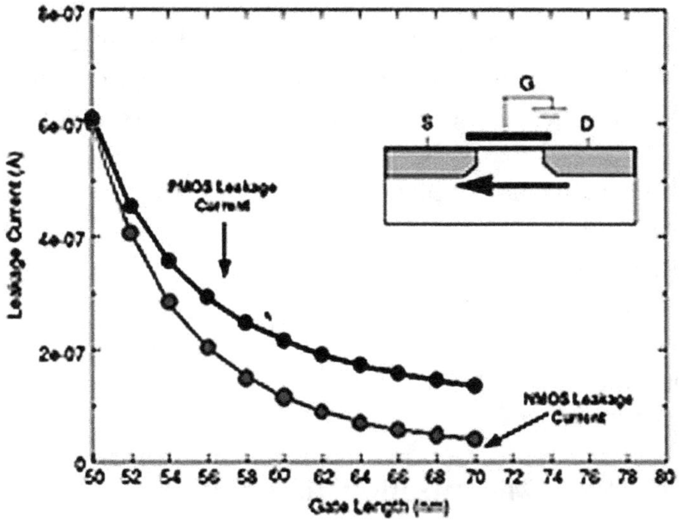

3.3 Band to Band Leakage Current (IBTBT)

A high electric field across a reverse biased p-n junction causes significant current to flow through the junction due to tunneling of electrons from the valence band of P-junction to the conduction band of n-region. It is evident that for such tunneling to occur the total voltage across the junction must be more than the band gap. Since silicon is an indirect band-gap semiconductor, The BTBT current in silicon involves emission and absorption of phonons (Taur &ing, 1998). The tunneling current density through a silicon p-n junction is given by

$$J_{b-b} = A\frac{EV_{app}}{\Sigma_g^{\frac{1}{2}}}\exp\left(-B\frac{\Sigma_g^{\frac{3}{2}}}{E}\right) \tag{5}$$

where,

$$A = \frac{\sqrt{2m^*q^3}}{4\pi^3\hbar^2} \text{ and } B = \frac{4\sqrt{2m^*}}{3q\hbar}$$

m* is the effective mass of electron, Σ_g is energy band gap, E is the electric field at the junction, q is electronic charge and ħ is the reduced plank's constant.

$$\eta = 1 + \frac{\text{Cdm}}{\text{Cox}}.$$

4. CIRCUIT LEVEL LEAKAGE MINIMIZATION TECHNIQUES

In this section, we review and classify some of the circuit level techniques for leakage minimization from the literature. All these techniques mainly reduce subthreshold leakage currents. Each technique has its own advantages and disadvantages. These techniques can be classified as follows:

- Subthreshold leakage minimization techniques
- Body Biasing techniques
- Transistor Stacking techniques
- Leakage Minimization with data retention
- Input Vector Control techniques
- Algorithmic Approaches
- Application oriented techniques

4.1 Subthreshold Leakage Minimization Techniques

Sub threshold leakage current is the feeble current that flows between source and drain of a MOSFET, when the transistor is still in OFF condition. In sub threshold region the gate to source voltage is below the threshold voltage of the device. Here we briefly describe the different techniques to minimize the subthreshold current. In each such technique, either a high-Vt transistor or an additional leakage control transistor is inserted in pull-up or pull down path.

4.1.1 Power Gating

Multi-threshold CMOS (MTCMOS) (Mutoh et al., 1995) is a popular power gating approach that uses high V_t (threshold voltage) device for power gating. High threshold transistors, which are slow and have low subthreshold leakage current, are used as sleep transistors (S and Sbar). The technique of MTCMOS consist of inserting an additional PMOS sleep transistor with control signal S (with high Vt) in series between pull up network (PUN) and supply, commonly known *as header* and one additional NMOS sleep transistor with control signal Sbar (with high Vt) between pull down network (PDN) and ground, known as *footer* as shown in Figure 6. During active mode both high V_t transistors are turned on, so low V_t transistors in pull-up and pull-down network operates normally according to logic inputs. The sleep transistors are turned off when the sleep signal goes high and the circuit becomes idle. Due to high threshold sleep transistors, it reduces the subthreshold leakage current as the path from power supply to ground is cut off by the sleep transistors. This techniques has however few drawbacks. First, the circuit will lose data when sleep transistors are turned off. Secondly, timing is critical to generate sleep signal, increased delay due to high Vt transistors and lastly, for large circuit transistor sizing is a difficult task.

4.1.2 Super-Cut Off CMOS Circuit

Super cutoff CMOS technique (Kawaguchi, Nose, & Sakurai, 2002) is an alternative to power gating technique. When the circuit is in standby mode, the sleep transistor is under driven, whereas, it is overdriven with a slight negative voltage. For example, an NMOS device would be turned off with a

Figure 6. Circuit representing MTCMOS (Mutoh et al., 1995)

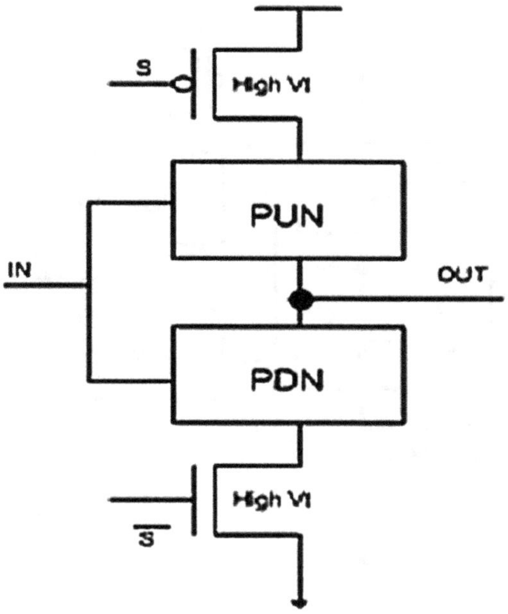

slight negative voltage instead of zero voltage. This means that the gate voltage is negative (Vg < 0). From equation (5) of sub threshold leakage current it is clear that if we apply a negative voltage at gate terminal then it will cause an exponential decrease in leakage current. However, this technique requires two additional power supplies and hence area increases.

4.1.3 Zig-Zag Approach

The zigzag approach (Min, Kawaguchi, & Sakurai, 2003) reduces the wake-up overhead caused by sleep transistor in power gating. It places an alternating sleep transistor assuming particular pre-selected input vector as shown in Figure 7. This approach uses fewer transistors than the original approach and removes the problem of floating node. However, designing the circuit through this approach is a complicated task, delay and area is an issue in this technique.

4.1.4 LECTOR

LECTOR uses two extra transistors called leakage control transistor (LCT's). These two transistors are inserted between pull-up and pull-down network within the logic circuit as shown in Figure 8. LCT's are connected in such a manner that transistor are always near the cutoff voltage (Hanchate & Ranganathan, 2004). Leakage control transistor cause the increase in path resistance from supply to ground, leading to significant leakage reduction. Lector comes under self leakage controlled technique because leakage control transistor are internally biased and not externally. However, Lector faces a signal quality problems and delay is a major bottleneck in this technique.

Figure 7. Circuit showing Zigzag approach (Min, Kawaguchi, & Sakurai, 2003)

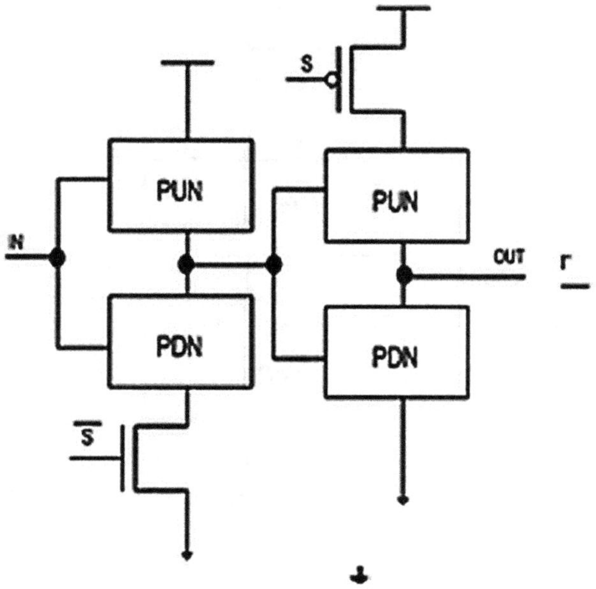

Figure 8. Concept of LECTOR (Hanchate & Ranganathan, 2004)

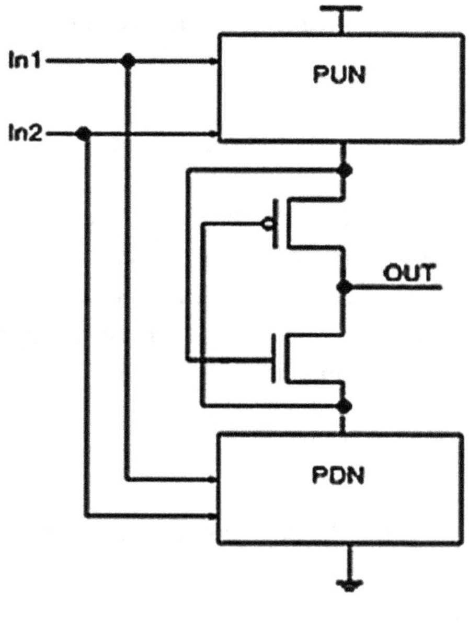

4.1.5 GALEOR

GALEOR technique (Srikanth & Dhireesha, 2008) uses the same structure as the LECTOR, except the location of extra gate leakage transistor (GLT) are swapped as shown in Figure 9. A PMOS GLT is located between pull-down network and output and an NMOS GLT is located between pull-up network and output. GALEOR technique however, suffers from a significant problem that the low signal level is very much higher than 0V and high signal level is very much lower than Vdd. Both GLT transistors are high threshold voltage transistors.

4.1.6 Drain Gating

In drain gating, leakage current can be reduced by inserting extra sleep transistor between pull-up and pull-down transistor (Chun & Chen, 2010) as shown in Figure 10. A PMOS (S) sleep transistor is inserted between pull-up network and output whereas an NMOS sleep transistor (S bar') is between pull-down network and output. During standby mode, both the sleep transistors are turned off and provide stacking effect which reduces the leakage current by increasing resistance of the path from power supply to ground. In active mode, both the sleep transistors are turned on to produce exact logic state. The disadvantage of this technique is that a separate control supply is needed to control sleep transistors.

4.1.7 ONOFIC

This approach uses the insertion of an extra logic circuit between the pull-up and pull-down network for minimizing the leakage current. This extra inserted logic circuit is called ONOFIC block as shown in Figure 11. When ONOFIC (Sharma, Pattanaik, & Raj, 2013) block is turned on, this block provides a conducting path while in its turn off condition the logic block has very high resistance which precisely controls the leakage current. The propagation delay penalty in this approach is less than LECTOR because of single NMOS transistor between pull-up and pull-down network. However, this technique suppresses less leakage current compared to LECTOR.

Figure 9. Concept of GALEOR (Srikanth & Dhireesha, 2008)

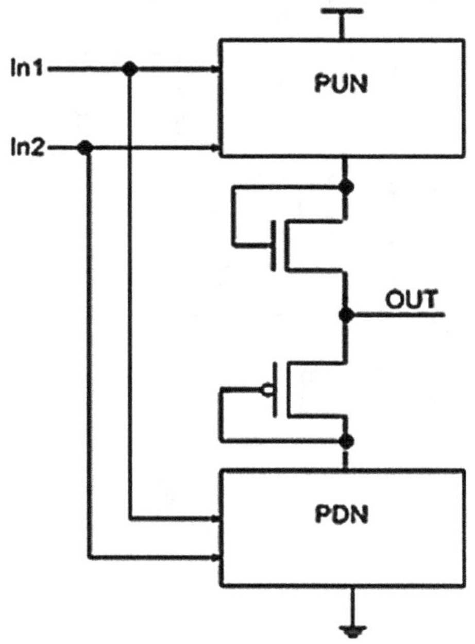

Figure 10. Circuit representing Drain Gating (Chun & Chen, 2010)

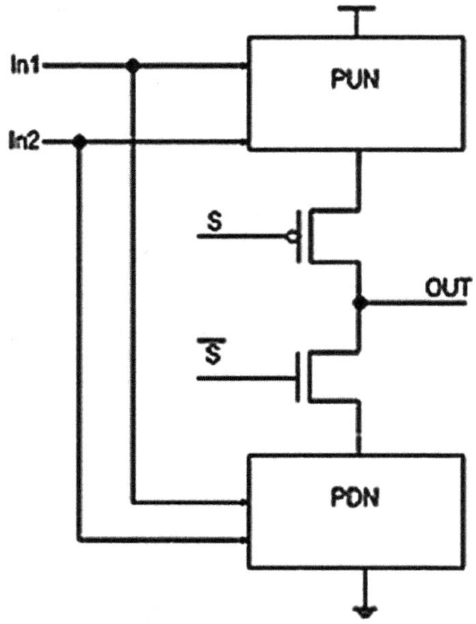

Figure 11. Circuit showing ONOFIC (Sharma, Pattanaik, & Raj, 2013)

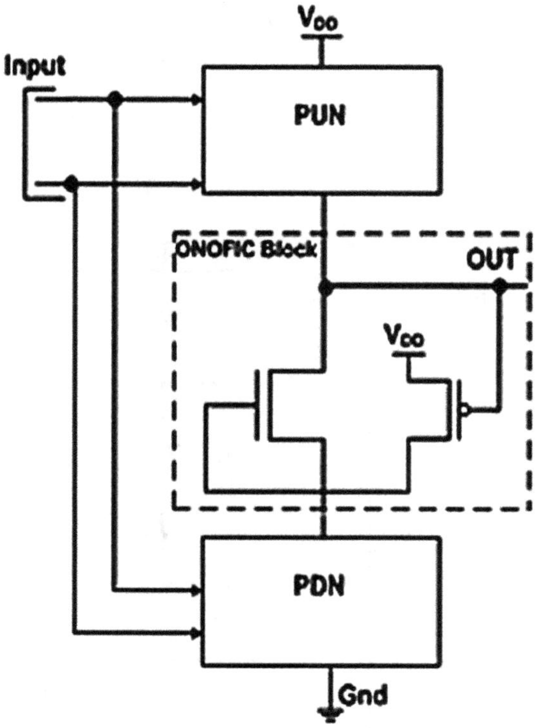

4.2 Body Biasing Technique

In this section, we discuss various leakage minimization schemes which are based on body biasing technique. In body-bias, either a positive or a negative voltage is applied across the source and the substrate (or body) which otherwise remains grounded. Body biasing techniques have been used to alter the threshold voltage in a MOS device in order to increase either the speed or to reduce the leakage current.

4.2.2 DTMOS

Figure 12 shows the dynamic threshold MOS (DTMOS) (Drake, Zamdmer, Nowka, & Brown, 2003). The structure of DTMOS inverter consists of a NMOS and PMOS transistor with body terminal and gate terminal tied together. In DTMOS, body voltage varies with the gate voltage. The ability of DTMOS is that it dynamically changes the threshold voltage (V_t) of MOSFET. It offers low Vt for high on-current during active mode and gives a high Vt for low off-current during standby mode. In this way it helps the designer to achieve system performance and power constraints more easily.

In active mode, body-source junction is forward biased and in standby mode body-source junction is reverse biased. DTMOS is a popular technique to achieve dynamic Vt in bulk CMOS. However, the drawback of this device is that the body resistance and body capacitance both increases resulting in higher RC time constant which may shift the gate-body signal level.

Figure 12. Circuit representing DTMOS (Drake, Zamdmer, Nowka, & Brown, 2003)

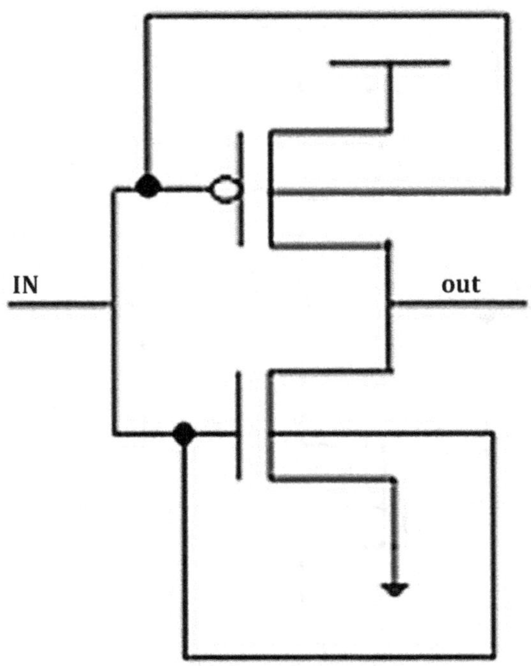

4.2.3 VTCMOS

The variable threshold MOS (VTCMOS) technique (Nukai, Hiromoto, & Sakurai, 2001) is based on adjusting the threshold voltage of the transistor in order to avoid the leakage in standby mode by controlling the substrate bias with a separate substrate bias control circuit as shown in Figure 13. The controlled voltage V_T is a function of its source to substrate bias V_{SB}. This way it automatically controls the leakage current. However, an external controller circuit is needed in this approach which will increase the total area.

4.2.4 Dual Threshold

Dual threshold CMOS is a technique often used at subsystem design level. The subsystem is implemented with low Vt or a high Vt transistor depending on weather they lie on critical path or non-critical path. If subsystem lie on critical path low V_t transistors are used, and if it lies on non-critical path then high Vt transistors are used. Normally, in such situations slack based algorithms (Ho & Hwang, 2004) are used to get better result. However, dual threshold scheme doubles the library size as compared to single threshold design, so the design efforts are increased.

4.2.5 Mixed Vth Scheme

(Wei, Chen, & Roy, 1999) has proposed a mixed threshold voltage static CMOS design scheme to reduce leakage power in low voltage, low power circuits at transistor level and claimed to have achieved 20% more leakage reduction compared to the corresponding gate level dual threshold technique.

Figure 13. Circuit of VTCMOS (Nukai, Hiromoto, & Sakurai, 2001)

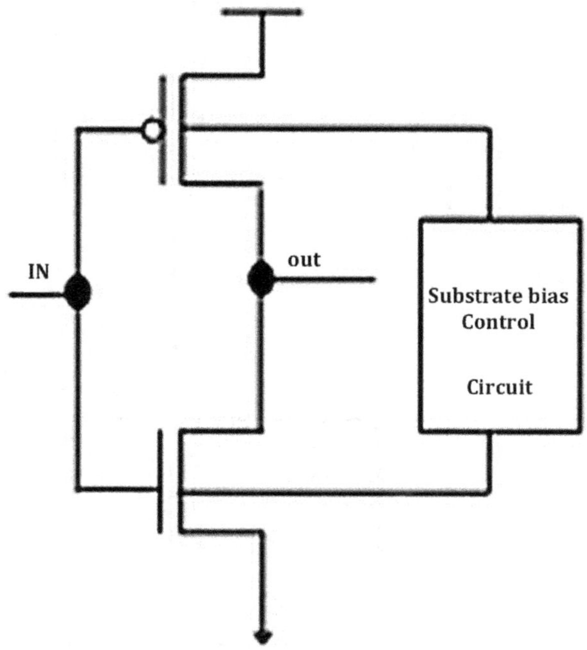

4.3 Transistor Stacking

Sub threshold leakage current flowing through a stack of series connected transistors reduces when more than one transistor of the stack is turned off. Following are the different schemes to minimize the sub threshold leakage using different types of transistor stacking.

4.3.1 Forced Stack

In forced stacking technique (Chandrakasan, Bowhill, & Fox, 2000), single transistor of width W is replaced by two transistor of equal width W/2. Figure 14 shows the concept of forced stacking. Two transistors are turned off simultaneously during off state. It is due to reverse gate source bias and transistor stacking effect the leakage current decreases. Transistor stacking technique exploits the dependence of I_{SUB} on the source terminal voltage Vs. If we increase the Vs of the transistor, the sub-threshold leakage current reduces exponentially. This would happen due to the following reasons:

First, the Gate-to-Source voltage Vgs would reduce and if the input applied is grounded, it would turn negative. This would reduce I_{SUB} exponentially. Secondly, the threshold voltage increases due to body effect. Third, the drain induced barrier lowering (DIBL) coefficient decreases due to lower drain to source potential thereby further reducing leakage. However, the main drawbacks of the technique are that the area and delay will go high.

4.3.2 Sleepy Stack

The sleepy stack approach (Park, 2005) is shown in Figure 15. It combines the advantage of sleep and stack approach. The sleepy stack technique divides the existing transistors into two half size transistors like stack approach. We can see that one of the sleep transistors is added in parallel to the divided transistor. During sleep mode, sleep transistors are turned off and stacked transistors suppress leakage current while saving state. Each sleep transistor, placed in parallel to the one of the stacked transistors, reduces resistance of the path, so delay is decreased during active mode. However, area penalty is a significant matter for this approach since every transistor is replaced by three transistors and since additional wires are added for S and Sbar, which are sleep signals.

Figure 14. Circuit representing Forced Stack (Chandrakasan, Bowhill, & Fox, 2000)

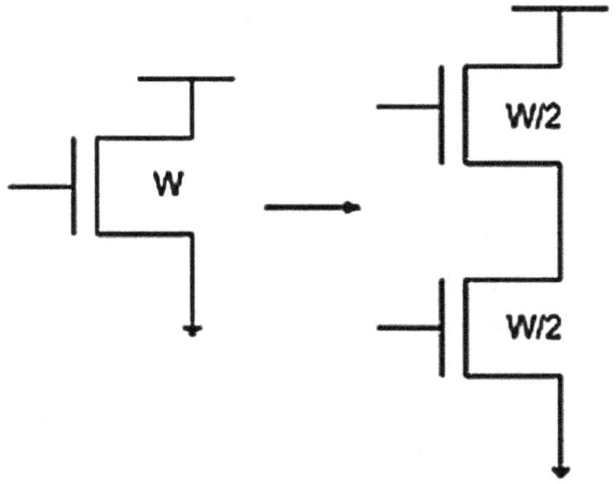

Figure 15. Circuit of Sleepy Stack (Park, 2005)

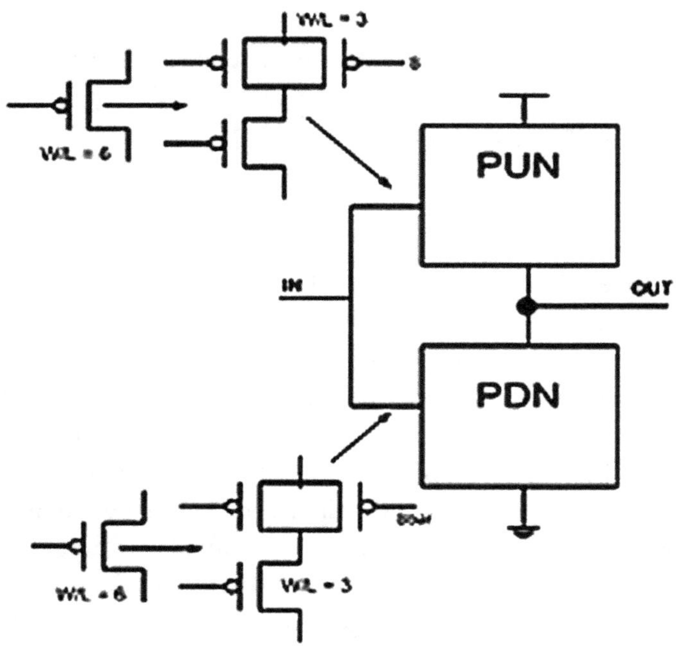

4.3.3 Control Point Insertion

This technique is based on insertion of a transistor called control point (Rahman & Chakrabarti, 2005) in a stack near the supply rail as shown in Figure 16. Control point insertion is based on whether the gate structure is NAND or NOR. The leakage current in NMOS devices is much greater than PMOS devices, and hence an effective solution for leakage reduction is to add NMOS as a control transistor at the bottom. However, an efficient algorithm is needed in this approach for the selection of control points. This method results in an average leakage reduction but the penalty in each case is the increase in dynamic power and area due to additional transistors.

4.4 Data Retention Techniques

In data retention techniques, apart from minimizing the sub threshold leakage, whenever the circuit goes into standby mode, the previous data (output) will be retained by the circuit. Following are the few schemes in this category.

4.4.1 Leakage Feedback Approach (LFA)

Leakage feedback approach is based on two sleep transistors (S and Sbar) and helper transistor (H1 and H2) as shown in Figure 17. These two helper transistors are used to maintain logic state during sleep mode and are driven by the output of an inverter utilizing leakage feedback (Kao & Chandrakasan, 2001). During sleep mode, sleep transistors are turned off and one of the helper transistors in parallel to the sleep transistors keep the connection with the appropriate power rail and thus retain the previous data. This approach solves the data retention problem but at the cost of area and delay.

Figure 16. Circuit showing Control Point Insertion in a Gate (Rahman & Chakrabarti, 2005)

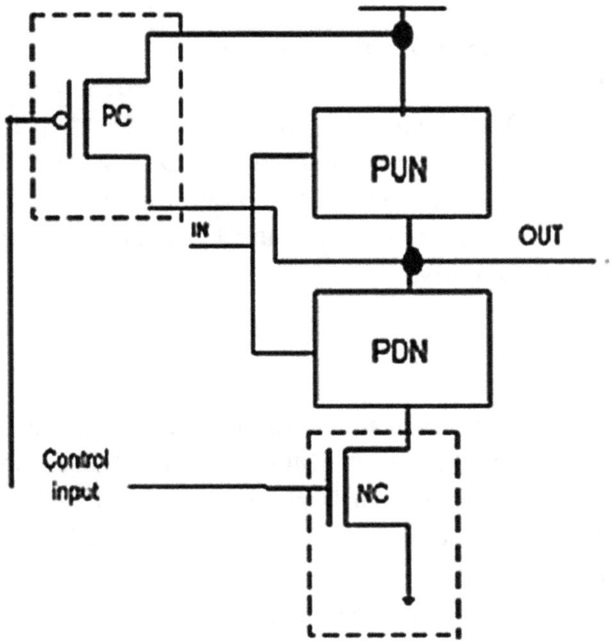

Figure 17. Circuit showing leakage feedback approach (Kao & Chandrakasan, 2001)

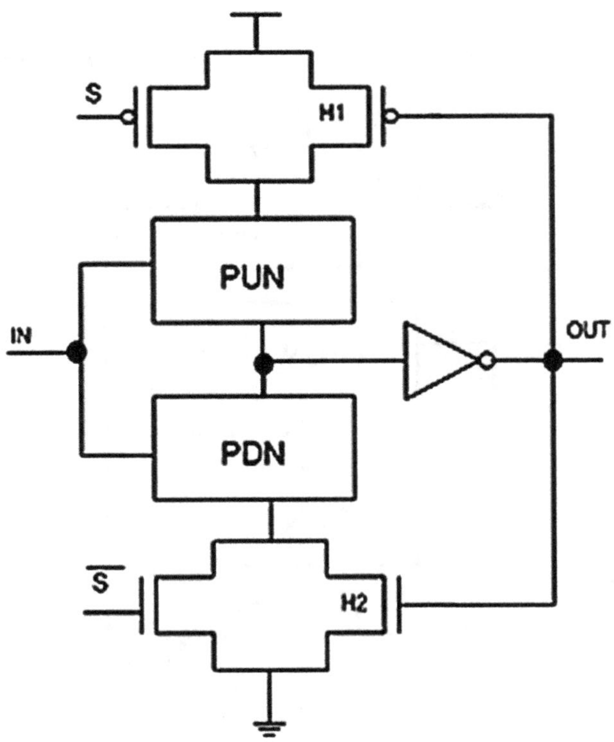

4.4.2 Sleepy Keeper Approach

Sleepy keeper approach (Hun & Mooney, 2006) is a modification of leakage feedback approach. As shown in Figure 18, this technique utilizes a NMOS helper (H1) transistor which is placed in parallel to the sleep transistor (S) and a PMOS helper (H2) transistor is placed in parallel to the sleep transistor (Sbar). The two transistors are driven by the output directly. During sleep mode, sleep transistors are turned off and one of the helper transistors in parallel to the sleep transistor keep the connection with the appropriate power rail. This approach also solves the data retention problem like leakage feedback approach but at the cost of area and delay.

4.4.3 Dual Stack

In dual stack technique (Islam, Nasrin, Mansur, & Tasneem, 2010), two additional PMOS transistors (helper) in the pull-down network and two other NMOS transistors (helper) in pull-up network are connected in parallel to sleep transistors as shown in Figure 19. In active mode, both the sleep transistors are on ON, as the sleep signal is low and so the circuit will operate normally. But in standby mode, as the sleep signal goes high, both the sleep transistors are now in OFF state. However, the pull-up and pull-down network are connected to virtual rail due to helper transistors which actually gets a (Vdd - 2Vt) as positive rail and (Vss + 2Vt) as negative rail. Thus overall potential difference in the Vdd to ground path is reduced causing a reduction in subthreshold leakage current. However, as compared to all other techniques discussed earlier, it has higher power delay product. Furthermore, the delay and area penalty is more in this approach.

Figure 18. Circuit representing sleepy keeper approach (Hun & Mooney, 2006)

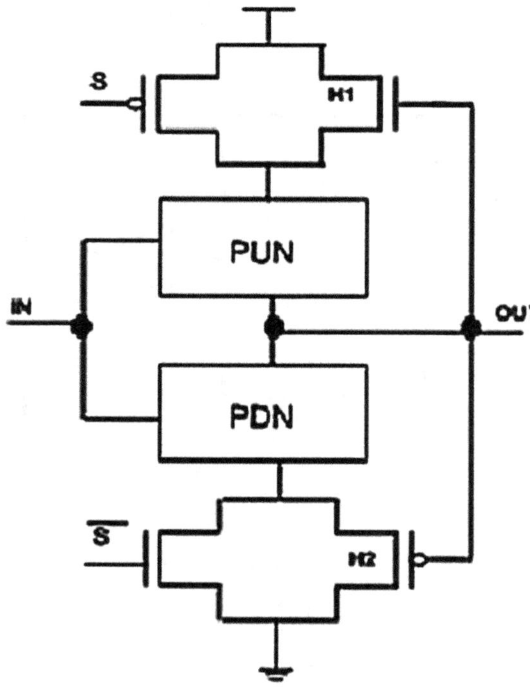

Figure 19. Circuit showing dual stack technique (Islam, Nasrin, Mansur, & Tasneem, 2010)

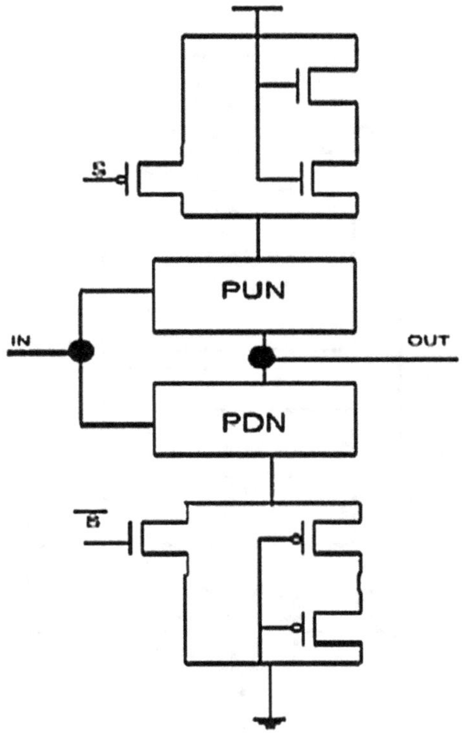

4.4.4 All NMOS Leakage Feedback Technique with Data Retention

In (Lorenzo & Chaudhury, 2013), a new structure for NMOS based leakage feedback is proposed which is shown in Figure 20. The proposed circuit technique includes NMOS only sleep transistors in parallel to both pull-up and pull-down paths which will reduce subthreshold current while saving the exact logic state in standby mode. The circuit of Figure 20 uses all NMOS parallel transistors in pull-up and pull down paths. Moreover, a single PMOS transistor is connected between lower NMOS (N2) transistor and the output. The parallel pull-up NMOS (N1) transistor is connected directly from the output. In standby mode, both the sleep transistors are turned off and one of the pull-up and pull down parallel transistor turns on. Therefore, there are two possibilities, first case - if the output is at logic high previously due to low input logic then this logic high output turns off P1 transistor so also the NMOS transistor N2, thereby completely eliminating any leakage path from ground power rail. As the N1 transistor turns on and so it maintains the output logic by connecting it to VDD (power rail). In second case, when the output is at logic low, it will turn on P1 transistor so also the NMOS transistor N2. The logic low output turns off the N1 transistor. In this way, pull up network is disconnected from VDD supply (i.e., power rail) and pull down network is now connected to ground (supply rail).

The result of simulation on Microwind with BSIM4 at 45nm technology shows that the proposed NMOS leakage feedback design suppresses more leakage power as compared to other existing techniques as shown in Table 1 with minimized area and delay.

Figure 20. Circuit for all NMOS leakage feedback technique with data retention (Lorenzo & Chaudhury, 2013)

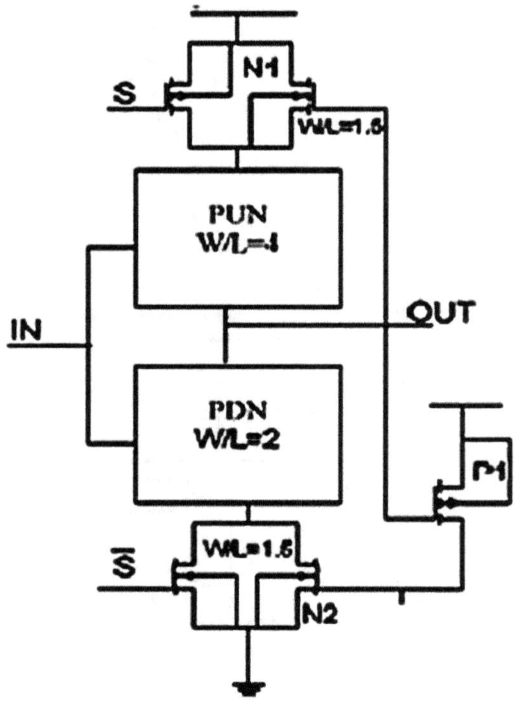

A similar design have been proposed in (Lorenzo & Chaudhury, 2014) wherein the PMOS (P1) transistor of Figure 20 has been replaced with an inverter which connects the output node to N2 transistor in pull down path. The transistor N1 in pull up path is fed directly with the output node voltage. The proposed NMOS leakage feedback approach (PNLFA) also achieves a substantial saving in static power dissipation as compared to other techniques while having a comparable area and delay performances (Figure 20 and Table 2).

Table 2. Power, delay and area estimation in a chain of 4 inverters

Chain of 4 Inverters at 45nm (BSIM4)	Static Power (uW)	Dynamic Power (uW)	Delay (ns)	Area (um)^2
Forced stack (FS)	1.546	3.28	0.1526	58.6
Leakage Feedback Approach (LFA)	0.567	3.748	0.1649	61.13
Leakage Feedback with stack (LFS)	0.517	7.281	0.5126	124.8
Sleepy Keeper Approach (SKA)	0.320	1.630	0.1274	49.05
Stacked Sleep Approach (SSA)	0.317	4.638	0.2048	63.9
Dual Stack (DS)	0.711	12.57	0.428	87.9
Proposed Single Transistor Leakage Feedback Approach (PSTLFA)	0.148	2.338	0.1272	60.2

4.5 Input Vector Control Techniques

It has been observed that leakage power consumption is highly dependent on the kind of pattern of inputs (commonly known as input vectors) that are present at the input of logic circuits. Selection of a suitable input vector is highly essential in designing any such technique for leakage minimization during standby mode.

4.5.1 Input Vector Control

Input vector technique is based on the selection of input vectors to reduce the leakage power (Abdollahi, Fallah, & Pedram, 2004). Such a technique uses an automated tool which will check for a combination of inputs that will produce minimum leakage and shut down the circuit with minimum leakage input vector in standby mode. However, this technique requires extensive computational time for finding out the minimal leakage input vector out of all possible input vectors. Also dynamic leakage is not taken into consideration in this approach. There are some other input vector schemes which are also found to be effective means of power minimization in VLSI circuits (Chopra, Sarma, & Vrudhula, 2004; Gao & Hayes, 2004; Lee, Blaauw, & Sylvester, 2005; Zhao, Jiangfang, Tong, & Cheng, 2006; Chen, Johnson, Wei, & Roy, 1998; Ho, Chang, & Chauchin, 2012).

4.5.2 INDEP Gates

INDEP approach uses two extra inserted transistors between pull-up and pull-down networks of the CMOS logic circuit as shown in Figure 21. These extra inserted transistors are called INDEP (input dependent) transistors (Sharma, Pattanaik, & Raj, 2014). The technique is called as INDEP approach because the benefit of low leakage and high speed depends on the inputs of INDEP transistors. INDEP transistors are connected in such a way to form transistor stacks with the pull-up/pull-down networks for each primary input combination. The operation of MP1 and MN1 are controlled by the Boolean logics at input terminals V0 and V1, respectively. Criteria for leakage power reduction in INDEP approach depends on the appropriate Boolean logic calculations for the input signals of the extra inserted transistors. The input voltage signal for INDEP transistor is selected by applying some algorithms; however propagation delay is the penalty in this approach because of extra INDEP transistors.

4.6 Algorithmic Approaches

Genetic algorithm based approaches for finding minimum leakage input vector has been presented in (Zhao, Yi, Tong, & Cheng, 2006; Chen, Johnson, Wei, & Roy, 1998). A heuristic method for finding a minimum leakage input vector during sleep state has been presented by (R.M. Rao et al, 2003). A mixed integer linear programming model is proposed (Lu & Agarwal, 2007) to minimize leakage power with specified timing yield under process variation. The MLP method is specifically devised with set of constraints whose size is linear in the number of gates. They have also shown that actual complexity depends on the nature of the problem and achieved 30% more reduction in leakage power by using the statistical approach when the result is compared to a deterministic approach. Rastogi et al. [52] propose an efficient technique for estimating major leakage components in a circuit by computing leakage using a logic state based analysis while incorporating the loading effect to improve accuracy of leakage current estimation.

Figure 21. Circuit showing INDEP gates (Sharma, Pattanaik, & Raj, 2014)

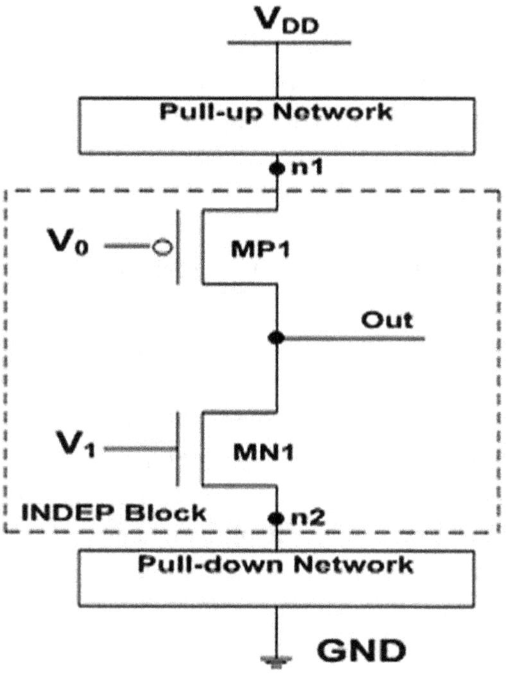

4.7 Application Oriented Techniques

A few leakage minimization schemes have been briefly discussed here which are specific to a particular application. Each of these techniques are based on the fundamental concept of leakage minimizing by increasing the threshold voltage either by substrate biasing, source biasing or by causing reduced DIBL effect with stacking. The clamping diode SRAM cell in (Zhang, Wu, Mao, & Zheng, 2012; Zhu, Bai, & Wu, 2013) works on source biasing technique to control the leakage current however, with a performance degradation. Again, the NC SRAM cell (Elakkumanan, Thondapu, & Sridhar, 2004) uses multiple grounds to control the leakage. It uses normal active mode, the source is connected to ground as usual (with 0 V) whereas, during idle mode, the sub-threshold and leakage currents would be reduced by the act of applying a positive voltage (Vs) to the source of transistors, thereby, raising the voltage level of ground. The reduction of sub-threshold currents is because of increasing in threshold voltage due to body effect and increase the source voltage. On the other hand, reduction of leakage current is due to reduction of the drain-source voltage and consequently the reduction in the DIBL effect. This technique, however, suffers from extra dynamic power consumption. Moreover, it also causes SNM and performance degradation. In PP-SRAM cell Sinha, & Caviglia, 2013; Razavipour, Afzali-Kusha, & Pedram, 2009), PMOS access transistors have been used instead of NMOS access transistors in order to reduce gate leakages as shown in Figure 22. This is because PMOS transistors have higher threshold voltages than NMOS. But it causes performance to degrade due to lesser mobility of holes compared to electrons. In order to compensate for the performance degradation in active mode, a body-bias voltage of $V_{DD}/2$ is being applied to all the PMOS transistors of the SRAM cell, which improves the timing characteristics although at the expense of increasing the power dissipation. A bootstrapped repeater circuit has been proposed in (Ho, & Su,

Figure 22. The circuit of PP SRAM cell (Razavipour, Afzali-Kusha, & Pedram, 2009)

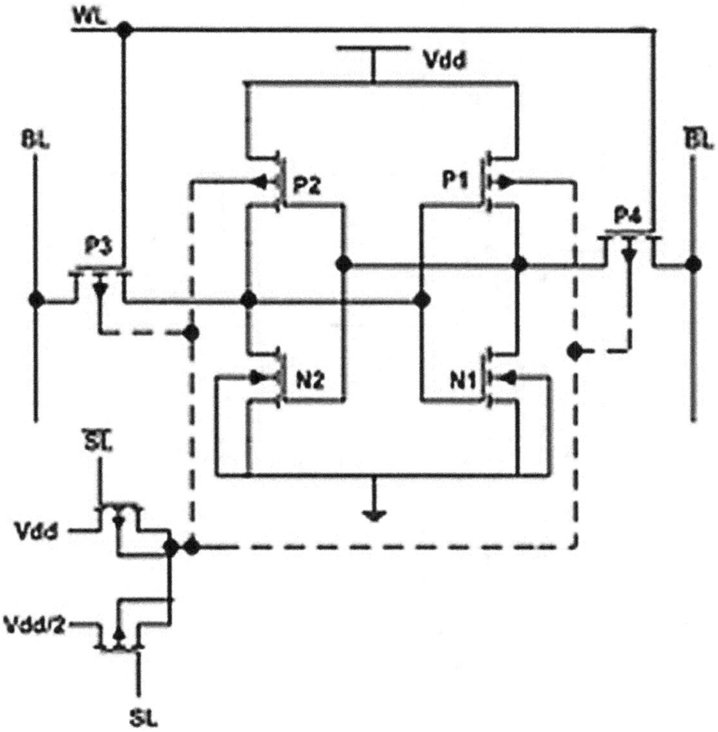

2012) in order to have higher I_{ON}/I_{OFF} ratio and higher speed of operation. This is because the bootstrap control circuit produces a boosted output swing from $-V_{DD}$ to $2V_{DD}$ to increase the driving current (with $2V_{DD}$) and turn off the transistor aggressively (by $-V_{DD}$). Moreover, the precharge enhancement scheme improves the precharge capability to achieve high-speed operation. A multi-bit retention register has been proposed in (Chen, Geng, Yu Lai, Shi, & Chang, 2014) for power gated circuits. This technique presents a method of designing multiple bit retention register (MBRR) as well as an efficient assignment algorithm to reduce the size of retention storage and the leakage power in sleep mode. The master flip flop designed with low V_t transistor for fast switching during active mode and the slave retention latch designed with high V_t transistor.

The Schmitt trigger inverter (Lotze & Manoli, 2012) has the same basic functionality as a normal inverter with the PMOS transistor P0/P1 pulling the output voltage to a high value, whereas NMOS inverters N0/N1 are off, though still conducting a leakage current which may potentially degrade the output level V_Z at low supply voltage as shown in Figure 23. The improvement introduced by Schmitt trigger structure is due to the effect of the feedback transistors P2/N2 (Lotze & Manoli, 2012). The feedback transistor is active in the block, where driving transistors are off, as the gates of P2/N2 are controlled by the output Z = Abar. For low input, N2 pulls middle node X to a high potential. This forces drain-source voltages of N1 close to zero and, more importantly, its gate-source voltage into the negative region, reducing the leakage through N1 exponentially. The result is an effective leakage quenching from the output node Z, consequently minimizing the output level degradation. A Novel low leakage and high speed triple threshold voltage buffers with skewed inputs and outputs are introduced in (Zhu & Kursan,2014).

Figure 23. Leakage quenching in Schmitt trigger inverter (Lotze & Manoli, 2012)

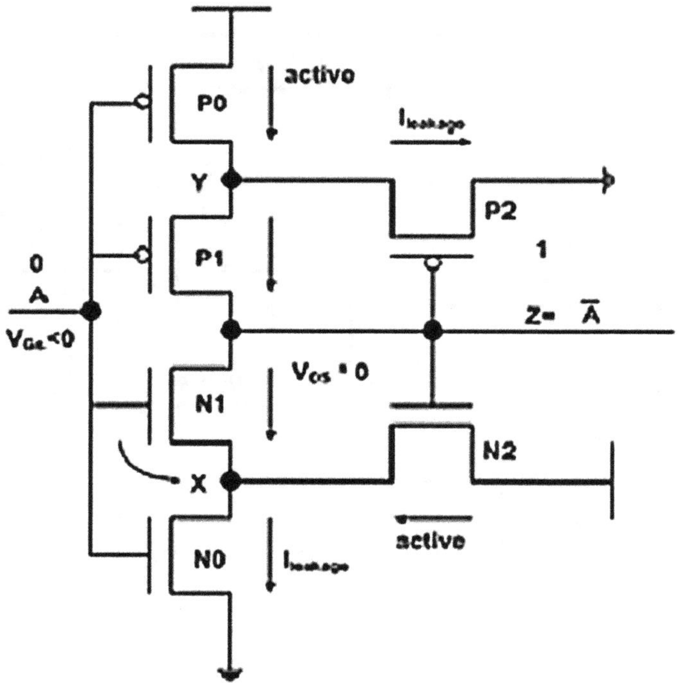

5. CONCLUSION

In this chapter a detailed review of the circuit level leakage minimization techniques in CMOS VLSI circuit has been presented. The techniques are then classified based on the fundamental concept of enhancing the threshold voltage, either by multi-Vt, transistor stacking, body-biasing or input vector control. All the techniques discussed earlier share the same objective of minimizing the leakage, however, most of the techniques have given emphasis on subthreshold leakage reduction only. Few schemes are there which apart from leakage minimization they also retain the data during sleep mode. We have also identified few techniques of leakage minimization using algorithmic approaches based on finding minimum leakage input vector. At last, we have reviewed some of the application oriented techniques where a specific concept leakage minimization is applied to one particular application. It is seen that circuit level design techniques improve the subthreshold operation in terms of performance and robustness. It can be also concluded that *MTCMOS technique* provides an efficient leakage reduction in idle mode. However, if data retention is the main criteria then *leakage feedback approach* is a better alternative. Whereas, the *sleepy stack technique* provides the fastest speed of operation apart from minimizing the leakage substantially. Table 3 summarizes various leakage reduction techniques with their advantages and limitations. While making such a review most of the referred techniques have been taken till 2014. All efforts were to give a comprehensive review on circuit level leakage minimization techniques, although power minimization is also possible at device level or at any other higher level in the design hierarchy, which is beyond the scope of this Chapter.

Table 3. Summary of leakage minimization techniques

Sl. No	Year of Publication	Techniques	Advantages	Limitation
1	1995	Power gating (Mutoh et al., 1995)	Very high leakage power saving, Industry preferred technique	Delay penalty, need for dual threshold transistors, data retention problem, controller design needed
2	2000	Forced stack (Chandrakasan, Bowhill, & Fox, 2000)	Substantial leakage saving, Ease of implementation, no need for controller design, single threshold transistor and easy to fabricate	Increases circuit delay
3	2000	Super-cut off CMOS circuit (Kawaguchi, Nose, & Sakurai, 2002)	Substantial leakage power saving, single threshold transistor and easy to fabricate	Complex control design and large delay penalty
4	2001	DTMOS (Drake, Zamdmer, Nowka, & Brown, 2003)	Dynamically tuned threshold voltage for low leakage and high speed	RC time constant shift the body signal, higher energy consumption
5	2001	VTCMOS (Nukai, Hiromoto, & Sakurai, 2001)	leakage and delay advantage by tuning the threshold voltage	Additional controller is needed to control the body bias
6	2001	Leakage feedback approach (Kao & Chandrakasan, 2001)	Solves data retention problem and minimizes leakage	Area and delay penalty due to additional transistor
7	2003	Zigzags Approach (Min, Kawaguchi, & Sakurai, 2003)	Removing the problem of floating node as in power gating approach and better leakage saving	Designing is complicated and delay is an issue
8	2004	Input Vector Control (Abdollahi, Fallah, & Pedram, 2004)	Leakage saving is based on input vectors	Extensive computational time requirement and dynamic power is not taken into consideration
9	2004	LECTOR (Hanchate & Ranganathan, 2004)	Better leakage reduction, no need of additional circuitry to monitor leakage	Signal quality problem
10	2004	Dual threshold (Ho & Hwang, 2004)	Leakage saving and speed advantage	Doubles the library size and increased design efforts
11	2005	sleepy stack (Park, 2005)	Lesser delay penalty compared to the forced stacking	Lower leakage saving with respect to forced stacking, increase in transistor count
12	2005	Control point insertion (Rahman & Chakrabarti, 2005)]	Average leakage reduction	dynamic power and area is a penalty
13	2006	Sleepy keeper Approach (Hun & Mooney, 2006)	Solving data retention problem	Control signals are needed for sleep transistors and having area penalty
14	2008	GALEOR (Srikanth & Dhireesha, 2008)	better leakage saving with no external control	Signal quality is poor than lector technique
15	2010	Drain gating (Chun & Chen, 2010)	Better leakage saving and logic levels are good	External controller is needed
16	2010	Dual stack (Islam, Nasrin, Mansur, & Tasneem, 2010)	Better leakage saving	Delay and area penalty due to additional helper transistors
17	2012	Bootstrapped circuit (Ho, & Su, 2012)	Active leakage current reduction	Additional circuit is needed to control the leakage
18	1999	Mixed Vth Scheme (Wei, Chen, & Roy, 1999)	More leakage reduction	This technique require extra library
19	2013	ONOFIC (Sharma, Pattanaik, & Raj, 2013)	Less leakage reduction than lector and less delay	leakage reduction is not more as compared to other techniques

continued on following page

Table 3. Continued

Sl. No	Year of Publication	Techniques	Advantages	Limitation
20	2014	INDEP (Sharma, Pattanaik, & Raj, 2014)	More leakage saving in this approach	As a separate algorithm is needed, so the computational complexity increases
21	2004	Algorithm (Chopra, Sarma, & Vrudhula, 2004)	More leakage saving	Heuristics optimization technique and so the computational complexity is more in order to reduce the leakage current
22	2012	Clamping diode SRAM cell (Zhang, Wu, Mao, & Zheng, 2012)	More leakage saving by clamping the source voltage to high value	This technique have delay degradation
23	2004	NC SRAM cell (Elakkumanan, Thondapu, & Sridhar, 2004)	leakage current reduction by DIBL effect and raising the source voltage	This technique also have the problem of delay degradation
24	2009	PP SRAM cell (Razavipour, Afzali-Kusha, & Pedram, 2009)	leakage reduction is more than NC SRAM, performance is also better	At the cost of additional circuitry
25	2012	Bootstrapped repeater circuit (Ho, & Su, 2012)	This technique reduce the leakage current, delay and noise	Additional circuit is controlling the leakage current
26	2014	Multibit retention register for power gated design (Chen, Geng, Yu Lai, Shi, & Chang, 2014)	It reduces more leakage power than single bit register	It require multiple bit retention register as well as efficient assignment algorithm
27	2012	Schmitt trigger logic (Lotze & Manoli, 2012)	This technique reduces the leakage current effectively	However, it requires additional quenching circuit
28	2014	Skewed IO with split input/output cell (Zhu & Kursan,2014)	It reduces the leakage current substantially.	But, this technique needs extra PMOS and NMOS transistors
29	2013	All NMOS leakage feedback with Data Retention (Lorenzo & Chaudhury, 2013)	This technique is an improvement over LFA with reduced area, delay and leakage	However, dynamic power increases because of the additional transistors

REFERENCES

Abdullah, A., Fallah, F., & Pedram, M. (2004). Leakage current reduction in CMOS VLSI circuits by input vector control. *IEEE Trans. Very Large Scale Integration*, *12*(2), 140–154. doi:10.1109/TVLSI.2003.821546

Bohr, M. (1996). A high performance 0.25m logic technology optimized for 1.8V operation. In *Proceedings of the International Electron Devices Meeting*, (pp. 847-850). Academic Press.

Borkar, S. (1999). Design challenges of technology scaling. *IEEE Micro*, *19*(4), 23–29. doi:10.1109/40.782564

Brews, J. (1990). *High Speed Semiconductor Devices*. New York: Wiley.

Chandrakasan, A., Bowhill, W. J., & Fox, F. (2000). *Design of High- Performance Microprocessor Circuits*. Wiley-IEEE Press. doi:10.1109/9780470544365

Chen, C. X. (2008). A cost effective 32nm high effective k/metal –gate technology for low power application with single metal/gate –first process. Symposium on VLSI Technology Digest of Technical Paper, (pp. 88-99). Academic Press.

Chen, Z., Johnson, M., Wei, L., & Roy, K. (1998). Estimation of standby leakage power in CMOS circuits considering accurate modeling of transistor stacks. In *Proc. of Int. Symposium on LPED,* (pp. 239-244). Academic Press.

Chopra, K., & Vrudhula, B. K. (2004). Implicit pseudo boolean enumeration algorithms for input vector control. In *Proc. of IEEE-ACM Design Automation Conference*. doi:10.1145/996566.996774

Chun, J. W., Roger, C. Y., & Chen. (2010). A Novel Leakage reduction technique for CMOS circuit design. *Proceedings of ISOCC.*

Drake, A. J., Zamdmer, N., Nowka, K. J., & Brown, R. B. (2003). Analysis of the impact of gate body signal phase on DTMOS Inverters in 0.13µm PD-SOI. IEEE International SOI Conference, (pp. 99-100). IEEE.

Elakkumanan, P., Thondapu, C., & Sridhar, R. (2004). A gate leakage reduction strategy for sub-70 nm memory circuits. In *Implementation of High Performance Circuits. (DCAS-04),Proceedings of the IEEE Dallas/CAS Workshop*. IEEE. doi:10.1109/DCAS.2004.1360446

Fallah, F., & Pedram, M. (2005). Standby and active leakage current control and minimization in CMOS VLSI circuits. *IEICE Transactions on Electronics*, *E88–C*(4), 509–519. doi:10.1093/ietele/e88-c.4.509

Gao, F., & Hayes, J. P. (2004). Exact and Heuristic approaches to input vector control for leakage power reduction. In *Proc. of IEEE Int. Conference on CAD*, (pp. 527-532). IEEE.

Guang, Y., Geng, H., Lai, K-Y., Shi, Y., & Chang, S-C. (2014). Multibit Retention Registers for Power Gated Designs: Concept, Design, and Deployment. *IEEE Transaction on Computer Aided Design of Integrated Circuit and System*, *33*(4).

Hanchate, N., & Ranganathan, N. (2004). LECTOR:ATechnique for leakage reduction in CMOS circuits. *IEEE Trans. On VLSI Integration System*, *22*(2), 196–205. doi:10.1109/TVLSI.2003.821547

Ho, Y., Chang, C., & Chauchin. (2012). Design of a subthrehold–supply bootstrapped CMOS inverter based on an active leakage current reduction technique. *IEEE Transaction on Circuit and System Express brief–II*, *59*(1), 55-59.

Ho, Y.-T., & Hwang, T-T. (2004). Low Power Design Using Dual Threshold Voltage. In *IEEE ASP Design Automation Conference.*

Hun, K., & Mooney. (2006). Sleepy Keeper: A New Approach to Low-leakage Power VLSI Design. In *Proc. IFIP International Conference on Very Large Scale Integration* (pp. 367-372). Academic Press.

International Technology Roadmap for Semiconductor. (2009). Retrieved from http//www.itrs.net

Inukai, T., Hiromoto, T., & Sakurai, T. (2001). Variable threshold voltage CMOS (VTCMOS). *Proceedings of ISPLED.*

Islam, M. S., Nasrin, M. S., Mansur, N., & Tasneem, N. (2010). Dual Stack Method: A Novel Approach to Low Leakage and Speed Power Product VLSI Design. In *Proc. ICECE2010.*

Kahng, A. B. (2007). Design challenges at 65nm and beyond. In *Proceedings of the Conference on Design, Automation and Test in Europe*, (pp. 1466-1467). Academic Press.

Kang, S., & Lebleblici, Y. (2003). *CMOS Digital Integrated Circuits*. New York: McGraw Hill.

Kao, J., & Chandrakasan, A. (2001). MTCMOS sequential circuits. In *Proceedings of European Solid-State Circuits Conference.*

Kao, J., & Chandrakasan, A. P. (2000). Dual-Threshold Voltage Techniques for Low-Power Digital Circuits. *IEEE Journal of Solid-State Circuits, 35*(7), 1009–1018. doi:10.1109/4.848210

Kao, J., Narendra, S., & Chandrakasan, A. (2002). Subthreshold Leakage Modeling and Reduction Techniques. In *Proceedings of the International Conference on Computer Aided Design.*

Kaur, S. (2013). On chip network. *IETE Technical Review, 30*(3), 168–172. doi:10.4103/0256-4602.113467

Kawaguchi, H., Nose, K., & Sakurai, T. (2002). A Super Cut-Off CMOS(SCCMOS) Scheme for 0.5-V Supply Voltage with Picoampere Stand-By Current. *IEEE Journal of Solid-State Circuits, 35*(10), 498–1501.

Kim, K. K., Kim, Y. B., Choi, M., & Park, N. (2007). Leakage minimization technique for nanoscale CMOS VLSI. *IEEE Design & Test of Computers, 24*(4), 322–330. doi:10.1109/MDT.2007.111

Lee, D., Blaauw, D., & Sylvester, D. (2005). Static leakage reduction through simultaneous Vt/Tox and state assignment. *IEEE Transactions on Computer-Aided Design of Integrated Circuits and Systems, 24*(7).

Lee, D., Kwong, W., Blaauw, D., & Sylvester, D. (2003). Analysis and Minimization Techniques for Total Leakage Considering Gate Oxide Leakage. In *Proceedings of the 40th design automation conference*, (pp. 175-180).

Lorenzo, R., & Chaudhury, S. (2013). A Novel all NMOS Leakage Feedback with Data Retention Technique. In *Proc. of CARE 2013, IEEE Int. Conference.*

Lorenzo, R., & Chaudhury, S. (2014). Analysis of leakage feedback technique. In *Electronics, Communication and Instrumentation (ICECI), 2014 International Conference on*. IEEE.

Lotze, N., & Manoli, Y. (2012). A 62mV 0.13 μm CMOS standard cell based design technique using Schmitt-trigger logic. *IEEE Journal of Solid-State Circuits, 47*(1), 47–60. doi:10.1109/JSSC.2011.2167777

Min, L-S., Kawaguchi, H., & Sakurai, T. (2003). Zig-zag super cut-off CMOS (ZSCCMOS) block activation and self-adaptive voltage level controller: An alternative to clock–gating scheme in leakage dominant Era. *IEEE International Solid–State Circuits Conference*, (pp. 400-401). IEEE.

Mutoh, S., Douseki, T., Matsuya, Y., Aoki, T., Shigematsu, S., & Yamada, J. (1995). 1-V Power Supply High-speed Digital Circuit Technology with Multithreshold-Voltage CMOS. *IEEE Journal of Solid-State Circuits, 30*(8), 847–854. doi:10.1109/4.400426

Park, J. (2005). *Sleepy Stack: a New Approach to Low Power VLSI and Memory*. (Ph.D. Dissertation). School of Electrical and Computer Engineering, Georgia Institute of Technology. Available http://etd.gatech.edu/theses

Pierret, R. F. (1996). *Semiconductor Device Fundamentals*. Reading, MA: Addison-Wesley.

Rahman, H., & Chakrabarti, C., (2005). An efficient control point insertion technique for leakage reduction of scaled CMOS circuits. *IEEE Transaction on Circuit And system Brief-II*, *52*(8), 496-500.

Rao, R. M. (2003) A heuristic to determine low leakage sleep state vectors for CMOS combinational circuits. In *Proc. of IEEE Int. Conf. CAD*, (pp. 689-692). IEEE.

Razavipour, G., Afzali-Kusha, A., & Pedram, M. (2009). Design and analysis of two low-power SRAM cell structures. *IEEE Transactions on Very Large Scale Integration (VLSI) Systems*, *17*, 1551–1555.

Roy, K., & Prasad, S. C. (1990). *Low-Power CMOS VLSI Circuit Design*. New York: Wiley Interscience.

Sakurai, T. (2004). Perspective of Low Power *VLSI. IEICE Transactions on Electronics*, *87*(4), 429–436.

Sharma, V.K., Pattanaik, M., & Raj. (2013). ONOFIC approach: Low power high speed nanoscale VLSI circuits design. *International Journal of Electronics*, *101*(1), 61-73.

Sharma, V.K., Pattanaik, M., & Raj, B. (2014). INDEP approach for leakage reduction in CMOS circuits. *International Journal of Electronics*, *102*(1), 200-215.

Sinha, A. K., & Caviglia, D. D. (2013). Designing high value resistive network using weak inversion region of PMOS device at the floating Gate of a sensor. *IETE Technical Review*, *30*(6), 473–482. doi:10.4103/0256-4602.125665

Srikanth, K., & Dhireesha, K. (2008). GALEOR: Leakage reduction for CMOS circuits. In *Proceedings of 15th IEEE International Conference on Electronics Circuits and Systems*. IEEE.

Taur, Y., & Ning, T. H. (1998). *Fundamentals of Modern VLSI Devices*. Cambridge University Press.

Wei, L., Chen, Z., & Roy, K. (1999). Mixed-Vth (MVT) CMOS circuit design methodology for low power applications. In *Proc. of IEEE-ACM Design Automation Conference*. doi:10.1145/309847.309974

Yingchieh, H., & Su. (2012). A 0.1–0.3 V 40–123 fJ/bit/ch On-Chip Data Link With ISI-Suppressed Bootstrapped Repeaters. *IEEE Journal of Solid-State Circuits*, *47*(5).

Zhang, L., Wu, C., Mao, L.-F., & Zheng, J. (2012). Integrated SRAM compiler with clamping diode to reduce leakage and dynamic power in nano-CMOS process. *Micro & Nano Letters*, *7*(2), 171–173. doi:10.1049/mnl.2011.0680

Zhao, X., Yi, J., Tong, D., & Cheng, X. (2006). Leakage power reduction for CMOS combinational circuits. In *Proc. of ICSICT*. doi:10.1109/ICSICT.2006.306330

Zhu, H., & Kursan, V. (2014). Novel low leakage and high speed triple threshold voltage buffers with skewed inputs and outputs. *IEEE Transactions on Circuits and Systems*, (61), 7.

Zhu, J., Bai, N., & Wu, J. (2013). A review of sense amplifier for static random access memory. *IETE Technical Review*, *30*(1), 72–81. doi:10.4103/0256-4602.107343

Section 2

Non-Traditional Low Power Methodologies

Chapter 5
Contemporary Low Power Design Approaches

Lini Lee
Multimedia University, Malaysia

ABSTRACT

This chapter describes three contemporary low power design approaches; a resistor-less bandgap reference circuit, a hybrid voltage level shifter with a diode connected NMOS and a modified dynamic comparator, each design with the objective to demonstrate the feasibility of contemporary approaches in achieving lower power VLSI design. All three designs are simulated in 0.18 µm CMOS technology using industrial simulation tool and the results are based on performance parameters defined in the chapter.

INTRODUCTION

Today's power-conscious climate has elevated low power design approaches even at contemporary level, infiltrating every aspect of building blocks in VLSI system especially with the shrinking circuit board sizes in portable and wearable electronics and the much talked-about Internet of Things devices which require wireless and energy-storage-based design. The growing market of portable, battery-powered devices demand circuits design and VLSI system with ultra-low power dissipation. As the integration, size and complexity of the chips continue to increase; the difficulty in providing adequate cooling might either add significant cost or limit the functionality of the computing system in those integrated circuits.

OVERVIEW FOR BANDGAP REFERENCE CIRCUIT (BGR), VOLTAGE LEVEL SHIFTER (VLS) AND DYNAMIC COMPARATOR

In today's new deep submicron CMOS technology especially in the System on Chip (SoC) products, the CMOS technology scaling and demand for high efficiency low power devices is higher than ever. One fundamental approach to low power design is low power supply: when the power supply is low, the total power consumption would be lowered subsequently. How about a circuit which needs to operate at certain

DOI: 10.4018/978-1-5225-0190-9.ch005

temperature condition? What happen if the temperature is higher than the selected testing temperature? Would the power consumption of the circuit increases? In order to overcome all these obstacles, a self-biased low power CMOS bandgap reference (BGR) circuit is designed. This is a circuit which uses two temperature-control techniques to provide flexibility and functionality.

How about speed in low power design? Having high speed in low power design would be considered as luxury in circuit design. In the second part of the chapter, a hybrid voltage level shifter (VLS) using the contemporary approach of a diode connected NMOS would be introduced. In this part, there would be some equations introduced in order to aid in achieving an optimized VLS.

The final approach uses a modified dynamic comparator which has been designed from a modified latch utilizing a NMOS transistor. The t_{latch} of the comparator shows improvement of approximately 23% compared to conventional comparator. This modified version of latch type voltage sense amplifier comparator uses a contemporary approach and it is able to dissipate only 118 pW at 1.5 V power supply with 200 MHz frequency as reported by Iffa S. and L. Lee (2014). Detailed simulation results are included to give an insight onto the actual architecture.

BANDGAP REFERENCE CIRCUIT (BGR)

Electronic devices require steady and constant power supply in order to function smoothly, without flaws. With growth in demand for more portable electronic devices, more and more circuits and systems would require low power, small area and low cost features. Bandgap reference is one of the essential units in analog circuit systems. It provides a constant voltage supply regardless of fluctuations or changes in temperature and power supply. Brokaw Paul (1981) invented the first BGR circuit which has a temperature compensated band-gap reference of the type employing two transistors operating at different current densities to develop a positive TC (Temperature Coefficient) current. As CMOS technology developing rapidly and bringing us into the era of high integration and ultra-low power consumption, the need for lower than the conventional bandgap voltage reference was created. Hence, it is critical to design low power consumption bandgap reference circuits which work under low voltage supply and yet compatible with other modern circuit blocks. There are a number of researchers or authors such as Bo Wanget al. (2015), Klimach H. et al. (2013) and Yat-Hei Lam et al. (2010). They have contributed new solutions based on the concept of using a fraction of the original bandgap voltage level.

Here in this chapter, a self-biased low power CMOS bandgap reference is introduced, working at 0.8V power supply and 0 to 100°C temperature range. Advanced startup circuit was adopted in the voltage reference to improve circuit precision and stability. The proposed BGR circuits are designed in 0.18 μm CMOS technology. A first proposed BGR with a resistor load (Figure 1) is introduced with minimum of 13 transistors and 3 passive resistors while second proposed BGR (Figure 2) would not have any resistor. Xin Ming and authors (2010) mentioned that using resistor increases a chip area size and causes variation on the resistor sheet from substrate. The resistor would be replaced by a transistor which acts as an active resistor. A typical BGR circuit consisted of three building blocks: self-biased differential amplifier, start-up circuit and bandgap core circuit.

A voltage reference is an essential sub-electronic circuit present in many Systems on Chip (SoC) product such as mobile electronics product, Analog-to-Digital (A/D) and Digital-to-Analog (D/A) converters. However, the heat dissipation issues raised from the circuit operation would create huge

Figure 1. Conventional BGR circuit

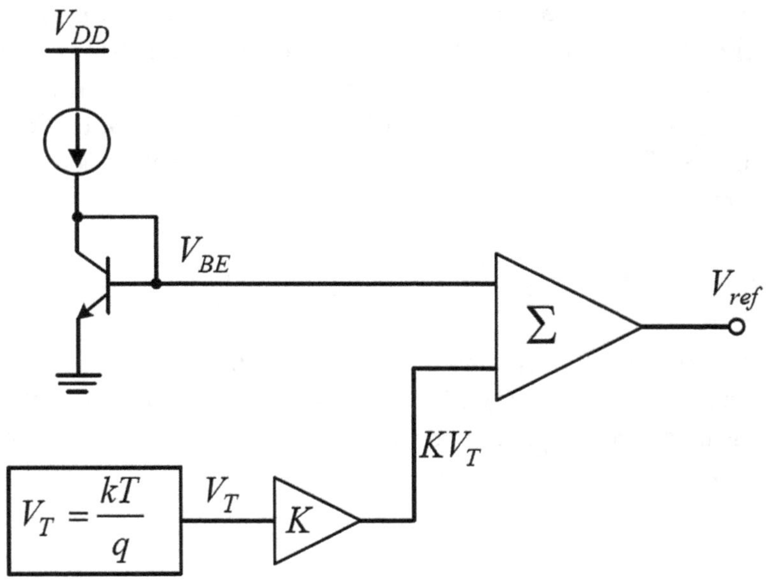

Figure 2. Proposed resistor-subdivision BGR circuit

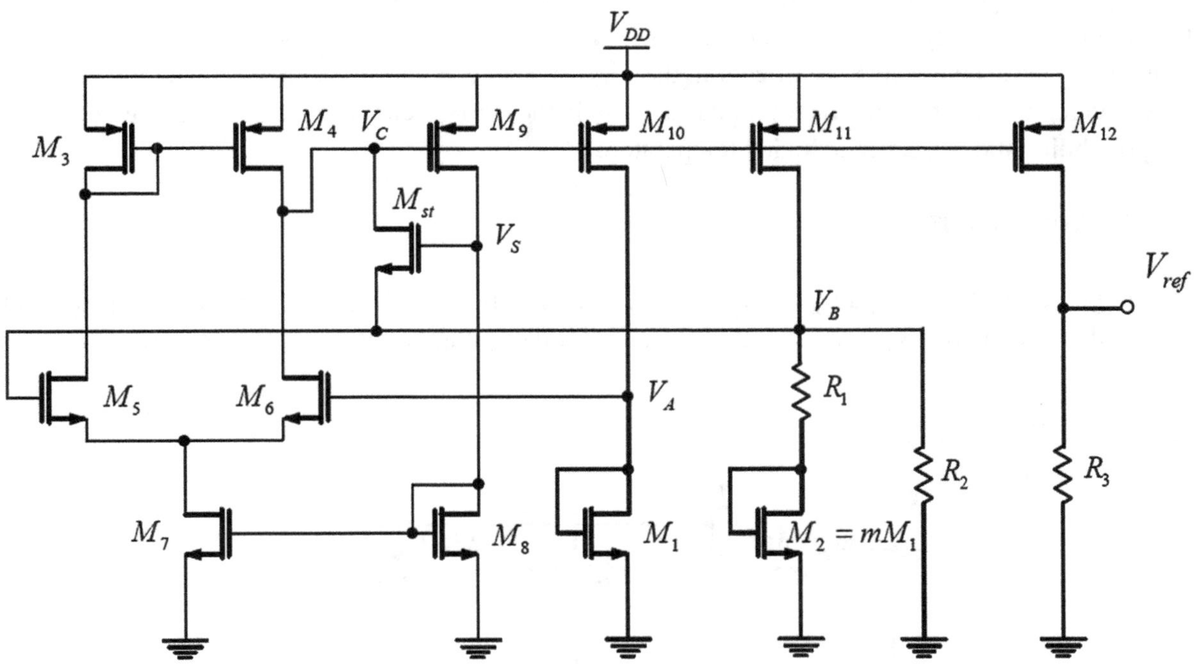

on-die temperature variation effect. The challenges of the on-chip temperature gradient will cause the inaccuracy of dc supply voltage. Hence, a voltage reference is required to operate at minimum constraint of parameters such as temperature and power supply variation. This feature is very important for the electronic products to be able to operate at extreme-case environment.

The basic principle of the BGR is to generate a zero temperature coefficient (TC) dc voltage based on two quantities which are CTAT(Complementary to Absolute Temperature)and PTAT(Proportional to Absolute Temperature). The first-order TC of bipolar junction transistor (BJT) generated from the base-emitter voltage V_{BE} has a CTAT characteristic of TC about -2 mV/^{0}C whereas PTAT effect is obtained from the thermal voltage $V_{THERMAL}$, which has TC of +0.085 mV/^{0}C. Based on Equation (1), the factor K is used to boost the positive TC term to balance the overall temperature performance for the output V_{REF} equivalent to zero.

$$V_{REF} = V_{BE} + KV_{THERMAL} \tag{1}$$

The BGR A circuit consists of an op-amp (M1 – M7), bandgap core circuit (M8-M12) and a start-up circuit (M13) as shown in Figure 2. The PMOS gate terminal is connected to the output node of the feedback op-amp to generate the V_{DD}-independent current. The differential input of the op-amp is tied to the PTAT and CTAT generator nodes respectively.

Circuit Implementation

Two different techniques are being used on the bandgap core temperature-control and these are the resistor-subdivision method and resistor-less method to control over the output voltage against the temperature and supply voltage variation as shown in Figure 2. A simplify version of start-up circuit is used to bias the op-amp at the early stage and should turn off once the op-amp has been biased in stable condition. In entire analysis and design specification, the transistor on core circuit need to be operating at weak inversion region, and the biased current particularly in the start-up circuit must be minimized to reduce the possibility of disturbance at higher temperature.

BGR Core Circuit

The proposed BGR core circuitry is shown in Figure 3, the differential input voltage of op-amp has been designed such that node V_a is equaled to node V_b. Hence the voltage across the diode-connected structure

Figure 3. BGR core circuit

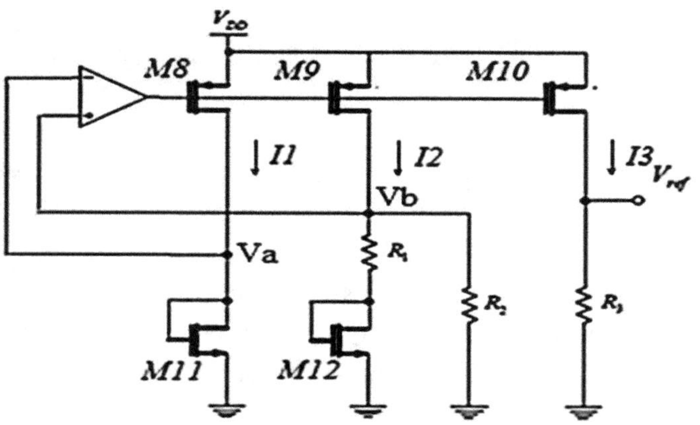

M_{11} will provide the CTAT behavior at weak-inversion region analysis. The biased current *I1, I2* and *I3* are connected by current mirror structure which has unity gain that allows same current value flow. The gate biased voltage of PMOS is controllable by the output amplifier. Assuming that the transistor operating at the weak inversion region; the V_{GS} voltage depends on the slope factor *n* which has direct relationship on temperature dependence as shown in Equations (2) and (3), as known in references of these authors; P.R. Gray et. al (2001) and P.E. Allen and D.R. Holberg (2002).

$$I_D = I_0 \frac{W}{L} e^{\frac{V_{GS}-V_{TH}}{nV_{THERMAL}}} \tag{2}$$

$$I_0 = 2n\mu_n C_{ox} (\frac{kT}{q})^2 \tag{2.1}$$

$$V_{GS} = nV_{THERMAL} \ln(\frac{I_D L}{I_0 W}) + V_{TH} \tag{3}$$

Substituting Equation (2) and Equation (2.1) into Equation 3 and then performing first-order differential would give V_{GS} a TC of about -2 mV/^{0}C on the V_{GS} drop as calculated using Equation (4). The CTAT generator will cause the same temperature effect on the voltage drop at R_2.

$$\frac{\partial V_{GS}}{\partial T} = -2\frac{nK}{q} + \frac{V_{GS}}{T} \tag{4}$$

Node V_b voltage can be written in term of Equation (5).Two different current-densities on the V_{GS11} and V_{GS12} generates the voltage drop at R_2 corresponds to PTAT behavior which contained $V_{THERMAL}$ and has TC of +0.085 mV/^{0}C as shown in Equation (6). Current-densities effect between transistor M_{11} and M_{12} is represented by constant M and is used to fine-tune the area size of transistor to give a desired temperature-control. M consisted of the dimensions of transistors M_{11} (L_{11}/W_{11}) and M_{12} (L_{12}/W_{12}).

$$V_b = V_{GS12} + I_2 R_2 \tag{5}$$

$$V_{GS11} - V_{GS12} = I_2 R_2 = nV_{THERMAL} \ln(M) \tag{6}$$

Since the current *I1, I2* and *I3* are mirrored, thus the output V_{REF} can be expressed as Equation (7);

$$V_{REF} = I_3 R_3 = (\frac{nV_{THERMAL} \ln(M)}{R_1} + \frac{V_b}{R_2}) R_3 \tag{7}$$

Temperature-compensation can be achieved by tuning appropriate resistor ratio value for R_1, R_2 and R_3 respectively and area ratio of transistor M_{11} and M_{12} as given in Equation (8). Combining Equation (4), the V_{GS} with TC of about -2 mV/^{0}C and Equation (6) with TC of +0.085 mV/^{0}C, Equation (8) is written.

$$\frac{\partial V_{GS}}{\partial T} = (\frac{0.085mV/^0 C\ln(20)}{R_1} + \frac{-2mV/^0 C}{R_2})R_3 = 0 \tag{8}$$

Self-Biased Op-Amp

A self-biased differential amplifier is shown in Figure 4 and it serves the purpose to reduce the minimum dependency on power supply variation. This is accomplished through a self-biased technique called bootstrap bias. By driving back the output current as the input current source itself through transistors M_5 and M_6, this makes the op-amp circuit consumed less power dissipation yet remains stable over a wide supply voltage range.

The symmetrical differential pair allows the geometry of transistors to be the same on M_5-M_6 and M_3-M_4 in order to conduct the same dc current flows. The input common-mode voltage range at transistor M_5 and M_6 are defined at around 900mV to provide sufficient headroom operating at strong inversion region. The voltage gain of the op-amp is found to be $Av = -g_m (r_{DS9} \parallel r_{DS8})$ with g_m as defined in Equa-

Figure 4. Self-biased op-amp

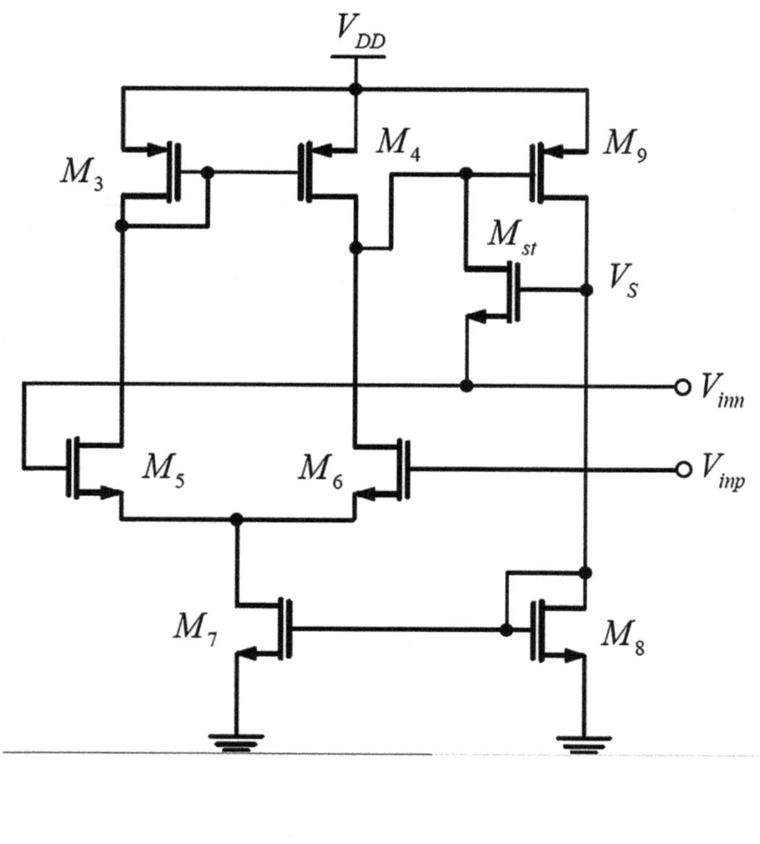

tion (9). Increasing the current in the transconductance transistor would increase the gain by square root of the current. The voltage gain is expected to be higher in order to stabilize the negative feedback loop which would help reduces the sensitivity of power supply.

$$g_m = \sqrt{K'(\frac{W}{L})I_D} \tag{9}$$

Start-Up Circuit

Only a single n-channel MOS is needed to drive the circuit out of the zero-current condition during the initial power-on stage. In Figure 5, the source terminal of M_{13} is tied together with V_b node while gate terminal voltage, V_G will pull the voltage until condition $V_{GS}>V_{TH}$ is met to turn on $M_{13.}$ Hence a small current at pico-ampere range will be injected into node V_b to generate the differential pair input of the op-amp. Once the different-pair input reach a stable voltage level, the M_{13} will turn off since $V_S> V_G$. Hence the geometry of the transistor M_{13} should be smaller to minimize such effect, ensuring normal operation.

Molithic Resistor

The active resistor is implemented by using an n-channel MOS transistor. The behavior of an active resistor is nonlinear, hence it is more challenging and difficult to fine tune to the desire resistivity value. By square law equation, the V_{GS} can be obtained through;

Figure 5. Modified MOS load-BGR circuit

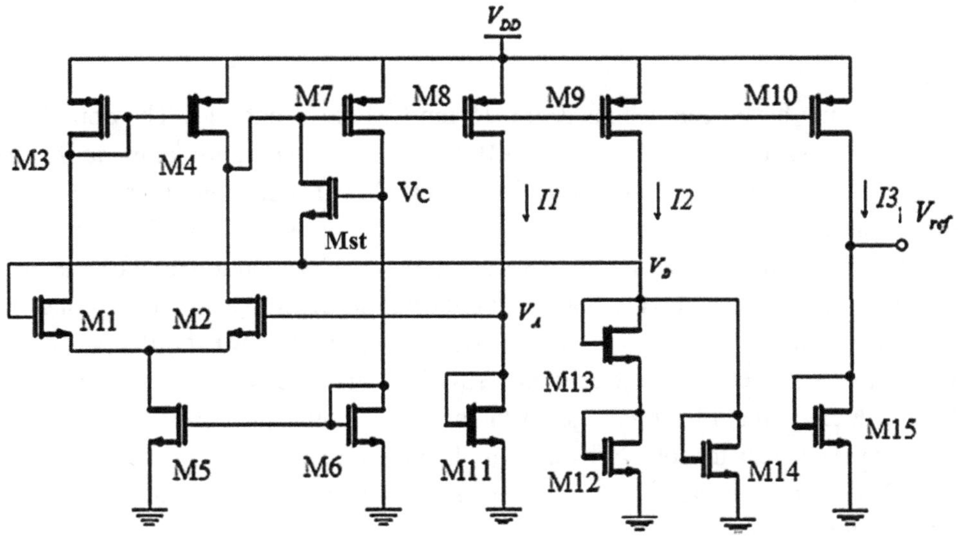

$$V_{GS} = V_{DS} = V_{TH} + (\frac{2I_D L}{K'W})^{1/2} \tag{10}$$

The small signal channel transconductance can be found by differentiating Equation (10) with respect to I_D. Referring to Equation (11), the L/W is defined as number of squares while second term, $[K(V_{GS} - V_{TH})]^{-1}$ is corresponding to sheet resistance, ohm per squares. Hence, the sheet resistor is controllable by the applied gate bias voltage.

$$r = \frac{1}{g} = \frac{\partial V_{GS}}{\partial I_D} = \frac{L}{W}\frac{1}{K(V_{GS} - V_{TH})} \tag{11}$$

The modification of the active resistor has been implemented in Figure 5. Transistors M_{13}, M_{14} and M_{15} are used to replace the passive resistors R_1, R_2 and R_3. The geometry size of transistor is fine tune based on the resistor value calculated using Equation (12). As mentioned, the zero TC dc voltage is based on CTAT and PTAT. Hence, a proper sizing on the transistors would generate an optimized CTAT and PTAT current for the core circuit.

$$\frac{W}{L} = \frac{1}{rK(V_{GS} - V_{TH})} \tag{12}$$

Now, the current path at *I2* is required to consume more current compared to previous BGR circuit. The passive resistor, R_1 is replaced by transistor M_{13} which requires the voltage level to be re-evaluated to drive the transistor to operate at a stable point. The output voltage reference V_{REF} has been deduced based on Equation(13). The I_{D3} is the temperature-independent bias current mirrored from I_1 and I_2.

$$V_{REF} = V_{GS} = \sqrt{\frac{2I_{D3}L}{K'W} + V_{TH}} \tag{13}$$

These BGR circuits have been designed in 0.18-µm CMOS technology. The optimized simulated reference voltage of resistor-less BGR is about 910.88 mV with maximum temperature coefficient of 12.99 ppm/^{0}C over the temperature range from 0°C to 100°C as presented in Figure 6. On another hand, a resistor-subdivision ratio BGR has been simulated with reference voltage of 529.24 mV with variation of 22.8 ppm/^{0}C as shown in Figure 7. In Figure 8, the supply voltage variation for the resistor-subdivision ratio BGR circuit is presented showing that the circuit has successfully produce a constant dc output voltage although the supply voltage (V_{DD}) varies from 0.9 to 10 V. However, the performance for the resistor-less BGR did not achieve the expected result. The main drawback could be the mismatch of transistor sizes inducing input offset voltage for the op-amp and the current-densities area size.

The layout size of the resistor-load BGR is around 25000 um^2 and the majority area size is occupied by the 3 NPOLY resistors, whereas the entire layout size of resistor-less BGR is only 1000um^2 area which is 25 times smaller compared to previous BGR circuit.

In Table 1, both BGR circuits have generated a constant low output voltage (<1V) as desired per specification. As the technology scaled down, the resistor-less BGR would be able to provide better

Figure 6. MOS load-BGR V_{REF} vs temperature

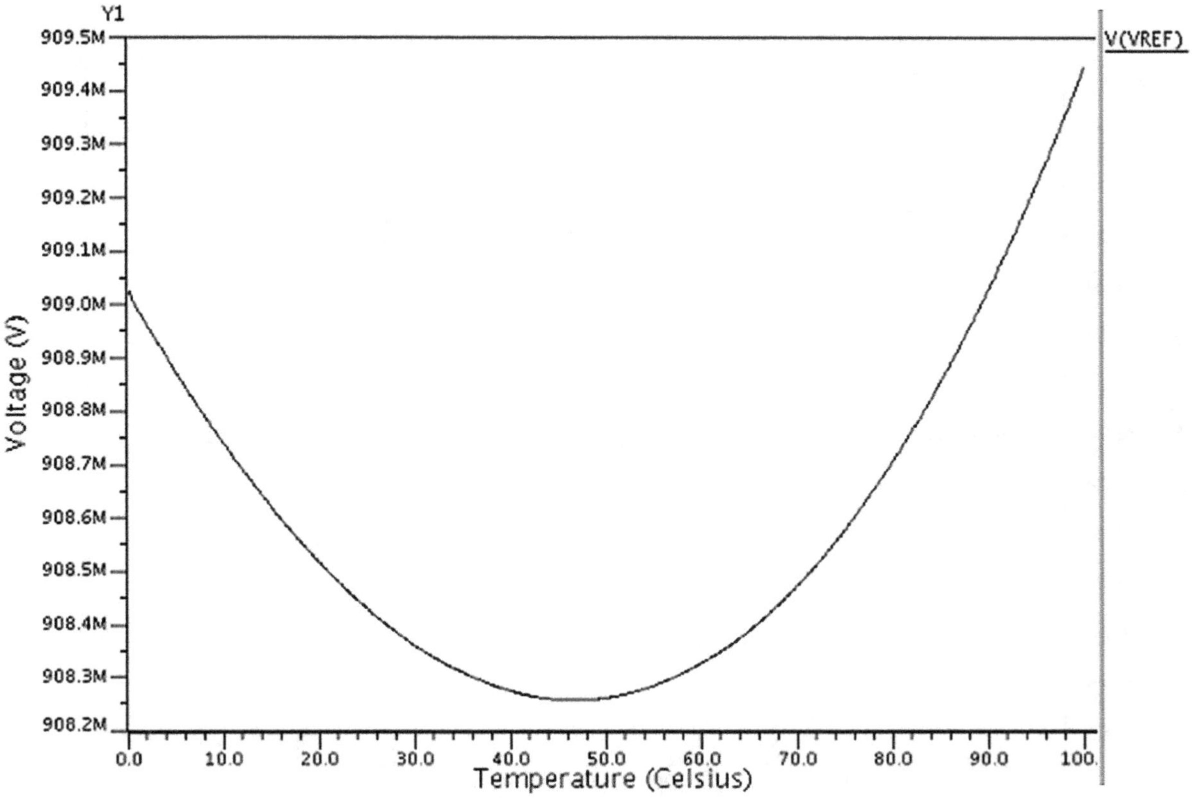

temperature control and smaller silicon die size. However, the resistor-subdivision BGR offers better performance due to low power consumption with the ability to operate at minimum of 0.9V supply voltage. Nonetheless, the poor temperature control is the main disadvantage of this circuit. High TC at 102 ppm/^{0}C would diminish the accuracy of the output result when there is a change in temperature.

A low voltage BGR circuit with resistor-subdivision has been designed in 0.18-µm CMOS technology. The minimum power supply voltage is 0.9 V and it produces an optimized 551.78 mV output reference voltage with TC of 101.9 ppm/^{0}C.The power consumption is only 96.67 µW at 1.8V. Meanwhile, the resistor-less BGR provides 910.88 mV with better TC at 12.2 ppm/^{0}C. However, its disadvantage is at the poor performance on power supply variation and high power dissipation. The start-up circuit has been modified into this simple version to reduce the complexity of the circuit. The simulation results show that the performance of the circuit is suitable for low-voltage and low-power electronic application which is capable of operating down to 1V output voltage.

There are some challenges in designing a low power BGR circuit. Transistors sizing is the most important parameter to be considered as it determines the amount of current driving the transistors. As a fresh designer or engineer, the hand-calculation based on equations (Equation 1 – 13) would be a good start. For advanced designers, the resistor-subdivision technique used in the resistor-load BGR circuit requires large area size in transistor in order to generate a large PTAT voltage for temperature variation cancellation.

Figure 7. Post & pre-simulated resistor-subdivision BGR V_{REF} vs temperature

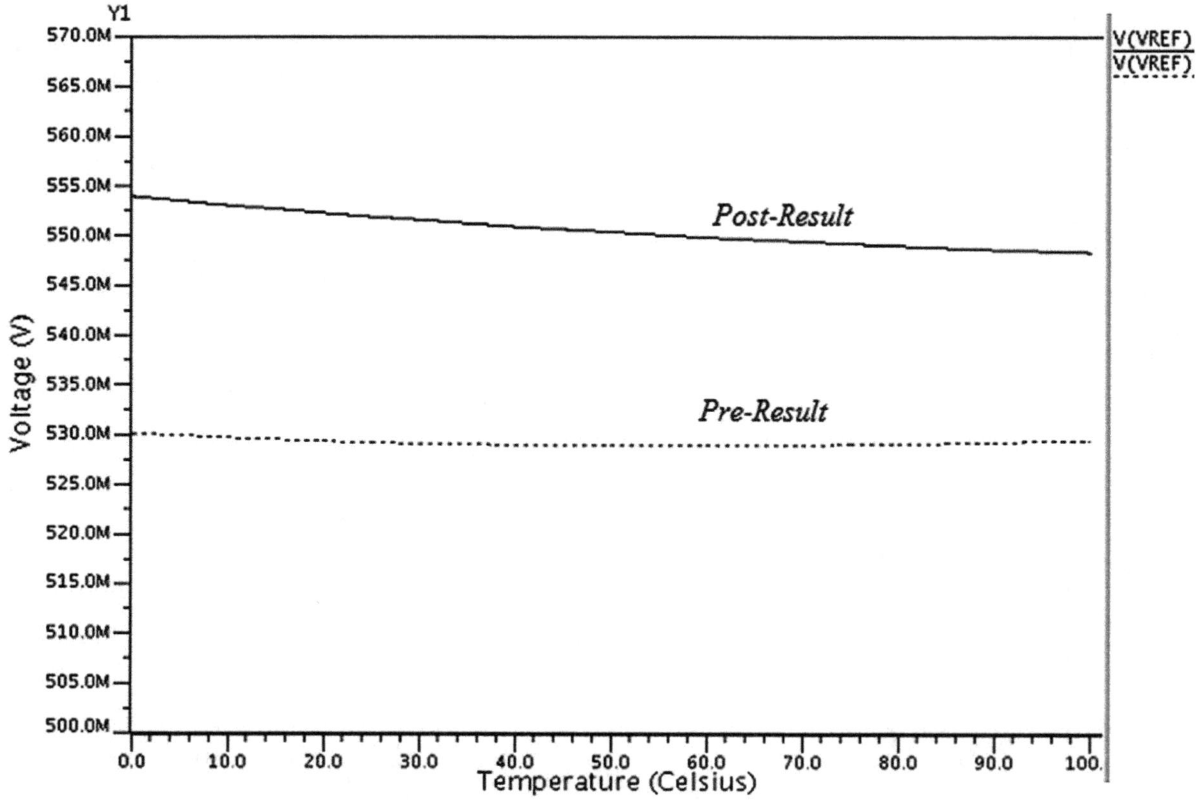

Table 1. Performance summary of two different loads BGR

Parameter	Resistor Load BGR	NMOS Load BGR
Technology	0.18um CMOS	0.18um CMOS
Size	71µ x 358.5µ	28.94µ x 35.88µ
Power Supply	0.85-10 V	1.8 V
Power Consumption @1.8V	96.67µ W	2.0215m W
Voltage Reference	551.78m V	910.88m V
TC	101.9ppm/⁰c	12.2ppm/⁰c

LEVEL SHIFTER

Low power application is highly sought after especially by battery-operated portable devices. The reduction in technology size and voltage power supply contributed to the decrease in power consumption. Voltage level shifter (VLS) allows selective components to work in several voltage blocks, optimizing speed and power which deemed as fit. Several researches include Wilson Current Mirror as described

Figure 8. Resistor-subdivision BGR V_{REF} vs V_{DD}

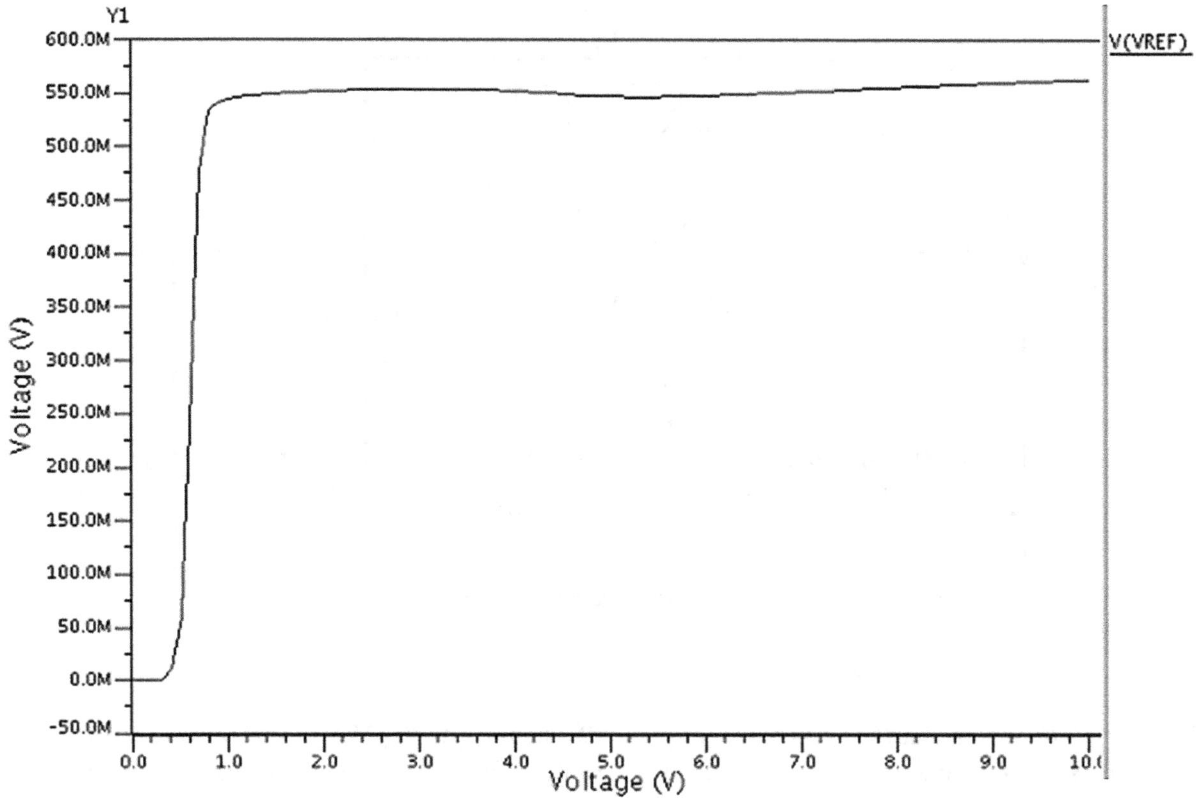

by Jun Zhou et al. (2015), subthreshold VLS as described by Hosseini S.R. et al. (2014), logic error correction as described by Osaki Y. et al. (2012) and also digital floating level shifters as described by T. Lehmann (2014).

A VLS with diode-connected NMOS as current-limiter has been designed in 0.18 μm CMOS technology. Both hand analysis and simulation results performed on this design showed an increase in low-to-high propagation delay (tpLH) when input signal amplitude (Vin) decreases. On the other hand, the high-to-low propagation delay (tpHL) is not affected by Vin. The maximum operating frequency for this VLS is 38.46MHz. The proposed VLS with diode-connected NMOS as active resistance is shown in Figure 9.

MN5 and MN6 are diode-connected NMOS instead of a V_{DD} connected NMOS and this is an enhancement to the circuit due to the simple notion of power saving design. When the gate of the NMOS is connected to V_{DD}, it is always "ON" hence leakage current is imminent. It is helpful during the charging up of MP1 and MP2 gate voltages, which ideally stop at ($V_{DD} - V_{thp}$) and at the same time MN5 and MN6 should turn "OFF". However both MN5 and MN6 are still "ON" as ($V_G = V_{DD}$) >>V_{thn}. There is still current flowing through the NMOS and this increases power consumption of the circuit. In Figure 9, V_G will follow V_D in both MN5 and MN6. When gate voltages of MP1 and MP2 reaches (V_{dd}-V_{thp}), $V_{GD} = V_{DS}$ will dwindle down to a maximum of V_{thp}, cutting down the current across the diode-connected NMOS significantly. Both of these NMOS would still work as active resistances to aid the charging and discharging of the PMOS. The difference is that both the gate voltages are drain-dependent instead of fixed on V_{DD} to decrease the current flowing during steady state. Current will start flowing through

Figure 9. Proposed VLS with diode-connected NMOS as active resistance

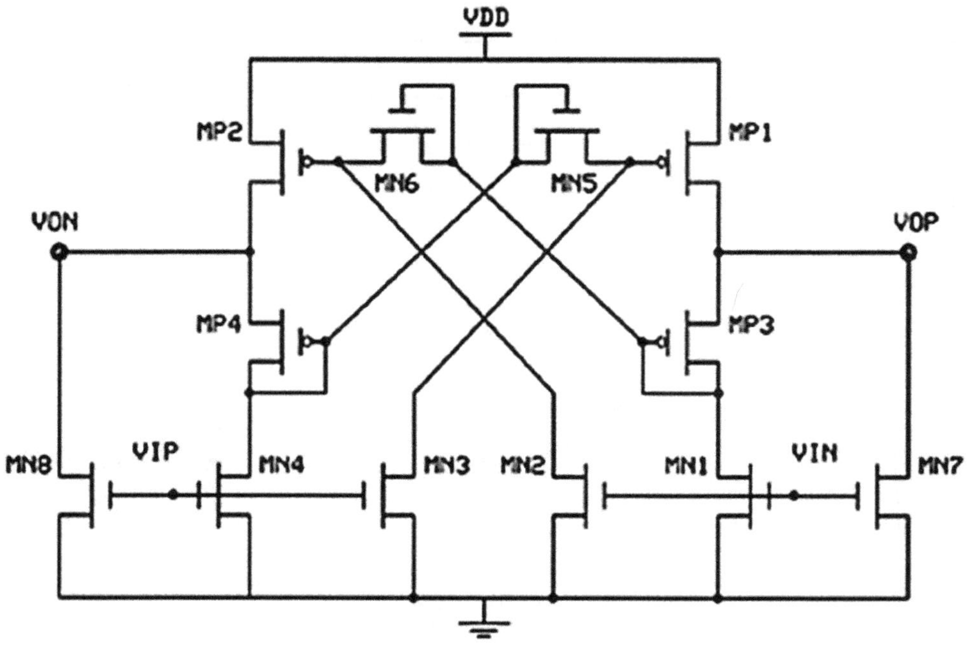

MN5 or MN6 during charging process as in Figure 9 when $V_{DS} > V_{GS}$-V_{thn}, comparatively in Figure 10 the NMOS starts conducting earlier as soon as $V_{GS} = V_{DS} > V_{thn}$, speeding up the charging up of PMOS gate voltages, in turns speeding up the discharging of V_{on} and V_{op}. This circuit is expected to consume less power and perform at a faster speed, i.e. faster discharge speed because PMOS turns OFF faster.

Small incremental time step is performed during steady state to observe the changes in each of the transistors. Half a period of a cycle is sufficient for the evaluation of t_p value since the VLS is symmetrical at vertical center. For simplicity purpose, NMOS and PMOS are considered either fully ON or fully OFF only. Both will turn ON immediately once $|V_{DS}| \geq |V_{GS}$-$V_{th}|$ and a resistor is assumed to be

Figure 10. Hybrid VLS

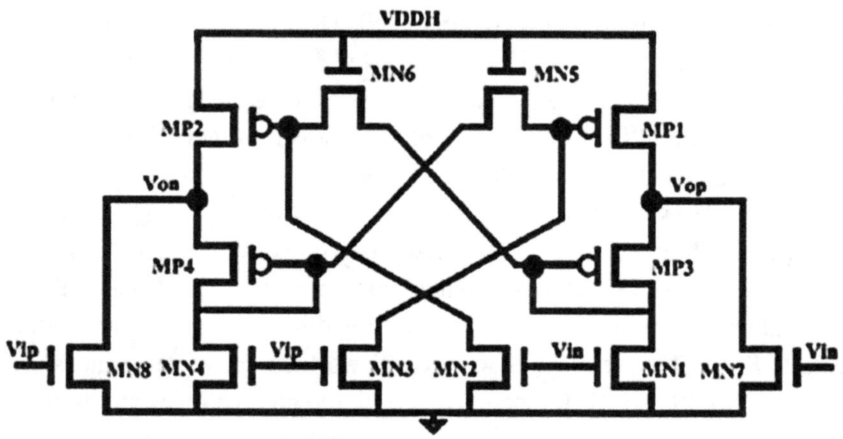

connected. They would turn OFF once $|V_{DS}|$ goes lower than $|V_{GS}-V_{th}|$ and the drain-source connection is open. In the diode connected NMOS and PMOS, both are assumed to be a diode with voltage drop of V_{th} and finite resistance when it's ON and open-circuit when it's OFF. V_{IN} and V_{IP} are assumed to be a perfect voltage source with zero rise time t_r and fall time t_f in this analysis. Figure 11 shows the expected circuit behavior in this analysis and power consumption is optimized by quickly turning OFF MP1, though minimizing t_{PLH}would influences charging of V_{ON}, cutting off the current pathway from V_{DD} through MP1 to ground. Diode-connected MN5 speeds up the charging of V_{GMP1}. Trade-off exists whereby increasing the pull-up strength of PMOS would minimize t_{pLH} and further minimizes power consumption; relatively the pull-down strength of NMOS decreases hence upsetting t_{pHL}, decreasing the overall speed of the circuit.

Figure 11. Timing diagram and expected equivalent circuit representation

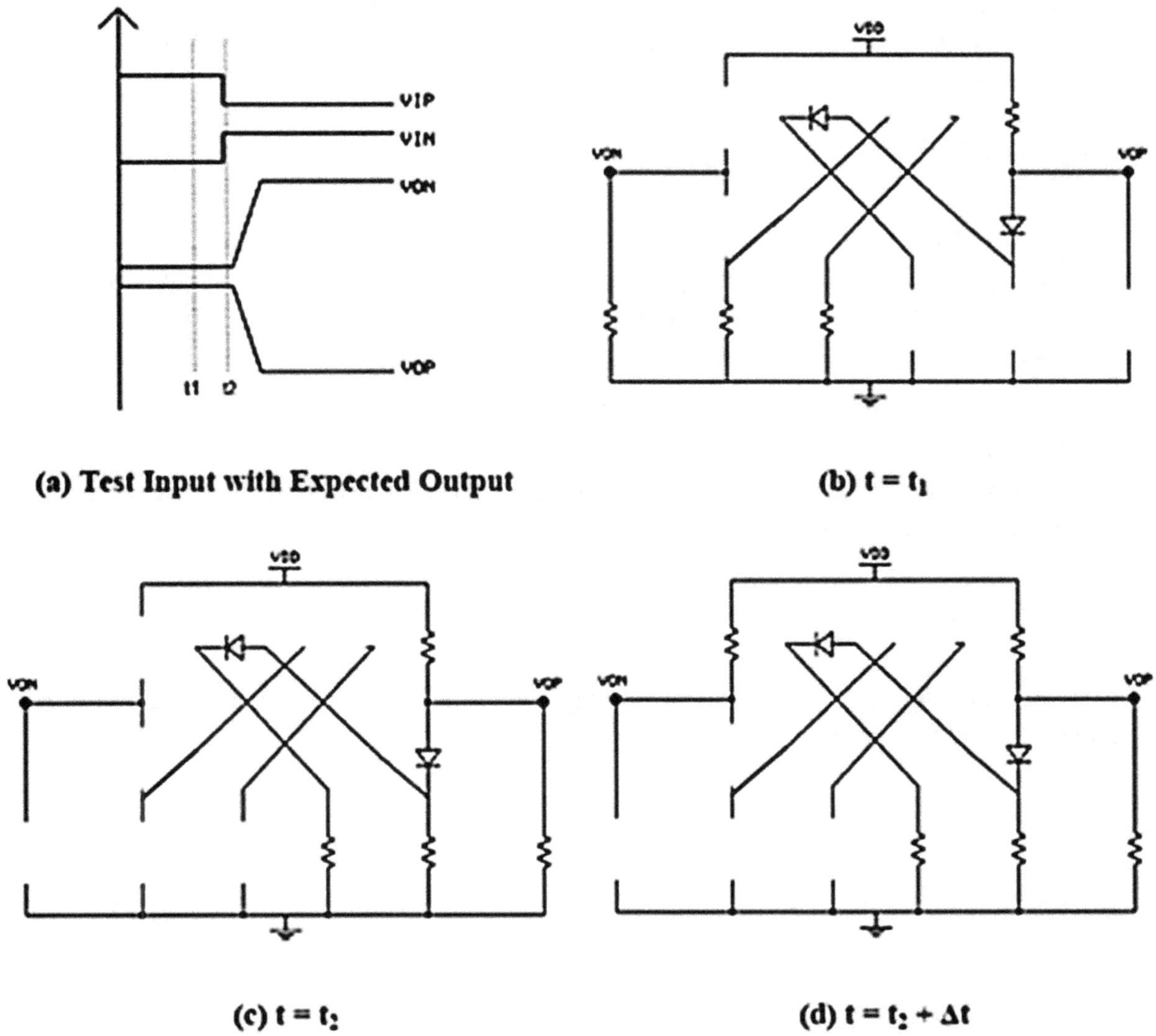

Several assumptions are made for this analysis and the assumptions are:

- $t_{pHL} = t_{pLH}$ as this is required for a symmetrical VTC as described in J.M. Rabaey (2012). This indicates that the change of rate in V_{OP} is the same as the change of rate in V_{ON} but in the opposite direction.The ratio of the PMOS and NMOS could be optimized for symmetrical t_{pHL} and t_{pLH}.
- The rate of change in V_S in all the diode-connected MOSFET would be the same as the rate of change of their $V_G(V_D)$ when they are turned ON. This suggests that the voltage drop across all of the diode-connected transistors remain the same as long as $V_{GS} > V_{th}$.
- The gates of MP1 and MP2 would be fully charged up or discharged down before input switches. Hence at the start of t_{pHL} or t_{pLH}, V_{GS} is either 0 or V_{DD}. This is for ease of the analysis before dwelling into complexity.

J.M. Rabaey (2012) mentioned that the propagation delay, t_p of an inverter is nothing but the average of t_{pHL} and t_{pLH} as shown in Equation (14). This is true since t_{pHL} and $t_{pLH}$of an inverter designed is such that they are as close as possible for a given configuration to optimize power dissipation and to achieve a symmetrical VTC.

$$t_p = \frac{t_{pHL} + t_{pLH}}{2} \quad (14)$$

However, t_{pHL} of a VLS with multiple input amplitudes is difficult to maintain since pull-down strength varies with input amplitude while t_{pLH} are relatively more stable to changes. Speed derivation through t_p value is more appropriate if the higher value of t_{pHL} or t_{pLH} is chosen for optimized operation. For an inverter, t_{pHL} is solely influenced by the discharging through one NMOS only and t_{pLH} is influenced by the charging through one PMOS. Refer to the analysis in Figure 11(d) for the VLS, t_{pHL} is dictated by not only the discharge through PDN, but constant pull-up of the PUN as well. By ignoring the discharge current through MP4 and MN5, Equation (15) can be used for t_{pLH}.

$$t_{pLH} = 0.69.C_L.R_{eq} \quad (15)$$

The charging voltage, V_c and discharging voltage, V_d of the circuit can be analyzed by using the equations (16) and (17) below:

$$V_c = V_o\left(1 - e^{-\frac{t}{\tau_c}}\right) \quad (16)$$

$$V_d = V_o\left(e^{-\frac{t}{\tau_d}}\right) \quad (17)$$

whereby V_o is the output voltage, while t is the charging and discharging time. Taking the derivative of the above expressions with respect to t, and substitute t with V_c and V_d expressions respectively, the overall rate of change involtage is obtained by adding both the rate of change of V_c and V_d. The equation can be written as:

$$\frac{dV}{dt} = \frac{V_o}{\tau_c} - V\left(\frac{1}{\tau_c} + \frac{1}{\tau_d}\right) \tag{18}$$

Substituting R_c with infinite resistance value in Equation(18), it's shown that it represent an open circuit, meaning that the circuit is discharging through R_d only. Inverse of $\tau_{c|Rc\to\infty}$ approaches zero and the overall discharge voltage would revert back to Equation (17). To obtain the expression needed, time to discharge the circuit, the left hand side (LHS) and right hand side (RHS) in Equation (18) are integrated with respect to t, from 0 to t. Rearranging and substituting $t = t_{pHL}$, $V = 0.5V_{DD}$ and $V_o = V_{DD}$, the expression of t_{pHL} can be written as:

$$t_{pHL} = -\frac{1}{A}\ln\left|\frac{B - A\frac{V_{DD}}{2}}{B - AV_{DD}}\right| \tag{19}$$

Where

$$A = \frac{1}{\tau_c} + \frac{1}{\tau_d} \tag{20}$$

$$B = \frac{V_o}{\tau_c} \tag{21}$$

As $\tau_d = C_L R_d$ and $\tau_c = C_L R_c$, assumption made in this derivation is that $R_d < R_c$.

As for the load capacitance, the entire capacitances model is based on Figure 12. The expression of C_L is expressed as below:

$$C_L = C_w + 2(C_{gdn} + C_{gdp} + C_{gsp}) + C_{bdp} + C_{bdn} + C_{bsp} + C_{gp} + C_{gn} \tag{22}$$

whereC_{gdn}, C_{gdp} and C_{gsp} are coupling capacitance of gate to drain and gate to source of the MOSFETS which are doubled due to Miller effect. C_w is the interconnect capacitance which can only be extracted at physical layout process. Fortuitously, $C_w << C_L$ hence it is negligible in the derivation of t_{pHL} and t_{pLH}.

In Figure 11, V_{OP} is discharging while V_{ON} is charging. Transistors involved in t_{pHL} are MP1, MP3, MN1, MN2, MN6 and MN7 while in $t_{pLH}$the transistors involved are MP2, MP4 and MN5. Since the

Figure 12. Components of C_L for Node V_{OP} in VLS

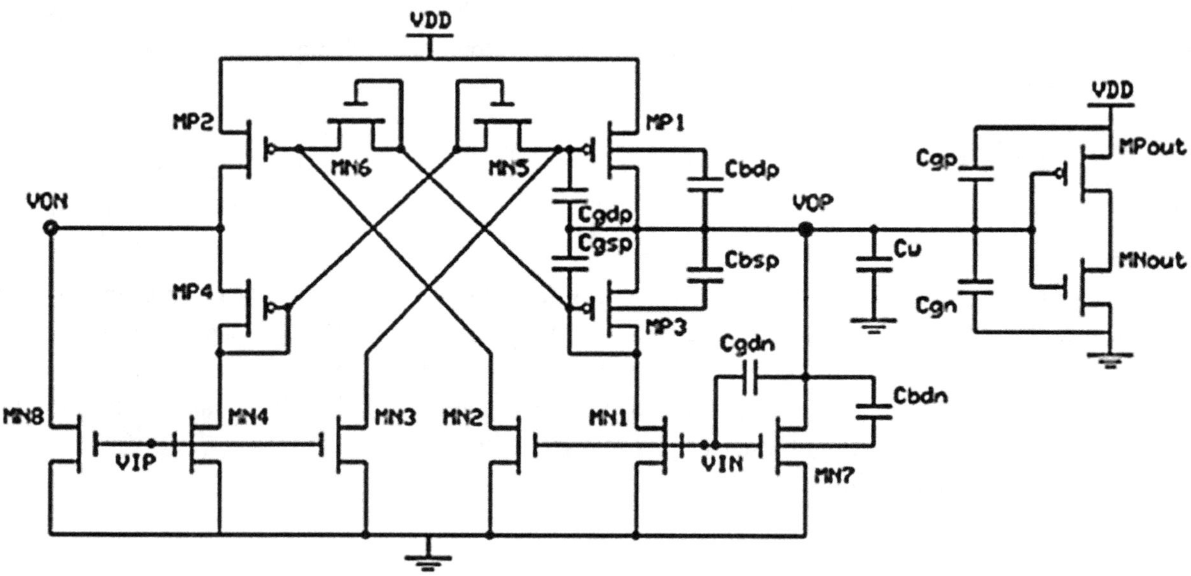

architecture is a symmetrical circuit, in t_{pLH} analysis, MP1, MP3 and MN6 would replace the role of MP2, MP4 and MN5 respectively.When V_{OP} is low, the resistance of MN6 is given by

$$R = \frac{\beta}{V_{DD} - V_{thn}} \tag{23}$$

where V_{DD} is V_{GS} of MN6 and β is a technology factor that is a function of the mobility and the gate capacitance C_g of the MOSFET. As V_{IP} goes high and turns on MP1, V_{OP} starts charging up. V_{OP} is equivalent to V_{DS} of MN6 and the resistance is dependent on this voltage. Equation (23) can be written as:

$$R = \frac{\beta}{V_{DD} - V_{thn} - V_{OP}} \tag{24}$$

Since the gate and drain of a diode-connected NMOS and PMOS are connected, hence $V_{GS} = V_{DS}$. Applying this into Equation (23), the resistance of the MOSFET is fixed at

$$R_{diode} = \frac{\beta}{-V_{thn}} \tag{24}$$

The diode connected MN6 and MP3 has a resistance relationship shown in Equation (24). R_{eq1}, R_{eq2} and R_{eq7} equations are derived based on the equivalent resistance of an inverter NMOS as shown in equation below:

$$R_{eq} = \frac{1}{V_{DD}/2} \int \frac{V_{DS}}{I_{DSAT}\left(1+\lambda V_{DS}\right)} dV_{DS} \tag{25}$$

where the I_{DSAT} is defined in equation below:

$$I_{DSAT} = k'\frac{W}{L}\left[\left(V_{GS}-V_{th}\right)V_{DSAT} - \frac{V_{DSAT}^2}{2}\right] \tag{26}$$

k' is the device conductivity while λ is the channel length modulation parameter dependent on factory specification. I_{DSAT} in Equation (26) is replaced by I_{Dlin} in Equation (27)for R_{eq1} since MP1 operates in linear region throughout t_{pHL}. Since both biasing voltages of MP1 change with time, it would increase the complexity of the resistance evaluation. Hence, the average value of V_{GS} is substituted for R_{eq1} derivation.

$$I_{Dlin} = k'\frac{W}{L}\left[\left(V_{GS}-V_{th}\right)V_{DS} - \frac{V_{DS}^2}{2}\right] \tag{27}$$

Based on related equations mentioned, possible W values for each of the transistors are inserted into the circuit in Figure 11, starting from the minimum transistor width to find the most suitable values of t_{pHL} and t_{pLH} for every V_{in} cases. t_{pHL} values would be affected by all 5 values, namely W_{MN7}, W_{MP1}, W_{MP3}, W_{MN2} and W_{MN1} while t_{pLH} would be affected by W_{MN7}, W_{MP1} and W_{MP3}. Since t_{pLH} doesn't involved PDN, V_{in} would only affect t_{pHL}. The relationship is described in Table 2.

Since the circuit is symmetrical, hence the transistors on the LHS are identical to the transistors on the RHS. Transistor MN6 is evaluated as diode-connected transistor and its size doesn't affect the speed of VLS. Economically, W_{MN6} should be as low as possible. From Table 2, W_{MP3} should be as low as possible to maximize the speed of VLS. Likewise, W_{MN1} and W_{MN2} would be as large as possible as long as it does not cause self-loading effect. Figure 13 shows the simulated result of V_{ON} and V_{OP} with increasing V_{in} at frequency of 5 MHz and $t_r = t_f = 0.1$ ns. t_{pHL} decreases as V_{in} increases which is in line with the previous discussion and t_{pLH} also decreases but with a smaller deviation. This might due to the increased V_G discharge rate of the charging transistors MP1 and MP2 through MN3 and MN2 respectively.

Table 2. Relationship between transistors' width and propagation delay

Increasing Parameters	t_{pHL}	t_{pLH}
W_{MN7}	Decrease	Increase
W_{MP1}	Increase	Decrease
W_{MP3}	Increase	Increase
W_{MN1}	Decrease	N.A.
W_{MN2}	Decrease	N.A.
V_{in}	Decrease	N.A.

Figure 13. Decreasing t_{pHL} with increasing V_{in} at 5 MHz

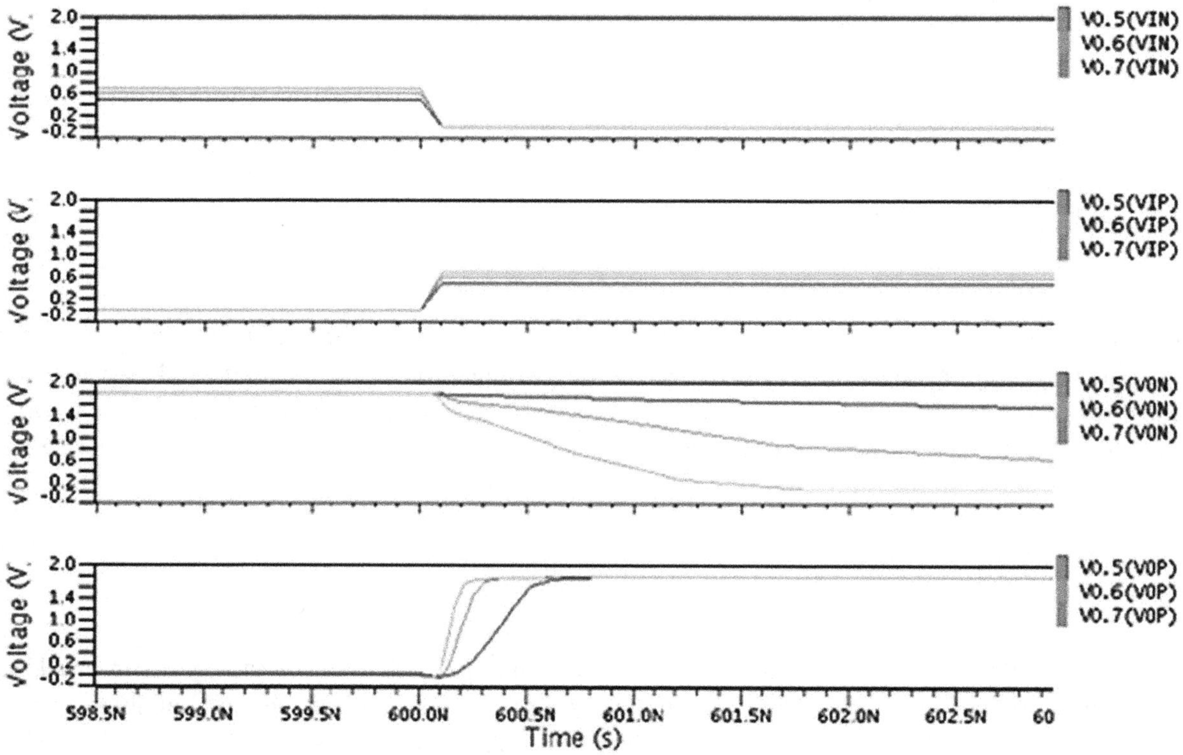

As shown in Figure 13, the difference between t_{pHL} and t_{pLH} is not as large as calculated. $t_{pHL}>t_{pLH}$ for $V_{in}< 1$ V, $t_{pHL}\cong t_{pLH}$ when $V_{in} = 1$V and $t_{pHL}<t_{pLH}$ for $V_{in}> 1$V. τ_c is progressively dominating as V_{in} increases during discharging process. The VLS is achieving $t_{pLH} = t_{pHL}$ when $V_{in} = 1$V, which is improving VTC and power consumption. Figure 14 shows the propagation delay (t_{pHL}, t_{pLH} and t_p) for increasing input signal amplitude V_{in}.

The propagation delay, t_p reflects a similar behavior with VLS in Ali S., Tanner S. &Farine P.A. (2011) and Shao H. &Tsui C. (2007). It indicates that the diode connected MN5 and MN6 does not affect the speed of the VLS. The VLS is reported to consume approximately 1.1 pJ energy consumption per transition with the increasing input signal amplitude from 0.4 to 1.4 V.Depending on the application of the VLS, t_p value can be decreased further by decreasing the size of MP1 (MP2) for low V_{in}. This would reduce the rift between t_{pHL} and t_{pLH} at low V_{in} but increase the difference at high V_{in}. It varies between a logic level shifter, unidirectional level shifter, bidirectional level shifter or application specific level shifter. Texas Instrument has more than a hundred products for different voltage level translation family product.

As a result, a 5 MHz VLS with diode-connected NMOS as an active resistive is proposed. The proposed VLS is expected to have smaller layout area. Moreover, the PDN sizes can still be further increased to facilitate the discharging process, thus improving t_{pHL}, without impending on the total cost. The maximum operating frequency is 38.46MHz. The proposed VLS still has a lot of rooms for improvement and this include derivation of equations for the discharge process, t_{pLH} and the value of V_{GS} in the pull-up PMOS should be considered as a variable instead of a constant for more accuracy in PUN strength. For

Figure 14. Propagation delays versus input signal amplitude, V_{in}

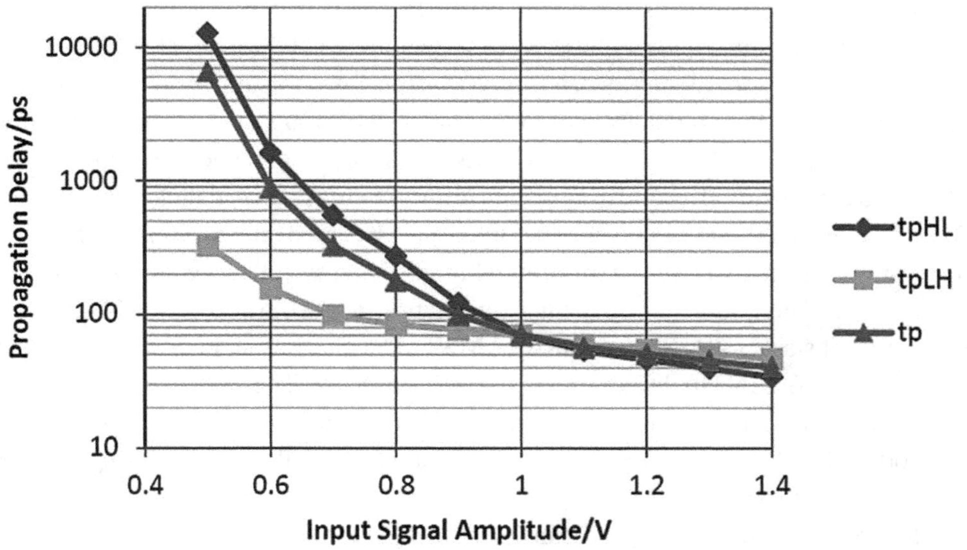

advance designers, one should also consider modelling an ideal voltage source by inserting t_{rise} and t_{fall} based on t_{pHL} and t_{pLH} values respectively rather than using the input signal period. The rate of change in all nodes of the circuit needs to be evaluated separately for more precise analysis. For improvement, the charge rate of pull-up PMOS gate need to be considered and added intothe analysis to further improve t_{pHL} and power consumption.

COMPARATOR

Comparators may be the underrated and underutilized monolithic linear component but this is very unfortunate as comparators are one of the most flexible and universally application components available. Comparators, in particularly the high speed comparators can be used to implement linear circuit functions which are as sophisticated as any op amp-based circuit. Combining a comparator with a low drift amplifier would form a high speed voltage to frequency converter, which can be further enhanced to work as a voltage controlled sine wave oscillator with applications in audio, shaker table driving and automatic test equipment. In this section, the comparator has been optimized to implement an analog-to-digital converter (ADC). Linear Technology (LT) has LT1016 which can be used to implement a fast 12-bit ADC.

A comparator represents a very important block in most ADC. It is used for converting analog signal into digital signal by comparing with the reference input signal. Nowadays, most of electronic device is invented to be able to work under ultra-low power consumption and smaller feature size. Comparator is the only analog block in the successive approximation register (SAR) ADC. The analog signal will be first sampled by sample and hold circuit (S/H) and then being compared with the reference voltage from digital to analog converter (DAC).

Dynamic comparator is composed of two phases; reset phase and regeneration phase. It consists of differential input pair and cross coupled inverter. Differential input pair circuit will feed current into cross coupled CMOS regenerative pair circuit. The cross coupled pair circuit provides a large gain and amplifies the differential input into near full logic level. The reset switch in the input pair helps to reduce comparator recovery time during reset phase, as described in Xiaolei Zhu et al.(2012).

Figure 15 is the improved version of latch type voltage SA comparator. When clock is low (reset phase), the node of the comparator, Di will charge to VDD through transistor reset transistor MP5. The transistor of MN6 and MN7 with transistor MN1 will turned off when the clock is low thus will completely cut the connection from cross couple inverter. There is no current activity in the differential pairs as Di node is isolated from the source of input pair by transistor MN6 and MN7. During reset phase there a no current activity in the differential amplifier where there is no charging to VDD and discharging to GND. Then, after clock is high (evaluation phase), transistors MN1, MN6, MN7 are turned on and current starts to flow into transistor MN1. The charged Di node would start to discharge through transistors MN6, MN7 to differential pair MN2 and MN3. The size of reset transistor MN6 and MN7 should be larger than transistors MN4 and MN5 in order to drive the high current density of the latch and Di node. Thereby, the current flow in MN2 and MN3 will increase during evaluation phase. The drain nodes of input transistors have large voltage variations causing a large kickback noise and the variation of the operation region of input transistors causes voltage variation at their gates producing kickback noise(RahelehHedayati, 2011). These transistors (MN6 and MN7) are used to avoid the clock feed through and kickback effects from the latch to the input.

Figure 15. Improved version of latch type voltage SA comparator

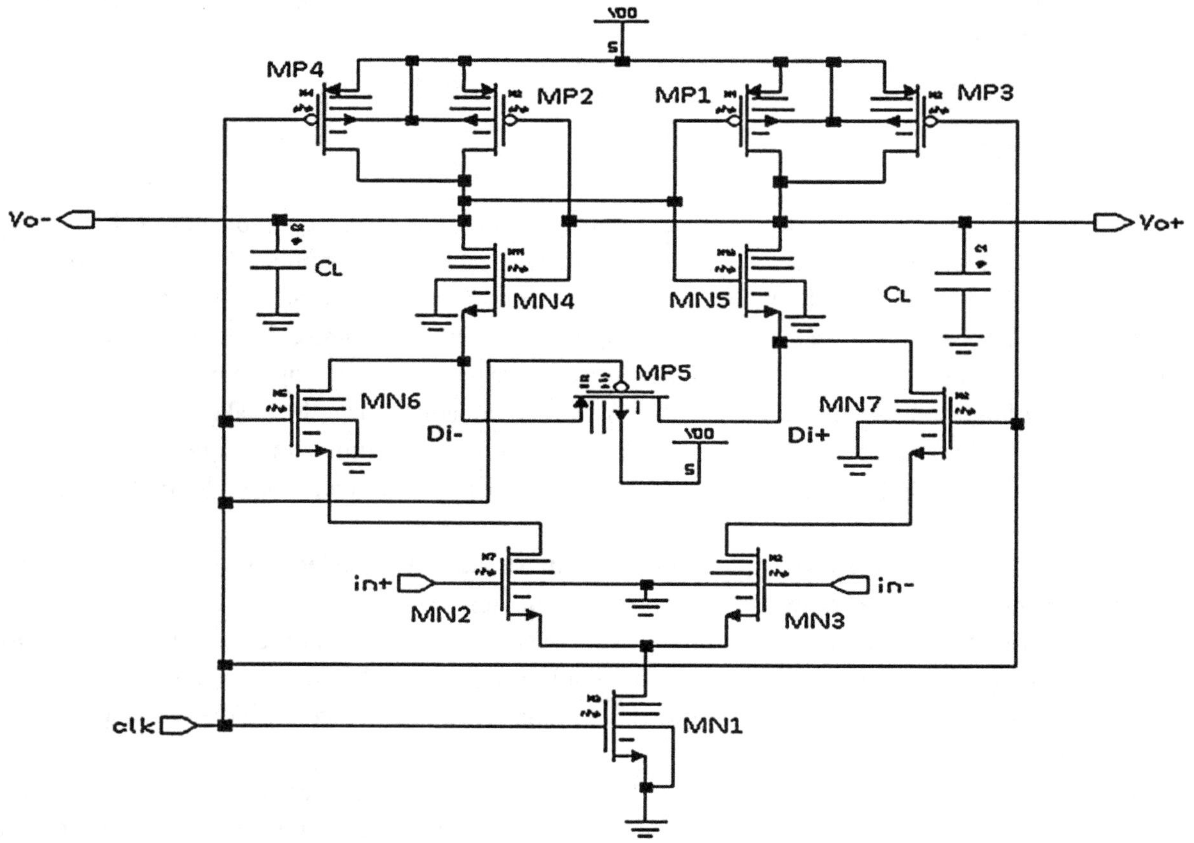

Comparator is the only analog component in SAR ADC. The modified comparator is designed to meet the SAR ADC requirement. The main criterion needed is resolution, power, and delay. The output node of the proposed comparator is connected to the output buffer and SR latch. The SR latch is used to hold the comparator output during reset. Otherwise, it will always pre-charge to VDD during reset mode and will draw unnecessary power. Figure 16 depicts the modified SR latch circuit. The NOR gates were used instead of NAND gate to hold the output value. SAR ADC required steady output signal instead of pulse signal. When the comparator produces output high or low, it needs to hold the value until the next comparison.

Figure 17 shows the simulation result for modified comparator with the inverter buffer and SR latch. The sinusoidal signal 1.5V amplitude with 20 MHz frequency is applied at the input terminal, Vin+ while reference voltage is set to dc voltage equaled to 0.75V at the other terminal, Vin-. One single clock has been used to drive the comparator with frequency of 0.1 MHz. Figure 18 shows the output voltage of the comparator with the use of NAND gate and with NOR gate type SR latch. When the clock is low, the output voltage reset to VDD and when the clock is high, the input and reference voltage are compared and subsequently, produced the output signal. If the input signal Vin+ is more than reference voltage Vin+, it produce logic level 1 and if Vin+ less than Vin- it works otherwise. Based on Figure 18, the NAND gate type SR latch exhibit pulse signal when the input voltage Vin+ is more than Vin- instead of constant zero level as in NOR gate type SR latch. The comparison between sampled input voltage and DAC voltage in the ADC operation should maintain constant level until the first clock cycle finished. Otherwise, it will result error in the ADC.

For 10-bit ADC with reference voltage of 1.5 V, the V_{LSB} value is equaled to 1.46 mV. The minimum resolution voltage, (ΔV_{IN})min for the comparator should be less than 0.5 V_{LSB} Thus, it must resolve voltages as small as 732μV. The sensitivity of the comparator can be tested by replacing the input terminal Vin+ with input pulse signals of value 732 μV. The previous design exhibit incorrect output at the first 10μS in the transient analysis. The output should be charged to VDD instead of VSS. In this design, only one tail current is used to drive the current out from differential pair transistor. When the clock is

Figure 16. Modified design schematic

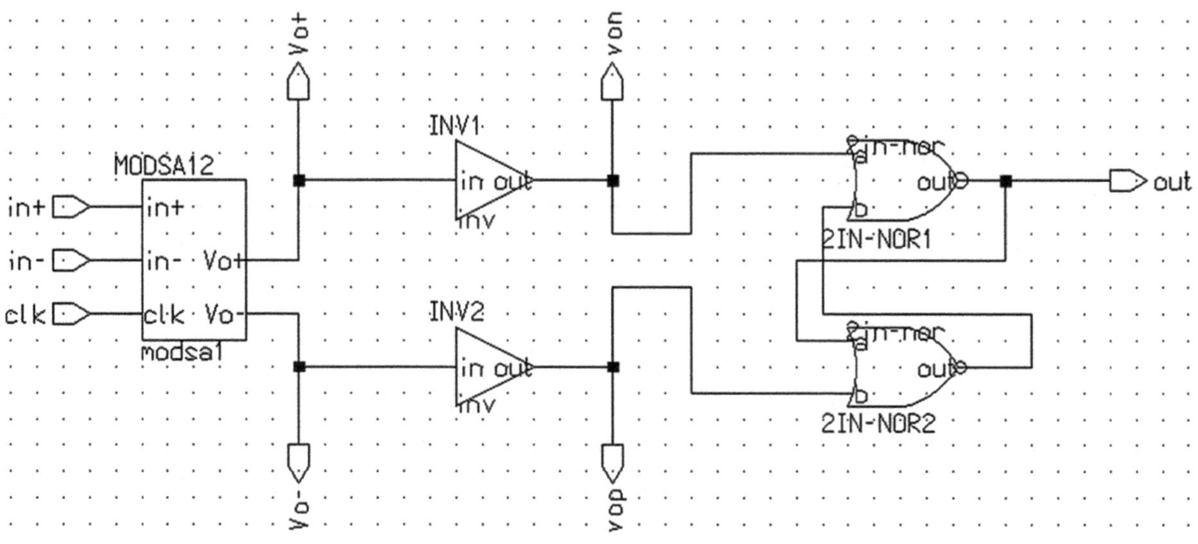

Figure 17. The simulation result for modified comparator with the inverter buffer and SR latch

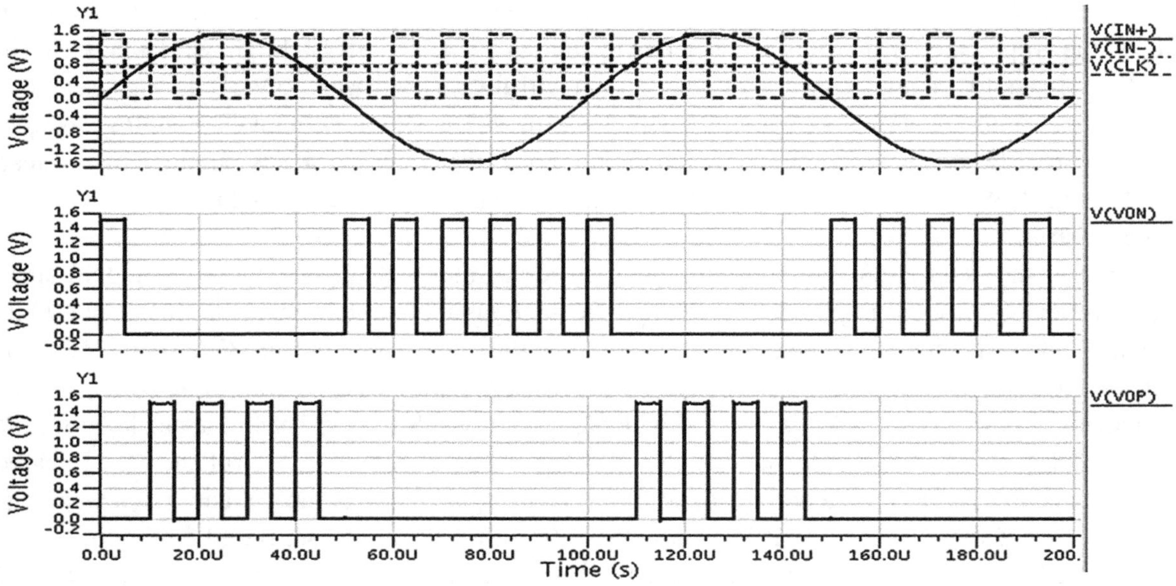

Figure 18. Output signal voltage, Vout using (a) 2 NAND gates and (b) 2 NOR gates

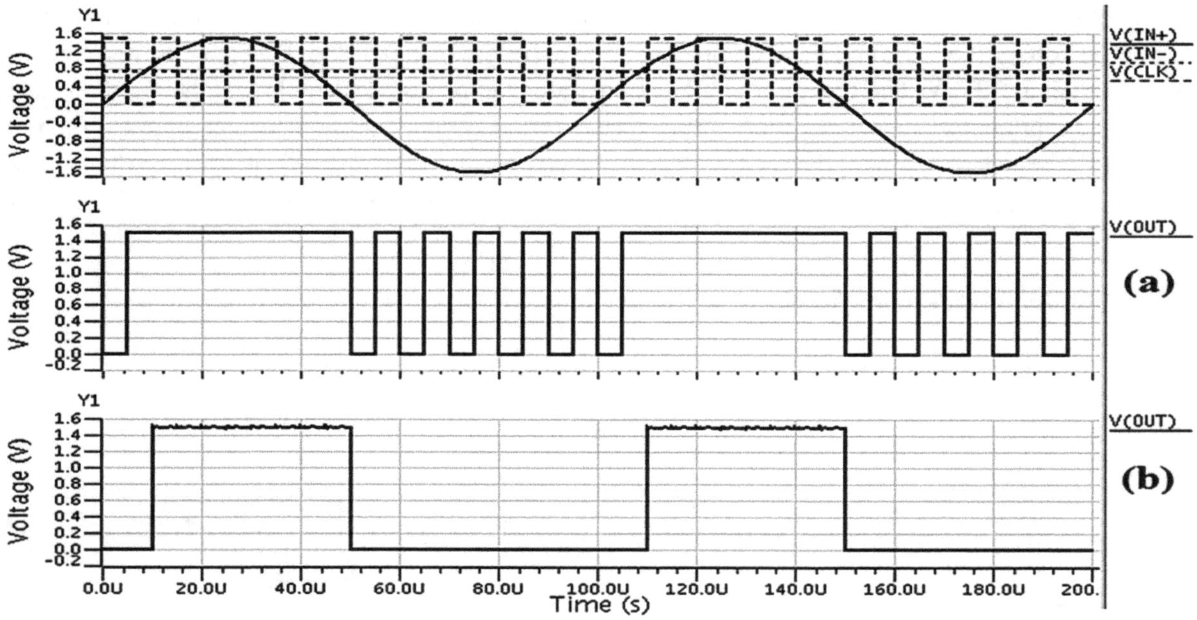

high the tail current will go into linear region and start the generation. The differential pair will go out from saturation and current start to flow through the tail transistor. In this design, the resolution and delay is considered to be the most important parameters. The modification is done in the tail transistor and differential pair transistors where we revised the sizing of the transistor. Table 3 shows the size of all transistors in previous design, Comparator A and modified design, Comparator B. The length of all transistors is set to be equaled to 0.18 μm.

Table 3. Transistors sizing parameters

Transistors	Comparator A	Comparator B
MN1	1.8 μm	11.52 μm
MN2,MN3	7.2 μm	2.88 μm
MN6, MN7	1.44 μm	1.44 μm
MN4-MN5,MP5	0.72 μm	0.72 μm
MP1-MP4	1.8 μm	1.8 μm

The simulations are carried out with $(\Delta V_{IN})_{min}$ equaled to 732μV for clock frequency 0.1 Mhz and 1.5 V power supply. Figure 19 presents the simulation of the output voltage for both comparator A and B. Comparator A was designed to have smaller area and faster latching delay. However, the circuit shows that it produced incorrect output level for the first 10 μs. This result can contribute error to the whole operation as it needs to settle within 10 μs. At the first clock cycle, when the clock is high the output terminal Vo+ should be charged to VDD instead of discharging to ground. The same goes to Vo- terminal where it should be discharged to ground. During regeneration phase, the current will start flowing through transistor MN1, MN2, MN3, MN6 and MN7. The back to back inverter will start the evaluation phase depending on the input voltage and current which flow through the differential pair. Since the Vin+ is more than Vin+, precharge Di+ node start to discharge. Transistor MN4 will turnon

Figure 19. Simulation of the output voltage with minimum resolution, ΔV_{IN} equaled to 732 μV of comparator A and comparator B

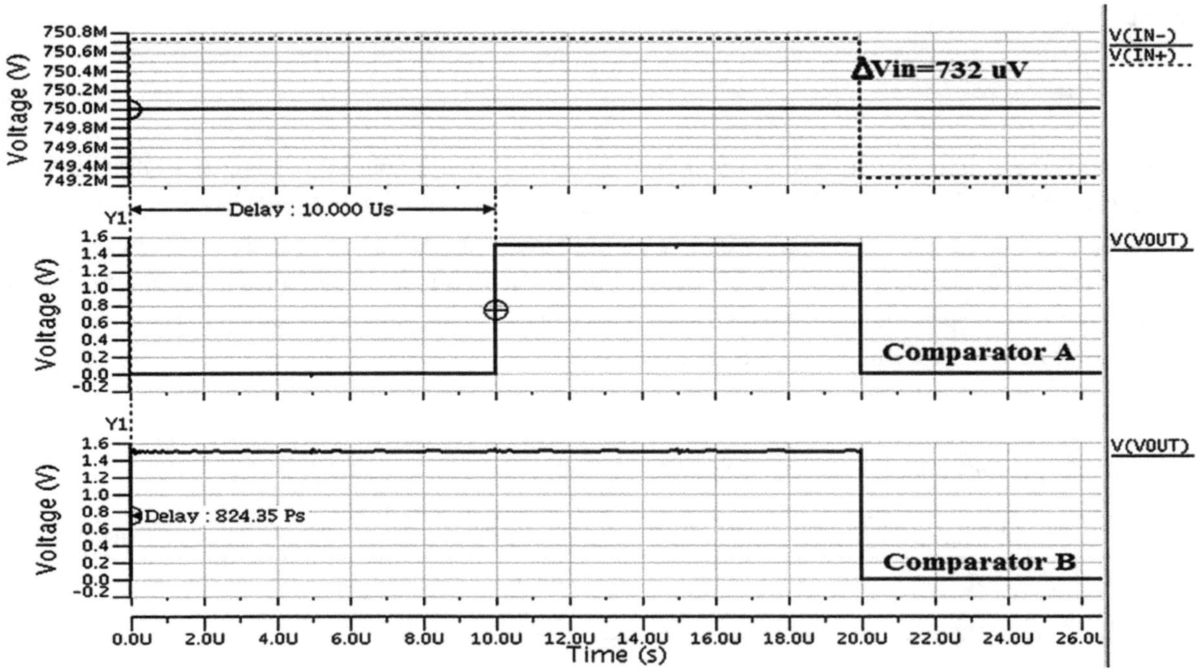

once Di+ node drop to V_{DD}-V_{thn}. When the output node is dropped to V_{DD}-V_{thp}, the transistor MP1 turn on and this would initiate a positive feedback. The Vo- terminal continued to discharge to 0V while terminal Vo+ should be charged to VDD.

In ADC design, the response time of the comparator must be less than one clock cycle. The response time of the comparator is the sum of the delay in the latch to generate full logic level. Sizing transistor in Comparator A has the worst response time as it produce delay equaled to one clock cycle. This will produce error during implementation of the ADC. The sensitivity of the comparator can be increased by increasing the gain of the comparator. The sizing in Comparator B shows that it has solved the previous problem by increasing the width of the tail transistor. However, the tradeoff of power consumption becomes high due to the high tail current supply. This is shown in Table 4. In ADC operation, the sensitivity of minimum resolution and the delay are the critical parameters.

Based on the ideal transfer curve of comparator, the output voltage level will change from 0 to 1 as Vp-Vn exceeds0. However, offset voltage, Vos would occur if the output level did not change until the Vp-Vn reach a value Vos. The output offset voltage of the dynamic comparator can be found by simply grounding both of the input terminals. The output voltage presents at this point is the Vos. The required Vos for this designed should be less than (ΔV_{IN})min.

For dynamic clocked comparator, the offset voltage is simulated by applying a triangle pulse varying from (0.75 - 1.46m)V to (0.75 - 1.46m)V at positive terminal Vin+. The reference voltage Vin- is connected to the 0.75V. The input and output are observed in Figure 20. The output voltage Vout should changes its output level for 0 to 1 once the value of Vin+ is more than the reference voltage Vin-. The offset value can be calculated from the change in the Vin+ value from the intersection of Vin+ and Vin- to the middle of the changing state output voltage. The target offset voltage value should be less than half of 1LSB to avoid error during the comparison stage. The input offset voltage is generally due to mismatch, process variations in the input and load devices at the differential stage. The minimum resolution can be defined as in Equation (28).

$$(\Delta V_{IN})_{min} = V_{OH}\text{-}V_{OL}/Av \tag{28}$$

where Av is the gain of the comparator and the V_{OH} and V_{OL} are the output of logic level high and low which is equaled to 1.5V. For (ΔV_{IN})min equal to 732 μV, the required gain is equaled to 66.2 dB. Thus, the design of the comparator targets a gain of at more than 70 dB. The gain of the comparator can be increase by increasing the tail current. The gain of the comparator can be obtained applying the derivative of Vout with respect of Vin from the range of (0 to 1.5V). The gain of the comparator is equaled to 112.55 dB which more than the required gain.

Table 4. The performance of the comparators

Parameter	Comparator A	Comparator B
Power Consumption	163 pW	181pW
Response time	10uS	511.63pS

Figure 20. Offset voltage for the modified comparator

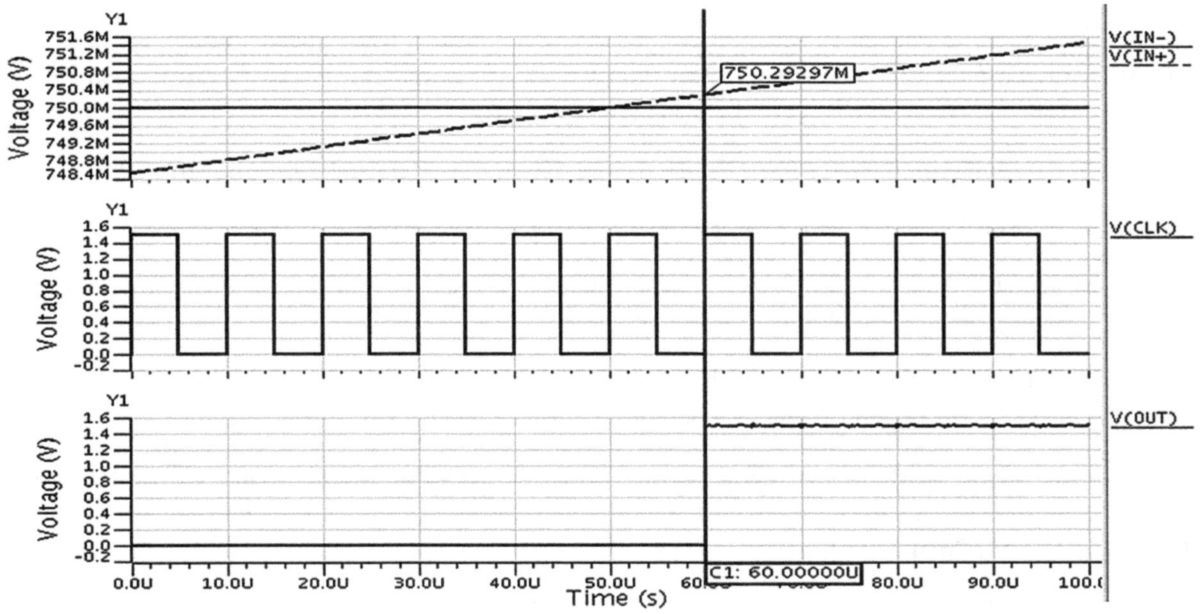

The delay and power consumption analysis of pre and post layout is presented. The layout of the dynamic comparator designed by using 0.18 µm CMOS process and the interdigitate method is used to reduce the dimension of the dynamic comparator. The VSS ground ring is drawn surrounding the comparator to prevent from noise from digital circuit in the SAR ADC layout.The delay and the power consumption in the post-layout simulation are higher than the post-layout simulation because of the parasitic capacitance and resistance present in the actual physical layout. However, both shared the same offset voltage. The output offset voltage of the dynamic comparator can be found by simply grounding both of the input terminals. The output voltage presents at the output terminal is the value of offset voltage.

Finally, a low power dynamic comparator has been designed to meet the requirement for ultra-low power SAR ADC. To achieve smaller resolution, a proper sizing of the each transistor can be revised. Otherwise, it will affect the performance of the comparator. The output comparator needs to be stable during the whole clock cycle as it will degrade the output of SAR ADC. By modifying the SR latch a stable output could be obtained at the output terminal. The total power consumption reported is equaled to 191 pWatts at 1.5V power supply and 0.1 MHz clock frequency. This low power dynamic comparator is shown to be suitable for ultra-low SAR ADC Design.

SUMMARY

This chapter has described three contemporary low power design approaches: a resistor-less BGR, a hybrid VLS with a diode connected NMOS and a modified dynamic comparator. Each design demonstrated how contemporary approaches can be effective in achieving low power VLSI designs.

REFERENCES

Ali, S., Tanner, S., & Farine, P. A. (2011). A Robust, Low Power, High Speed Voltage Level Shifter with Built-in Short Circuit Current Reduction. *20th European Conference on Circuit Theory and Design (ECCTD)*, (pp. 142-145). doi:10.1109/ECCTD.2011.6043302

Allen, P. E., & Holberg, D. R. (2002). CMOS Analog Circuit Design (2nd ed.). Academic Press.

Brokaw, P. (1981). *Band-gap Voltage Reference with Curvature Correction*. United States Patent.

Gray, P. R., Hurst, P. J., Lewis, S. H., & Meyer, R. G. (2001). *Analysis and Design of Analog Integrated Circuits* (4th ed.). New York: Wiley.

Hedayati. (2011). *A Study of Successive Approximation Registers and Implementation of an Ultra-Low Power 10-bit SAR ADC in 65nm CMOS Technology*. (Master's thesis). Linköping Institute of Technology.

Hosseini, S. R., Saberi, M., & Lotfi, R. (2014). A Low-Power Subthreshold to Above-Threshold Voltage Level Shifter. *IEEE Transactions on Circuits and Wystems. II, Express Briefs*, *61*(10), 753–757. doi:10.1109/TCSII.2014.2345295

Klimach, H., Costa, A. L. T., Monteiro, M. F. C., & Bampi, S. (2013). Resistorless switched-capacitor bandgap voltage reference with low sensitivity to process variations. *Electronics Letters*, *49*(23), 1448–1449. doi:10.1049/el.2013.1761

Lam, Y.-H., & Ki, W.-H. (2010). CMOS Bnadgap References with Self-biased Symmetrically Matched Current-Voltage Mirror and Extension of Sub-1-V Design. *IEEE Transactions on Very Large Scale Integration (VLSI) Systems*, *18*(6), 857–865.

Lehmann, T. (2014). Design of fast low-power floating high-voltage level-shifters. *Electronics Letters*, *50*(3), 202–204. doi:10.1049/el.2013.2270

Ming, X., Ying-qian, M., Ze-kun, Z., & Zhang, B. (2010). A high precision compensated CMOS bandgap voltage reference without resistors. IEEE Transaction on Circuits and System, 57(10).

Osaki, Y., Hirose, T., Kuroki, N., & Numa, M. (2012). A Low-Power Level Shifter with Logic Error Correction for Extremely Low-Voltage Digital CMOS LSIs. *IEEE Journal of Solid-State Circuits*, *47*(7), 1776–1783. doi:10.1109/JSSC.2012.2191320

Rabaey, J. M., Chandrakasan, A., & Nikolic, B. (2012). *Digital Integrated Circuits*. Prentice Hall.

Shao, H., & Tsui, C. (2007). A Robust, input voltage adaptive and low energy consumption level converter for sub-threshold logic. 33rd ESSCIRC, (pp. 312-315).

Sharuddin & Lee. (2014).An Ultra-low Power and Area Efficient 10 bit Digital to Analog Converter Architecture. *IEEE Int. Conference of Semiconductor Electronics 2014, IEEE-ICSE2014*.

Wang, B., & Law, M. K. (2015). A Precision CMOS Voltage Reference Exploiting Silicon Bandgap Narrowing Effect. *IEEE Transactions on Electron Devices*, *62*(7), 2128–2135. doi:10.1109/TED.2015.2434495

Zhou, J., Wang, C., Liu, X., Zhang, X., & Je, M. (2015). An Ultra-low Voltage Level Shifter Using Revised Wilson Current Mirror for Fast and Energy-Efficient Wide-Range Voltage Conversion from Sub-threshold to I/O Voltage. *IEEE Transactions on Circuits and Systems I*, *62*(3), 697–706. doi:10.1109/TCSI.2014.2380691

Zhu, X., & Chen, Y. (2012). A 9-bit 100MS/s tri-level charge redistribution SAR ADC with asymmetric CDAC array. *2012 International Symposium onVLSI Design, Automation, and Test (VLSI-DAT).*

Chapter 6
Low Power VLSI Circuit Design using Energy Recovery Techniques

V. S. Kanchana Bhaaskaran
VIT University, India

ABSTRACT

With the rapidly evolving silicon technology, the power density becomes increasingly high. Quadratically related to power, the voltage scaling offers a means of minimizing energy. However, power supply scaling demands less threshold voltage, which rises leakage current. Several low power techniques have been devised. This chapter deals with the non-conventional low power design solutions, based on adiabatic switching theory. In such circuits, the energy rather than getting dissipated during every cycle, is transferred back and forth between the logic and power-clock sources. A brief discussion on the reversible logic circuits will be presented followed by the fully adiabatic and quasi-adiabatic circuits. The use of power-clock sources for operating the adiabatic circuits will also be introduced. The generalized energetics of an adiabatic circuit followed by the typical loss models of the adiabatic families are presented. Some of the adiabatic circuits employing CMOS transistors are introduced in the chapter. A short comparison for the adiabatic circuit leakage models follows.

INTRODUCTION

Need for Low Power

With the rapid evolution in silicon technology, the transistor size continues decreasing remarkably, and the chips grow up in functionality with the switching frequencies increasing in parallel. The power dissipation in the chip thus becomes increasingly high. In 1980s, the growing power dissipation in the microprocessors, the memory and a multitude of ASICs prompted an industry-wide shift from the bipolar and NMOS technologies to CMOS to assuage the problem of heat dissipation. This attempt greatly reduced the average power dissipation resulting in a provisional solution. However, higher packing

DOI: 10.4018/978-1-5225-0190-9.ch006

densities necessitated during the 1990s brought in the requirement for low power design methodologies to the fore. While the design parameters were performance and area previously, *power* was recognized in 1990s. Presently, the power management of integrated circuits has attained the primary status and proves one of the most important performance metrics for SoCs than the throughput, latency or frequency.

The need for low power operation arises for the following reasons:

1. Low power computing applications are small systems, where battery weight is crucial. The battery technology does not advance in line with the portable systems of high performance and complexity, and its energy density improves only by about 6% annually.
2. Large and high performance systems require huge power and necessary cooling.
3. The environmental concerns insist low power design with electricity generation being a major source of air pollution.

Thus, it can be concluded that designing power efficient electronic systems design is more prudent than the research on power generation itself.

CONVENTIONAL METHODOLOGIES OF LOW POWER VLSI DESIGN AND THE CURRENT STATUS

The commonly used three degrees of freedom available in the low-power design arena are the supply voltage, the switched capacitance and the switching activity. Being quadratically related to power, voltage reduction directly offers a means of minimizing the energy consumption. However, an unrelenting voltage reduction demands appropriate reduction in threshold voltage across each technology node resulting in increasing leakage and this makes it harder to continue voltage scaling envisaged as shown in Figure 1(a) (Piguet, 2007). The reduction of the supply voltage leads to reduced speed performance also. The energy consumed per logic operation does not drop as it had been occurring with older technologies as shown in Figure 1(b) with the energy required per operation saturating beyond 130 nm. Additionally, the low voltage operation makes the threshold voltage characteristics more vulnerable to temperature variations.

The transistors working at GHz frequencies suffer from exponentially increasing leakage. It is more difficult to scale V_{dd} down than the other parameters. This is mainly due to the fundamental limit of the threshold slope of approximately *60mV/decade*. It can be more severe when supply voltage reaches down to *0.6V*. International Technology Roadmap of Semiconductors (ITRS) (ITRS, 2013) finds that the rising dynamic power density per unit area will sooner than later reach an acceptable level. On the other hand, the static power dissipation is expected to become more difficult to control, while working for high performance as the target. Hence, the ITRS report given in Table 1 foresees that the innovations in circuit design and architecture for performance and power management are imperative. This includes parallelism for improved performance at the cost of increased transistor count and silicon area, and aggressive design of power gating the circuits. A trade-off between speed performance and other factors, namely, lower leakage current or low standby power, device count and chip area is the goal of low power solutions.

Table 2 shows various abstraction levels and related design technologies. With focus on power management, the electronics community has been employing new techniques and materials on several fronts. The foundries introduce multi-threshold, multi-voltage transistors, FinFET devices, SoI (silicon-

Figure 1. (a) Supply voltage scaling and (b) Energy per logic operation across technology generations

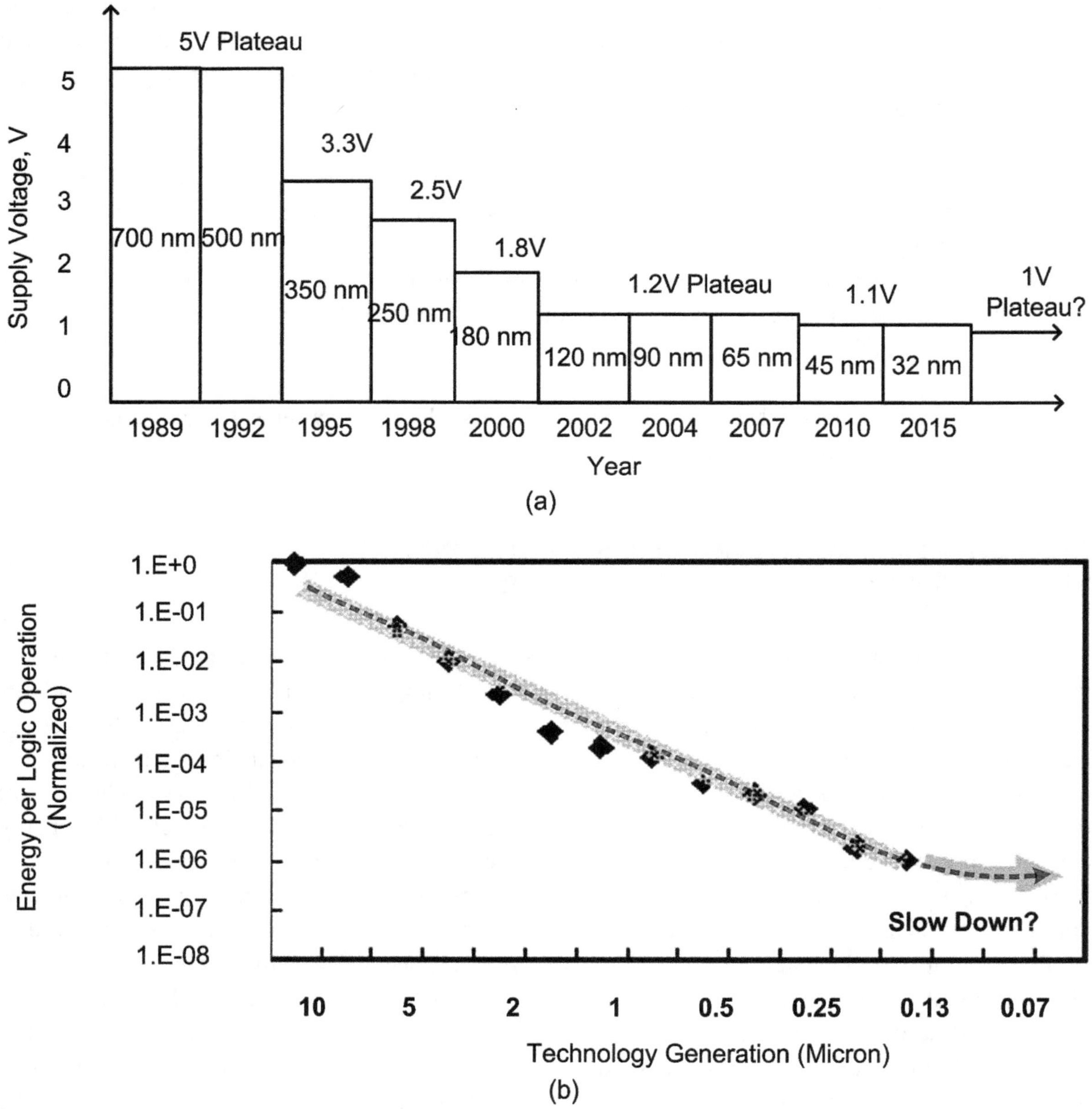

on-insulator) and low-k materials, body or back biasing, copper-metal and SiGe (silicon-germanium) substrates. In parallel, the chip architects and software designers focus on hardware versus software trade-offs, power-aware operating systems, power gating and devising hibernation modes, and by selective memory access. IC designers employ several techniques to lower power in their designs.

The various techniques that have been devised employ different methodologies, involving area overhead, speed and performance overhead. The choice of a particular technique is determined by the related dynamic or leakage power reduction. Thus, the innovation that is needed at all levels is to continue

Table 1. Overall design technology challenges (ITRS, 2013)

Challenges > 22 nm	Summary of Issues
Design productivity	System-level: high level of abstraction (HW/SW) functionality spec, platform based design, multi-processor programmability, system integration, AMS co-design and automation Verification: executable specification, ESL formal verification, intelligent test bench, coverage-based verification Logic/circuit/physical: analog circuit synthesis, multi-objective optimization Logic/circuit/physical: SiP and 3D (TSV-based) planning and implementation flows Heterogeneous component integration (optical, mechanical, chemical, bio, etc.)
Power consumption	Logic/circuit/physical: dynamic and static, system- and circuit-level power optimization
Manufacturability	Performance/power variability, device parameter variability, lithography limitations impact on design, mask cost, quality of (process) models ATE interface test (multi-Gb/s), mixed-signal test, delay BIST, test-volume-reducing DFT
Interference	Logic/circuit/physical: signal integrity analysis, EMI analysis, thermal analysis
Reliability and resilience	Logic/circuit/physical: MTTF-aware design, BISR, soft-error correction
Challenges <22 nm	**Summary of Issues**
Design productivity	Verification: complete formal verification of designs, complete verification code reuse, complete deployment of functional coverage Tools specific for SOI and non-static logic, and emerging devices Cost-driven design flow
Power consumption	Logic/circuit/physical: SOI power management Logic/circuit/physical: Reliability and resilience- and temperature-constrained 3D physical implementation flows
Manufacturability	Uncontrollable threshold voltage variability Advanced analog/mixed signal DFT (digital, structural, radio), "statistical" and yield-improvement DFT Thermal BIST, system-level BIST
Interference	Interactions between heterogeneous components (optical, mechanical, chemical, bio, etc.)
Reliability and resilience	Autonomic computing, robust design, SW reliability and resilience
EMI—electromagnetic interference ESL—Electronic System-Level HW/SW—hardware/software MTTF—mean time to failure SOI—silicon on insulator	

Table 2.

Abstraction Level	Design Techniques
System	Partitioning, Power Down, Power States
Algorithm	Complexity, Concurrency, Regularity, Locality
Architecture	Parallelism, Pipelining, Data Representation, Redundancy, Instruction Set Selection
Circuit/Logic	Logic Styles, Logic Manipulation/Optimization, Transistor Sizing, Adiabatic, Clock/Power Gating
Fabrication Processes	Threshold Reduction, Multi/Variable-Threshold and V_{DD}

performance scaling, even while maintaining tolerable power. The optimization so realized with respect to power, performance, delay and the use of parallel cores would thus control the power dissipation.

Power consumption and design robustness being the primary constraints taken into account in every level of design, the choice of circuit/logic style is significant. The mostly prevalent and conventional low power logic styles can be broadly categorized as follows:

1. CMOS circuit for its energy efficiency and robustness.
2. Domino logic, faster logic design using less device count.
3. Pass Transistor Logic and related enhancements.
4. Static DCVS logic followed by its variations.

REVERSIBLE AND ADIABATIC COMPUTING FOR LOW POWER

The computing performance and highly optimized logic design of IC depends primarily on the energy efficiency. The conventional methodologies do not include logical reversibility and adiabatic computing. The logical reversibility substantiates the physical reversibility of the system (Landauer, 1961). Any loss of information implies energy loss and in the reversible logic, every input vector is distinctively recovered from output vectors, and this fact ensures that no information is lost.

The adiabatic computing process can be said to represent the practical reversible circuits. The term adiabatic (meaning *impassable* in Greek) is also called *energy* or *charge recovery*. It portrays the nature of the circuits which reclaim part or whole of the energy stored in the nodal capacitances during computation, and reuse the energy for subsequent computations. The energy recovery circuits employ ramped or sinusoidal power clocks in place of DC voltage. The power clock is so called, since it powers the circuit while donning the role of the timing clock also across the adiabatic pipeline. Section 3.1 discusses in brief some of the reversible logic circuits found in the literature, followed by the adiabatic CMOS logic circuits in Section 3.2.

Reversible Logic for Low Power Operation

The low power design methods and the hierarchy of limits assume that a certain amount of electrical energy gets dissipated during each binary switching transition. Any such irreversible technology remains subject to the Von Neumann-Landauer bound of the minimum amount of kTln(2) energy dissipation per operation (Landauer, 1961). He proposed that logical irreversibility is associated with physical irreversibility and needs a minimal heat generation of the order of *kTln2* every cycle for each irreversible function. The principle was confirmed (Berut et al. 2012) through an experiment using a generic one-bit memory. These class of circuits avoid energy loss by doing *uncomputation* on the computed information (Vieri, 1999). The computing system avoids energy dissipation if the information erasure is avoided. Bennet and Landauer (1985) demonstrated that zero power dissipation in circuits is possible when the circuit comprises of reversible logic gates. This approach has resulted in logically reversible gates such as Toffoli or Fredkin gates (Dueck, 2012).

The reversible logic has several applications including Fault Tolerant Designs, Optical Information Processing, Quantum Computing and low power circuit design. However, the major problems associated with the reversible logic gates are:

1. The fan-out gates are not allowed.
2. The feedback to input from output is not possible.

Reversible Logic Gates

A circuit can be considered reversible if each of the input vectors can be mapped into an exclusive output vector and each output vector can be mapped into an exclusive input vector. Hence, the reversible circuits possess an inherent structure of one-to-one correspondence between the input and output vectors. While synthesizing a logical function, it may so happen that one or more of the input or output bits are to be kept constant at either logic 1 or 0 and these are called garbage outputs. The quantum cost is a measure that indicates the cost of the circuit. It is calculated based on the number of basic reversible logic gates required to realize the circuit. Comparison across the reversible logic gates is made through the parameter of quantum cost. Brief introduction on some reversible logic gates follows:

1. **Feynman Gate:** Feynman. R. (1985) presents 2*2 Feynman gate shown in Figure 2(a), consisting of input vectors (A, B) and the output vectors P=A and Q=A⊕B. The quantum cost of a Feynman gate is 1.
2. **Double Feynmann Gate (F2G):** The double Feynmann gate of Figure 2(b) has been demonstrated in Parhami (2006), which consists of a 3*3 structure with input and output vectors defined by I(A, B, C) and O(P, Q, R) respectively, where P=A, Q=A⊕B and R=A⊕C. The quantum cost of F2G is 2.
3. **Toffoli Gate:** The Toffoli gate is shown in Figure 2(c) with the input vector I(A,B,C) and the output vector O(P, Q, R), where the output bits are given by P=A, Q=B and R=AB⊕C. The Toffoli gate consists of a 3*3 gate structure with quantum cost of 5.
4. **Fredkin Gate:** The Fredkin gate with input vector I(A,B,C) and the output vector O(P, Q, R) is shown in Figure 2(d). The quantum cost of Fredkin gate is 5 and input and output are related by P=A, Q=A'B⊕AC and R=A'C⊕AB.
5. **Peres Gate:** Figure 2 (e) shows the 3*3 Peres gate with input and output vectors defined by I(A,B,C) and O(P, Q, R). The output is identified by P=A, Q=A⊕B and R=AB⊕C. The quantum cost of the Peres gate is 4.
6. **Double Peres Gate:** The input and output vectors are defined by I(A, B, D) and O(P, Q, R, S) for the 4*4 double Peres gate as shown in Figure 2(f).
7. **TS Gate:** A 4*4 reversible gate called TS gate is proposed by Himanshu, (2005), which can implement all Boolean functions and it can be used a reversible full adder cell. Figure 2(g) shows a 4*4 TSG reversible full adder. Figure 2(h) shows TSG gate operating as reversible full adder. The input vector I(A, B,C D) generates the output vector O(P, Q, R, S), where P=A, Q=A'C'⊕B, R= (A'C'⊕B')⊕D and S=(A'C'⊕B').D⊕(AB⊕C). A reversible 8x8 Wallace tree multiplier has been designed using TSG (Himanshu, 2005).
8. **Reversible Bennett Clocking:** The second type of logic reversibility realized by Bennett-clocking (Lent, C.S. et al. 2006), has timing so managed that computing involves two steps, namely, forward computing step followed by de-compute or reverse computing step. The major concerns of these circuits are reduced throughput and pipelined stages. The schematic of two inverters connected in cascade operated by Bennett-clocking is shown in Figure 3(a). Figure 3(b) shows the power clock transients and the output nodal voltages. Initially, all the power clock signals and the input are set at $+V_{dd}/2$. The three possible states of the circuit are logic 0 $(-V_{dd}/2)$, logic 1 $(+V_{dd}/2)$ and steady or reset state represented by *0V*. The circuit operation is briefly discussed. With input steady at logic 1 $(+V_{dd}/2)$, when PC1 ramps up, MN1 turns *on* and MP1 turns *off*. Node V_{o1} gets discharged

Figure 2. Reversible logic gates (a) Feynman Gate, (b) Double Feynman Gate, (c) Toffoli Gate, (d) Fredkin Gate, (e) Peres Gate, (f) Double Peres Gate and (g) TS Gate and (h) TS Gate as a full adder

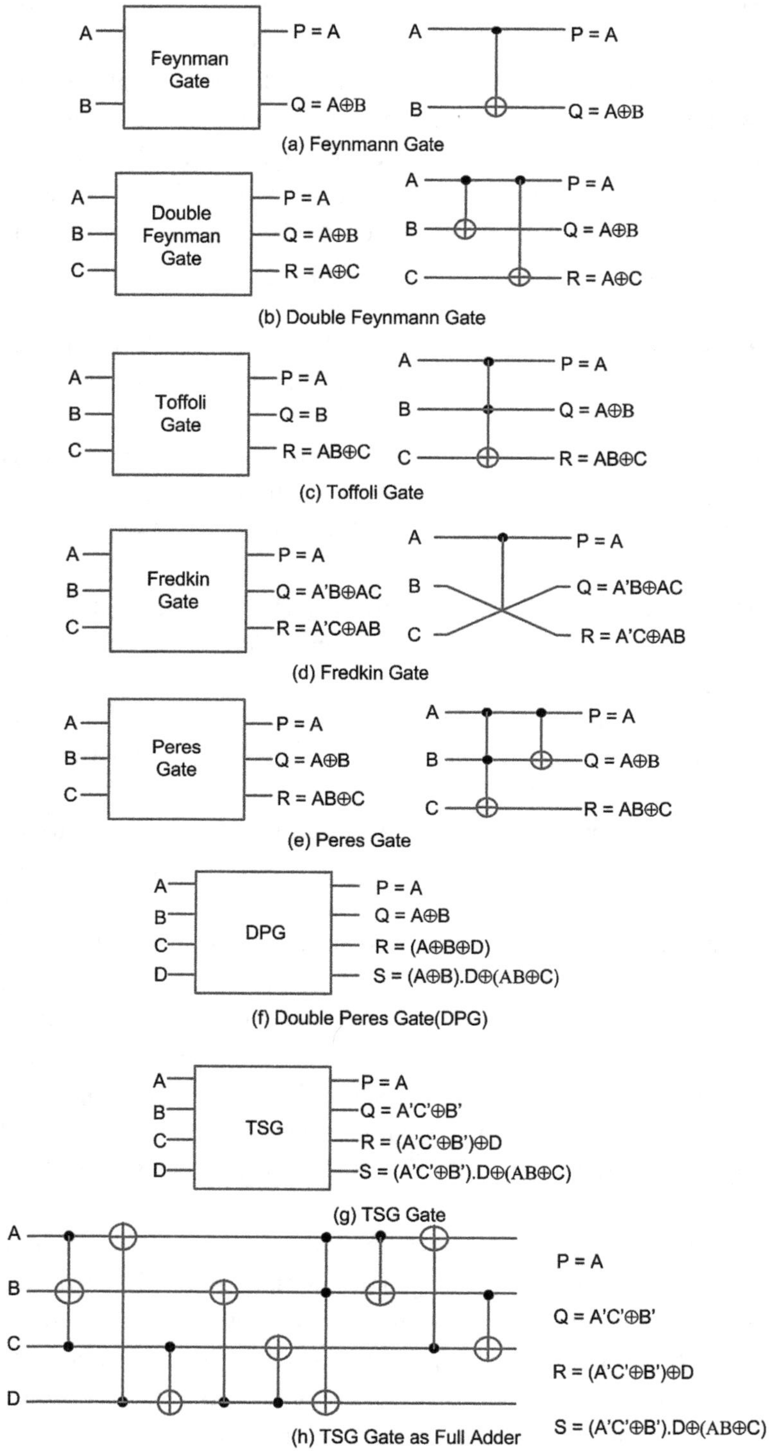

Figure 3. (a) Two inverter cascade and (b) Bennett clocking and operation Sequence

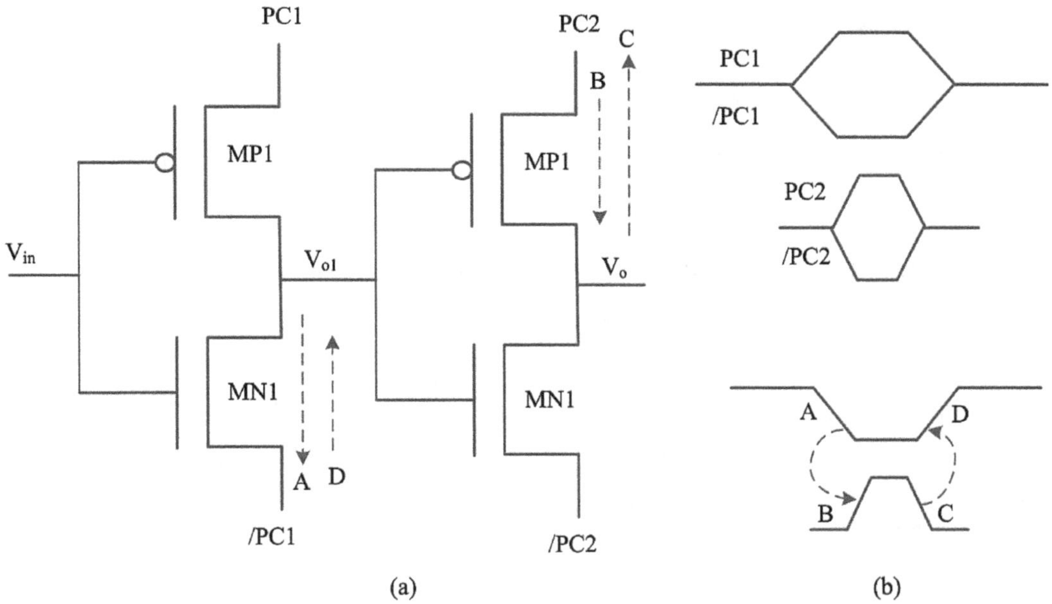

through MN1 gradually until PC1 and /PC1 become constant. When a constant output is reached at node V_{o1}, the PC2 and /PC2 starts ramping as shown. Then, MP2 turns *on* and MN2 turns *off*. Output V_o starts charging through MP2 gradually until PC2 and /PC2 remains constant and steady logic value of 1 is received at V_o.

Decomputation happens when PC2 and /PC2 ramps back to *0V* and the charge at output capacitance at node V_o gets recovered through MP2. When PC2 and /PC2 becomes constant at *0V*, the first level clocks PC1 and /PC1 commences relaxing back to *0V*, letting V_{o1} rise back to *0V*. As the last step, input is relaxed.

Figure 4 shows typically connected logic blocks and dual rail power clock transients employed in the reversible Bennett clocking for 4 level circuit cascades (Lent, 2006). The circuit computes the logic commencing from the block 1 operated by PC1. When the output is held by the block 1, which acts as input to block 2, block 2 operates with PC2, and this goes on. As output of stage 4 is obtained, the circuit then de-computes from block 4 to block 1 in reverse order.

Interested readers can find (Ismo, 2014) presenting an adiabatic arithmetic logic block and realization of Booth's multiplier in (Jakia, 2015). Current investigations in the reversible logic using quantum computer algorithms permit innovation in computer architecture and arithmetic logic unit designs (Morrison, 2013). The programmable reversible logic is an up-and-coming logic design for implementation in quantum computing with reduced heat dissipation characteristics.

Adiabatic (a.k.a. Energy Recovery) Logic Circuit Design

While the discussions made so far focused on logical reversibility, the newer theory based on the adiabatic switching asserts that the energy rather than getting fully dissipated can get reclaimed partly or in

Figure 4. Reversible Bennett-clocking for a 4-stage circuit

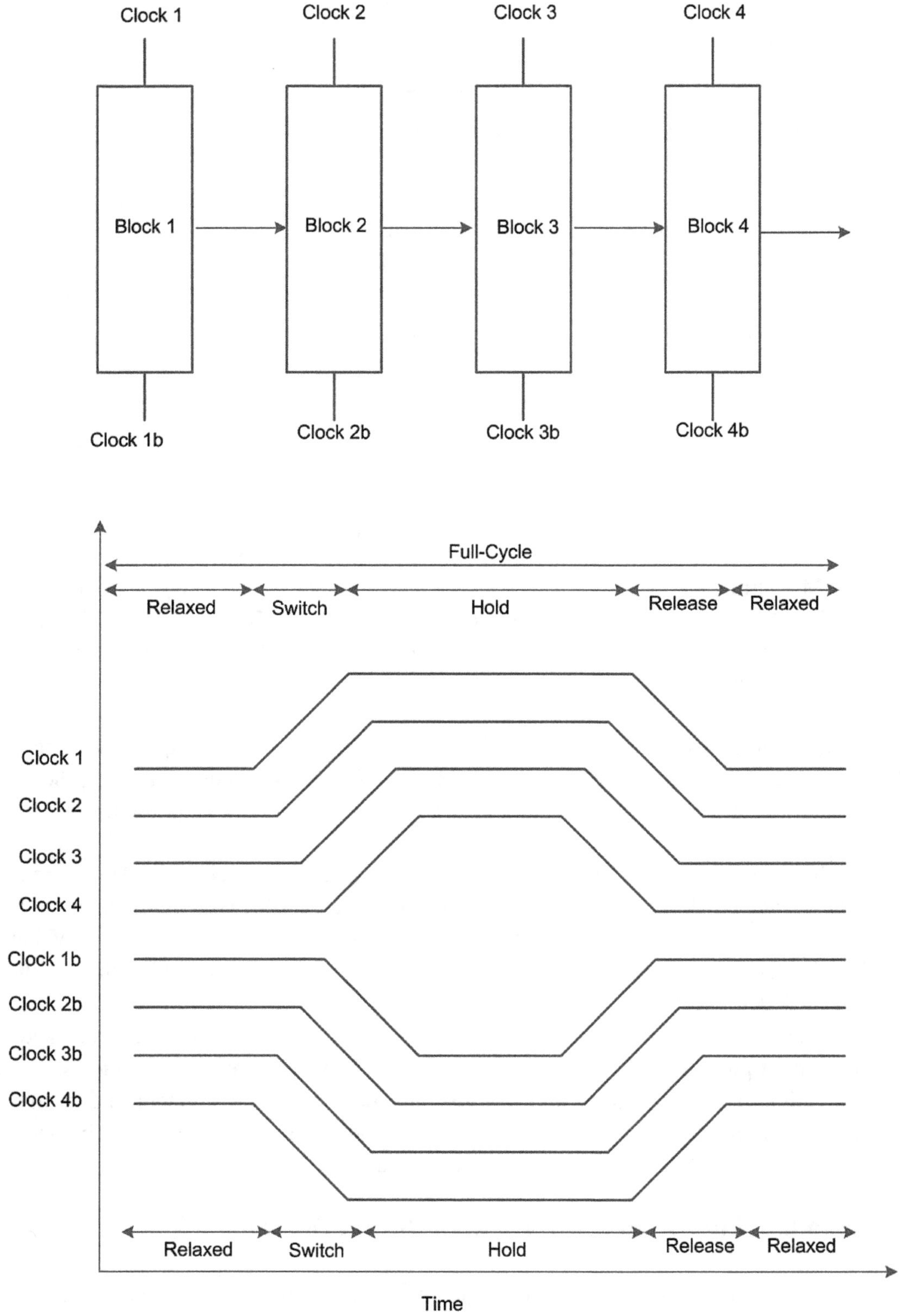

full, the fact based on the second law of thermodynamics (Athas, 1994). The $1/2(CV^2)$ energy rather than getting dissipated during every cycle is transferred back and forth between the logic and power-clock sources, thus recycling the energy. In the process, it indeed loses some negligible energy, as an inadvertent circuit overhead. This makes the circuits partial or quasi-adiabatic. The literature present adiabatic circuits that can operate at 100s of MHz. Microcontrollers, ASICs, LCD drivers, SRAMs and I/O line drivers have been implemented using adiabatic design. It may be noted that the CMOS circuits fall into utterly non-adiabatic circuits.

The adiabatic literature continue proposing new developments. The encryption standard hardware architectures where security is the primary goal, e.g., low-power secure S-box circuits, employ charge-sharing symmetric adiabatic logic (Cancio, 2014). The radio frequency identification tags (RFID) tags and Wireless sensor networks (WSN) which make use of cryptographic modules and thus vulnerable for differential power analysis attacks (DPA) use novel adiabatic logic methods (Moshe 2015). They prove low-power operating and also DPA immune.

Types of Adiabatic Circuits

Based on the energy recovery performance the adiabatic circuits are categorized (Denkar, 1994) as

1. Fully adiabatic, a.k.a. asymptotically adiabatic circuits, which, when operated slowly would dissipate arbitrarily little energy per operation, so as to incur no non-adiabatic power dissipation. However, the asymptotically zero energy is achieved at the cost of increased circuit complexity
2. Quasi-adiabatic circuits, which realize partial energy recovery with some energy lost due to irreversible energy spent in operating the circuit

Fully Adiabatic Circuits

The CRL and SCRL reversible logic styles (Younis and Knight, 1993), reversible Pendulum computer architecture (Vieru, 1995) and fully reversible nMOS Reversible Energy Recovery Logic (RERL) (Lim et al, 2000) (Patra, 1995) are some of the fully adiabatic logic families. The Pass transistor Adiabatic logic (PAL) (Vojin et al, 1997) with simple circuit uses complementary power-clock phases for operation. However, it suffers from the capacitive coupling effect and floating output node during the charging phase. This limits its operating frequency to 100 MHz even for logic circuit as small as a multiplexer. Operated by two-phase power-clocks its latency becomes twice that of the four-phase clocked circuits.

The Split Level Charge Recovery Logic (SCRL) based NAND gate (Younis, 1994) is shown in Figure 5. It is very similar in structure to that of conventional static CMOS, except the fact that the two rails of the gate are driven by ramp power clocks *PC* and */ PC* in place of V_{dd} and *Gnd* of CMOS. To discuss briefly, initially the circuit is set to $V_{dd}/2$ at *PC* and */PC,* and P1 and /P1 are made *Gnd* and V_{dd} respectively. This makes the transmission gate T1 non-conducting. T1 becomes *on* when P1 and /P1 gradually rises *high.* At the same time, as *PC* and */PC* are split to V_{dd} and *Gnd* as shown in Figure 6, the gate computes A NAND B. When the output is used by the next gate in the pipeline, T1 is turned *off* gradually and *PC* and */PC* returns to $V_{dd}/2$. The additional PMOS device in the pull-down network avoids energy dissipation in the NMOS device *A* for the case *A=1* and *B=0.*

Figure 5. SCRL NAND Gate

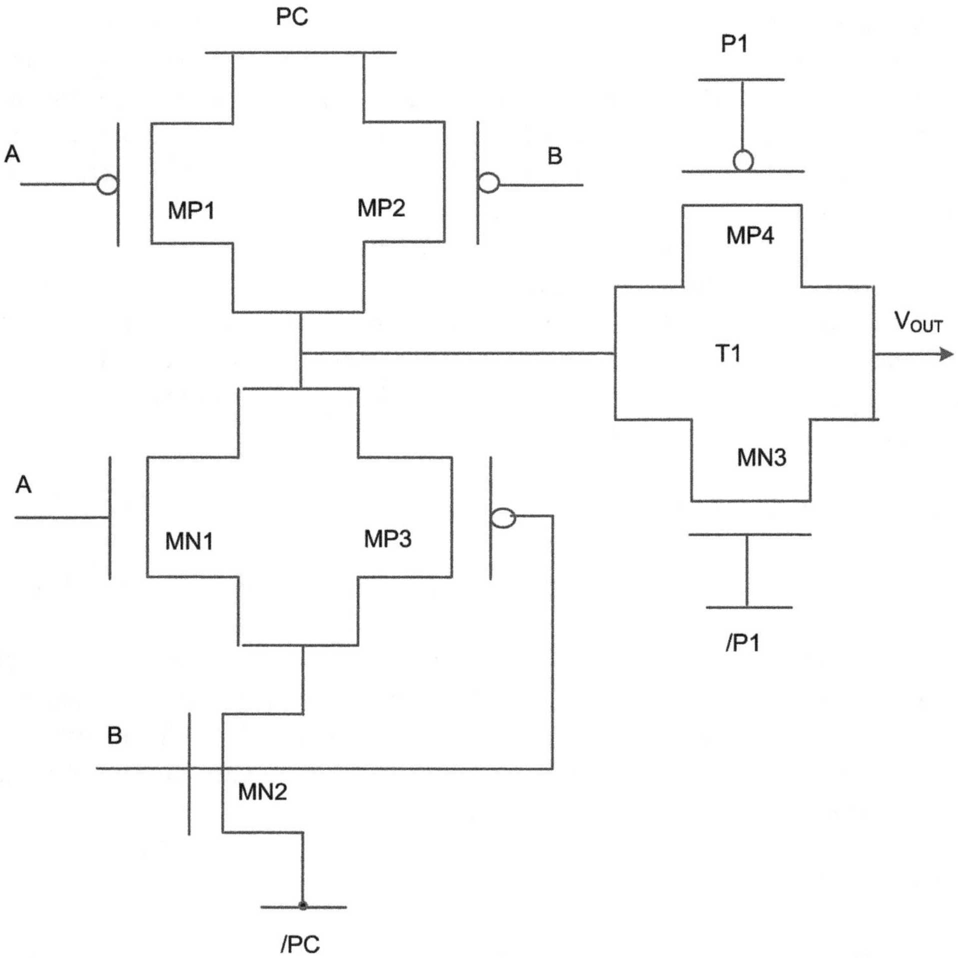

Figure 6. (a) Conventional CMOS Circuit and (b) Adiabatic CMOS Circuit

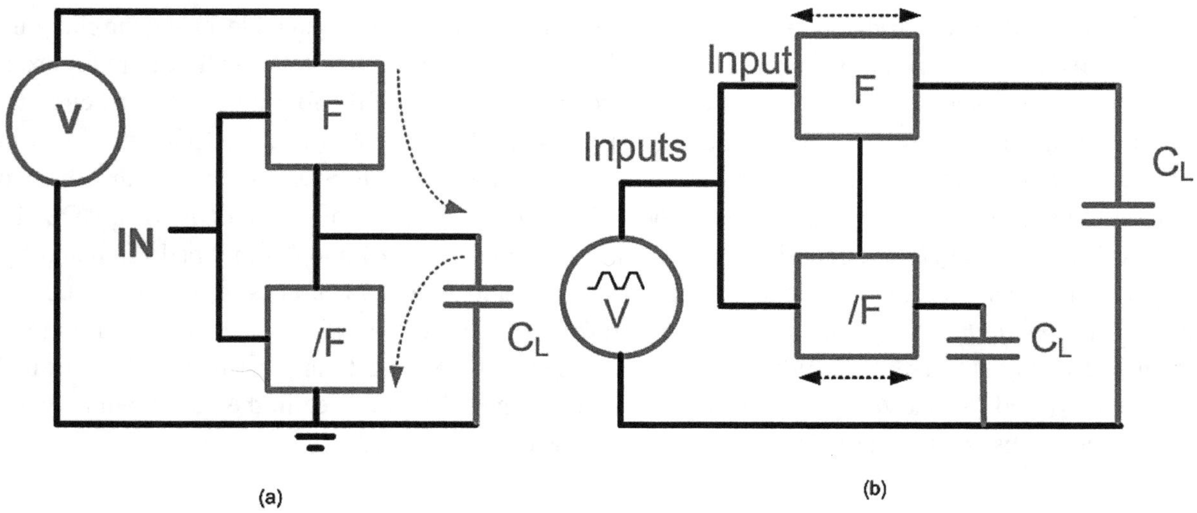

Quasi-Adiabatic Circuits

The quasi-adiabatic circuit can function with minimum energy dissipation and tolerable circuit complexity. They employ diodes or transistors to steer the charge and choice of the device characterizes the non-adiabatic power to a larger extent. Most of such circuits are operated by two or four-phase power-clock (represented as PC hereafter) signals with the literature offering some single-phase, three-phase and six-phase fully adiabatic circuits also. The timing of the positive/negative ramp edges of the PC are so timed that the logic is evaluated and retained for some time followed by charge recovery. The quasi-adiabatic circuits presented in the literature can be categorized into

1. Those using diodes
2. Those employing transistors

Hot-clock nMOS circuit proposed in (Seitz et al, 1985) with bootstrapped devices have been fabricated and demonstrated by (Younis and Knight, 1993). The earlier circuits using diodes are 2N-2N2D (Yibin and Roy, 1996), adiabatic dynamic logic (ADL) (Dickinson and Denker, 1995) using four-phase PC, the adiabatic pseudo domino logic (APDL) (Wang and Lau, 1995) and the 2N-2N2P2D (He at al, 2006). The major limitations of these designs are the use of diode, which inherently suffers from dissipation due to cut-in voltage V_γ . The quasi-adiabatic circuits using transistors overcome this limitation.

The energy recovery logic primitives were proposed in (Athas et al, 1994) with trades off between power and performance using adiabatic logic circuits and clock powered microprocessor in (Tzartanis and Athas, 1997). Some of the transistor based quasi-adiabatic structures are the Efficient Charge Recovery Logic (ECRL) (Moon and Jeong, 1996), a RISC microprocessor using ECRL (Youngjoon et al, 2004), 2N-2N2P logic (Kramer et al, 1995), Positive Feedback Adiabatic Logic (PFAL) (Vetuli et al, 1996), the modeling for PFAL (Blotti et al, 2000) and semi-custom design of PFAL arithmetic circuits proposed by (Blotti and Saletti, 2004). While most of the above configurations were of differential and dynamic, the Quasi-Static Energy Recovery Logic (QSERL) (Yibin and Roy, 2001) using the diode-connected transistors followed the earlier work (Yibin and Roy, 1996).

(Kim et al, 2005) proves that the adiabatic circuits have the potential to operate with reduced electromagnetic interference. The clock network in a system dissipates significant portion of total power in today's processors. (Supreeth. M.S, 2014) and (Jennifer, 2015) present the opportunity for reduction of power in clock distribution network by energy recovery technique and adiabatic flip-flops. (Sandeep, 2014) demonstrates how graphene based p-n junctions can perform as adiabatic logic gates with an ability to reuse the dissipated energy.

The four-phase PC operated circuits employing latch and dual-rail encoded families, namely, the 2N-2P, 2N-2N2P, PFAL, DCPAL and PSAL exhibit the best energy savings (Kim et al, 2005,) (Kanchana and Raina, 2008), (Kanchana and Raina, 2010) and (Kanchana and Raina, 2012), and they have been included in this chapter for implementation and comparison.

ENERGETICS OF CONVENTIONAL STATIC CMOS VS. QUASI-ADIABATIC CMOS CIRCUITS

This section presents the energy dissipation components of a conventional CMOS circuit and the quasi-adiabatic circuits. In the static CMOS circuits shown in Figure 6 (a), the energy dissipation occurs when the charging current is driven through devices with finite *on* resistance. Such resistive losses are proportional to the voltage drop and hence the energy dissipation in the device. During each switching event between logic 1 and 0, either the pull-up or pull-down network dissipates energy.

The switching energy E_{CMOS} of the conventional CMOS circuit is

$$E_{CMOS} = E_{sw} + E_{sc} + E_{Leak} = C_L V_{dd}^2 + E_{sc} + E_{Leak} \tag{1}$$

Here, E_{sc} and E_{Leak} are the energy losses due to the short circuit and the leakage current respectively.

On the other hand, the quasi-adiabatic CMOS circuit shown in Figure 6(b) allows the computation with partial energy recovery. These structures follow differential logic style generally. When the power clock PC rises, and when output node warrants logic 1 in response to input, the respective output derives charge from PC through the conducting path created by functional block *F* or its complement */F*. When the node warrants logic 0, PC ramps *low* and the charge on the output node is reclaimed by the PC through the recovery path F or /F. Note that the power source is made capable of pumping and reclaiming energy from the circuit. This is normally realized by the use of LC resonant power-clock generators. Stepwise capacitor charging circuits for powering the adiabatic circuits have also been proposed (L. J. Svensson and J. G. Koller, 1994) (Dmitry, 2012).

ENERGETICS OF ADIABATIC CIRCUIT

Adiabatic Power Dissipation Model

A generalized adiabatic power dissipation model is discussed using Figures 7(a) to (c). Figure 7(b) shows the circuit equivalent for determining the energy dissipation of the node with load capacitance C_L that is adiabatically charged with ramp power supply shown in Figure 7(a). The resistance R_{on} indicates the device channel resistance and V_{DD} is the ramped voltage peak in time T. The slope is slow enough that $V_C(t)$ is able to follow the supply voltage $V(t)$ instantly to maintain $V_C(t) \approx V(t)$. Then, the current drawn into the circuit is

$$i(t) = C_L \frac{dv(t)}{dt} = \frac{C_L V_{DD}}{T}$$

The energy dissipation E_{ch} during charging can be calculated as

Figure 7. (a), (b) & (c) Adiabatic CMOS operation

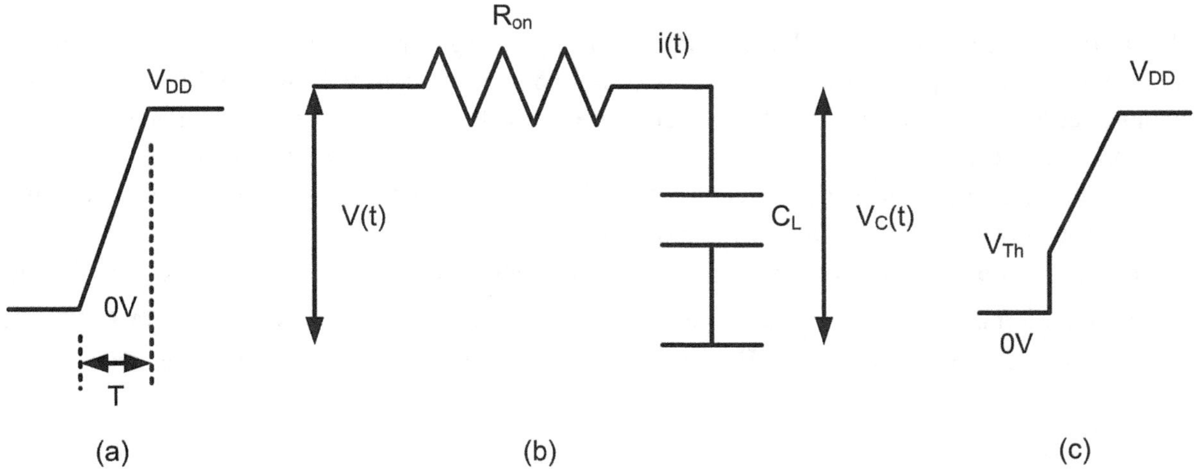

$$E_{ad} = \int_0^T p(t)dt = \int_0^T v(t).(i(t)dt = \int_0^T \left(v_R(t) + v_C(t).(i_C(t)\right) dt$$

The second component being zero for a complete cycle with charging and recovery operations, the adiabatic energy dissipation is

$$E_{ad} = \int_0^T R_{on} \frac{C_L^2 V_{DD}^2}{T} dt = \frac{R_{on} C_L}{T} C_L V_{DD}^2$$

More detailed analysis of the adiabatic energy E_{ad} of a quasi-adiabatic circuit taking into account the threshold voltage component can be found at (Yibin and Roy, 1996) which expresses

$$E_{ad} = \left(R_{on} C_L / T \right) C_L V_{dd}^2 \left(1 - R_{on} C_L / T + \left(R_{on} C_L / T \right) e^{-T / R_{on} C_L} \right)$$

When $T >> R_{on} C_L$, the dissipation is

$$E_{ad} = \left(R_{on} C_L / T \right) C_L V_{dd}^2$$

For values of T much less than the time constant $R_{on} C_L$ of the circuit, the energy dissipation becomes $(1/2) C_L V_{dd}^2$.

Non-Adiabatic Power Dissipation Models

This section presents the non-adiabatic power loss models of a generalized adiabatic circuit with differential architecture for pull-down networks and a cross-coupled latch using PMOS pull-up network. The main power loss components in adiabatic circuits are, 1) the loss due to incomplete charge recovery, 2) dissipation due to leakage currents, 3) capacitance coupling effect at the output nodes and power dissipation due to charge-sharing during HOLD state. These components along with the adiabatic family specific components will be discussed in more details in later section.

Refer to Figure 7(c). The conducting PMOS device cannot be made *on* unless *v(t)* reaches V_{tp} and goes beyond. Similarly, while recovering, the recovery path will be *on* only as long as the PMOS source is at level V_{tp} above the gate voltage. Hence, the energy loss due to the threshold voltage factor is

$$E_{th} = \left(\frac{1}{2}\right) CV_{th}^2$$

Here, the PMOS device resistance is approximated using standard BSIM3V3 model parameters as

$$R = \left[\mu_{eff,p} C_{ox} \frac{W_{eff,p}}{L_{eff,p}} \left(V_{dd} - \left|V_{tp}\right|\right)\right]^{-1}$$

where $\mu_{eff,p}$, $W_{eff,p}$ and $L_{eff,p}$ are the effective hole mobility, width and length of the device (Kanchana and Raina, 2008).

The second component of the non-adiabatic energy dissipation E_{Leak} is due to the current leakage components. It has become all the more severe for the UDSM nodes, with the disproportional rate of device size scaling and the threshold voltage scaling. One of the major components of leakage current is due to the subthreshold leakage

$$I_D = I_{D0} \exp\left(\frac{V_{GS} - V_{th}}{\eta v_T}\right)\left(1 - \exp\left(\frac{-V_{DS}}{v_T}\right)\right)$$

The third component of power dissipation is due to the energy E_{opn} spent in operating the 2N2P latch of the adiabatic circuit. The energy E_{opn} spent in a sense amplifier (Sathe, 2007) is

$$E_{opn} = \left(\pi^2/2T\right) R_{on} V_{dd}^2 C_L^2$$

Hence, the total energy dissipation of an adiabatic circuit can be expressed (kanchana and Raina 2012) by

$$E_{diss} = \left(2R_{on}C_L/T\right) C_L V_{dd}^2 + \left(1/2\right) C_L V_{th,p}^2 + V_{dd} I_{Leak} kT + E_{opn}$$

THE FOUR-PHASE POWER CLOCK SOURCES

Most of the adiabatic circuits employ resonant power-clock generator that alternately charge and discharge the capacitive load at resonance (Meimand and Kusha, 2001) and (Athas et al. (1996)). The oscillator frequency is determined by the capacitive load offered by adiabatic circuit and the external inductor L with the frequency of oscillation $f = 1/2\pi\sqrt{LC_L}$. The PC sources are divided into (1) asynchronous and (2) synchronous power-clock generators. While the former does not need any external control signals, the later needs external timing control signals for varying the operating frequency of adiabatic circuits. The power clock switching transistors can be integrated along with the same CMOS chip. The resonant-tank inductor can be added as an external component with the resonant circuit powered by a low-voltage dc power source, such as a battery. The typical schematic of an adiabatic system is shown in Figure 8(a) and the generation of a two phase power clock using an external inductor is depicted in Figure 8(b).

Power clocking and its integration in adiabatic circuits have the following benefits:

1. One of the salient features of an adiabatic circuit is the constant load presented to the power clocks.
2. The ability of providing both output and its complement on the same clock phase.
3. The PCs drive a constant capacitive load offered by the circuit, irrespective of the switching activities of the data. Thus, they operate with substantially reduced electromagnetic interference.
4. The PC network distribution is improved since the relatively large and detrimental parasitic inductances are removed from the distribution network, due to the ramp power clock signals.
5. The temporal and spatial incoherence due to the vast clock network of a CMOS circuit and the related skew and jitter effects are avoided.
6. The on-chip power-clock control circuitry can also easily be implemented.

QUASI-ADIABATIC LOGIC STRUCTURES

General Guiding Principle

The common guidelines followed in the design of most of the adiabatic circuits (Athas et al, 1994) are as follows:

1. The adiabatic pipeline stages are powered by multiphase and sequential PC sources. Figure 9(a) shows the cascade of four adiabatic buffers powered by PC1 to PC4 respectively. The four phases are called 1) Evaluate, 2) Hold, 3) Recovery and 4) Wait phases as shown in Figure 9(b). Each power clock lags behind the previous by 90°.
2. The ramp rate of the PC must be as low as possible. The source-drain voltage V_{ds} is changed only when the device is *off* and the device turns *on* only when V_{ds} is zero or only a small potential exists between the source and the drain, and any voltage change is done gradually.

This section describes the widely discussed quasi-adiabatic circuits, namely, 2N-2P, 2N-2N2P, PFAL, DCPAL and PSAL.

Figure 8. (a) Schematic of an adiabatic system (b) Typical two phase power clock generator

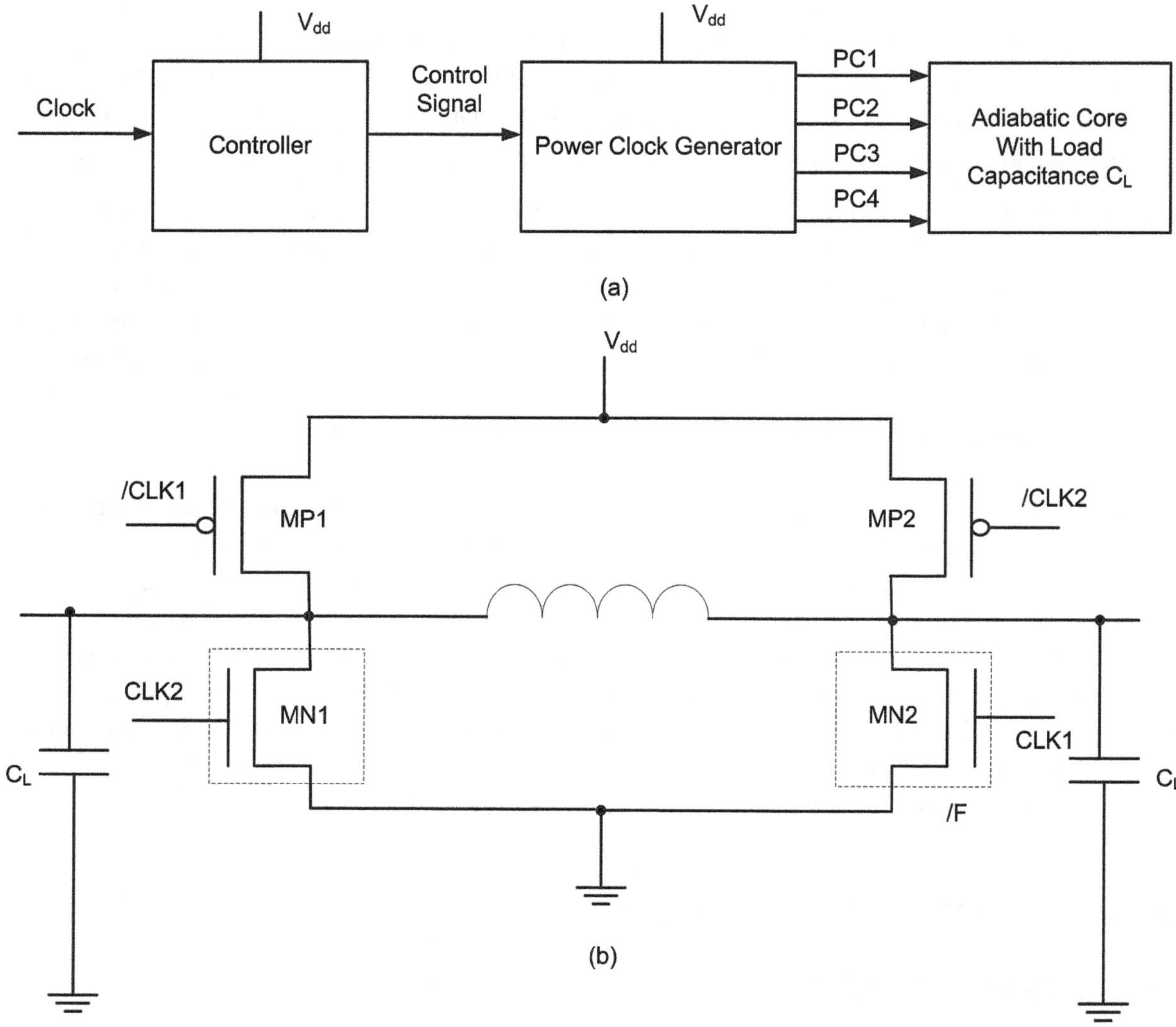

2N-2P or ECRL Buffer/Inverter

The 2N2P or the Efficient Charge Recovery logic, which is one of the widely published quasi-adiabatic logic families, is shown in Figure 10 and it employs the four-phase PC. The dotted blocks in the pull-down arm of the differential structure, namely, F and /F represent the logic functional blocks. Any desired logic function can be realized by replacing these blocks with the corresponding function and its complement. For simplicity, Figure 10 represents single transistors for discussing the inverter operation. The 2N-2P circuit has a latch formed by MP1 and MP2 in the pull-up arms.

To discuss the operation of the circuit, consider Figure 11(a). The input IN leads the power-clock PC by 90° as shown in Figure 11(b). When IN is at logic 1 during T1, the NMOS device MN1 conducts and pulls node /OUT to ground. During this time, the power clock PC is rising and this phase is called the *Evaluate* phase. With the logic 0 realized from /OUT at its gate and with PC rising, MP2 becomes ON. This operation connects node *OUT* to PC, and OUT starts following PC. During T2, IN falls low, and

Figure 9. (a) Adiabatic pipeline using buffers and (b) Four-phase power-clocks

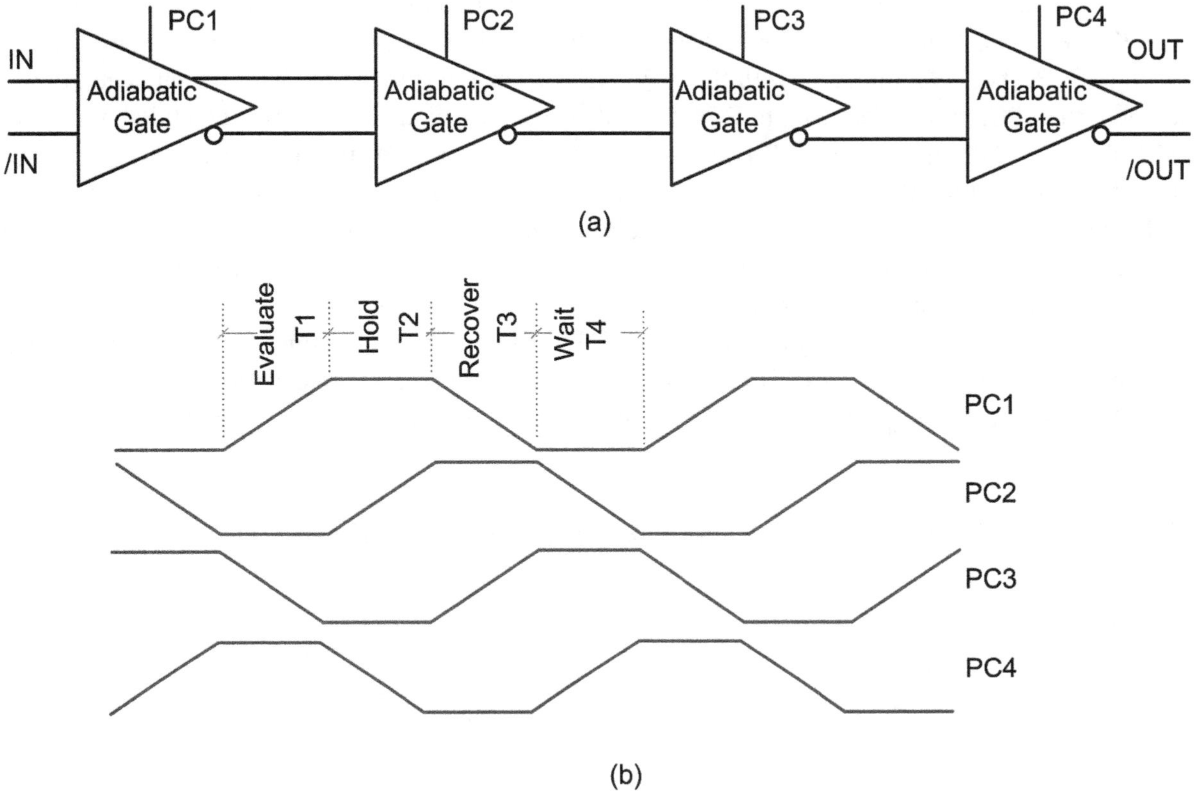

Figure 10. 2N2P Buffer/Inverter Circuit

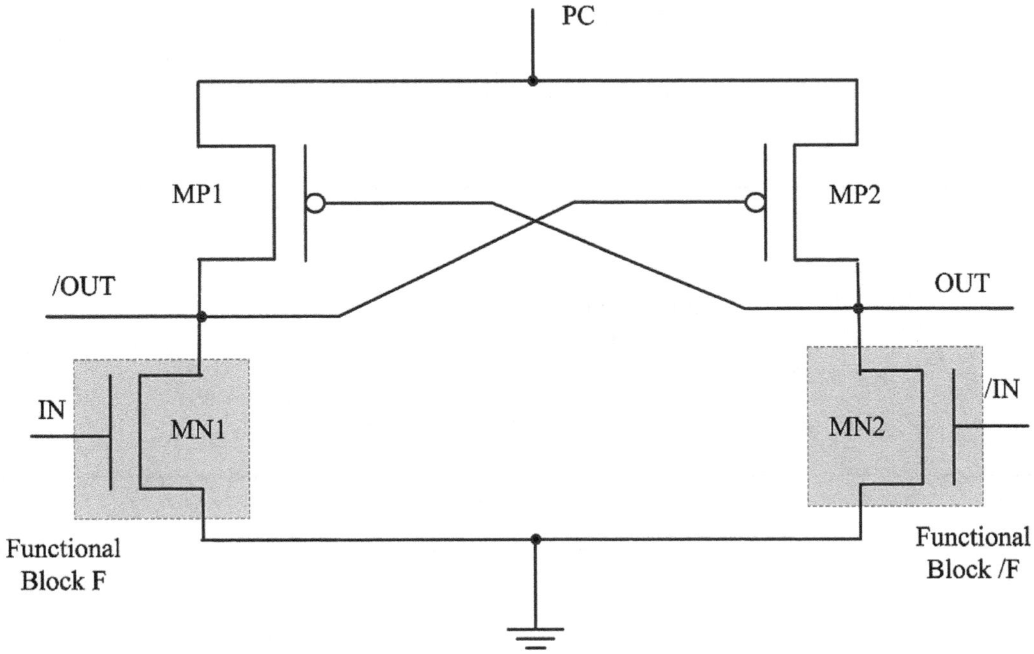

PC remains steady at logic 1, and node OUT follows the same. At this time, the current stage is said to maintain its output in its *Hold* phase, and the next stage in the pipeline enters its *evaluate* phase. Note that PC1, PC2, PC3 and PC4 are applied sequentially to the adiabatic pipeline to realize this feature. During T3, PC begins falling and with the MP2 ON, the charge from node OUT, which is at a higher potential than PC, is recycled back to PC or, in other words, reclaimed by power clock PC and hence OUT also starts falling. This is the *Recovery* phase. This is followed by the *Wait* phase during T4. This phase helps in resetting the state of the current adiabatic stage and awaits the next evaluation stage. This cycle of operation repeats across stages of the adiabatic pipeline. The input applied through successive evaluation stages reaches the output of the last stage, and it may be noted that it will be in phase with the power clock operating the last stage.

Figure 11 (c) depicts the energy transient of the 2N2P circuit. The rising edge of the waveform depicts the energy extracted from the power clock, the near flat edge represents the energy transient during the hold phase, and the falling edge represents the energy recovered from the node by the power clock. The

Figure 11. Signal transients (a) IN and /IN (b) power clock phases (c) Energy

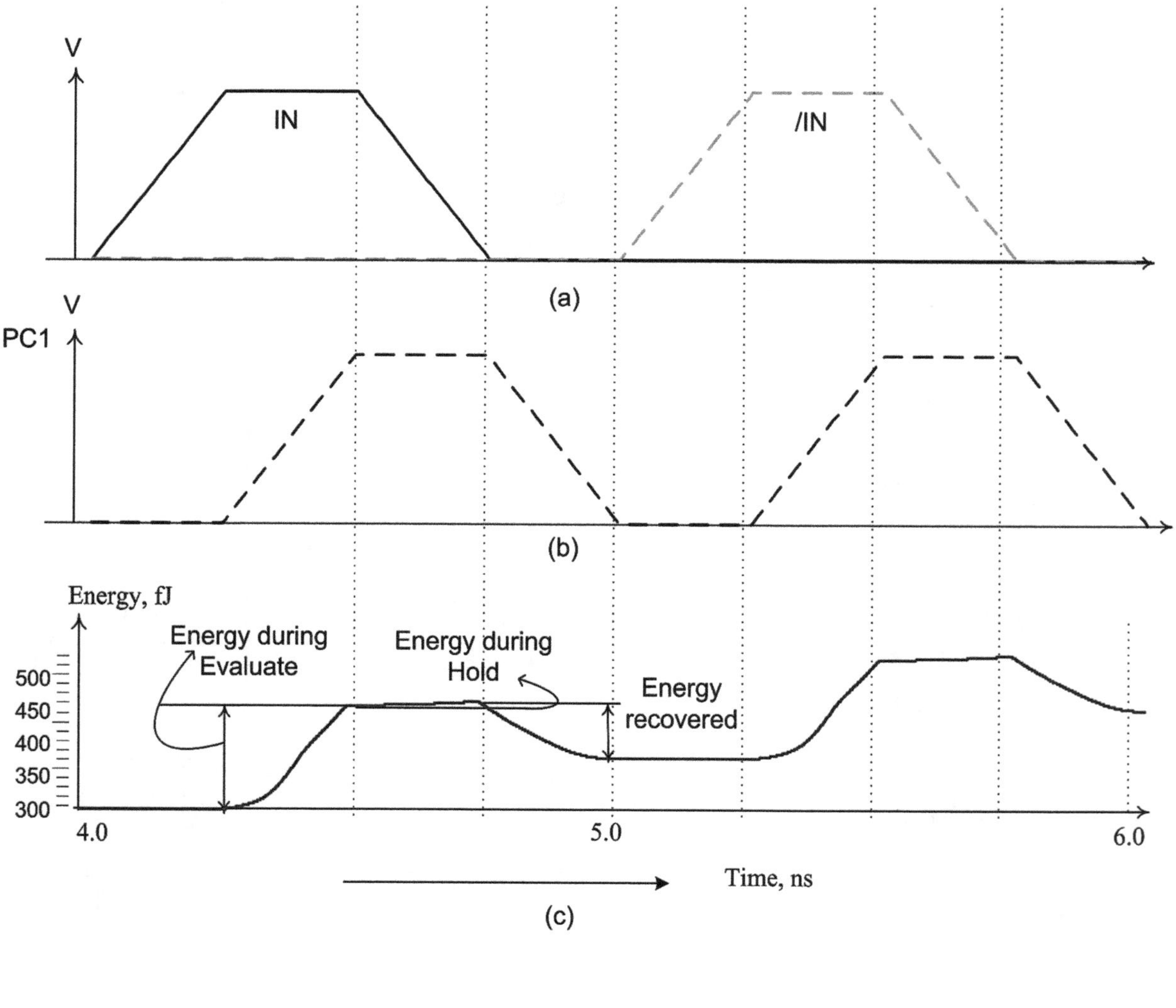

difference between the energy values depicts the adiabatic energy spent in the process of the adiabatic switching of the node. It may be noted for discussion that the energy dissipation in the case of CMOS would successively be adding up every time the switching of the node happens.

Hence, it can be said that though the 2N-2P enjoys the advantage of reduced device count than the other adiabatic logic circuits to be discussed and as a result enjoys the advantage of low power operation at optimal frequency ranges and pipeline length, it incurs a major shortcoming in the form of floating output nodes. This occurs during the charge recovery process, when the power clock voltage drops down below the threshold voltage V_{tp} of the PMOS device. In such a condition, the node OUT finds itself floating and the charge then available at its node cannot find path to discharge to ground.

2N-2N2P Buffer/Inverter

Figure 12 (a) shows the 2N-2N2P inverter which consists of a cross-coupled latch formed by MN1-MP1 and MN2-MP2 inverter structures. The dotted blocks represented by MN3 and MN4 constitute the logic functional block F and its complement /F respectively. The desired logic can be obtained by substituting these blocks. This energy recovery logic operates in the same manner as that of the 2N-2P logic. Its added advantage is the prevention of floating output node when neither MP2 nor MN2 conducts with respect to OUT node, or when neither MP1 nor MN1 conducts with respect to node /OUT. Figure 12(c) depicts the energy dissipation during *evaluate* and *hold* phases, and the energy recovered per cycle. It may also be noted that the *Hold* phase incurs some energy loss, which is due to the leakage energy spent by the 2N-2P latch in holding up the states. The total energy consumed over the two cycles is found to be 24.2fJ as against the 22.9fJ of the 2N-2P logic.

Figure 12 (b) shows the structure of the Positive Feedback Adiabatic Logic (PFAL) realized by connecting MN3 and MN4 as the functional blocks, along with the 2N2P latch formed by MN1-MP1 and MN2-MP2 pairs. Connecting the functional blocks in parallel with MP1 and MP2 makes R_{on} of the PFAL buffer less than that of 2N-2P and 2N-2N2P buffers. However, the functional blocks in the pull-up network account for higher C_L at output while designing for complex logic functions. To briefly explain the operation of the circuit using Figure 12(b), whenever the functional block *F* is true (with false */F*), the node */OUT* becomes *high* following the rising power clock. This in effect makes *MN2* conduct beyond its V_{tn} and *OUT* reaches ground potential. This aids in making MP1 conduct more, speeding up the charging of */OUT*. This circuit is hence called the positive feedback adiabatic logic. When input changes with */F* high, the reverse case happens with respect to node *OUT*. It may be noted that when PC falls below V_{tp}, the recovery process stops. This factor remains the cause of increased energy dissipation in comparison with the counterparts as shown in Figure 12(d).

Differential Cascode Pre-Resolved Adiabatic Logic (DCPAL) Buffer/Inverter

The dynamic CMOS circuits are known for their lower device count, comparatively less standby leakage current, use of an NMOS footer device and their reduced switching capacitance due to reduced device count, and more specifically, their non-vulnerability to false transitions. The important feature of the dynamic logic family, namely, the *pre-charging* phase has been employed in the DCPAL. This enhances the adiabatic gain. DCPAL include a *pre-resolving* phase before the evaluate phase of previously discussed adiabatic circuits. This Pre-resolving phase (as against the pre-charging of the dynamic circuits)

Figure 12. Buffer/Inverter Circuits (a) 2N2N-2P and (b) PFAL, (c) 2N2N-2P energy transient and (d) PFAL energy transient

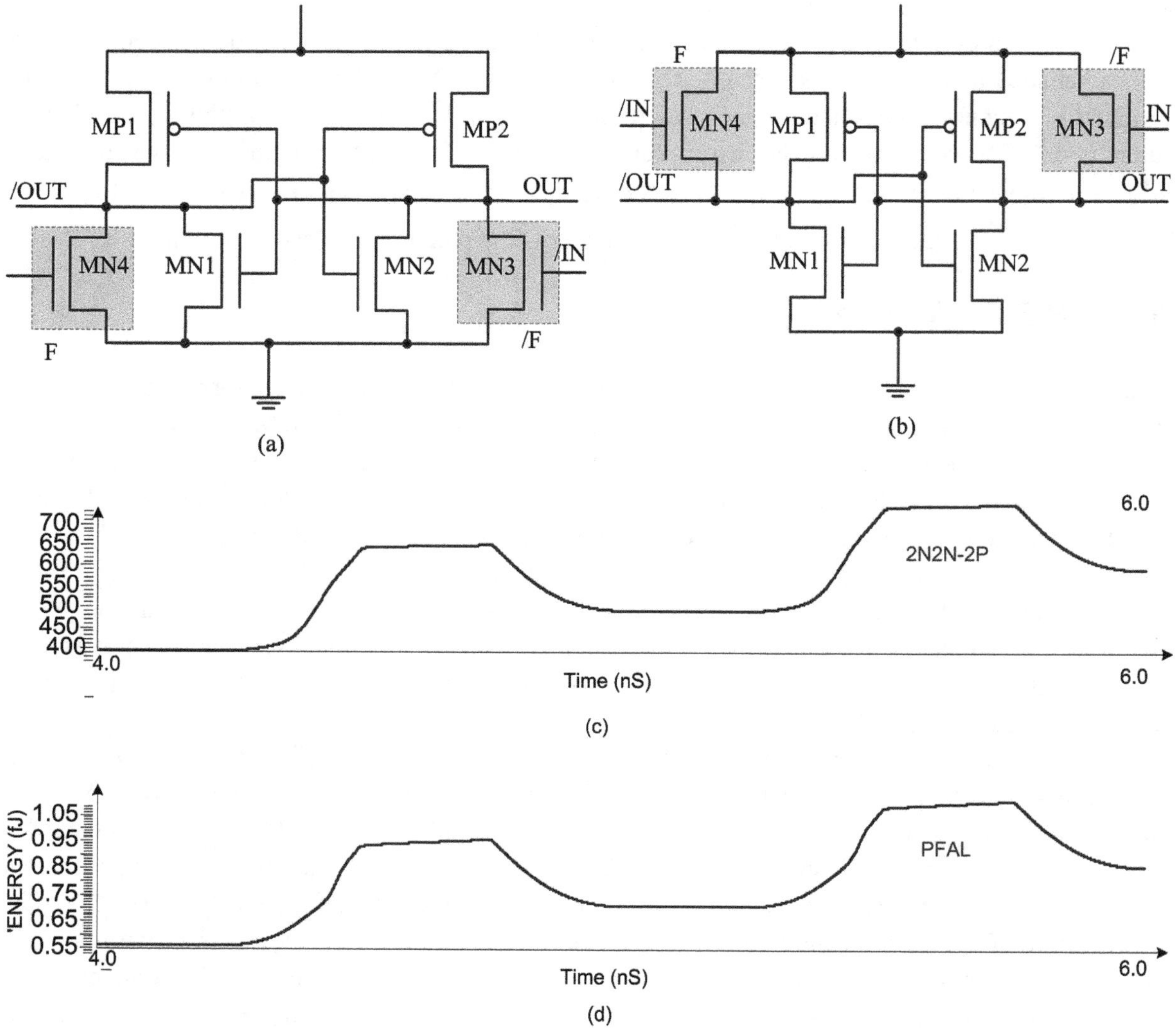

pre-resolves the input to reflect on the output nodes before evaluation. Quintessentially, it dissipates zero adiabatic energy during the pre-resolving process.

Figure 13 (a) shows the basic buffer/inverter circuit of the DCPAL buffer/inverter. It is structured as a dynamic, differential and a dual rail pre-resolve logic. The basic circuit has devices MN1 and MN2 in the pull-down arms. A pre-resolving footer MN3 is placed at the common source terminal of MN1 and MN2. The required logic function is obtained by substituting MN1 and MN2 with the structured logic tree. The use of structured tree makes the complex logic functional blocks such as those involving XOR realizable using shared transistors. The power-clock phase PC3, so named for lucidity, at the gate of MN3 acts as pre-resolve enabling signal. It is 180° ahead of PC1 and 90° ahead of the input signal *IN*. MP1 and MP2 constitute the PMOS latch and they provide the charging/discharging paths for the output nodes *OUT* and */OUT*.

Figure 13. Differential Cascode Pre-resolved Adiabatic Logic Buffer/Inverter Circuit (a) Circuit diagram, (b) Input, Output and Power clock transients and (c) Energy transient

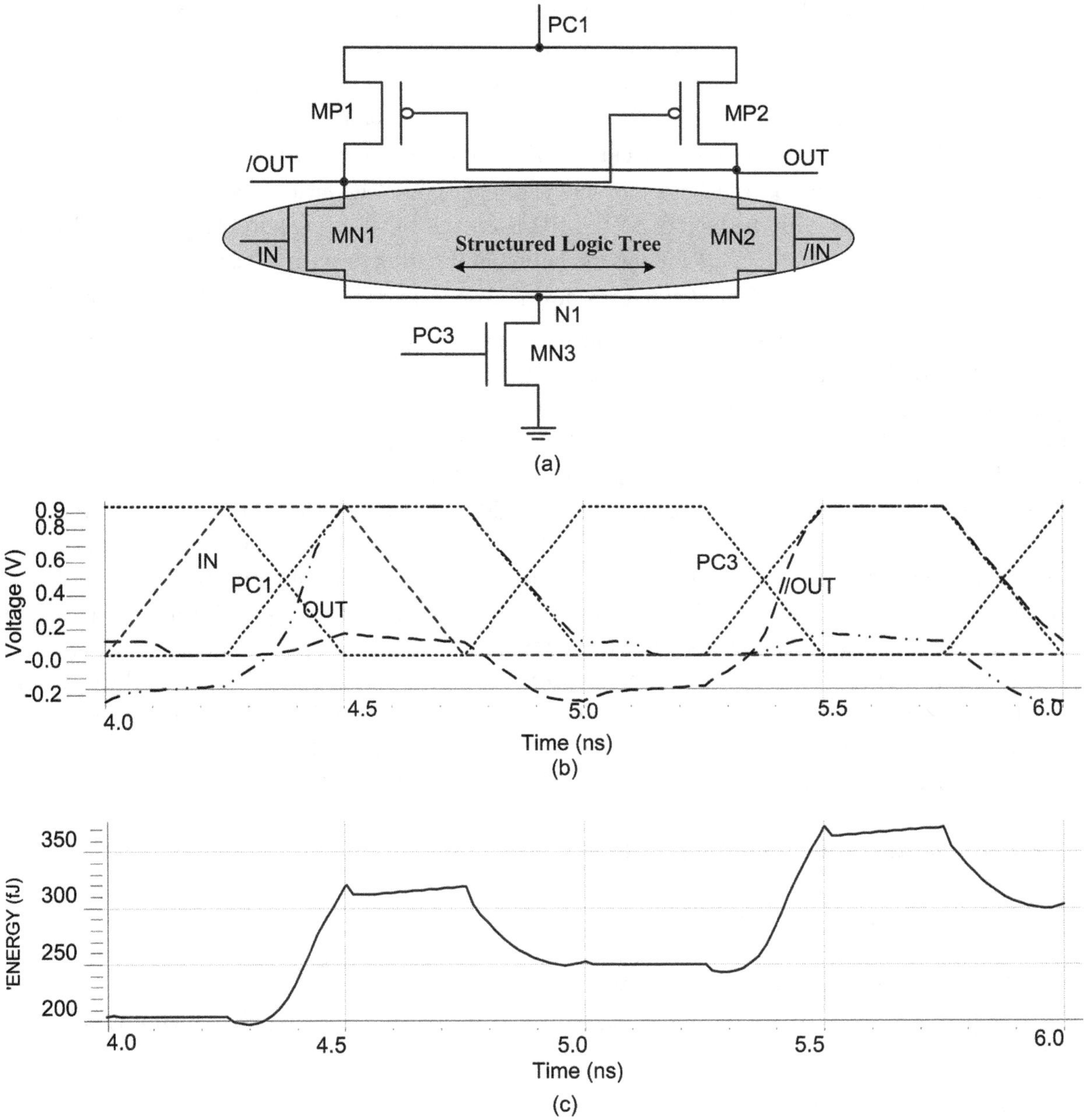

The four stages of operation of the pre-resolved logic family are

1. Pre-Resolve,
2. Evaluate,
3. Hold, and
4. Recover.

Refer to Figure 13(b). During the pre-resolving phase T1, with rising IN and PC3 high, the device MN3 conducts and connects ground to node N1. As *IN* reaches the threshold voltage V_{tn} of MN1, it conducts and the pre-resolving operation makes the node /OUT low. During T2, PC3 falls, MN3 stops conducting below V_{tn} and */OUT* retains its *low* state by its nodal capacitance. At the same time, the evaluating PC1 ramps up and turns MP2 *ON*. This makes node *OUT* rising high following PC1. At this time, MN2 with *low* input at its gate (Ref. Figure 13(b)) does not conduct and passes negligible leakage current. The body effect posed by the stacking nMOS transistor MN3 which is *off* when $PC_3 < V_{tn}$ further adds to the reduction of leakage current. The input goes *low* during T3 and nodes *OUT* and / *OUT* retain their values with the help of MP1-MP2 latch and MN3 goes entirely *off*. During T4, PC1 enters *recovery* phase and node *OUT* follows the falling PC1 letting the charge be recovered through the recovery path of MP2. At this time, */OUT* goes down below zero due to C_{gd} of MP1. However, this has the favorable effect of allowing enhanced charge recovery during T4, through the conducting MP2. Such a potential of -0.5V occurring because of the capacitive coupling effect poses no threat in terms of any signal issues to the next stage, since this occurs only during the *WAIT* state of the subsequent stage. Similarly, during T4, the negative voltage at NMOS input transistors of the next stage do not get affected. The negative voltage on the other hand assists the charge recovery process.

Figure 13 (c) shows the energy transients of the DCPAL buffer/inverter. The pre-resolve, evaluate, hold and recovery phases are denoted by TP, TE, TH and TR respectively. Note the individual energy dissipation components due to PC1 and PC3. The total energy dissipation during the two cycle duration as done for the 2N2P, 2N-2N2P and PFAL circuits is 20.8fJ. The enhanced energy recovery characteristics and reduction of leakage components make this circuit family less energy consuming.

The pipelined operation of the circuits are realized by the use of the successive power clocks operating with 90° phase difference as shown in Figure 14. For example, the following stage is operated by evaluating power clock PC2 and pre-resolving PC4. Figure 14 shows the chain of inverters with the power clock sequence PC1/PC3, PC2/PC4, PC3/PC1 and PC4/PC2.

Power-Loss Models of DCPAL Buffer

The power loss models of DCPAL are due to:

1. The use of pre-resolving transistor MN3,
2. The power dissipation due to charge-sharing and capacitance coupling effect in HOLD state,

Figure 14. DCPAL Cascade stages operated by PC1 to PC4

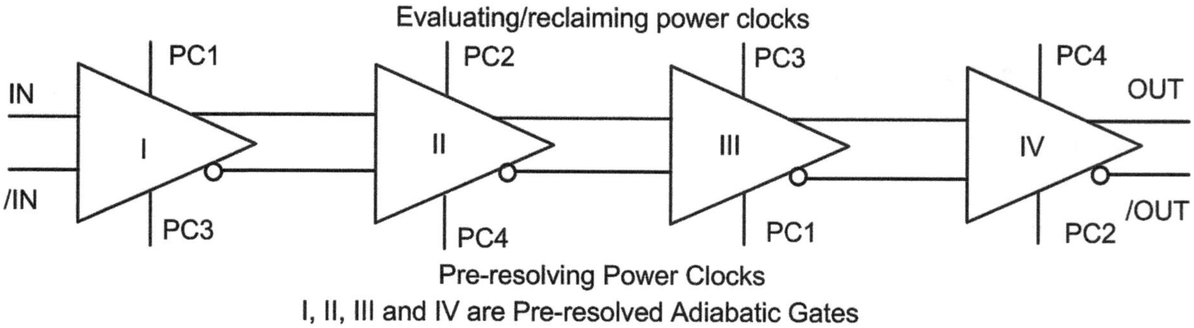

3. The loss due to incomplete charge recovery and
4. The leakage currents.

When footer MN3 operates in linear mode, the gate capacitance is approximated as $C_g = WLC_{ox}$. Then, the non-adiabatic dissipation during the pre-resolving is $\frac{1}{2} C_g V_{dd}^2$ for the ramp power clock, where V_{dd} is the peak voltage of PC3 and C_g, the gate capacitance of the footer device.

When the input warrants the same output during the subsequent cycle, it effects an energy dissipation of $\frac{1}{2} C |V_{tp}|^2$. This energy component added with E_{opn} is depicted by the climbing waveform for every cycle in the energy transient.

1. The non-adiabatic dissipation during the HOLD state is due to the charge sharing between the *OUT* node and N1 through the conducting MN1. Due to the capacitive coupling effect, *OUT* node drops below *zero*, whose value is determined by the size of MP1. The advantage is reduced leakage current due to the body effect.
2. Leakage current in DCPAL is vastly reduced by the rise in threshold due to stack effect. This causes source of MN2 to be at higher potential than the bulk node which reduces the sub-threshold leakage current.

Thus, the total energy dissipation of the DCPAL gate is the sum of the above four components.

Pre-Resolved Sense Amplifier Adiabatic Logic (PSAL) Buffer/Inverter

Figure 15 (a) shows the basic PSAL buffer/inverter. It operates in the same manner as DCPAL, with its 2P latch replaced by the 2N-2P latch, which handles the *Evaluate, Hold and Recovery* operations. The advantage of the PSAL over DCPAL is its excellent energy performance characteristics over a wide range starting from kilohertz of frequency to 500MHz (Kanchana and Raina, 2012). The input and output transients of the PSAL logic shown in Figure 15(b) depict this characteristics. While operating at 100kHz, the recovery process of the PSAL is near complete with the OUT following the falling PC1 most effectively. On the other hand, the 2N-2P, 2N2N-2P and PFAL circuits incur high energy dissipation at low frequencies due to leakage current components (Blotti and Saletti, 2000)].

The PSAL possesses the following advantages:

1. The presence of 2N-2P latch avoids floating output nodes. It enables low frequency operation of PSAL circuits.
2. The presence of pre-resolving footer avoids leakage and introduces stacking effect resulting in reduced leakage current and minimum overall power at low frequencies.

PERFORMANCE METRICS OF ADIABATIC CIRCUITS

The generally followed performance metrics for adiabatic circuits include the following:

Figure 15. (a) Pre-resolve and Sense Adiabatic Logic (PSAL) Buffer/Inverter Circuit (b) Input, Output and Power clock transients

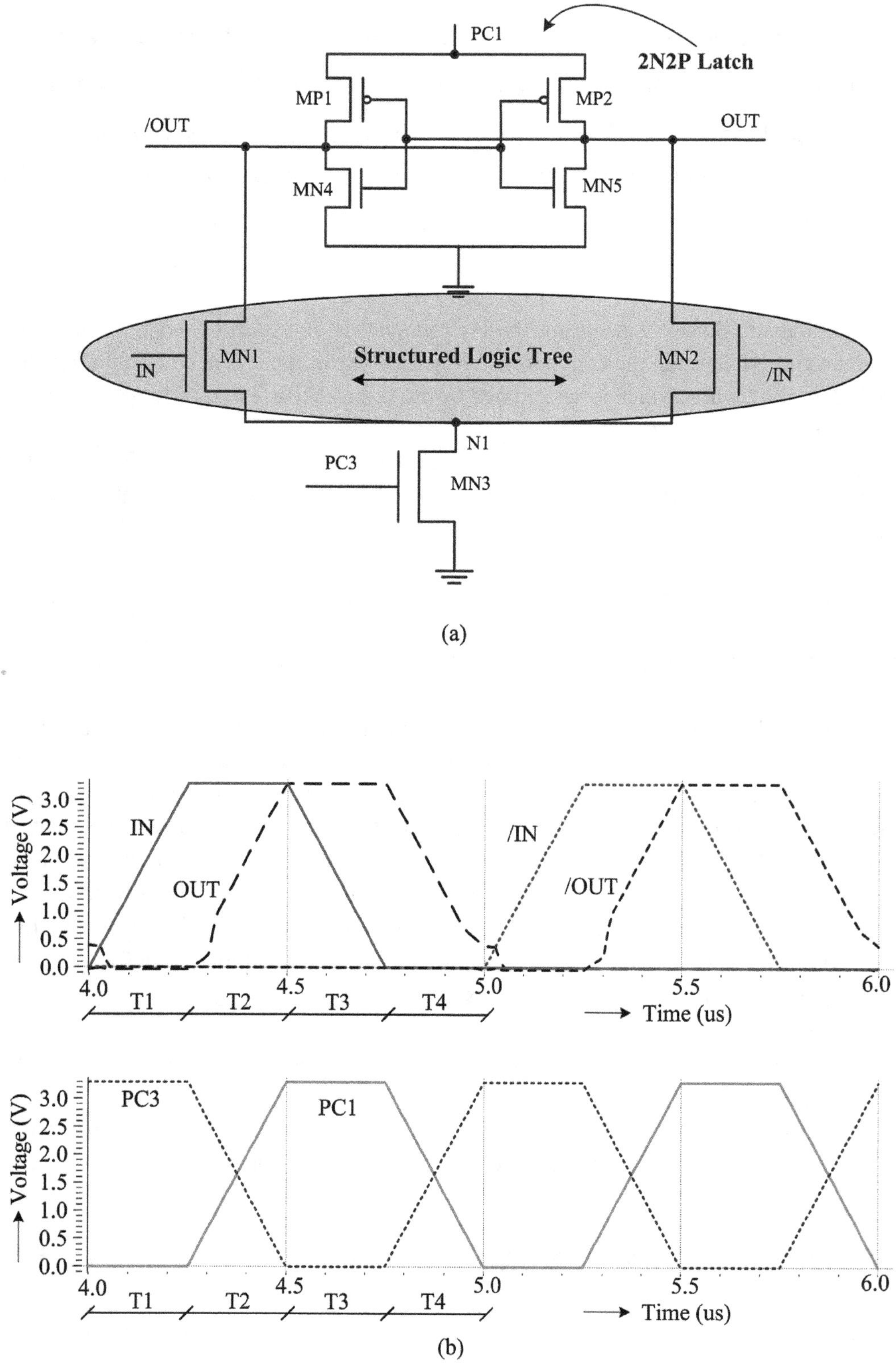

1. **Adiabatic Gain**: It is the ratio between the power dissipated by a conventional CMOS and the adiabatic logic counterpart. Normally, it is measured across the minimum to maximum range of operating frequencies. It validates the advantage of the use of energy recovery circuits. Note that the computation takes into account the power incurred in generation of power clock, while considering a self-powered adiabatic system.
2. **Operating Frequency**: The frequency range of operation of the adiabatic circuit is determined by the feasibility of operating the circuit with optimal gain values.
3. **Transistor Count**: The energy recovery circuits involving charging and recovery paths introduces inadvertent component overhead. However, the major advantage enjoyed by structures discussed in this chapter is the availability of the complementary outputs. This avoids the necessity of introducing inverters for realizing the output complement. It also eliminates the timing related issues such as the generation of glitches, skew and jitter in digital integrated circuits.
4. **Latency**: Latency indicating the computation time delay incurred in adiabatic pipeline, is determined by the number of stages involved in the system. This can be a stumbling block for high speed circuits. While using a four phase PC, it amounts to a latency of $T = \left(4/f\right)$, where f is the frequency of operation.

However, improved designs are realized by the use of:

1. Structured logic trees with large fan-in and
2. The use of shared transistors and reduced gate capacitance and minimum circuit complexity.

LEAKAGE MODELS OF THE INVERTERS

The leakage power models depicting their impact on power dissipation are shown in Figure 16. The dashed arrowed lines in colored boxes show the subthreshold leakage current paths. These lines also identify the effective widths of the leakage paths. It illustrates the minimum leakage realized by DCPAL and PSAL circuits. Kanchana (2011) details the energy characteristics of lower technology nodes such as 65nm, 45nm and 32nm of Predictive Technology Models obtained from of www.eas.asu.edu/~ptm, through various phases of adiabatic operation.

To discuss impact of technology nodes, Figures 17 (a) and (b) show the energy dissipation of the inverters, while employing 350nm models from Austria Microsystems and 32 nm FF corner models from PTM. The evaluate, hold and recovery phase transients for Low V_{th} corner models for 200 kHz to 1 GHz are shown for comparison. The domination of energy due to increased leakage current for lower technology nodes is apparent. While the 350nm technology node demonstrates various adiabatic circuits performing with approximately flat energy dissipation for lower frequency, the 32nm low V_{th} PTM models on the other hand, show highly obvious leakage effects at lower frequency. This fact has been widely acknowledged in the literature, and this feature impedes the use of 2N-2P, 2N-2N2P and PFAL adiabatic circuits for lower operating frequencies. It may further be noted that the pre-resolved adiabatic family overcomes this limitation due to its modified circuit architecture.

Figure 16. Leakage models of the adiabatic buffer/inverters

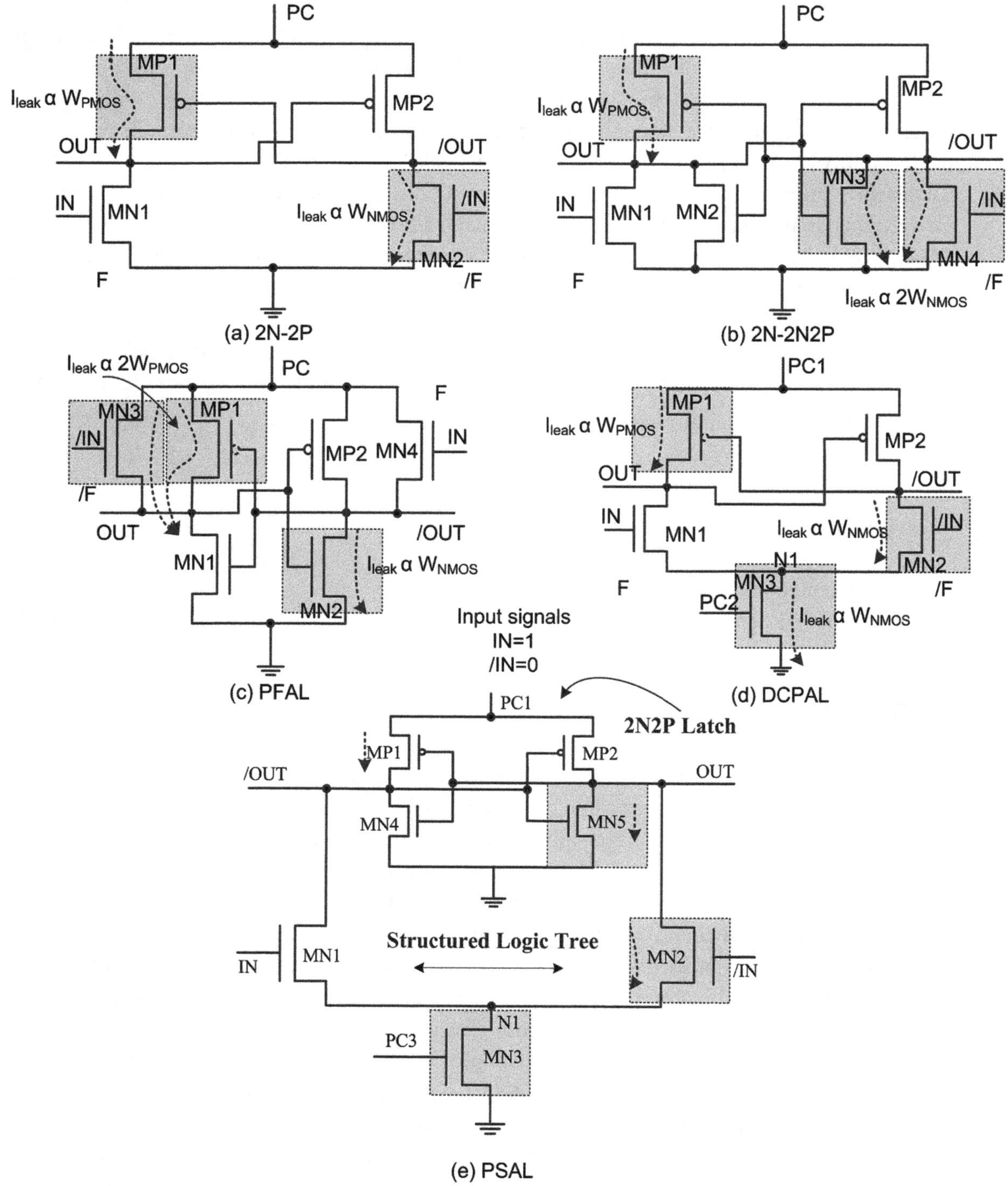

Figure 17. Leakage current characteristics of the buffers (a) 350nm AMS (b) 32nm PTM models

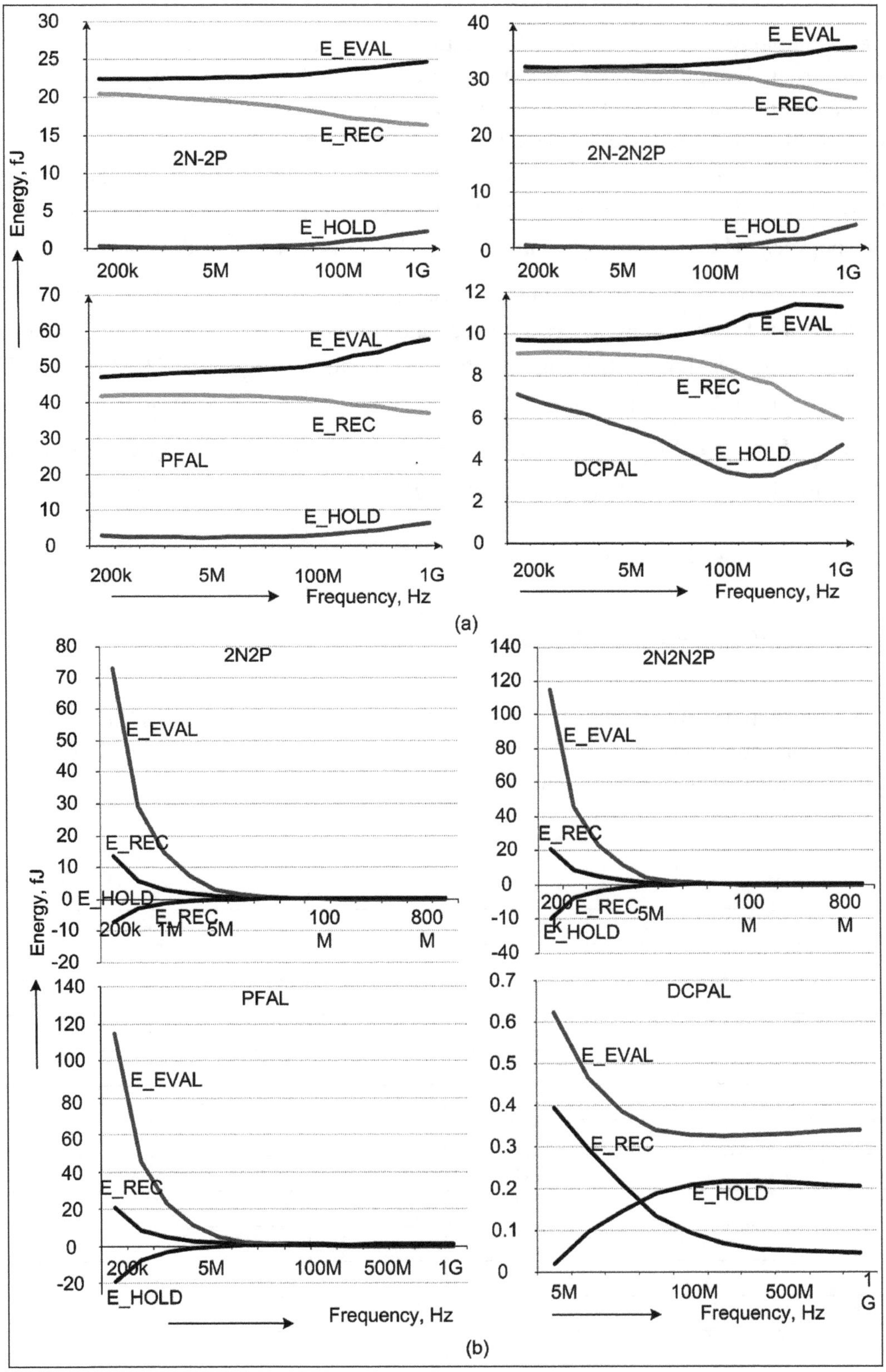

PERFORMANCE COMPARISONS AMONG THE ADIABATIC CIRCUITS

Predictive Technology Model (PTM) files for 32nm have been employed in the simulations on industry standard Electronic Design Automation (EDA) tools. Table 3 depicts the energy dissipation values incurred while operating at 1 GHz of frequency for two cycles across 2ns duration. The adiabatic gain is also shown for the adiabatic logic families. Figure 18 depicts the energy curves for the adiabatic and CMOS circuits. Note that while the adiabatic energy curves show rising and partial recovery phases every cycle, that of the CMOS circuit show only rising transient, as can be expected, with the logic transitions between '1' and '0' always claiming $\frac{1}{2} C_L V_{dd}^2$ energy. The peak voltage V_{dd} and the node capacitance C_L determine the adiabatic energy dissipation. The energy recovery property of adiabatic circuit is determined by charging path resistance, load capacitance, leakage and charge sharing. Typically considering the PSAL circuit, C_L is contributed by drain-bulk C_{db} of MN2, MN5 and MP2, gate-drain C_{gd} of MN4 and MP2, gate-source C_{gs} of MN4, gate-bulk C_{gb} of MP1, gate-drain overlap C_{gdo} of MN2 and MN5, C_{gdo} and C_{gso} of MP1, and C_o, the sum of wire and input capacitances of next stage. The overall capacitance C_L at */OUT* node with high *IN* is

$$C_L = C_{gdo,N2} + C_{db,N2} + C_{gdo,N5} + C_{db,N5} + C_{gd,P2}$$

$$+C_{db,P2} + C_{gdo,P1} + C_{gso,P1} + C_{gb,P1} + C_{gs,N4} + C_{gd,N4} + C_o$$

The node capacitances obtained 32nm are *0.277fF, 0.407fF, 0.405fF, 0.272fF* and *0.303fF* respectively for the adiabatic inverters 2N-2P, 2N2N-2P, PFAL, DCPAL and PSAL. The modeling for the capacitances are made using HPICE manual.

Table 3.

Adiabatic Logic	Energy in Joules
2N2P	8.02E-17
2N2N2P	9.87E-17
PFAL	15.0E-17
DCPAL	6.93E-17
PSAL	9.68E-17
CMOS	58.2E-17
Adiabatic Gain = E_{CMOS}/E_{Adiab}	
2N-2P	7.26
2N2N-2P	5.90
PFAL	3.88
DCPAL	8.40
PSAL	6.02

Figure 18. Energy transients of 2N-2P, 2N2N-2P, PFAL, DCPAL, PSAL and CMOS logic buffers/inverters

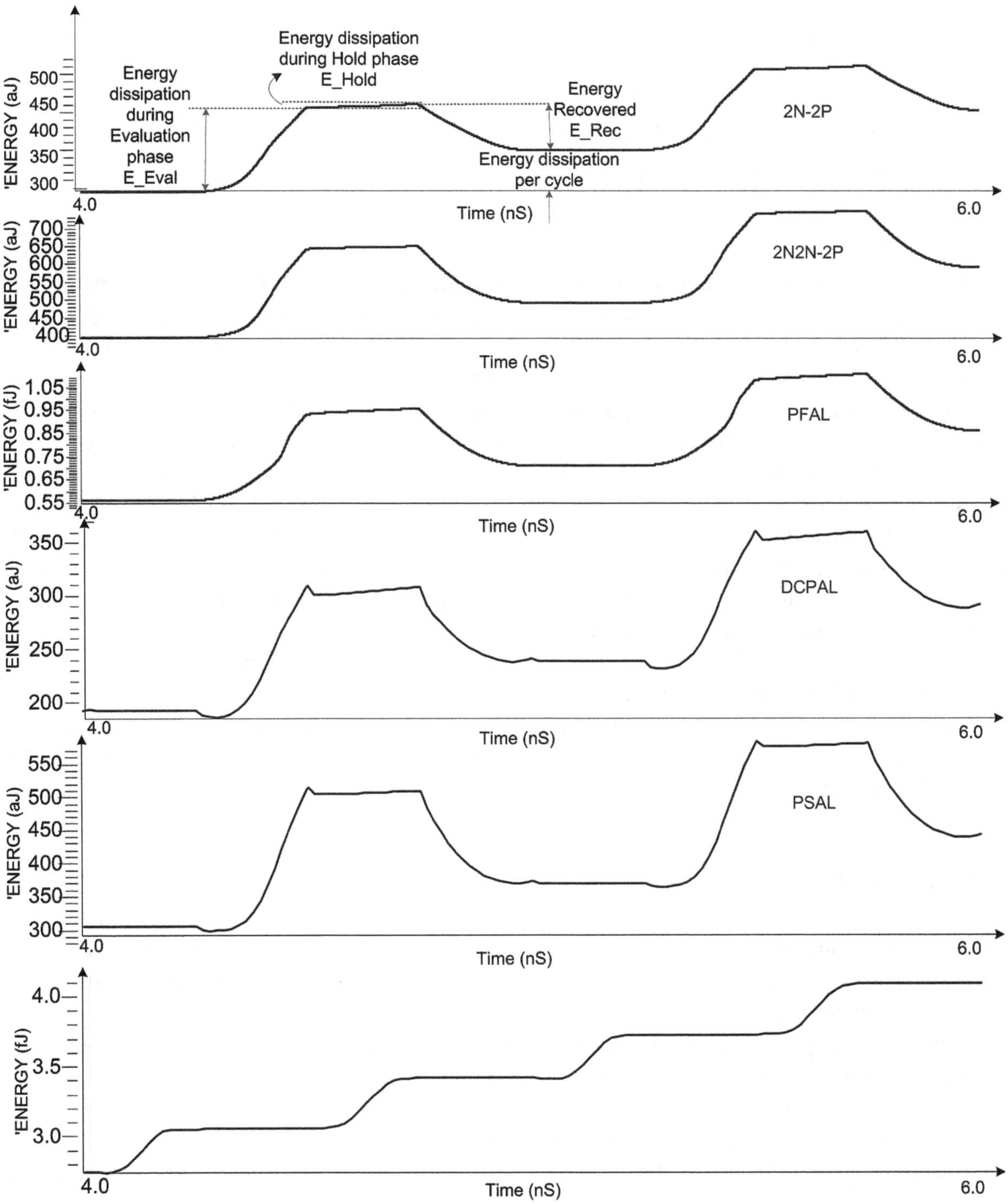

$$C_{gs} = \frac{C_j.AS}{\left(1 - {V_{BS}}/{\varphi_o}\right)^{Mj}} + \frac{C_{jw}.PS}{\left(1 - {V_{BS}}/{\varphi_o}\right)^{Mjsw}}$$

$$C_{gd} = \frac{C_j.AD}{\left(1 - {V_{BD}}/{\varphi_o}\right)^{Mj}} + \frac{C_{jw}.PD}{\left(1 - {V_{BD}}/{\varphi_o}\right)^{Mjsw}}$$

The reduced device count achieves considerably less energy (*125aJ*) of 2N-2P inverter. However, with floating output nodes, the unrecovered energy is more than *50aJ* per cycle as shown in Figure 18. Corresponding values for 2N2N-2P are *240aJ* and *140aJ* and for PFAL, *0.37fJ* and *0.18fJ* respectively. The higher device count and hence increased nodal capacitance contribute for increased energy for both the logic types. Additionally, higher resistance in the discharging path of the PFAL contributes to increased energy for PFAL. On the other hand, minimum nodal capacitance and reduced leakage current due to pre-resolving device contribute for *110aJ* and *47aJ* for DCPAL. Additional devices in the sense amplifier structure of PSAL contribute for slightly increased energy values, *212aJ* and *63aJ* than DCPAL.

Figure 19 depicts the charging and recovery current transients for the adiabatic logic families and the CMOS counterpart. While the negative current values indicate the current spent in charging the output nodes, the positive values indicate the values when the power clock reclaims charge from the nodes. The comparative maximum current values of around *1uA* to *-2uA* for 2N-2P, *1uA* to *-3uA* for 2N2N-2P, *1uA* to *-6uA* for PFAL, *0.75uA* to *-1.5uA* for DCPAL, *1.75uA* to *-3uA* for PSAL and *-4uA* for CMOS counterpart justify the above argument.

Figure 20 shows a 16-bit CLA employing DCPAL, PC sequence and module level powering. More application circuits and comparisons are available in Kanchana Bhaaskaran V. S. & Raina J. P. (2008), (2012) (Visvesh, 2007).

CONCLUSION

The adiabatic circuits and reversible computing techniques are the fundamental transformations to circuits with their radically different architecture, which can lead to ultimate limits of energy efficiency. This chapter can help an electronics circuit designer interested in adiabatic circuit in learning the nuances of the energy recovery property of the circuits. The reversible logic for low power operation is presented with typical reversible gates such as Feynmann, Toffoli and Fredkin gates. The reversible Bennett-clocking has been discussed with a 4-stage circuit.

The circuit topology in lieu of the charging and recovery paths decides the adiabatic efficiency and the non-adiabatic loss of quasi-adiabatic circuits, and the control schemes decide the operational energy dissipation E_{opn}. The overhead incurred in the adiabatic CMOS is the need of transistors in the charging/recovery paths. However, the quasi-adiabatic circuits are proved to function with minimum energy and tolerable circuit complexity. Typical buffer/inverter circuit architectures of 2N-2P, 2N2N-2P, PFAL, DCPAL and PSAL have been discussed and the comparative energy performance along with the leakage models have been discussed. The influence of technology on energy recovery is analyzed using 350nm

Figure 19. Current transients of 2N-2P, 2N2N-2P, PFAL, DCPAL, PSAL and CMOS logic buffers/inverters

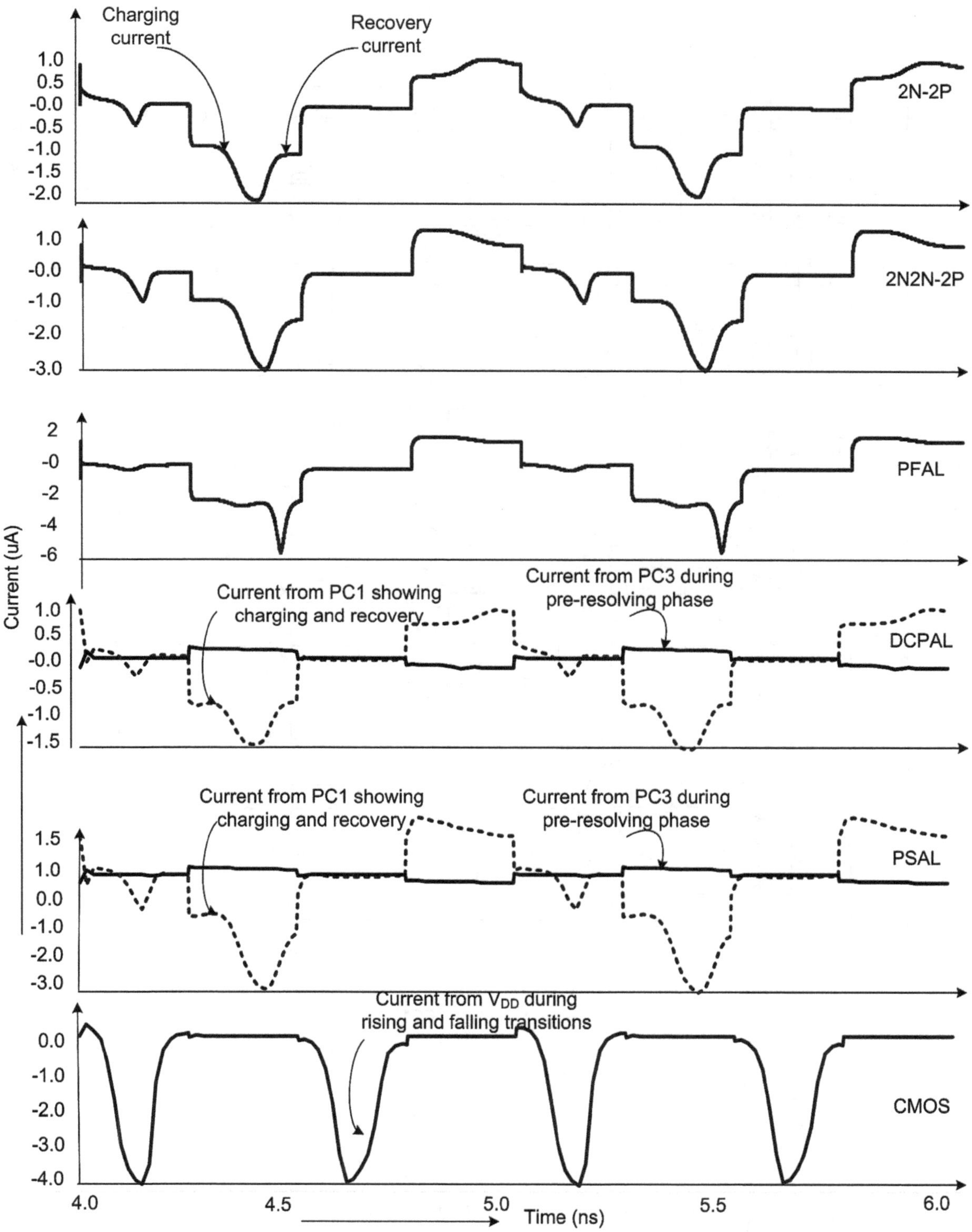

Figure 20. Typical structure of 16-bit adiabatic CLA adder

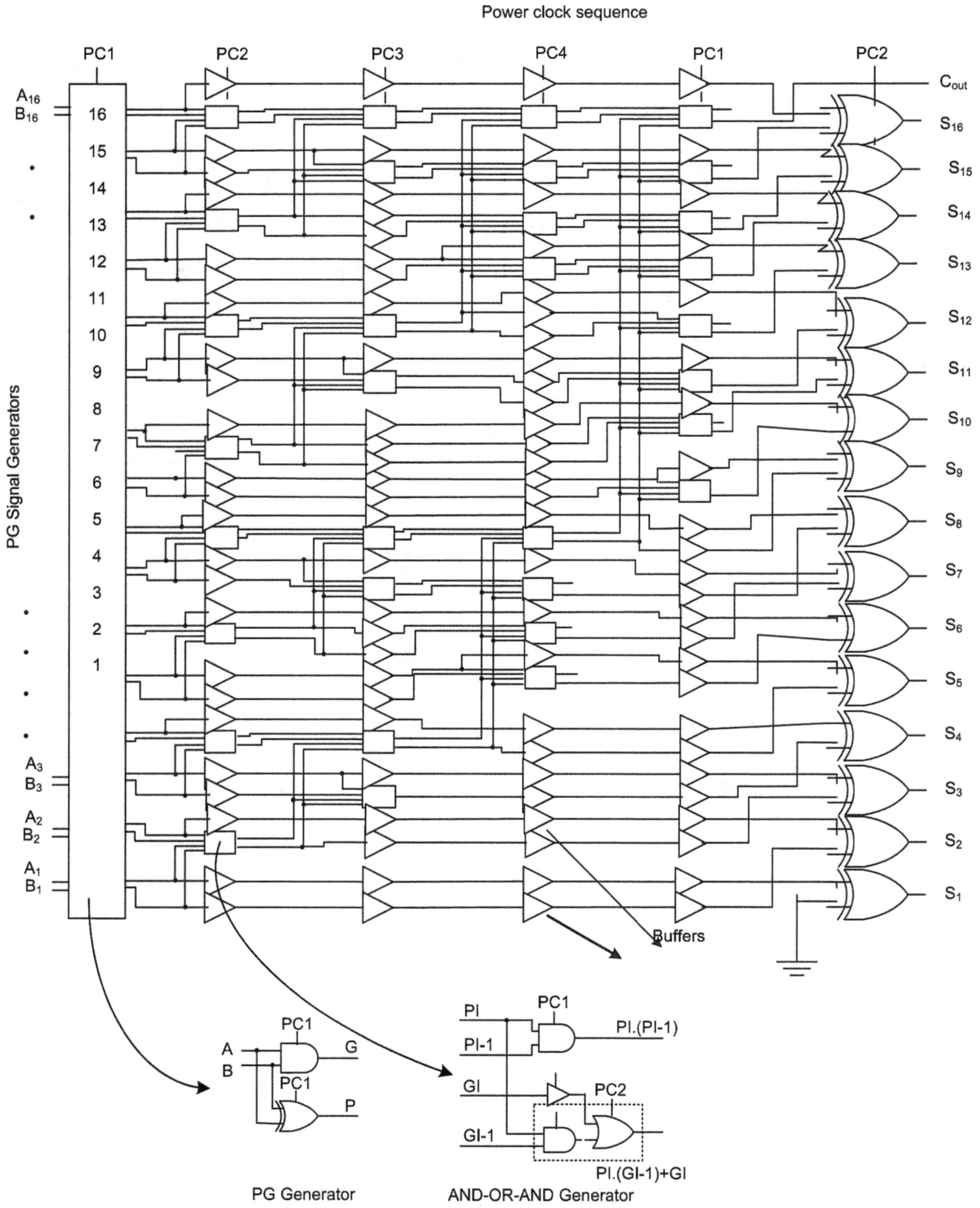

and 32nm technology nodes. The pre-resolved adiabatic logic family of circuits overcomes these limitations of lower adiabatic frequency limits and lower gain at high frequencies. DCPAL with 69.3aJ against 0.15fJ of PFAL for two cycles of operation proves more efficient. PSAL can operate at kHz range of frequencies with optimal gain, unlike any other adiabatic circuits found in the literature.

REFERENCES

Athas, W. C., Svensson, L. J., Koller, N., Tzartzanis, N., & Chou, E. (1994). Low power digital system based on adiabatic switching principles. *IEEE Trans. on VLSI Systems*, *2*(4), 398–407. doi:10.1109/92.335009

Athas, W. C., Svensson, L. J., & Tzartzanis, N. (1996). A resonant signal driver for two-phase, almost-non-overlapping clocks.*IEEE International Symposium on Circuits and Systems*. IEEE. doi:10.1109/ISCAS.1996.541916

Avital, M., Dagan, H., Levi, I., Keren, O., & Fish, A. (2015). DPA-Secured Quasi-Adiabatic Logic (SQAL) for Low-Power Passive RFID Tags Employing S-Boxes. IEEE Transactions on Circuits and Systems—I: Regular Papers, 62(1), 149-156.

Bennet, C. (1982). The Thermodynamics of Computation – a Review. *International Journal of Theoretical Physics*, *12*(12), 905–940. doi:10.1007/BF02084158

Bennett, C. (1973). Logical reversibility of computation. *IBM Journal of Research and Development*, *17*, 525–532.

Bennett, C. H., & Landauer, R. (1985). The Fundamental Physical Limits of Computation. *Scientific American*, *253*(1), 38–46. doi:10.1038/scientificamerican0785-48

Berut, A., Arakelyan, A., Petrosyan, A., Ciliberto, S., Dillenschneider, R., & Lutz, E. (2012). Experimental verification of Landauer's principle linking information and thermodynamics. *Nature*, *483*(7388), 187–189. doi:10.1038/nature10872 PMID:22398556

Blotti, A., Pascoli, S.D., & Saletti. (2000) Simple model for positive-feed-back adiabatic logic power consumption estimation. Electronic Letters, 36(2).

Blotti, A., & Saletti, R. (2004). Ultralow-Power Adiabatic Circuit Semi-Custom Design. *IEEE Trans. on VLSI Systems*, *12*(11), 1248–1253. doi:10.1109/TVLSI.2004.836320

Borkar. (2008). *Gigascale integration - challenges and opportunities*. Intel® Software Network.

Chernichenko, D., Kushnerov, A., & Ben-Yaakov, S. (2012). *Adiabatic Charging of Capacitors by Switched Capacitor Converters with Multiple Target Voltages*. 2012 IEEE 27th Convention of Electrical and Electronics Engineers, Israel.

Denkar, J. S. (1994). A Review of Adiabatic Computing. AT & T Bell laboratories. *1994 IEEE Symposium on Low Power Electronics*. doi:10.1109/LPE.1994.573218

Dickinson, A. G., & Denker, J. S. (1995). Adiabatic dynamic logic. IEEE Solid State Circuits, 30(3), 311-315. doi:10.1109/4.364447

Dueck, G. W. (2012). Synthesis of Toffoli Networks: status and challenges. In *International Symposium on Electronic System Design (ISED)*. doi:10.1109/ISED.2012.26

Feynman, R. (1985). Quantum Mechanical Computers. *Optics News*, *11*(2), 11–20. doi:10.1364/ON.11.2.000011

Feynman, R. (1986). Quantum Mechanical Computers. *Foundations of Physics*, *12*(6).

Hamid, M.-M., & Afzali, K. A. (2001). Efficient power-clock generation for adiabatic logic. *The 2001 IEEE Int.Symp. on Circuits and Systems, ISCAS 2001*.

Hänninen, I., Lu, H., Blair, E. P., Lent, C. S., & Snider, G. L. (2014). Reversible and Adiabatic Computing: Energy-Efficiency Maximized. *Field-Coupled Nanocomputing, LNCS*, *8280*, 341–356.

ITRS. (2013). Available from http://www.itrs.net

Johnson, M. C., Somasekhar, D., Chiou, L. Y., & Roy. (2002). Leakage Control with Efficient Use of Transistor Stacks in Single Threshold CMOS. *IEEE Trans. on VLSI Systems*, *10*, 1-5.

Kaeting, Flynn, Aitken, Gibbons, & Shi. (2007). *Low Power Methodology Manual for system-on-chip design*. Springer.

Kanchana Bhaaskaran V. S. (2011). Energy Recovery Performance of Quasi-Adiabatic Circuits using Lower Technology Nodes. *IICPE2010*.

Kanchana Bhaaskaran, V. S., & Raina, J. P. (2008). Differential Cascode Adiabatic Logic Structure for Low Power. *Journal of Low Power Electronics*, *4*(2), 178–191. doi:10.1166/jolpe.2008.264

Kanchana Bhaaskaran, V. S., & Raina, J. P. (2010). Two-Phase Sinusoidal Power Clocked Quasi-Adiabatic Logic Circuits. *Journal of Circuits, Systems, and Computers*, *19*(2), 335–347. doi:10.1142/S0218126610006098

Kanchana Bhaaskaran, V. S., & Raina, J. P. (2012). Pre-resolve and Sense Adiabatic Logic for 100 kHz to 500 MHz Frequency Classes. *Journal of Circuits, Systems, and Computers*, *21*(5).

Keating, M., Flynn, D., Aitken, R., & Alan, G. K. S. (2007). *Low Power Methodology Manual For System-on-Chip Design*. Springer Publications.

Kim, S., Ziesler, C. H., & Papaefthymiou, M. C. (2005). Charge-recovery computing on silicon. *IEEE Transactions on Computers*, *54*(6), 651–659. doi:10.1109/TC.2005.91

Kramer, J. S., Denker, B., & Moroney. (1995). 2nd order adiabatic computation with 2N-2P and 2N-2N2P logic circuits. *Proc. Int. Symp. Low Power Design*. doi:10.1145/224081.224115

Landauer, R. (1961). Irreversibility and heat generation in the computing process, IBM Jl. *Research for Development*, *5*(3), 183–191.

Laszlo, V., Kovacs, F., & Hosszu, G. (2001). An Improved Pass-Gate Adiabatic Logic.*Proceedings, 14th Annual IEEE Int. ASIC/SOC Conference*.

Lent, C. S., Liu, M., & Lu, Y. (2006). Bennett clocking of quantum-dot cellular automata and the limits to binary logic scaling. *Nanotechnology*, *17*(16), 4240–4251. doi:10.1088/0957-4484/17/16/040 PMID:21727566

Lim, J., Kim, & Chae. (2000). NMOS reversible energy recovery logic for ultra-low-energy applications. *IEEE J. of Solid-State Circuits, 35*(6), 865-875.

Meindl, J. D. (1995). Low Power Microelectronics: Retrospect and Prospect. *Proceedings of the IEEE*, *83*(4), 619–635. doi:10.1109/5.371970

Miryala, S., Calimera, A., Macii, E., & Poncino, M. (2014). Ultra Low-Power Computation via Graphene-Based Adiabatic Logic Gates. *2014 17th Euromicro Conference on Digital System Design*.

Monteiro, C. (2015). Yasuhiro Takahashi & Toshikazu Sekine. (2014). Low-power secure S-box circuit using charge-sharing symmetric adiabatic logic for advanced encryption standard hardware design. *IET Circuits Devices Syst.*, *9*(5), 362–369. doi:10.1049/iet-cds.2014.0150

Moon, Y., & Jeong, D. K. (1996). An Efficient Charge Recovery Logic Circuit. *IEEE Journal of Solid-State Circuits*, *31*(4), 514–522. doi:10.1109/4.499727

Morrison, M., & Ranganathan. (2013). Novel optimization method for reversible logic circuit minimization. *2013 IEEE Computer Society Annual Symposium on VLSI* (ISVLSI).

Parhami, B. (2006). Fault Tolerant Reversible Circuits.*Proceedings of 40th Asilomar Conf. Signals, Systems, and Computers.*

Patra, P. (1995). *Approaches to Design of Circuits for Low-Power Computation.* (Ph.D dissertation). Univ. Texas at Austin.

Piguet, C. (2007). Low power design in deep submicron 65 and 45nm technology. *14th IEEE Intl. Conf. on Electronics, Circuits & System 2007, ICECS, 2007*. doi:10.1109/ICECS.2007.4511140

PTM. (2013). Retrieved from http://www.eas.asu.edu/~ptm

Roy, K., & Prasad, S.C. (2000). *Low Power CMOS VLSI Circuit Design*. Wiley Interscience.

Sathe, V. S., Chuch, J.-Y., & Papaefthymiou, M. C. (2007). Energy-efficient GHz-class charge-recovery logic. *IEEE Journal of Solid-State Circuits*, *42*(6), 38–46. doi:10.1109/JSSC.2006.885053

Sathe, V. S., Chuch, J.-Y., & Papaefthymiou, M. C. (2007, January). Energy efficient GHz-class charge-recovery logic. *IEEE Journal of Solid-State Circuits*, *42*(6), 38–46. doi:10.1109/JSSC.2006.885053

Seitz, C. L., Frey, A. H. Mattisson Sven, S. D., Rabin, D. A., & Snepscheut, J (1985). *Hot Clock nMOS.* California Institute of Technology. Retrieved from http://resolver.caltech.edu/CaltechCSTR:1985.5177-tr-85

Smoline, J., & DiVincenzo. (1996). Five two-qubit gates are sufficient to implement the quantum Fredkin gate. *Physics Review A, 53*(4), 2855-2856.

Sultana, J., Mitra, S. K., & Chowdhury, A. R. (2015). On the Analysis of Reversible Booth's Multiplier. *28th International Conference on VLSI Design* (VLSID).

Supreeth, M. S., Sagar, D., & Kanchana Bhaaskaran, V. S. (2015). Low Power Energy Recovery Clock Distribution.*International Conference on Communications and Signal Processing (ICCSP'14).*

Supreeth, M.S., & Bhaaskaran, V.S. (2015). Self-gated resonant-clocked flip-flop optimized for power efficiency and signal integrity. *IET Circuits, Devices & Systems.*

Svensson, L. J., & Koller, J. G. (1994). Adiabatic charging without inductors.*International Workshop on Low-Power Design.*

Thapliyal & Srinivas. (2005). A Novel Reversible TSG Gate and Its Application for Designing Reversible Carry Look-Ahead and Other Adder Architectures.*Tenth Asia-Pacific Computer Systems Architecture Conference (ACSAC05).*

Tzartzanis, N., & Athas, W. C. (1997). Clock-powered logic for a 50 MHz low-power RISC Data path. In *Proc. of ISSCC.*

Vetuli, A., Pascoli, S. D., & Reyneri, L. M. (1996). Positive Feedback in Adiabatic Logic. *Electronics Letters, 32*(20), 1867–1869. doi:10.1049/el:19961272

Vieri, C. J. (1999). *Reversible Computer Engineering and Architecture.* (Thesis). Massachusetts Institute of Technology.

Vieru, C. (1995). *Pendulum: A Reversible Computer Architecture.* (SM thesis). Massachusetts Inst. of Technology.

Vojin,, G., Oklobdˇzija, F., Maksimovi'c, D., & Fengcheng. (1997). Pass-transistor adiabatic logic using single power-clock supply. *IEEE Transactions on Circuits and Systems. 2, Analog and Digital Signal Processing, 44*(10).

Wang, W.Y., & Lau, K.T. (1005). Adiabatic pseudo-domino logic. *Electronics Letters, 31*(23), 1982-83.

Ye & Roy. (2001). QSERL: quasi-static energy recovery logic. *IEEE Jl. of Solid-State Circuits, 36*(2), 239–248.

Ye, Y., & Kaushik, R. (1996). Energy Recovery Circuits Using Reversible and Partially Reversible Logic. *IEEE Trans. on Circuits and Systems-I: Fund Theory and Applications, 43*, 769–778.

Youngjoon, S., Chanho, L., & Yong, M. (2004). A low power 16-bit RISC microprocessor using ECRL circuits. ETRI Journal, 26(6).

Younis, S., & Knight, T. (1994). *Asymptotically zero energy computing using split-level charge recovery logic.* Technical Report AITR-1500. MIT AI Laboratory.

Younis, S. G., & Knight, T. F. (1993). Practical implementation of charge recovering asymptotically zero power CMOS.*Proc. of the 1993 Symposium on Integrated Systems.*

Section 3

Low Power Applications and Studies

Chapter 7
State-of-the-Art Master Slave Flip-Flop Designs for Low Power VLSI Systems

Kunwar Singh
Netaji Subhas Institute of Technology, India

Satish Chandra Tiwari
Netaji Subhas Institute of Technology, India

Maneesha Gupta
Netaji Subhas Institute of Technology, India

ABSTRACT

This chapter presents a comprehensive overview of the conventional fully static master slave flip-flops used in low power VLSI systems where power budget is critical. In addition, the chapter also presents alternative realization of fully static master-slave flip-flops utilizing a modified feedback strategy. The flip-flops designed on the basis of modified architecture have been explained in detail and compared with state-of-the-art master slave flip-flop designs available in the literature. Extensive capacitance calculations have been performed in terms of clock load and capacitance at internal nodes has also been estimated for all the flip-flop configurations. This is executed in order to compare their relative power and delay characteristics which are well supported by simulation results.

INTRODUCTION

Major technological advancements in low power domain during the last decade have been associated with phenomenal demand for portable electronic devices. Flip-flops (FFs) and latches are considered as indispensable components for design of synchronous digital VLSI systems. The maximum operating speed of clocked digital systems is determined by the FFs. From a low power design perspective, nearly 30 - 70% of the total chip power dissipation is due to FFs and the clock distribution network (Yeap, 1998; Kawaguchi & Sakurai, 1998). Reducing the clock power of FFs is thus a serious bottleneck for digital IC

DOI: 10.4018/978-1-5225-0190-9.ch007

designers. Various attempts have been reported in the past to reduce the clock load and associated power dissipation in the clocking network (Oklobdzija, Stojanovic, Markovic, & Nedovic, 2003; Giacomotto, Nedovic, & Oklobdizija, 2007; Mesgarzadeh, Hanson, & Alvandpour, 2007). Moreover, a design with a high transistor count occupies larger area on chip and leads to an increase in the overall manufacturing cost. Hence, it has always been a challenge to explore more efficient FF designs in terms of enhanced speed, reduced power dissipation and minimum transistor count.

FFs are broadly classified into the following four major categories depending on the type of application viz:

- Master-slave FFs
- Pulse triggered FFs (implicit pulse triggered and explicit pulse triggered)
- Differential FFs
- Dual-edge triggered FFs (Alioto, Consoli, & Palumbo, 2015; Alioto, Consoli, & Palumbo, 2011b).

Master-slave FFs are generally utilized for low power systems whereas pulse triggered FFs find their use in high speed applications. Differential FFs are most suitable for deeply pipelined systems where the FF must supply complementary output signals to the subsequent logic. As a consequence, both the outputs need to be generated at the same time to improve performance by equalizing worst case delays (Myjak, Frias, & Jeon, 2006; Nikolic et al., 2000). On the other hand, a double edge triggered flip-flop (DETFF) is useful for maintaining a constant throughput while operating at only half the clock frequency.

Master-slave FFs are further classified into non clock-gated FF structures and clock-gated topologies as shown in Figure 1. Amongst the non-clock gated FFs, Transmission gate flip-flop (TGFF) represents the most efficient design in terms of power-delay product (PDP) (Consoli, Palumbo, Rabaey, & Alioto, 2014; Alioto, Consoli, & Palumbo, 2010) while modified C^2MOS based flip-flop (mC^2MOSff) (Chao & Johnston, 1989) is considered to be area efficient. Write port master-slave FF (WPMS) (Markovic, Tschanz, & De, 2003) and pass transistor logic style FF (PTLFF) (Hossain, Wronski, & Albicki, 1994) show degraded power-delay tradeoff characteristics. This is due to adoption of NMOS pass transistors in the critical path and presence of partially non-gated keepers. Gated master-slave latch (GMSL) (Strollo, Napoli, & Caro, 2001) and Data-transition lookahead FF (DTLA) (Nogawa & Ohtomo, 1998) represent the FFs belonging to the clock-gated class. However, the obvious limitation of clock gated FFs in terms of enhanced delays and increased area overhead, restricts their use to applications where power budget is critical and input switching probability is sufficiently low (Strollo & Caro, 2000). In this chapter, an alternative design approach for both types of master-slave FFs viz. TGFF and mC^2MOSff has also been introduced based on a new architecture with reduced transistor count while simultaneous improvements are obtained in power-delay-area product (PDAP). The proposed configurations modified transmission gate FF (MTGFF), hybrid master-slave FF (HMSFF), mC^2MOSff1 and mC^2MOSff2 fall under the non-clock gated FF category.

This chapter is organized as follows. Section 1.2 explains the basic building blocks of sequential digital systems. Section 1.3 differentiates between dynamic and static latches and FFs. Section 1.4 presents a comprehensive review of conventional FF design techniques and introduces a novel architecture for the design of master-slave FFs. Based on this architecture, four new FF circuits have been implemented viz. MTGFF, HMSFF, mC^2MOSff1 and mC^2MOSff2 (Singh, Tiwari, & Gupta, 2014). Section 1.5 describes the simulation parameters and test bench used for executing this work. Section 1.6 illustrates the simula-

Figure 1. Classification of master-slave flip-flops

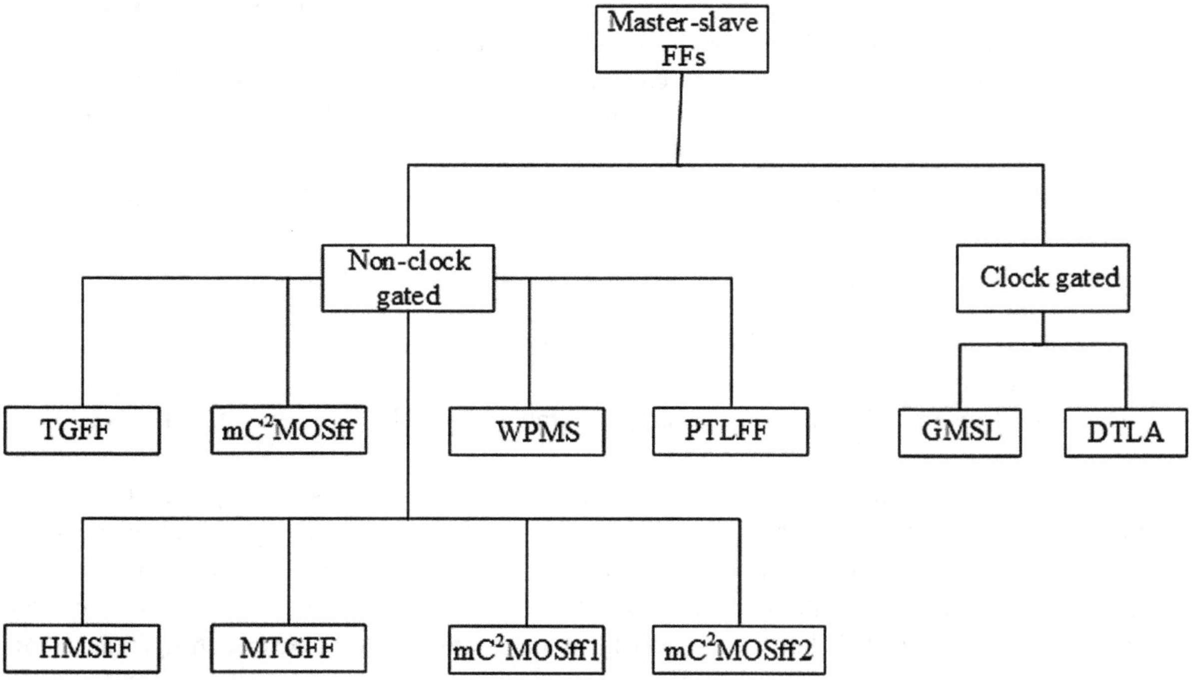

tion results and relates them mathematically with the expected theoretical behaviour based on logical effort (LE) theory. In addition, proposed configurations are also compared with state-of-the-art master-slave FFs in this section. The FFs selected for comparison are among the best-known in the literature in terms of power-delay tradeoffs. Finally, the conclusion is summarized in Section 1.7.

SYNCHRONIZATION CIRCUITS IN DIGITAL SYSTEMS

Level Sensitive Latches

Latches refer to the most basic synchronizing circuit in a sequential design. A latch is a level sensitive device that is transparent for one of the signal (clock) level while opaque for the other. A simple schematic of a positive level sensitive transmission-gate latch is shown in Figure 2 (a). When the clock signal 'CLK' is high, the latch lets the input 'D' pass to the output 'Q', while if the clock is low the output 'Q' will hold the previous input data on the output. A level-sensitive latch which is transparent for a high clock signal, samples the data on the falling clock-edge and stores it at the output during the low clock phase.

Edge Sensitive Flip-flops

An edge-triggered FF samples the data input on either the rising or falling edge of the clock, but in contrast to a level-sensitive latch, maintains the sampled data on the output during the remainder of the

Figure 2. (a) Level sensitive latch (b) Edge triggered flip-flop

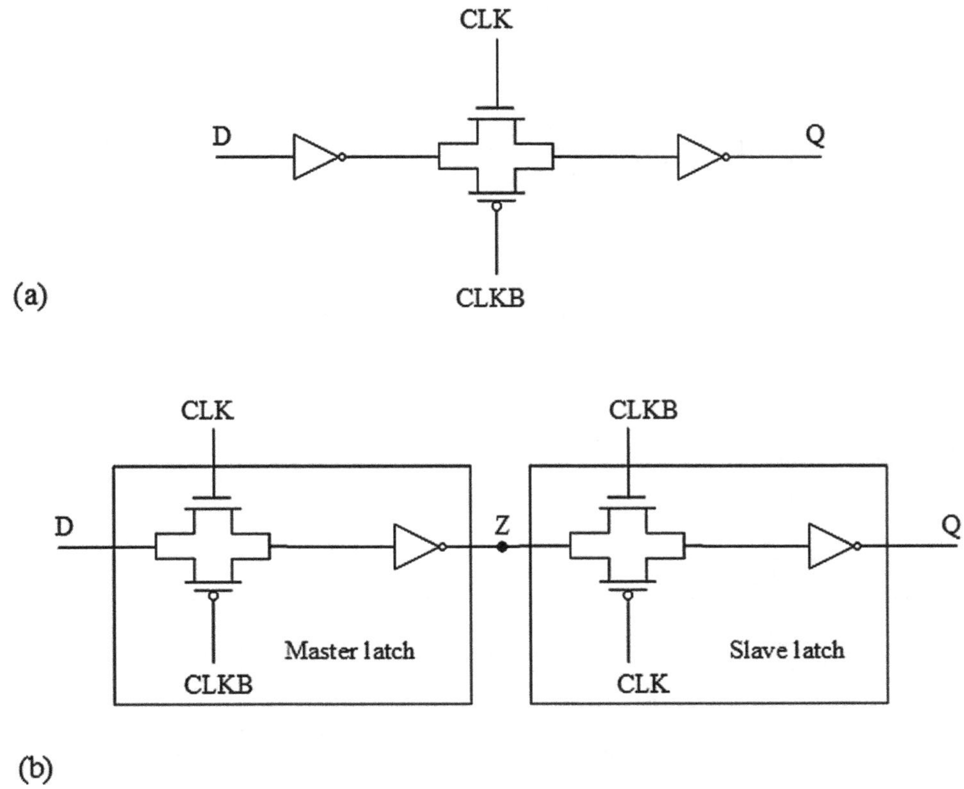

clock period. A simple master-slave FF is realized from two cascaded level-sensitive latches, as shown in Figure 2 (b). The master latch is transparent and the input is transferred to the intermediate node 'Z' when the clock signal 'CLK' is HIGH whereas the second latch called the slave latch is opaque so that the output is held at its previous logic state. When the clock signal makes a HIGH-to-LOW transition the master latch is insensitive to the changes in the input data while the intermediate data at 'Z' is transferred to the output 'Q'.

DYNAMIC AND STATIC LATCHES AND FLIP-FLOPS

Both the circuits illustrated in Figure 2 pertain to dynamic CMOS logic since the logic state at the output is retained only for the time period when clock is ON. As soon as the clock is OFF, the logic level at the output node is no longer reliable and may be lost due to subthreshold leakage at modern process nodes. To overcome this problem, regenerative looping action is provided by using inverters in the feedback paths as shown in Figure 3 to hold the logic levels correctly (Weste & Harris, 2011). Hence, the resulting circuit configurations become static in nature.

Figure 3. (a) Static latch circuit configuration (b) Static edge triggered flip-flop configuration

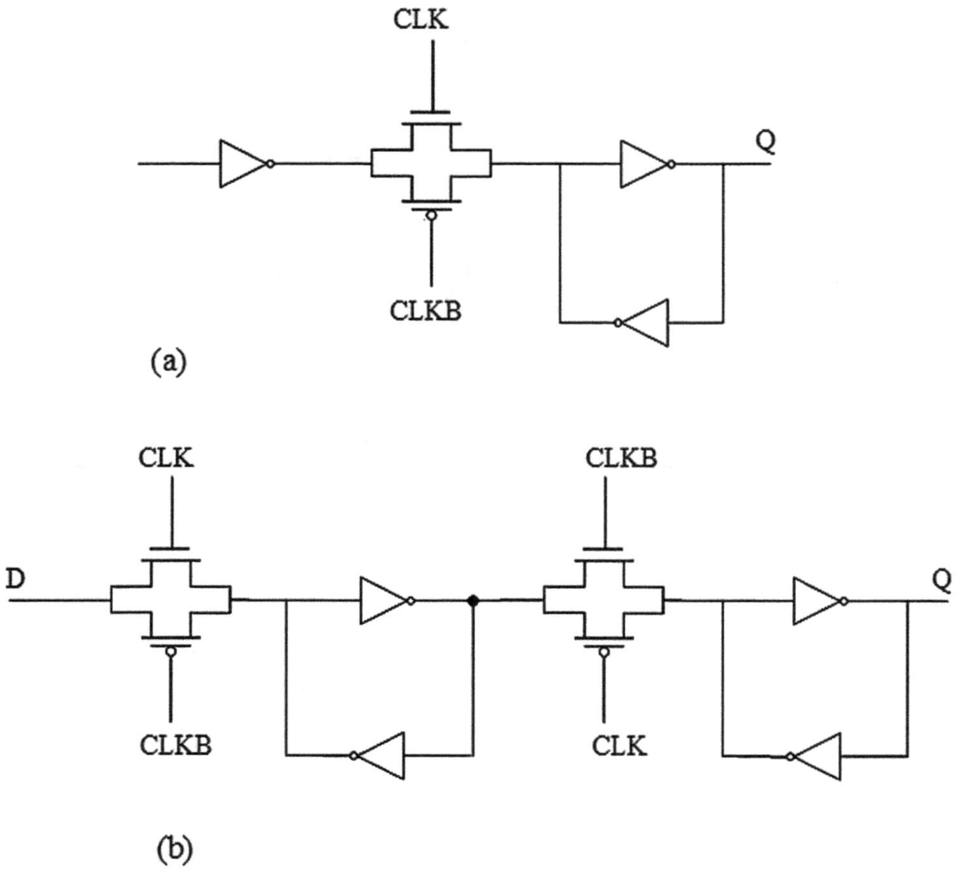

REVIEW OF STATE-OF-THE-ART FLIP-FLOPS AND PROPOSED DESIGNS

Conventional Static Flip-flop Design Techniques

In the conventional master-slave FF architecture as shown in Figure 4, complementary outputs Q and QB (Gerosa et al., 1994) are obtained by using two regenerative loops, (L1, L2 in Figure 4(a) and L3, L4 in Figure 4(b)) one each in the master and the slave sections to maintain a static functionality.

However, both the loops operate independently and become functional on complementary clock signals CLK and CLKB respectively. Regenerative loops generally require two cross coupled inverters. In the traditional design technique, for each loop, one inversion takes place in the forward (critical) path while the other (clocked) inversion takes place in the feedback path and there is no shared component between both the loops.

TGFF based Proposed Flip-Flop Architectures

In the proposed architecture, as reported in Figure 5, both the inversions occur in the forward (critical) path and the regenerative action is achieved through a clocked switch connected in the feedback path for

Figure 4. Conventional master-slave FF architecture with (a) clocked inverters in feedback paths, (b) clocked switches in feedback paths

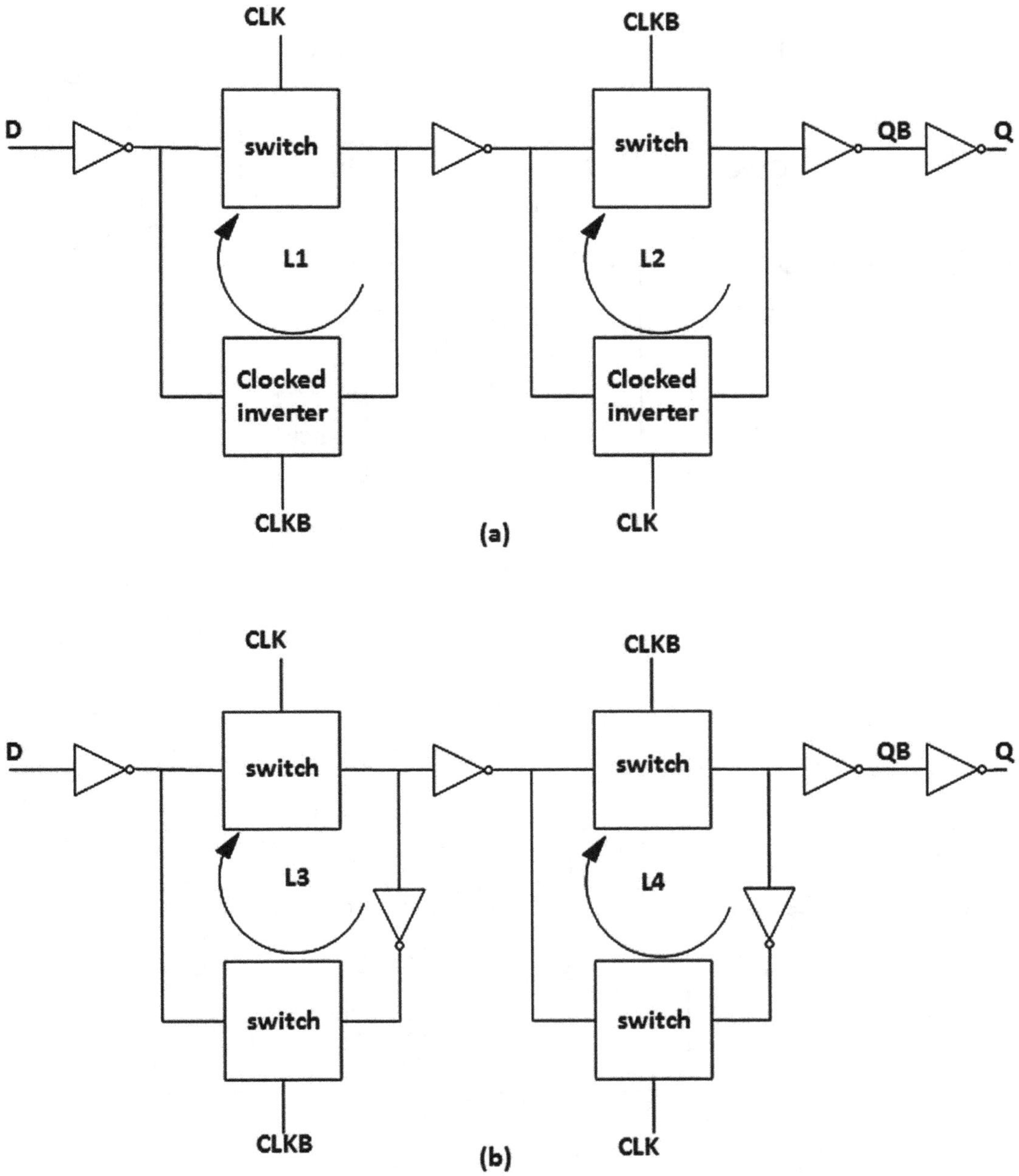

both the loops 'L5' and 'L6'. Note that the inverter INVZ is common to both the loops which is contradictory to the realization of conventional master-slave FF architectures as mentioned previously. Again, both the loops are active on complementary clock signals such that the output node is always driven and never floating thus ensuring a static functionality. The size of transistors in feedback paths marked by dots (.) is kept at minimum technology widths so as to resolve the races at nodes 'X' and 'Y' (Varma, 2004; Raychaudhuri, 2004). It is to be observed that Figure 5 (b) represents the actual circuit design

Figure 5. (a) Proposed architecture, (b) Modified transmission gate flip-flop (MTGFF)

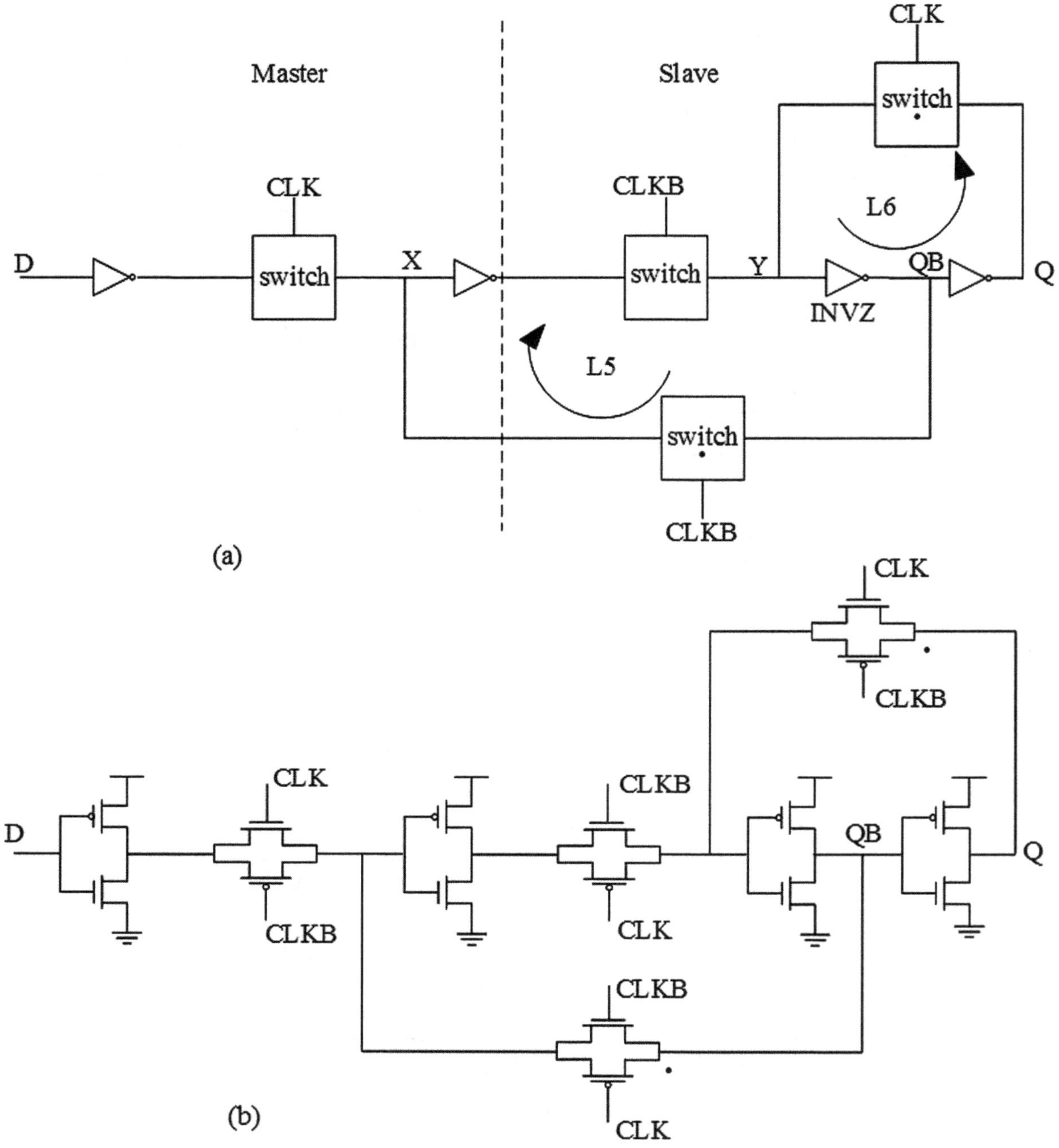

based on the proposed architecture termed as MTGFF, while TGFF (refer Figure 6) is implemented by utilizing transmission gates as switches in the conventional architecture as demonstrated in Figure 4(a).

Yet another variation in the proposed design, HMSFF, is illustrated in Figure 7 where transmission gates in the feedback paths are replaced by pass transistors which further reduces the transistor count to only fourteen. However, using NMOS pass transistors in the feedback path leads to increased short circuit power dissipation at nodes 'X' and 'Y' (refer Figure 5(a)) because of threshold voltage drop

Figure 6. Transmission gate flip-flop (TGFF)

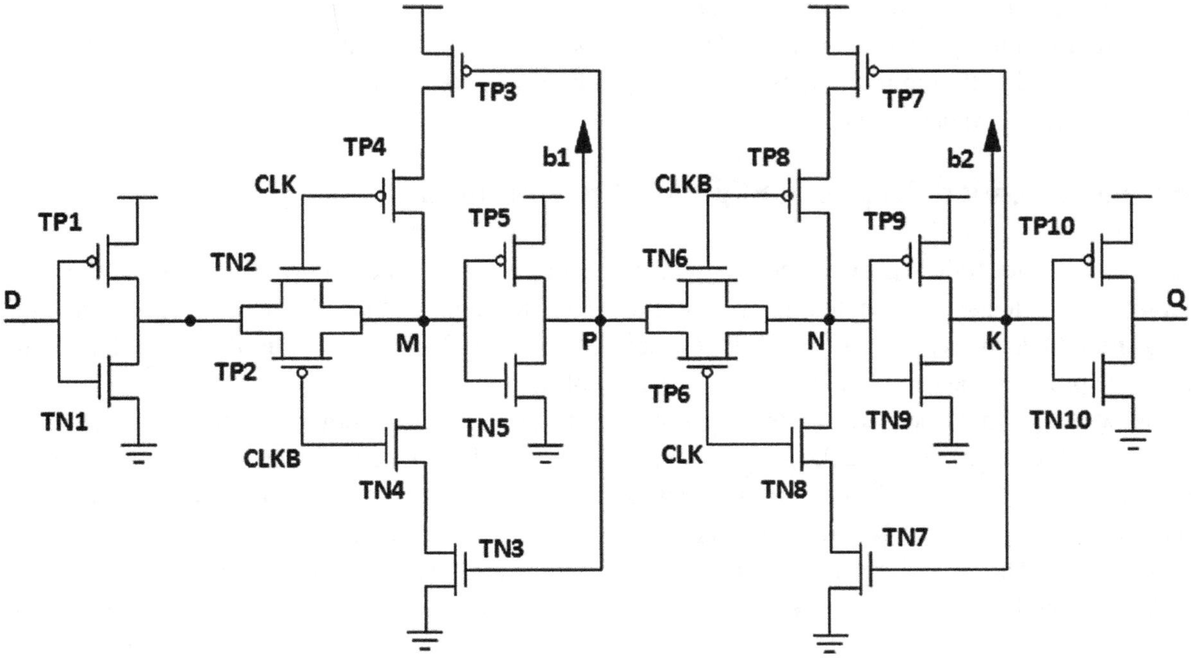

Figure 7. Hybrid master-slave flip-flop (HMSFF)

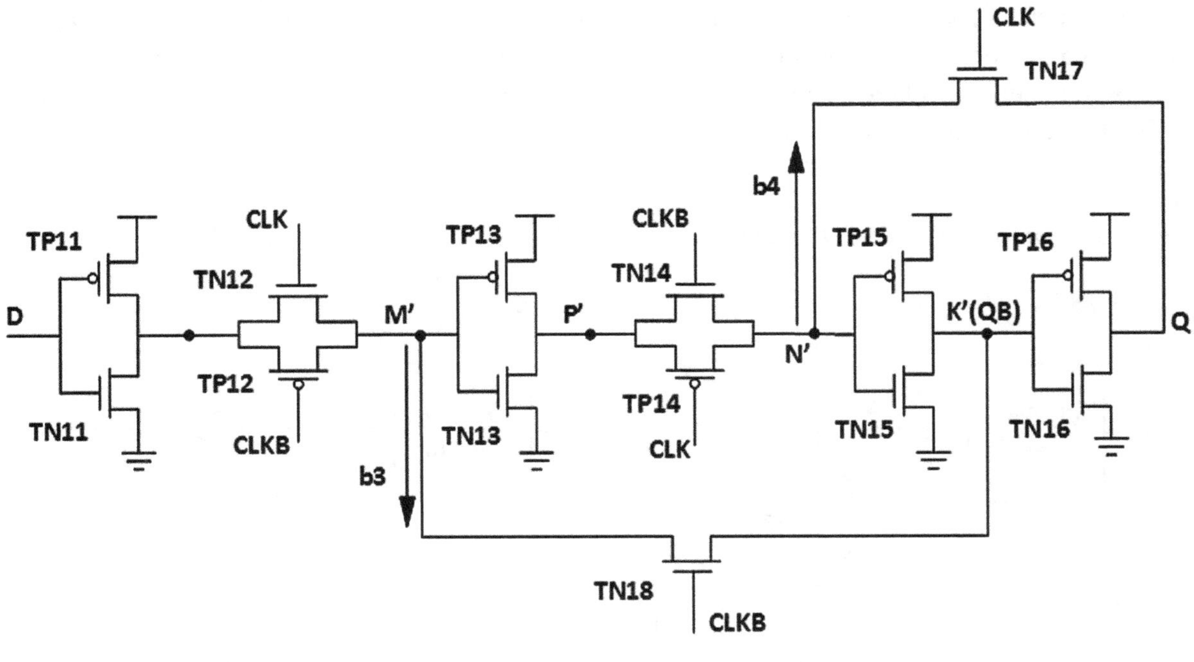

(V_{DD} - V_t) whenever logic HIGH is regenerated in loop 'L5' or 'L6'. On the contrary, the clock power is reduced due to lesser number of clocked transistors. Short circuit power is dependent on the ratio $v = V_t/V_{DD}$. For v = 0.3 or 0.2, short circuit power is around 2% or 10% of switching power respectively for sharp edges. In the nanometer processes, as the value of V_t is normally less than 0.3 V, therefore, the short circuit power dissipation is also non-significant (Weste & Harris, 2011). However, the overall effect is to reduce the total power dissipation of the circuit in comparison to TGFF as revealed by the simulation results in later sections. Interestingly, using the proposed architecture, a fully static DFF can be realized with minimum transistor count (twelve) if all the switches are replaced by NMOS pass transistors (refer Appendix 1).

mC^2MOSff based Proposed Flip-Flop Architectures

Since an inverter followed by transmission gate is equivalent to a clocked inverter, the combination is replaced by a clocked inverter to form a conventional mC^2MOS based FF architecture as shown in Figure 8 (Suzuki, Odagava, & Abe, 1973). Two regenerative loops 'L7' and 'L8' are used in a similar manner as in the previous case to maintain the static nature of the FF.

However, in the proposed architecture as reported in Figure 9 (a), both the inversions take place in the forward (critical) path and the loop is completed by a clocked switch for loop 'L10' while loop 'L9' is completed by using an inverter in the feedback path. Again, it is clearly noticed from Figure 9 (a) that the output node is always driven and never floating so that no compromise is made with the static functionality. The size of transistors in the feedback path marked by dots (.) is kept at 360 nm (minimum technology width) to eliminate race conditions at nodes 'U' and 'V'. Yet another implementation is shown in Figure 9 (b) which uses inverter INVX in the critical path and a clocked switch to form a regenerative loop 'L11'. It is to be noted that INVX is common to both the regenerative loops 'L11' and 'L12' which is contrary to the realization of previous clocked inverter based circuit designs.

Figure 10 shows the circuit diagrams for mC^2MOSff1 and mC^2MOSff2. It can be clearly observed that both mC^2MOSff1 and mC^2MOSff2 are realized using sixteen and fourteen transistors respectively. As a result, the area occupied by the proposed designs is significantly lesser than the conventional designs. Moreover, the number of clocked transistors in case of mC^2MOSff1 is six as compared to eight in case of TGFF or conventional mC^2MOSff.

Figure 8. Conventional mC^2MOSff architecture with clocked inverters in the critical path

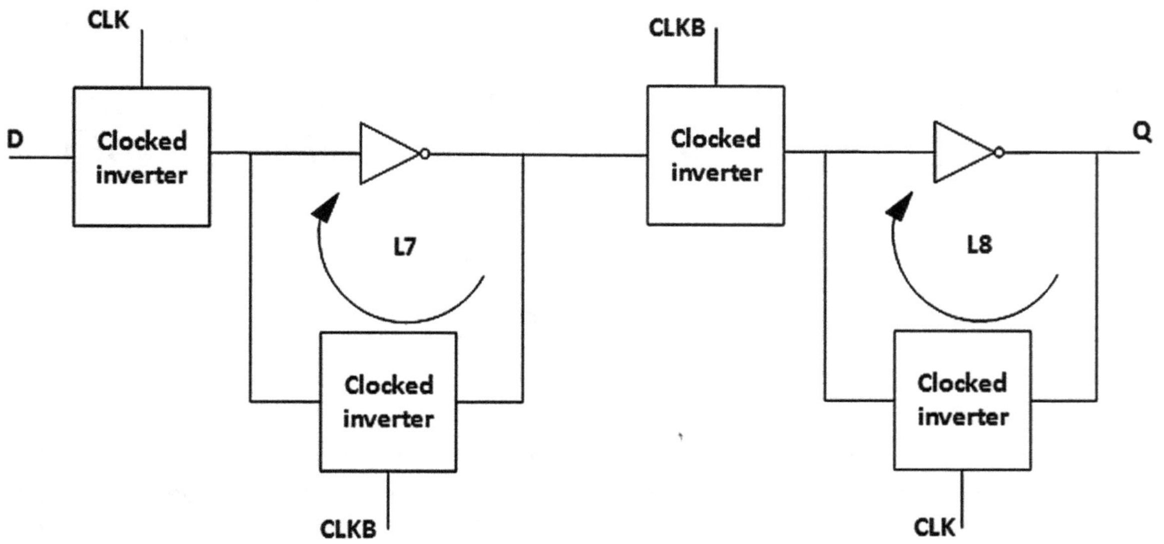

Figure 9. Proposed mC²MOSff architecture (a) and (b)

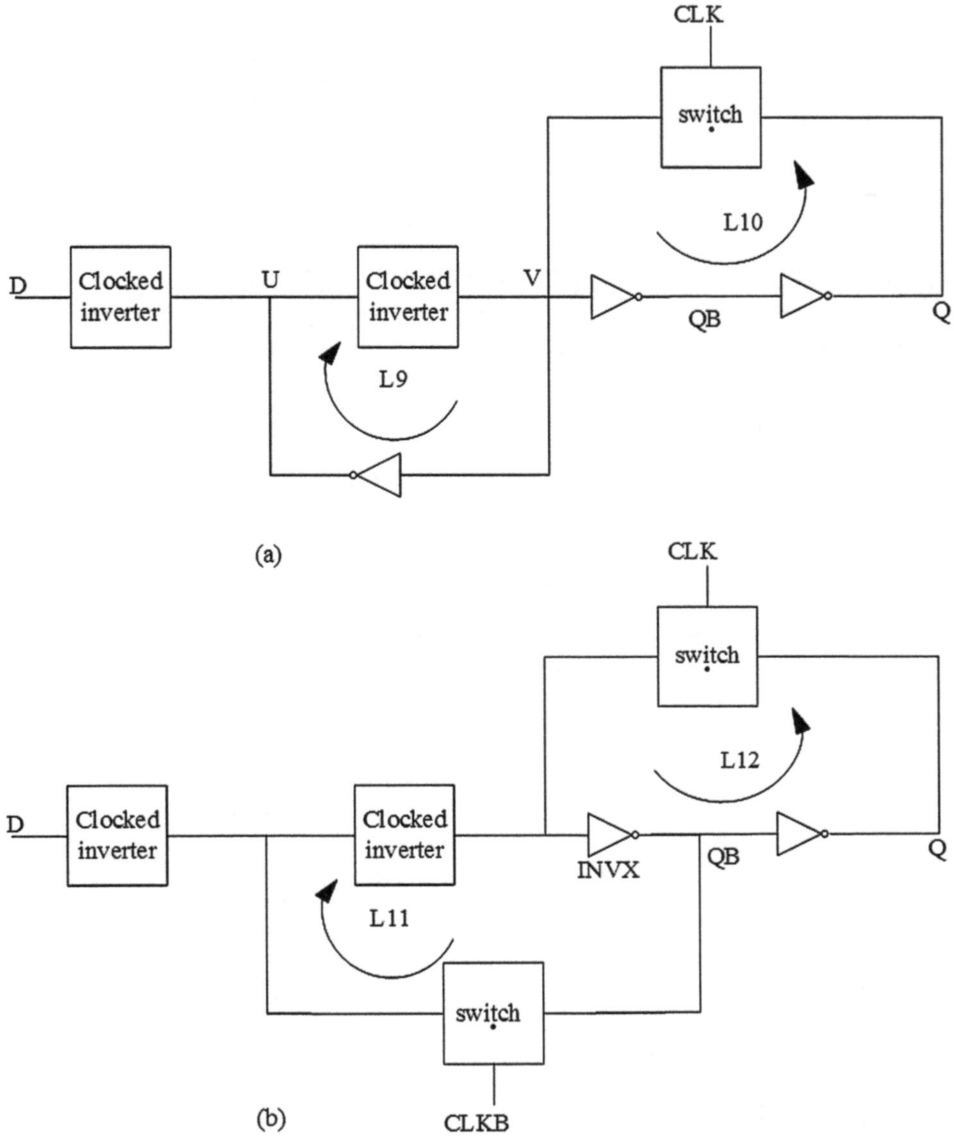

SIMULATION PARAMETERS, TEST BENCH AND METHODOLOGY

Table 1 lists the CMOS parameters used for simulation of the proposed and conventional FF designs. All the FFs have been designed using 180 nm CMOS process technology and a supply voltage of 1.8 V at 250 MHz clock frequency. The minimum technology width in case of feedback transistors has been fixed at 360 nm while the rising and falling transitions of data and clock signals are limited by a slope of 100 ps. To demonstrate the superiority of proposed designs, the performances of all the designs have been evaluated using post layout simulation results with parasitics extracted and backannotated in the circuit simulation environment. HSPICE has been used as the platform for schematic level simulations while layouts have been implemented using IC station (Mentor Graphics).

Figure 10. Proposed circuits (a) mC²MOSff1 and (b) mC²MOSff2

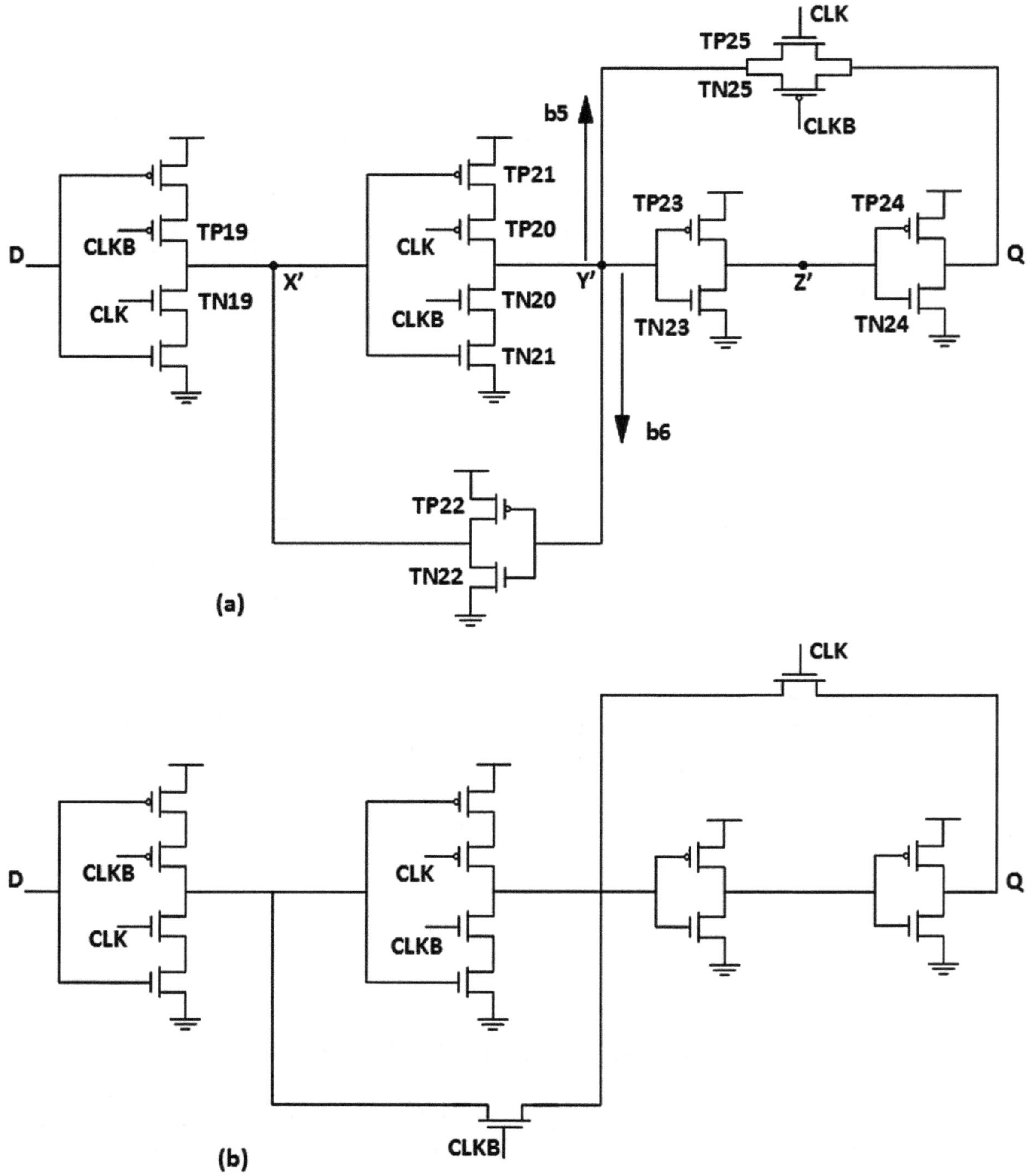

Table 1. Simulation parameters

Parameter	W_{min}	L_{min}	C_{min}	V_{DD}	Frequency	Signal slope
Value	360 nm	140 nm	1.24 fF	1.8 V	250 MHz	100 ps

The simulation setup used to characterize and compare the FF designs has been adopted from the previously available literature (Stojanovic & Oklobdzija, 1999). Data and clock buffers have been used to provide realistic clock and data signals. Data-to-output delay ($T_{DQ,min}$) has been used as the performance parameter. The delay sensitivity factor introduced by (Alioto, Consoli, & Palumbo, 2010) based on LE theory has been used for speed optimization.

A 16 clock cycle long pseudorandom data sequence with a switching activity $\alpha = 0.5$ is supplied at the data input for average power estimations (Stojanovic & Oklobdzija, 1999). Since the delay and power characterization is strongly dependent on the capacitive load offered to FFs (Heo & Asanovic, 2001), different loading conditions namely $4C_{min}$, $16C_{min}$ and $64C_{min}$ (C_{min} being the input capacitance of a symmetrical minimum sized inverter with $W_p = 2W_n = 2W_{min}$ have been used to test the FF response. Transistor sizing methodology adopted is the same as that in (Alioto, Consoli, & Palumbo, 2010). PDP and PDAP are the chosen figures of merit.

The relation between absolute gate capacitance (C_{GATE_fF}) in fF (femtofarads) and absolute transistor width (W) in nanometers (nm) has been derived at 180 nm process node by fitting simulation data (Weste & Harris, 2011; Palumbo & Pennisi, 2008) and is found to be:

$$C_{GATE\_fF} = (1.15\times10^{-9})\times W \tag{1}$$

LE method states that the optimized delay '*D*' of a path composed of '*N*' cascaded stages is:

$$D = N\sqrt[N]{GBH} + P \tag{2}$$

$$D = N\sqrt[N]{F} + P \tag{3}$$

where, G, B, H (= C_L/C_{in}), P, F (= GBH) and C_L are LE, branching effort along the critical path, electrical effort, parasitic delay, path effort and the applied load capacitance respectively.

$$D = P(1+t) \tag{4}$$

From Equations (2) and (4)

$$t = \frac{N\sqrt[N]{GB}\sqrt[N]{C_L}}{P\sqrt[N]{C_{in}}} \tag{5}$$

where 't' is the relative delay increment with respect to parasitic delay. Equations (4) and (5) indicate that increasing C_{in} to larger values leads to a reduction in the optimized delay but increasing it beyond a particular value has no significant effect on the FF latency. At this point, delay for the circuit is considered to be saturated. Based on the above analysis, the delay sensitivity factor introduced by Alioto, Consoli, & Palumbo (2010) is used to obtain the upper bound on transistor widths to explore the power-delay design space with minimal computational effort, and is expressed as:

$$S_D^{C_{in}} = \frac{\partial D}{\partial C_{in}} \frac{C_{in}}{D} = -\frac{1}{N} \frac{t}{t+1} \tag{6}$$

where, $S_D^{C_{in}}$ is the delay sensitivity factor and can be obtained using Equations (3 - 5).

Optimization is performed by varying input capacitance C_{in} of the FF using LE theory. For fixed values of C_L and C_{in}, LE theory optimizes the delay in the critical path. Since the transistors in the critical path affect the FF speed hence these widths are determined for different values of C_{in} to optimize the FF for minimum delay using Equation (1). The upper bounds on the normalized transistor widths wi (representing absolute transistor widths Wi normalized with respect to W_{min} are obtained such that the delay sensitivity tends to converge towards minimum value of S_{min} which indicates the delay saturation with respect to a particular FF topology. The S_{min} value for our analysis is selected to be -5%.

The input capacitance C_{in} (absolute value expressed in femtofarads) of the FF is determined from normalized width w1 of transistors in the first stage of the D-Q path as shown below:

$$C_{in} = (w1 \times 360 + 2 \times w1 \times 360)(1.15 \times 10^{-3}) \tag{7}$$

Figure 11 shows the conventional TGFF design. The size of transistors in the critical path are assumed to be independent design variables (IDVs) which are optimized for maximum performance using LE theory in congruence with the technique used in (Alioto, Consoli, & Palumbo, 2010). An inverter is added in the first stage before transmission gate in the critical path to protect the input terminal from noise variations. This approach is adopted for all the designs hence maintaining a fair level of comparisons (Consoli, Palumbo, & Pennisi, 2012).

Figure 11. Transistor sizing methodology for TGFF using LE theory

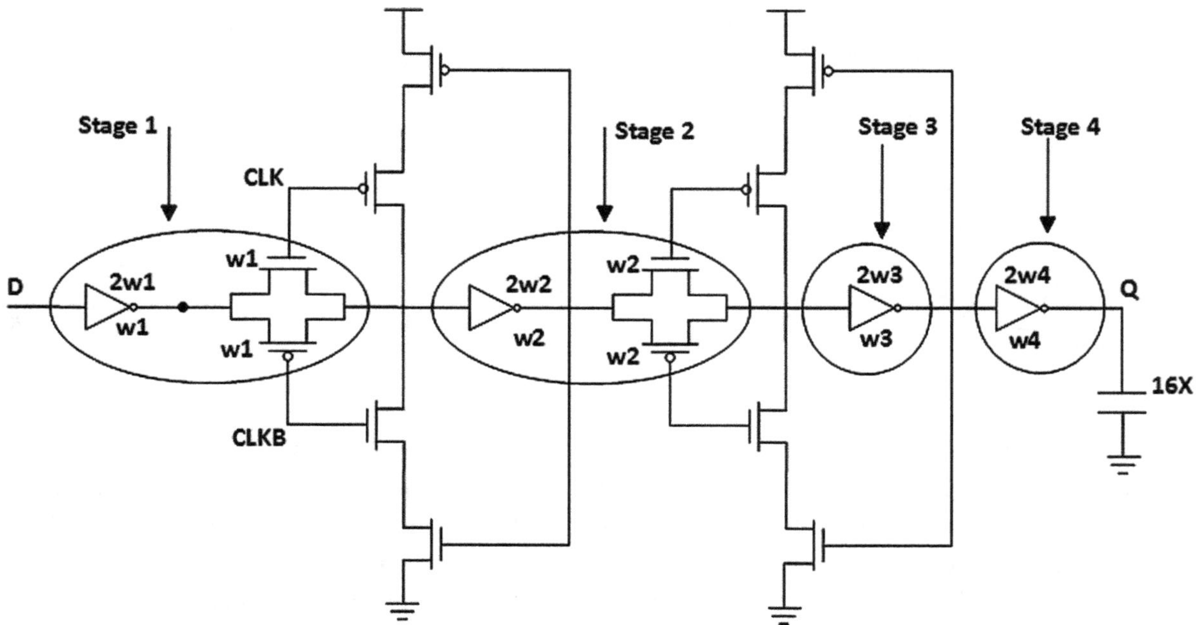

Table 2 shows the variation of delay for different values of C_{in}. It is to be noted that the delay saturates at 153 ps for C_{in} = 24.8 fF. This leads to determination of the upper bound on transistor widths early in the design cycle and hence defines the limits of power (energy)-delay design space (Alioto, Consoli, & Palumbo, 2012). The table also lists the corresponding power measured along with the PDP. It is readily observed that minimum PDP is obtained at C_{in} = 9.92 fF. The technology parameters listed in Table 3 are used for capacitance calculations at 180 nm CMOS process obtained from PTM (Predictive Technology Model) model files (Cao, Sato, Sylvester, Orshansky, & Hu, 2000).

In Table 3, C_{gdo} is the gate-to-drain overlap capacitance per unit transistor width, C_{gso} is the gate-to-source overlap capacitance per unit transistor width, C_{jsw} is the junction capacitance per unit parameter and C_j is the junction capacitance per unit area.

The method for estimation of C_{sb} (source-body diffusion capacitance), C_{db} (drain-body diffusion capacitance), C_{gd} (gate-to-drain overlap capacitance), C_{gs} (gate-to-source overlap capacitance) and C_g (total gate capacitance) is elaborated next.

The following relations have been used to determine the absolute value of capacitances throughout the chapter:

$$C_g = C_{gb} + C_{gd} + C_{gs}$$

$$C_g = \mathrm{W} \times \mathrm{L}_{\mathrm{eff}} \times \mathrm{C}_{\mathrm{ox}} + \mathrm{W} \times \mathrm{C}_{\mathrm{gdo}} + \mathrm{W} \times \mathrm{C}_{\mathrm{gso}}$$

$$C_{db} = C_j \times AD + C_{jsw} \times PD$$

Table 2. Conventional TGFF at 19.92 fF load (16X)

C_{in} (fF)	w1	w2	w3	w4	$T_{DQ,min}$ (ps)	Power (uW)	PDP (fJ)
2.48	2	2.35	2.79	6.65	226	554	125.2
4.96	4	3.95	3.95	7.91	191	585	111.7
7.44	6	5.35	4.84	8.76	173	599	103.6
9.92	8	6.65	5.59	9.41	166	615	102
12.4	10	7.86	6.25	9.95	162	632	102.3
14.8	12	9.01	6.85	10.4	159	648	103
17.3	14	10.1	7.40	10.8	157	665	104.4
19.8	16	11.1	7.91	11.2	155	675	104.6
22.3	18	12.2	8.39	11.5	154	682	105
24.8	20	13.2	8.84	11.8	153	689	105.4

Table 3. Technology parameters used for capacitance calculations

Parameter	C_{gdo} (F/m)	C_{gso} (F/m)	C_{jsw} (F/m)	C_j (F/m²)
NMOS	2.78E-10	2.78E-10	7.9E-10	0.00365
PMOS	2.78E-10	2.78E-10	1.44E-9	0.00138

$$C_{sb} = C_j \times AS + C_{jsw} \times PS$$

where, L_{eff} is the effective channel length, W is transistor width, C_{ox} is oxide capacitance per unit area, AD is area of the drain, AS is area of the source, PD is parameter of the drain and PS is the parameter of the source.

SIMULATION RESULTS AND DISCUSSION

The proposed master-slave FF circuits viz. MTGFF, HMSFF, mC^2MOSff1 and mC^2MOSff2 have been simulated and compared using standard metrics with the benchmark master-slave FFs available in the literature namely TGFF, WPMS, PTLFF, GMSL and DTLA. Post layout simulation results with the extraction of parasitics verify the correct FF operation at 250 MHz clock frequency.

The variation in absolute data-to-output delays $T_{DQ,min}$ with input capacitance (C_{in}) for 16X (19.92 fF) capacitive load is illustrated in Figure 12. TGFF, MTGFF and HMSFF utilize transmission gates in

Figure 12. Data-to-output delay as a function of FF input capacitance

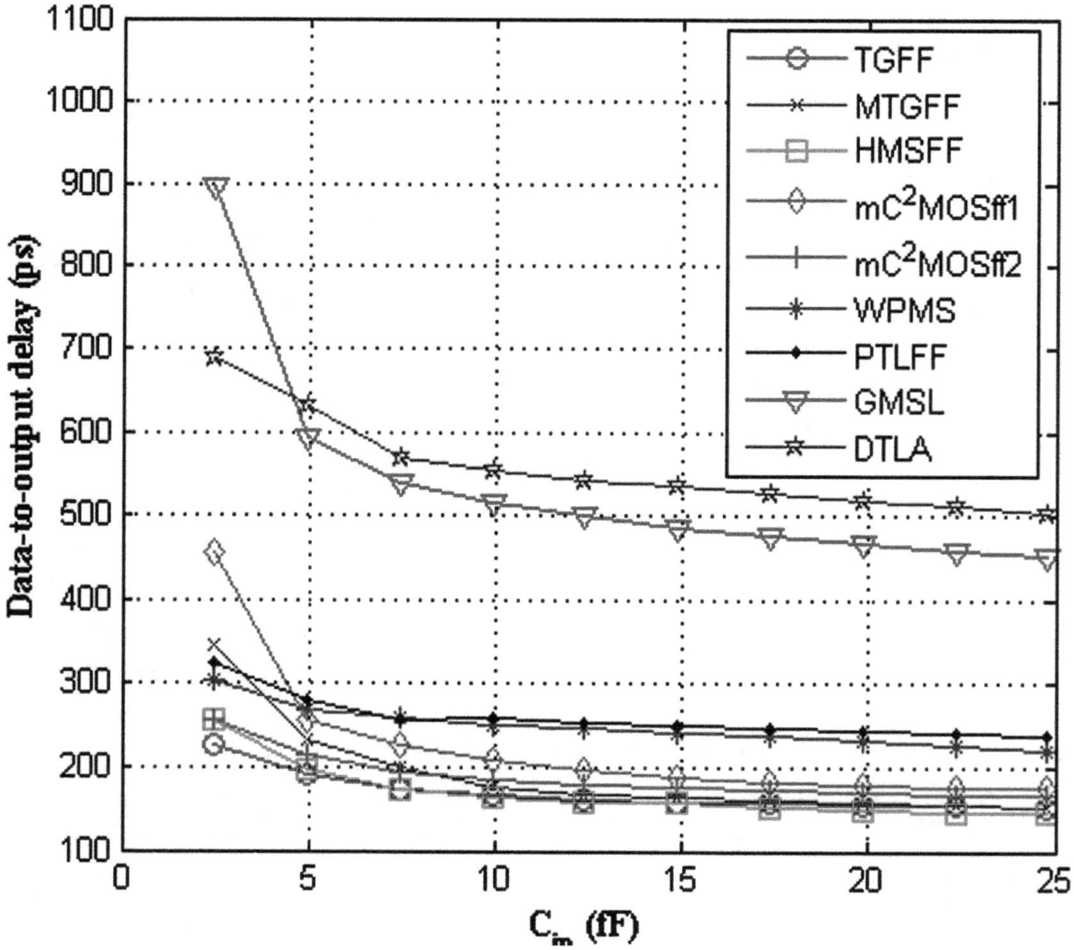

the critical path and are faster than their rival designs. The reason behind the degraded performance of WPMS and PTLFF designs is the presence of pass transistors in the critical path. The speed of clock-gated structures is worst among all the other topologies mainly due to the fact that gating circuit is inserted between the clock and FF terminals which deteriorates the timing characteristics. However, MTGFF and HMSFF are marginally faster than TGFF because of lesser branching effort as established by the following mathematical analysis based on LE theory. The calculations are performed assuming C_{in} = 12.4 fF and C_L = 19.92 fF (16X) where C_L represents the FF load capacitance.

There are exactly the same number of stages in the critical path of TGFF and mC^{2}MOSff1 with the only difference being that the latching circuit in case of TGFF is an inverter followed by a clocked transmission gate (inverting latch) whereas a clocked/tristate inverter is present in mC^{2}MOSff1. LE of both the latches is considered to be two, however it is apparent that an inverter followed by a transmission gate is faster because the output node is driven by both the transistors of the transmission gate in parallel and this behaviour is reflected in Figure 12 (data-to-output delay curves). From the above discussion, it is obvious that the value of LE for an inverting latch can be assumed to be two for most theoretical purposes, but as compared to a C^{2}MOS latch it must be slightly less than two if delays are to be modelled precisely.

It can be seen from Equation (2) that lesser branching effort leads to a faster circuit operation. Due to feedback paths, some of the drive current is directed along the path under analysis whereas rest of the drive current is directed off the path giving rise to internal fanout within the FF, accordingly the branching effort is:

$$b = \frac{C_{on-path} + C_{off-path}}{C_{on-path}} \tag{8}$$

where, $C_{on\text{-}path}$ is the load capacitance along the path under analysis, while $C_{off\text{-}path}$ is the capacitance of the connections that lead off the path.

The branching effort along the critical path is expressed as the product of individual branching efforts as shown below:

$$B = \prod b_i \tag{9}$$

There are two branches each in TGFF, HMSFF and mC^{2}MOSff1 represented as {b1; b2}, {b3; b4} and {b5; b6} in Figure 6, Figure 7 and Figure 10 respectively. The branching effort corresponding to branches b1, b2, b3, b4, b5 and b6 is calculated as follows:

Branching Effort in Case of TGFF

Calculation of b1:

$$C_{on\text{-}path} = C_{gd}(TN5) + C_{db}(TN5) + C_{gd}(TP5) + C_{db}(TP5) + C_{gd}(TN6) + C_{db}(TN6) + C_{gd}(TP6) + C_{db}(TP6) = 21.68 \text{ fF (refer Figure 6)} \tag{10}$$

$$C_{off\text{-}path} = C_g(TN3) + C_g(TP3) = 1.12 \text{ fF} \tag{11}$$

$b1 = 1.04$ (from Equation (8))

Calculation of b2:

$$C_{on\text{-}path} = C_{gd}(TN9) + C_{db}(TN9) + C_{gd}(TP9) + C_{db}(TP9) + C_{g}(TN10) + C_{g}(TP10) = 22.9 \text{ fF (refer Figure 6)} \quad (12)$$

$$C_{off\text{-}path} = C_{g}(TN7) + C_{g}(TP7) = 1.12 \text{ fF} \quad (13)$$

$b2 = 1.05$

$B = b1 \times b2 = 1.1$

Branching Effort in Case of HMSFF

Calculation of b3:

$$C_{on\text{-}path} = C_{gd}(TN12) + C_{db}(TN12) + C_{gd}(TP12) + C_{db}(TP12) + C_{g}(TN13) + C_{g}(TP13) = 19.67 \text{ fF (refer Figure 7)} \quad (14)$$

$$C_{off\text{-}path} = C_{gd}(TN18) + C_{db}(TN18) = 0.47 \text{ fF} \quad (15)$$

$b3 = 1.02$

Calculation of b4:

$$C_{on\text{-}path} = C_{gd}(TN14) + C_{db}(TN14) + C_{gd}(TP14) + C_{db}(TP14) + C_{g}(TN15) + C_{g}(TP15) = 15.55 \text{ fF (refer Figure 7)} \quad (16)$$

$$C_{off\text{-}path} = C_{gd}(TN17) + C_{db}(TN17) = 0.47 \text{ fF} \quad (17)$$

$b4 = 1.03$

$B = b3 \times b4 = 1.05$

Branching Effort in Case of mC²MOSff1

b5 calculation:

$$C_{on\text{-}path} = C_{gd}(TN20) + C_{db}(TN20) + C_{gd}(TP20) + C_{db}(TP20) + C_{g}(TN23) + C_{g}(TP23) = 21.01 \text{ fF (refer Figure 10)} \quad (18)$$

$$C_{off\text{-}path} = C_{gd}(TN25) + C_{db}(TN25) + C_{gd}(TP25) + C_{db}(TP25) = 1.47 \text{fF} \quad (19)$$

$b5 = 1.06$

b6 calculation:

$$C_{on\text{-}path} = C_{gd}(TN20) + C_{db}(TN20) + C_{gd}(TP20) + C_{db}(TP20) + C_{g}(TN23) + C_{g}(TP23) = 21.01\text{ fF}$$
(refer Figure 10) (20)

$$C_{off\text{-}path} = C_{g}(TN22) + C_{g}(TP22) = 0.828\text{ fF} \quad (21)$$

$b6 = 1.04$

$B = b5 \times b6 = 1.10$

where, C_{gd} is gate to drain capacitance; C_{db} is drain to body capacitance; and C_g is the gate capacitance of respective transistors.

It needs to be observed that TGFF, MTGFF and HMSFF comprise of exactly the same critical path consisting of four stages. Using Equation (2) and putting G = 4, B = 1.1, H = 19.92/12.4 = 1.60, N = 4 and P = 6 we have D = 12.51 (absolute delay 162.7 ps) for TGFF whereas putting G = 4, B = 1.05, H = 19.92/12.4 = 1.60, N = 4 and P = 6 for HMSFF we get, D = 12.44 (absolute delay 161.72ps). Absolute delay D_{abs} is obtained by multiplying parameter 'D' with parameter 'τ' as shown in the equation below:

$$D_{abs} = D \times \tau \quad (22)$$

In addition, putting G = 4, B = 1.10, H = 19.92/12.4 = 1.60, N = 4 and P = 6 for mC²MOSff1, D = 12.51 (absolute delay 162.7 ps) is obtained. It is clearly observed that the theoretical delay of mC²MOSff1 is nearly equal to the delay of TGFF. Now keeping other parameters to be same and assuming the LE of inverting latch to be 1.8, the updated value of TGFF is evaluated as D = 12.35 (absolute delay 160.55 ps).

The value of process dependent parameter 'τ' is determined as approximately 13 ps based on the calibration technique mentioned by (Sutherland, Sproull, & Harris, 2004) using the delay versus fanout curve (refer Appendix 2). The absolute values of the delay obtained through simulation are 162 ps for TGFF, 159 ps for HMSFF and 196 ps for mC²MOSff1 which are in close agreement with the theoretical values 165.1 ps, 161.72 ps and 165.1 ps respectively (typically within 15% error). The aforementioned delay approximations also hold good for MTGFF and mC²MOSff2.

A similar analysis performed at C_{in} = 24.8 fF and C_{in} = 4.98 fF (4X) shows that for smaller loads the relative difference in branching effort is even higher leading to about 21% improvement in $T_{DQ,min}$ whereas almost similar performance is exhibited by the FFs at 64X capacitive load which is comparatively larger. This behaviour can be understood by considering Equations (8 - 17) which clearly indicate that $C_{on\text{-}path}$ term in the numerator depends on width of transistors in the critical path whereas $C_{off\text{-}path}$ term which depends on the feedback structure, remains constant for a given FF configuration and its effect certainly becomes dominant whenever a FF is driving a relatively smaller load. This is mainly because smaller sized transistors if used in the critical path for driving smaller loads will further reduce the magnitude of $C_{on\text{-}path}$. Furthermore, smaller number of transistors in the feedback path lead to lesser interconnections and as an immediate consequence the delays due to these interconnects are also reduced.

Figure 13 depicts the variation of average power dissipation with C_{in} for 16X loading condition. WPMS and PTLFF display worst power dissipation characteristics mainly because of short circuit power dis-

Figure 13. Power dissipation as a function of FF input capacitance

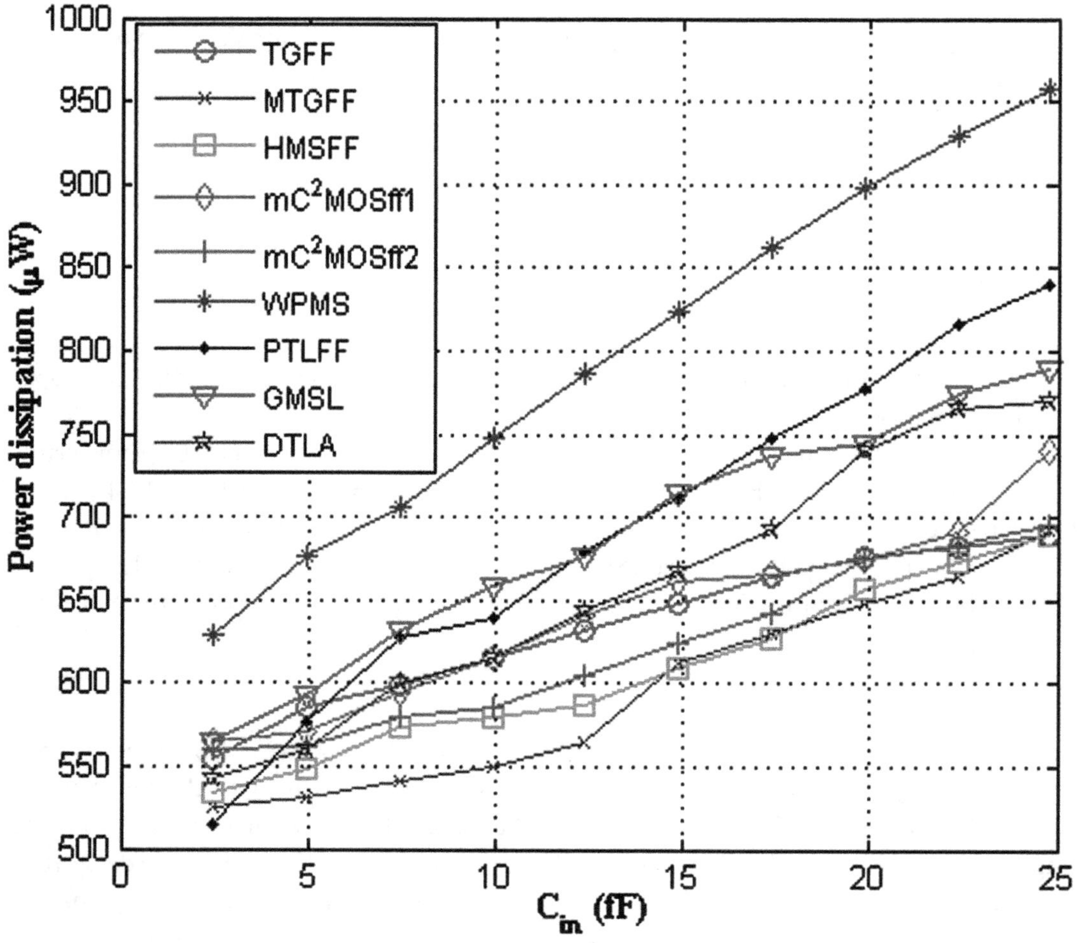

sipation at internal nodes of the FF in the critical path due to threshold voltage drop. Since the switching activity is typically high at $\alpha = 0.5$ for pseudorandom sequence, the power dissipation of GMSL and DTLA FFs is much higher than their non-clock-gated counterparts which also have transmission gates in the critical path viz. TGFF, MTGFF and HMSFF. MTGFF (refer Figure 5 (b)) represents a modified version of TGFF based on the proposed architecture, wherein the feedback transistors (TN3, TP3, TN7, TP7) connected in the feedback structure of TGFF (refer Figure 6) are entirely eliminated (preserving the static functionality of the FF) thereby leading to reduction in the internal node capacitances. Hence, MTGFF exhibits lesser dynamic power dissipation than TGFF even though clock load still remains the same because lesser capacitance has to be charged or discharged during every clock cycle. Since transmission gates in the feedback path of MTGFF have been replaced by pass transistors for implementing HMSFF, clock load decreases and the value of capacitance at internal nodes of the FF is further reduced leading to lesser dynamic power consumption. However, the short circuit power dissipation increases at these nodes due to threshold voltage drop but the overall effect is to reduce the total power dissipation.

Based on the simulations, PDP characteristics are derived for all the FFs at 16X loading condition as shown in Figure 14. The optimum PDPs of clock-gated structures, GMSL and DTLA are respectively 3.30× and 3.34× greater than optimum PDP of TGFF. Among the non-clock-gated structures, pass transistor logic based designs WPMS and PTLFF exhibit 1.77× and 1.57× enhancement in the PDP with respect to the benchmark flip-flop TGFF. However, both MTGFF and HMSFF show 7.7% and 8.8% improvement in the optimal PDP.

Moreover, to provide a detailed insight into the PDP of MTGFF and HMSFF with reference to TGFF, simulations have been carried out at varying capacitive loads and the corresponding results are listed in Table 4. It is worth noting that while MTGFF with 5.3% improvement represents a better alternative at larger loads, HMSFF is best suited for smaller loads with 21.1% improvement in the optimal PDP. It is also to be noted that both the mC^2MOS based FFs $mC^2MOSff1$ and $mC^2MOSff2$ are not efficient in terms of optimal PDPs but are more suited for high density applications since they perform better in terms of minimal PDAP as will be seen later.

Figure 14. Variation of PDP with input capacitance of FF at 16X load

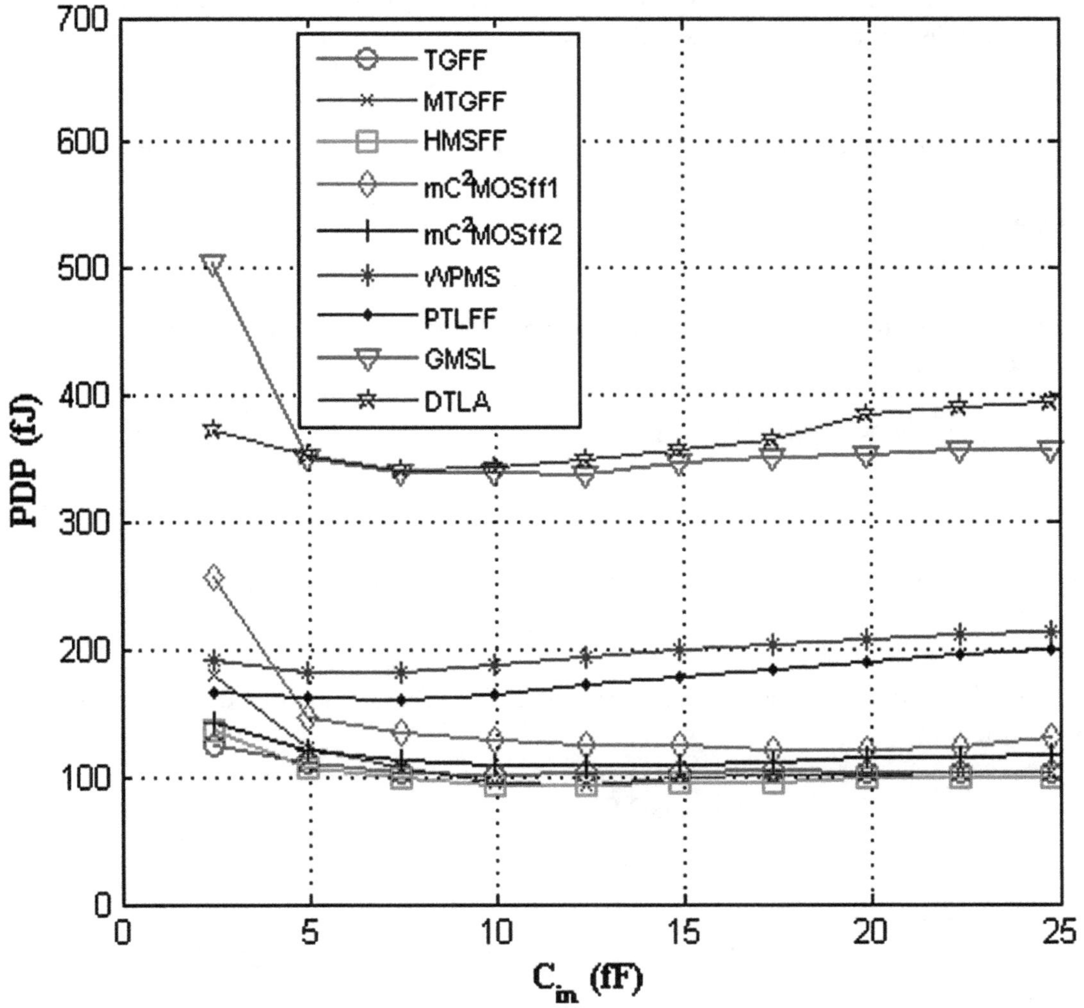

Table 4. Optimum PDPs of flip-flop structures at different output loads

	TGFF	MTGFF	HMSFF	mC²MOSff1	mC²MOSff2	WPMS	PTLFF	GMSL	DTLA
4X	85.2 fJ	85.7 fJ	67.2 fJ	115.7 fJ	100.4 fJ	161.2 fJ	142.6 fJ	271.3 fJ	277.6 fJ
16X	102 fJ	94.1 fJ	93 fJ	120.8 fJ	108.4 fJ	181.4 fJ	161.5 fJ	337 fJ	341.2 fJ
64X	130.1 fJ	123.2 fJ	129.8 fJ	147.6 fJ	135.6 fJ	257 fJ	206.4 fJ	409.2 fJ	421.1 fJ

Clock Load Calculations

The clock load estimation is done assuming C_{in} = 12.4 fF and C_L = 19.92 fF. The clock load is assumed to be composed of two components as observed from Equations (23) and (24). The first component is the capacitance offered by transistors in the critical path while the second component is the capacitance offered by transistors in the feedback structure. The capacitive component corresponding to feedback structure remains constant irrespective of the driving capabilities of transistors in the critical path because transistors in the feedback paths have minimum technology widths. Thus, indicating that the second component is independent of C_L and C_{in}. In contrast, the first component is directly dependent on the driving capabilities of transistors in the critical path.

TGFF:

$$[C_g(TN2) + C_g(TP2) + C_g(TN6) + C_g(TP6)] + [C_g(TN4) + C_g(TP4) + C_g(TN8) + C_g(TP8)]$$

(refer Figure 6)

[Clock load offered due to transistors in the critical path] + [Clock load offered due to transistors in the feedback structure]

$$C_{clock\_load\_TGFF} = 14.78\text{ fF} + 1.66\text{ fF} = 16.44\text{ fF} \tag{23}$$

HMSFF:

$$[C_g(TN12) + C_g(TP12) + C_g(TN14) + C_g(TP14)] + [C_g(TN17) + C_g(TN18)]$$ (refer Figure 7)

[Clock load offered due to transistors in the critical path] + [Clock load offered due to transistors in the feedback structure]

$$C_{clock\_load\_HMSFF} = 14.78\text{ fF} + 0.84\text{ fF} = 15.62\text{ fF} \tag{24}$$

mC²MOSff1:

$$[C_g(TN19) + C_g(\text{TP19}) + C_g(TN20) + C_g(TP20)] + [C_g(TN25) + C_g(TP25)]$$ (refer Figure 10)

[Clock load offered due to transistors in the critical path] + [Clock load offered due to transistors in the feedback structure]

$$C_{clock\_load\_mC^2MOSff1} = 22.18 \text{ fF} + 0.84 \text{ fF} = 23.02 \text{ fF} \quad (25)$$

Despite clock load reduction of a mere 5% in case of HMSFF as compared to TGFF, significant impact is created on clock power dissipation because the eliminated clocked feedback transistors are switched at highest frequency in the circuit which leads to decreased levels of total power consumption as will be shown later.

In particular, overall reduction in clock load using the above mentioned technique becomes much more significant at smaller loads. This is apparent from Equations (23) and (24) because the contribution of second component becomes even more pronounced as compared to the first component mainly due to the requirement of smaller sized transistors in the critical path. However, it is to be noticed that the clock load increases in case of mC^{2}MOSff1 and mC^{2}MOSff2 due to the fact that sizes of PMOS transistors in the clocked inverter are on the higher side as compared to the PMOS transistor sizes of transmission gates in the TGFF.

Calculation of Capacitances at the Internal Nodes of TGFF

Internal capacitance at nodes P, K, M and N:

Node P: $C_g(TN3) + C_g(TP3) + C_{gd}(TN5) + C_{db}(TN5) + C_{gd}(TP5) + C_{db}(TP5) + C_{gd}(TN6) + C_{db}(TN6) + C_{gd}(TP6) + C_{db}(TP6) = 22.53$ fF (refer Figure 6)

Node K: $C_g(TN7) + C_g(TP7) + C_{gd}(TN9) + C_{db}(TN9) + C_{gd}(TP9) + C_{db}(TP9) + C_g(TN10) + C_g(TP10) = 19.58$ fF

Node M: $C_{gd}(TN2) + C_{db}(TN2) + C_{gd}(TP2) + C_{db}(TP2) + C_{gd}(TN4) + C_{db}(TN4) + C_{gd}(TP4) + C_{db}(TP4) + C_g(TN5) + C_g(TP5) = 18.41$ fF

Node N: $C_{gd}(TN6) + C_{db}(TN6) + C_{gd}(TP6) + C_{db}(TP6) + C_{gd}(TN8) + C_{db}(TN8) + C_{gd}(TP8) + C_{db}(TP8) + C_g(TN9) + C_g(TP9) = 14.8$ fF

Calculation of Capacitances at the Internal Nodes of HMSFF

Internal capacitance at nodes P', K', M' and N':

Node P': $C_{gd}(TN13) + C_{db}(TN13) + C_{gd}(TP13) + C_{db}(TP13) + C_{gd}(TN14) + C_{db}(TN14) + C_{gd}(TP14) + C_{db}(TP14) = 21.7$ fF (refer Figure 7)

Node K': $C_g(TN16) + C_g(TP16) + C_{gd}(TN18) + C_{db}(TN18) + C_{gd}(TN15) + C_{db}(TN15) + C_{gd}(TP15) + C_{db}(TP15) = 19.23$ fF

Node M': $C_{gd}(TN12) + C_{db}(TN12) + C_{gd}(TP12) + C_{db}(TP12) + C_g(TN13) + C_g(TP13) + C_{gd}(TN18) + C_{db}(TN18) = 17.69$ fF

Node N': $C_{gd}(TN14) + C_{db}(TN14) + C_{gd}(TP14) + C_{db}(TP14) + C_g(TN15) + C_g(TP15) + C_{gd}(TN17) + C_{db}(TN17) = 14$ fF

Calculation of Capacitances at the Internal Nodes of mC²MOSff1

Internal capacitance at nodes X', Y' and Z':

Node X': $C_g(TN21) + C_g(TP21) + C_{gd}(TN19) + C_{db}(TN19) + C_{gd}(TP19) + C_{db}(TP19) + C_{gd}(TN22) + C_{db}(TN22) + C_{gd}(TP22) + C_{db}(TP22) = 28.24$ fF (refer Figure 10)

Node Y': $C_g(TN22) + C_g(TP22) + C_g(TN23) + C_g(TP23) + C_{gd}(TN25) + C_{db}(TP25) + C_{gd}(TN20) + C_{db}(TN20) + C_{gd}(TP20) + C_{db}(TP20) + C_{gd}(TN25) + C_{db}(TP25) = 23.31$ fF

Node Z': $C_g(TN24) + C_g(TP24) + C_{gd}(TN23) + C_{db}(TN23) + C_{gd}(TP23) + C_{db}(TP23) = 22.92$ fF

It can be easily concluded from calculations done above that a total of 2.7 fF capacitance has been reduced from the four internal nodes in the critical path of HMSFF in comparison to TGFF. This leads to reduced internal power dissipation at these nodes as lesser capacitance has to be charged or discharged per clock cycle. Again, the relative advantage of decreased internal node capacitance becomes predominantly high whenever HMSFF is driving smaller loads because width of transistors in the critical path is comparably small and hence result in relatively lower power dissipation. The overall effect due to reduced clock load as well as internal capacitance leads to about 12% improvement in power dissipation at $C_{in} = 2.48$ fF and $C_L = 4.96$ fF (4X).

It is also worth noting that a total of 19.34 fF capacitance has been reduced from the internal nodes in the critical path of mC²MOSff1 in comparison to TGFF. This leads to reduced internal power dissipation at these nodes as lesser capacitance has to be charged or discharged per clock cycle. However, reduction in the clock load of mC²MOSff1 due to transistors eliminated from the feedback structure is nullified due to PMOS transistors TP19 and TP20 whose size is twice that of transistors TP2 and TP6 in case of TGFF and as a result the total power dissipation of both the FFs is nearly the same as can be clearly observed from Figure 13.

Following a similar procedure as mentioned previously, the clock load of various FFs is obtained and listed in Table 5 along with number of clocked transistors and power consumption values. It is clear from Table 5 that MTGFF and HMSFF represent the most efficient designs in terms of power consumption having 10.8% and 7.1% lesser power dissipation than TGFF at $C_{in} = 12.4$ fF and $C_L = 19.92$ fF (16X).

TGFF also shows 20% improvement over mC²MOSff1 in terms of minimum PDP. However, despite the fact that TGFF represents a better alternative in terms of performance and optimum PDP, the area requirements also remain a major concern. It has been observed in the literature that conventional mC²MOSff is upto 20 - 25%more efficient in terms of occupied chip area. This stems mainly from the fact that at layout level (i) in comparison to TGFF, diffusion areas of most transistors can be shared in mC-

Table 5. Comparison of flip-flops at C_{in} = 12.4 fF and 16X load

Design (fF)	MTGFF	HMSFF	TGFF	mC²MOSff1	mC²MOSff2	WPMS	PTLFF	GMSL	DTLA
Transistor count	16	14	20	16	14	24	16	31	46
Number of clocked transistors	8	6	8	6	6	6	4	2	3
Clock-to-output delay, T_{CQ} (ps)	97	89	92	116	100	206	204	419	683
Optimum setup time (ps)	70	70	70	80	80	40	50	80	-140
Hold time (ps)	-19	-19	-19	-21	-21	-33	-32	-23	25
$T_{DQ,min}$ (ps)	167	159	162	196	180	246	254	499	543
Clock load (fF)	16.44	15.61	16.44	23.02	22.59	9.05	8.22	7.76	7.31
Total power dissipation (uW)	564	587	632	640	605	786	679	676	643
Leakage power (uW)	56.19	54.72	59.38	57.51	56.2	72.64	69.83	74.91	76.73
Layout area (um²)	165	153	175	125	122	228	134	272	316
PDAP [aJ.(um²]	15.54	14.28	17.91	15.67	13.28	44.08	23.1	91.75	110.33

The layouts were implemented using C_{in} = 12.4 fF, indicating almost similar transistor sizes throughout the critical path with the exception of TP19 and TP20 belonging to mC²MOSff1 which are twice in size compared to TP2 and TP6 in accordance with the LE theory.

²MOSff (Chao & Johnston, 1989) (ii) the number of contact holes can be reduced in the layout pattern (Suzuki, Odagava, & Abe, 1973) (iii) less complicated feedback structure leads to fewer interconnections.

It can be observed that MTGFF and HMSFF have the least transistor count while GMSL and DTLA use maximum number of transistors. Since only sixteen transistors are used for circuit realization of MTGFF, power dissipation is lower than TGFF. However, in case of HMSFF although transistor count is further reduced to fourteen and number of clocked transistors is reduced to six, power consumption is still greater than MTGFF mainly due to short circuit power dissipation at internal nodes since pass transistors are used in the feedback. It is noteworthy that clock load is minimum in case of both GMSL and DTLA, as a result these topologies exhibit minimum power dissipation at lower switching activities (as indicated by the simulation results at different switching activities later on). The reason for extended clock-to-output delays of GMSL and DTLA is the insertion of clock gating circuitry. DTLA has a pulsed operation and hence shows negative (positive) set-up (hold) time requirements. Leakage power at 180nm is mainly due to subthreshold conduction which is not a major issue and represents 5 - 10% of the total power consumption as confirmed by the simulation results. In particular, the leakage power of MTGFF and HMSFF are 5.3% and 7.8% less as compared to TGFF re-establishing the fact that leakage power is proportional to the area for any specific FF topology as mentioned in (Alioto, Consoli, & Palumbo, 2011a). Table 5 also lists area occupied by the layouts sketched for individual FFs. Due to a simplified feedback structure, MTGFF and HMSFF occupy 5.7% and 12.5% lesser area than TGFF because of reduced transistor count and interconnections and hence the PDAP of MTGFF, HMSFF is improved by 13.2% and 20.2% respectively.

Table 5 clearly shows that while TGFF is better in terms of PDP by 18.4%, mC²MOSff1 and mC-²MOSff2 show 12.4% and 25.8% improvement respectively in the PDAP making it suitable for high density applications where performance in terms of power and/or speed can be compromised.

The power dissipation results shown in Figure 15 are determined using C_{in} = 12.4 fF. As expected, clock-gated structures contribute towards maximum power savings at zero switching activity. GMSL shows 32.5% improvement while DTLA shows 46.3% reduction in power in case of input staying constant at logic HIGH whereas in case of logic LOW at the input, the power reduction is up to 19.2% and 35.4% respectively. Both MTGFF and HMSFF also show up to 17.6% improvement in power dissipation compared to TGFF at zero switching activity and input staying HIGH. At $\alpha = 0.25$ switching activity, MTGFF and HMSFF are most suitable since they consume about 9.5% and 7.2% lesser power. The trend continues even at higher switching probabilities where the power savings are up to 10.7% and 7.1% respectively at $\alpha = 0.5$. The layout of HMSFF is shown in Figure 16 for C_{in} = 12.4 fF and 16X capacitive load. Again, it can be clearly observed that there is only a slight difference in the power dissipation of TGFF and $mC^2MOSff1$ at different switching activities.

To address scalability issues at modern process nodes, the FFs were designed and simulated at 130 nm, 90 nm and 65 nm CMOS processes. The test bench and optimization methodology remain the same as mentioned earlier. Furthermore, to account for process, voltage and temperature (PVT) variations the simulations were carried out at all process corners viz. TT, FF, SS, FS, SF with voltages scaled from 0.9 - 1.1 V while the temperatures varied from 0 – 125 degrees for different process corners as shown in Table 6. Various simulation and technology parameters used for the simulation of FFs are also represented in Table 7, where, C_{GATE} represents the capacitance per unit gate oxide and was found to be 1.25 fF/um through simulations.

For illustration purposes, the delay and power variations with the FF input capacitance at 65 nm CMOS technology for HMSFF are demonstrated in Figure 17 and Figure 18 respectively at 16X loading condition. Both MTGFF and HMSFF showed correct circuital behaviour at the aforementioned process nodes which indicates that no internal noise violations exist especially due to the fact that logic levels are retained even at FF process corner. However, it is to be pointed out that HMSFF in a manner similar

Figure 15. Power dissipation at different switching activities

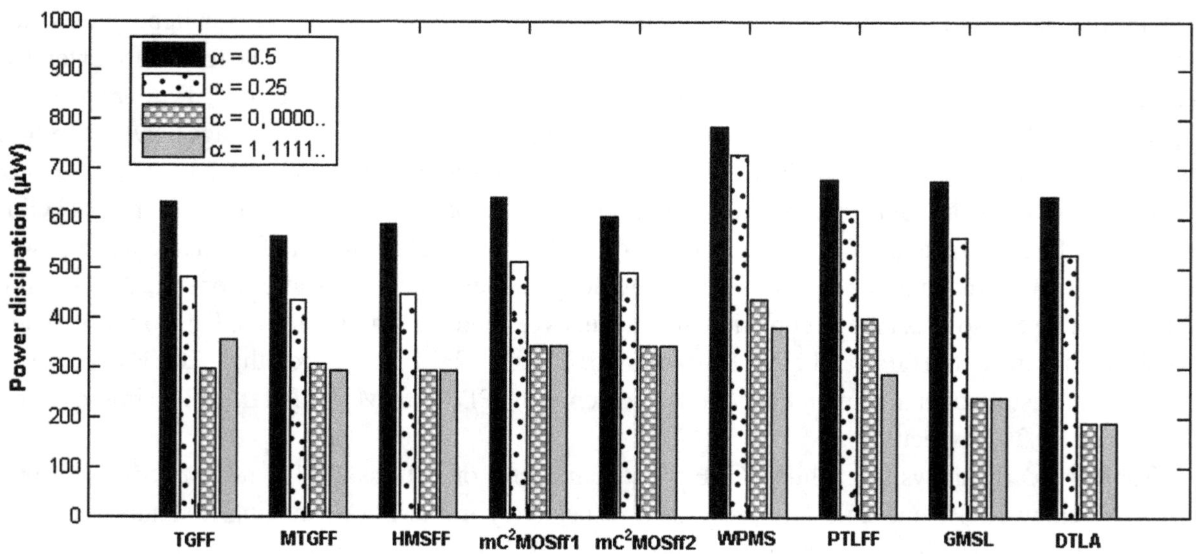

Figure 16. Implementation of HMSFF at layout level

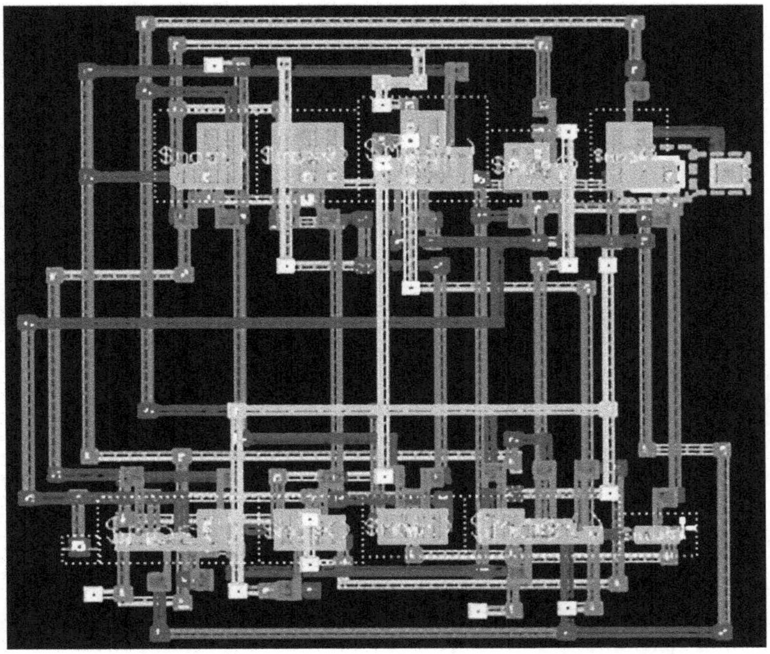

Table 6. Parameters for process corner simulation at 65 nm

Process corner	TT	FF	SS	FS	SF
Temperature (þC)	70	0	125	70	70
V_{DD}	1	1.1	0.9	1	1

Table 7. Parameters used for simulation of flip-flops at 65 nm

L_{min}	W_{min}	C_{min}	Frequency	Signal slope	C_{GATE}
60 nm	120 nm	507 aF	2 GHz	20 ps	1.25 fF/um

to TGFF starts to fail at SS corner for lower values of C_{in} (Dao, Nowka, & Oklobdizija, 2001). This is essentially due to the presence of transistors in the critical path with extremely weak driving strengths and lower supply voltage.

Implementation of 8-bit Asynchronous Counter

A mod-256 counter is designed by converting the TGFF (Data FF) into a Toggle FF using EXOR logic realized with transmission gates and inverter as shown in the first stage of the FF design in Figure 19.

Figure 17. Delay variations for HMSFF at 16X loading for different process corners

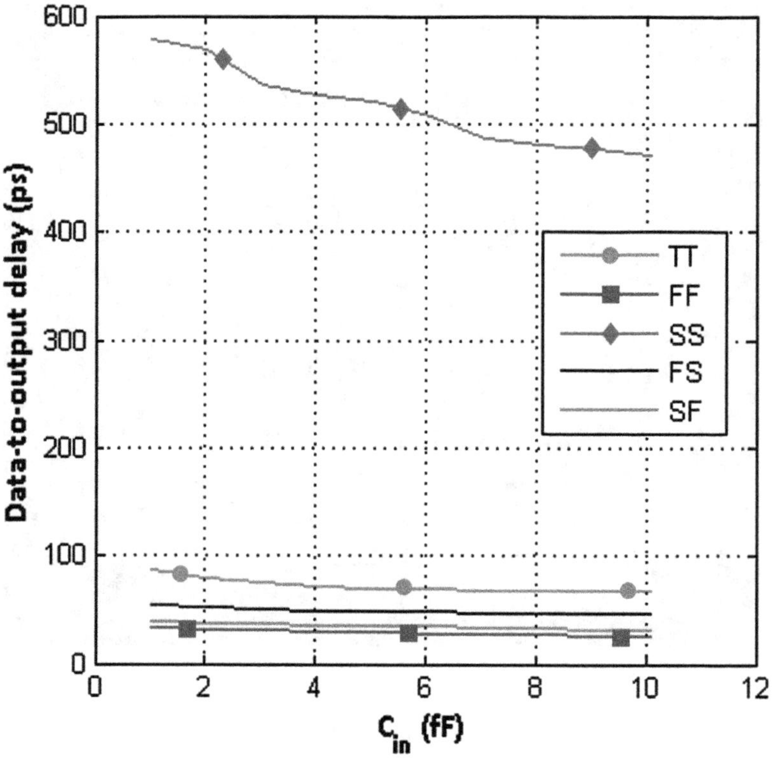

The sizing technique for a Toggle FF effectively consisting of five stages is also demonstrated in Figure 19 with the normalized transistor widths shown against each stage.

An 8-bit asynchronous counter has been implemented from an application point of view such that the output of each FF is connected to clock terminal of the subsequent FF by inserting two intermediate inverters acting as a buffer. As a result, the input gate capacitance of the first inverter in series acts as the load capacitance for the FF in the previous stage ensuring uniform loading of the FFs. The transistor sizes of inverters have been selected such that the input capacitance of the inverter is equivalent to 4.98 fF (4X). A similar technique is used to design 8-bit counters based on MTGFF, HMSFF, $mC^2MOSff1$ and $mC^2MOSff1$. Again, LE theory has been adopted for sizing individual FFs in each counter for optimum performance.

The correct functionality of the proposed FFs is further validated by designing the 8-bit ripple counter at 4X and 16X capacitive loads to layout level. The power dissipation was estimated through simulations over 256 clock cycles. When compared with an 8-bit counter realized using TGFF, MTGFF and HMSFF based counters show power savings up to 44.5% and 34.2% for 4X loading condition at 250 MHz whereas the reduction in power dissipation is recorded at 20.2% and 25.2% for $mC^2MOSff1$ and $mC^2MOSff2$ respectively at 250 MHz as indicated in Figure 20. In addition, it can also be observed that power savings are obtained on a consistent basis for capacitive loads up to 16X. The power dissipation of HMSFF based counter is higher in comparison to MTGFF based counter primarily because of short circuit power dissipation. The relatively lower reduction in power dissipation at elevated frequencies is

Figure 18. Power variations for HMSFF at 16X loading for different process corners

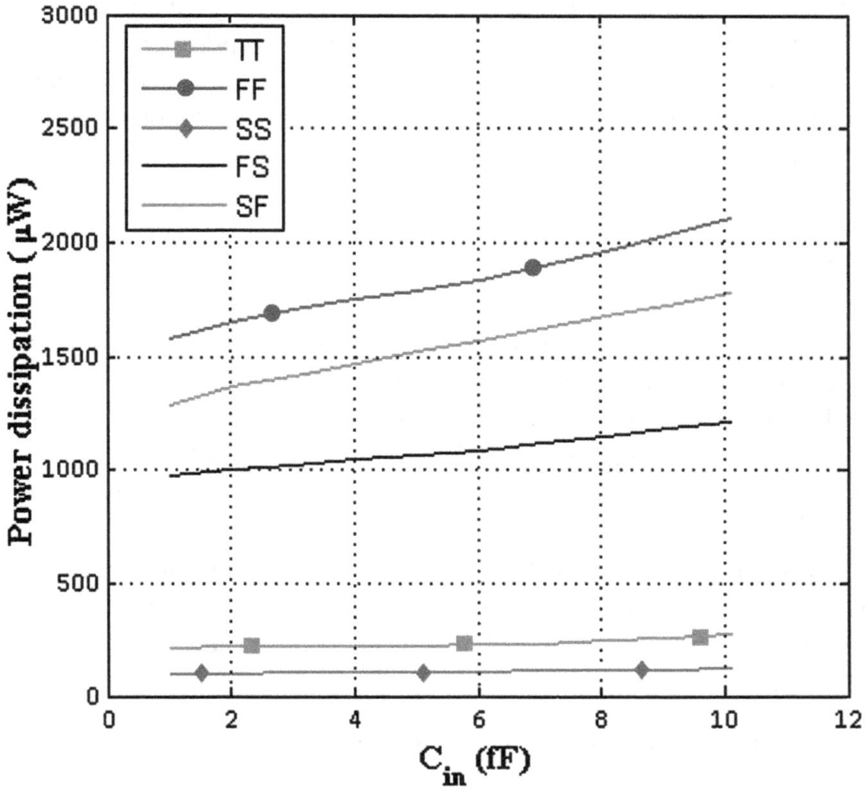

Figure 19. Toggle flip-flop realized using TGFF

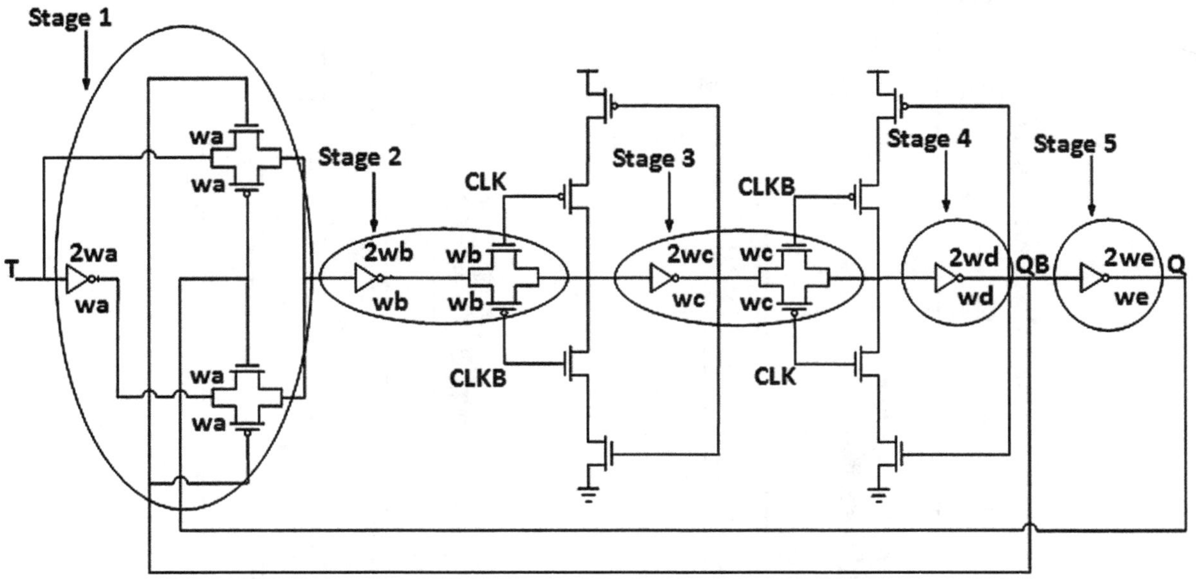

attributed to the fact that for the same transistor sizes and hence the same intrinsic capacitance, switching activity at internal nodes becomes the dominant factor for determining the average power consumption.

In a similar manner, the correct functionality of the proposed flip-flop mC²MOSff1 is validated by designing an 8-bit ripple counter at 16X capacitive load and the average power measurements were carried out over 256 clock cycles. It was noticed that the power consumption of the mC²MOSff1 based counter is comparable to the TGFF at varying frequencies.

CONCLUSION

In this chapter, an extensive simulation supported coverage of the most popular master slave flip-flop circuits is provided. Additionally, alternative architectures for the design of TGFF and mC²MOSff based master slave FFs have been presented with a modified feedback strategy while preserving the fully static operation. Using the new feedback approach, four modified topologies MTGFF, HMSFF, mC²MOSff1 and mC²MOSff2 were discussed which exhibit reduced clock line capacitance and decreased capacitance at internal nodes in comparison to the conventional master-slave FFs. It was demonstrated with proper reasoning that improved power and delay characteristics were obtained mainly at comparatively smaller output loads. Comparisons were carried out with state-of-the-art FFs in the master-slave class. The simulation results show appropriate correlation with mathematical analysis based on LE theory within

Figure 20. Power dissipation of counters at 4X and 16X capacitive loading

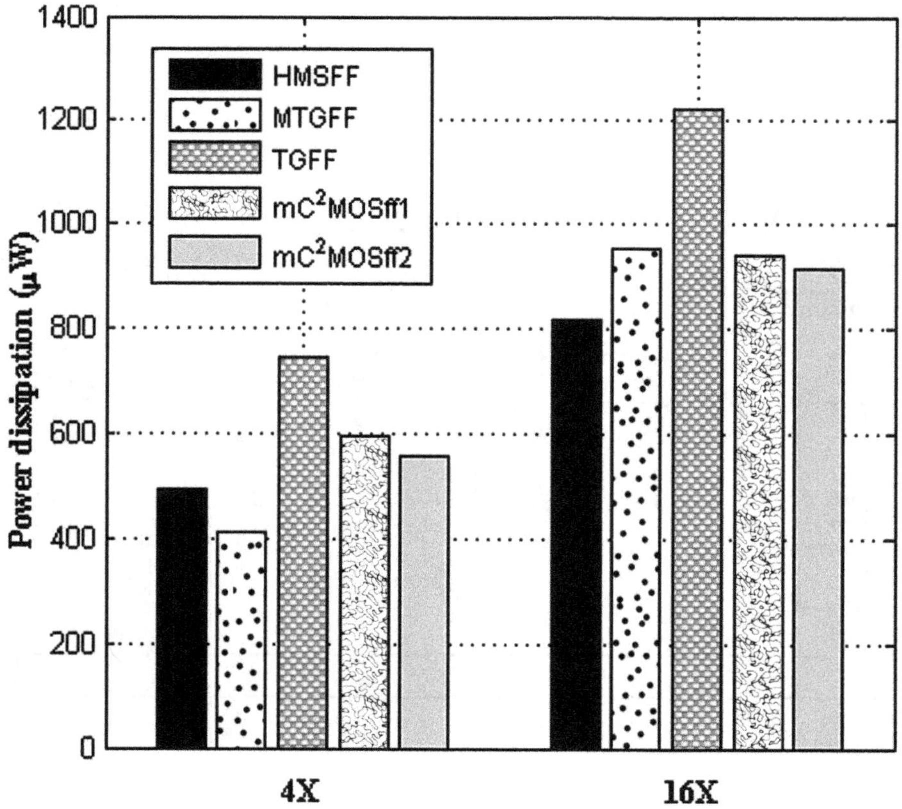

acceptable error (typically less than 15%). From an application point of view, an eight-bit counter was implemented using TGFF, MTGFF, HMSFF, mC^{2}MOSff1 and mC^{2}MOSff2. The power dissipation results showed that MTGFF and HMSFF based counters exhibit substantial power savings (up to 44.5%) on a consistent basis in comparison to TGFF based counter for a wide range of capacitive loads (4X-64X) and operating frequencies. Overall, MTGFF and HMSFF based designs represent the best choice in terms of PDP amongst the traditional master-slave FF configurations, whereas HMSFF and mC^{2}MOSff2 represent the most preferred alternatives for low power high performance and high density applications.

REFERENCES

Alioto, M., Consoli, E., & Palumbo, G. (2010). General strategies to design nanometer flip-flops in the energy-delay space. *IEEE Transactions on Circuits and Systems-Part I*, *57*(7), 1583–1596. doi:10.1109/TCSI.2009.2033538

Alioto, M., Consoli, E., & Palumbo, G. (2011a). Analysis and comparison in the energy- delay-area domain of nanometer CMOS flip-flops: Part I- Methodologies and design strategies. *IEEE Transactions on VLSI Systems*, *19*(5), 725–736. doi:10.1109/TVLSI.2010.2041376

Alioto, M., Consoli, E., & Palumbo, G. (2011b). Analysis and comparison in the energy-delay-area domain of nanometer CMOS flip-flops: Part II- Results and figures of merit. *IEEE Transactions on VLSI Systems*, *19*(5), 737–750. doi:10.1109/TVLSI.2010.2041377

Alioto, M., Consoli, E., & Palumbo, G. (2012). From energy-delay metrics to constraints on the design of digital circuits. *International Journal of Circuit Theory and Applications*, *40*(8), 815–834. doi:10.1002/cta.757

Alioto, M., Consoli, E., & Palumbo, G. (2015). Variations in Nanometer CMOS Flip-Flops: Part I—Impact of Process Variations on Timing. *IEEE Transactions on Circuits & Systems I*, *62*(8), 2035–2043. doi:10.1109/TCSI.2014.2366811

Cao, Y., Sato, T., Sylvester, D., Orshansky, M., & Hu, C. (2000). New paradigm of predictive MOSFET and interconnect modeling for early circuit design. In *Proceedings of IEEE Custom Integrated Circuits Conference* (pp. 201-204). IEEExplore Digital Library. Retrieved Sep. 24, 2013, from http://ptm.asu.edu/

Chao, J., & Johnston, C. A. (1989). Behaviour analysis of CMOS D flip-flops. *IEEE Journal of Solid-State Circuits*, *24*(5), 1454–1458. doi:10.1109/JSSC.1989.572637

Consoli, E., Palumbo, G., & Pennisi, M. (2012). Reconsidering high-speed design criteria for transmission-gate-based master-slave flip-flops. *IEEE Transactions on VLSI Systems*, *20*(2), 284–295. doi:10.1109/TVLSI.2010.2098426

Consoli, E., Palumbo, G., Rabaey, J. M., & Alioto, M. (2014). Novel class of energy-efficient very high speed conditional push-pull pulsed latches. *IEEE Transactions on VLSI Systems*, *22*(7), 1593–1605. doi:10.1109/TVLSI.2013.2276100

Dao, H. Q., Nowka, K., & Oklobdzija, V. G. (2001). Analysis of clocked timing elements for dynamic voltage scaling effects over process parameter variation. *InProceedings of IEEE International Symposium on Low Power Electronic Design* (pp. 56-59). IEEExplore Digital Library. doi:10.1145/383082.383094

Gerosa, G., Gary, S., Dietz, C., Pham, D., Hoover, K., Alvarez, J., & Kahle, J. et al. (1994). A 2.2 W, 80 MHz superscalar RISC microprocessor. *IEEE Journal of Solid-State Circuits*, *29*(12), 1440–1454. doi:10.1109/4.340417

Giacomotto, C., Nedovic, N., & Oklobdzija, V. (2007). The effect of the system specification on the optimal selection of clocked storage elements. *IEEE Journal of Solid-State Circuits*, *42*(6), 1392–1404. doi:10.1109/JSSC.2007.896516

Heo, S., & Asanovic, K. (2001). Load-sensitive flip-flop characterization. In *Proceedings of IEEE Computer Society Workshop on VLSI* (pp. 87–92). IEEExplore Digital Library.

Hossain, R., Wronski, L. D., & Albicki, A. (1994). Low power design using double edge triggered flip-flops. *IEEE Transactions on Very Large Scale Integration (VLSI) Systems*, *2*(2), 261–265. doi:10.1109/92.285754

Kawaguchi, H., & Sakurai, T. (1998). A reduced clock-swing flip-flop (RCSFF) for 63 percent power reduction. *IEEE Journal of Solid-State Circuits*, *33*(5), 807–811. doi:10.1109/4.668997

Markovic, D., Tschanz, J., & De, V. (2003). *Transmission-gate based flip-flop*. U.S. Patent 6642765.

Mesgarzadeh, B., Hansson, M., & Alvandpour, A. (2007). Jitter characteristic in charge recovery resonant clock distribution. *IEEE Journal of Solid-State Circuits*, *42*(7), 1618–1625. doi:10.1109/JSSC.2007.896691

Myjak, M. J., Frias, J. G. D., & Jeon, S. K. (2006). An energy-efficient differential flip-flop for deeply pipelined systems. In *Proceedings of 49th IEEE International Midwest Symposium on Circuits and Systems* (pp. 203-207). IEEExplore Digital Library. doi:10.1109/MWSCAS.2006.382032

Nikolic, B., Oklobdizija, V. G., Stojanovic, V., Jia, W., Chiu, J. K., & Leung, M. M. (2000). Improved sense-amplifier based flip-flop: Design and measurements. *IEEE Journal of Solid-State Circuits*, *35*(6), 876–884. doi:10.1109/4.845191

Nogawa, M., & Ohtomo, Y. (1998). A data-transition look-ahead DFF circuit for statistical reduction in power consumption. *IEEE Journal of Solid-State Circuits*, *33*(5), 702–706. doi:10.1109/4.668984

Oklobdizija, V. G., Stojanovic, V., Markovic, D., & Nedovic, N. (2003). *Digital system clocking: High performance and low power aspects*. Wiley. doi:10.1002/0471723703

Palumbo, G., & Pennisi, M. (2008). Design guidelines for high-speed transmission-gate latches: Analysis and comparison. In *Proceedings of IEEE International Conference on Electronics, Circuits and Systems* (pp. 145-148). IEEExplore Digital Library. doi:10.1109/ICECS.2008.4674812

Raychaudhuri, A. (2004). *Edge-triggered D-flip-flop circuit*. U. S. Patent 6762637.

Singh, K., Tiwari, S. C., & Gupta, M. (2014). A modified implementation of tristate inverter based static master-slave flip-flop with improved power-delay-area product. The Scientific World Journal.

Stojanovic, V., & Oklobdzija, V. (1999). Comparative analysis of master-slave latches and flip-flops for high-performance and low-power systems. *IEEE Journal of Solid-State Circuits*, *34*(4), 536–548. doi:10.1109/4.753687

Strollo, A., & Caro, D. (2000). Low power flip-flop with clock gating on master and slave latches. *Electronics Letters*, *36*(4), 294–295. doi:10.1049/el:20000268

Strollo, A., Napoli, E., & Caro, D. (2001). Low power flip-flops with reliable clock gating. *Microelectronics Journal*, *32*(1), 21–28. doi:10.1016/S0026-2692(00)00072-0

Sutherland, I. E., Sproull, B. F., & Harris, D. L. (2004). *Logical Effort: Designing fast CMOS circuits* (1st ed.). Morgan Kaufmann Publishers.

Suzuki, Y., Odagava, K., & Abe, T. (1973). Clocked CMOS calculator circuitry. *IEEE Journal of Solid-State Circuits*, *8*(6), 462–469. doi:10.1109/JSSC.1973.1050440

Varma, P. (2004). *Double edge-triggered flip-flops*. U.S. Patent 6794916.

Weste, N. H. E., & Harris, D. M. (2011). *CMOS VLSI Design: A circuits and systems perspective* (4th ed.). Addison-Wesley.

Yeap, G. K. (1998). *Practical Low power Digital VLSI Design* (1st ed.). Springer. doi:10.1007/978-1-4615-6065-4

KEY TERMS AND DEFINITIONS

Clock-Load: Clock load with respect to a flip-flop is defined as the value of capacitance seen from the clock terminal.

Clock-to-Output Delay: Clock-to-output delay with respect to a flip-flop is defined as the delay between the 50% transition points of output signal and the triggering clock edge.

Data-to-Output Delay: Data-to-output delay with respect to a flip-flop is expressed as the sum of setup time and clock-to-output delay.

Flip-Flop: A flip-flop is a bistable element used in sequential systems for data storage and processing. It is an edge triggered device.

Interconnect Capacitance: Interconnect capacitance is the capacitance offered by the metallic wires used for routing of an IC layout.

Latch: A latch is used as a basic building block in sequential systems. It is a level triggered device.

Logical Effort: Logical effort is the relative ability of a digital circuit to drive a capacitive load as compared to an inverter.

Power-Delay Product: Power-delay product is the figure of merit which explains the tradeoff between power and delay of a CMOS circuit.

APPENDIX 1

The circuit level realization of DFF by using NMOS pass transistors as switches both in the critical path and feedback structure of the proposed architecture leads to an implementation which requires only twelve MOSFETs (excluding the MOSFETs used for generating complementary clock signals). The DFF proposed by Raychaudhuri (2004) is claimed to be realized with 9 MOSFETs (excluding the MOSFETs used for generating complementary clock signals and MOSFETs used for SET and RESET operations). However, it is quite clear from the FF design that only the slave section is static in nature and no provision has been incorporated to implement a static master section. As a result, the intermediate node X is prone to leakage and noise variations. The absence of inverter at the data input terminal also makes the FF susceptible to noise effects.

Additionally, a feedback mechanism similar to the proposed architecture has been reported by Varma (2004) in case of a double edge triggered FF design. In this case, both master and the slave sections are claimed to have static functionality. A weak PMOS transistor is used in the slave section which serves the dual purpose of producing complementary output logic level and reducing the transistor count of the circuit. But, this scheme is fully static only in case of a double edge triggered FF since the weak PMOS transistor when used in the slave section of a single edge triggered FF will not offer a solution in terms of complete static functionality. This is because the output will be retained only when the clock is HIGH (LOW) in case of a positive (negative) edge triggered FF.

APPENDIX 2

The delays in this work for various FFs have been modelled using LE theory in terms of a basic delay unit 'τ' which is process dependent parameter. The absolute delay has been represented as the product of a unit less delay of the gate and the delay unit 'τ'.

The delay for a multistage path is represented by 'D' as explained earlier whereas 'd' corresponds to the delay of a single stage logic gate. To determine the absolute delays, parameter 'τ' needs to be estimated. This is achieved by plotting a delay vs fanout curve for an inverter by fitting simulation data. Since LE theory is based on a linear delay model, the curve is approximated as straight line where the slope of the line is represented as 'τ' since $d = (gh + p)\tau$ and logical effort of an inverter is 1. In our case the value of τ is estimated as 13 ps.

Chapter 8
Signal-Adaptive Analog-to-Digital Converters for ULP Wearable and Implantable Medical Devices:
A Survey

Nabi Sertac Artan
New York Institute of Technology, USA

ABSTRACT

The mission of this chapter is to introduce the reader the recent developments in the design of ultra-Low Power ADCs for Wearable and Implantable Medical Devices (WIMDs). The focus of this chapter will be on Signal-Adaptive Successive Approximation Register (SAR) ADC architectures and their derivatives, since the majority of the ULP medical devices rely on these architectures. The proposed chapter first provides an overview of the WIMDs, and electrophysiological signals. Then, basic SAR ADCs are introduced followed by the study of adaptive SAR ADCs. The chapter concludes with a brief summary of the other prevalent ADC architecture for WIMDs, namely the Level-Crossing ADCs.

1. INTRODUCTION

There is a surge in wearable devices for physiological monitoring and treatment (Medtronic MiniMed insulin pump (Medtronic 2015), Second Sight Argus II Retinal Prosthesis (Second Sight 2015)) and for activity monitoring for healthy living (FitBit 2015). The wearable market is expected to have a 35% compound annual growth rate in the next five years (Danova 2015) and to reach to a global retail revenue of $53.2B by 2019 (Juniper Research 2014).

Similarly, implantable medical devices (IMDs) have become an indispensable option for patient monitoring and treatment, while their size has continuously shrunk. IMDs came a long way from the early pacemakers. For instance, the latest pacemakers such as Nanostim (St. Jude Medical 2015) can fit

DOI: 10.4018/978-1-5225-0190-9.ch008

inside the heart eliminating the need for a surgical pocket, and can stimulate the heart without leads. There are over 4 million people with devices including pacemakers for managing cardiac rhythms worldwide, and 700,000 new pacemakers are implanted every year (Strickland 2013). Furthermore, modern IMDs offer new functionality for monitoring and treatment of various diseases and disorders ranging from Deep Brain Stimulation for Parkinson's disease with Medtronic ActiveRC (Medtronic 2015a) to weight loss treatment for obesity with EntroMedics Maestro System (EntroMedics 2015).

Together, Wearable/Implantable Medical Devices (WIMDs) provide exciting opportunities for significantly improving the health of the general population. On the other hand, the design and implementation of the WIMDs have unique engineering challenges.

1.1. Stringent Power Constraints for WIMDs

As the demand for achieving more functionality in smaller volume for wearable and implantable medical devices grow, the requirements for power consumption become more and more restricted. An important hurdle for the adoption of these devices is the need for large batteries or frequent battery recharging. The frequent battery recharging can easily become a hassle for the patients. More importantly, the limited battery life for implantable devices makes battery replacement surgeries a necessity. An ideal solution is to have these systems rely solely on energy harvesting to avoid the need for battery replacement or recharging, altogether. Yet, for most medical applications, there is a large gap between the energy demand and the energy that can be harvested.

A satisfactory compromise for the short-term can be wearable or implantable systems, which only require recharging infrequently, such as once a month. This extended period of use without recharging also opens the possibility of the recharging to take place only at routine visits to the physician's office. Thus, the risk of relying on the patients to recharge their own devices can be circumvented.

For making wearable and implantable devices ubiquitous, power consumption should be considered at every layer of the design. In this context, it is no longer feasible to design subsystems in isolation without intimate knowledge of the use patterns, power consumption, and capabilities of the surrounding systems. Analog-to-Digital Converters (ADCs) are a key building block in these Ultra-Low Power (ULP) systems since they dictate the accuracy and speed of the analog-to-digital conversion. As a result, ADCs also dictate the size of the resulting bit stream.

The mission of this chapter is to introduce to the reader the state-of-the-art methods, the challenges, and the opportunities of designing ultra-low power ADCs for wearable and implantable medical devices. The focus of this chapter will be on Successive Approximation Register (SAR) ADC architectures and their derivatives (e.g. adaptive SAR ADCs), since the majority of the ULP medical devices rely on these architectures. The SAR-ADC architecture is preferred for ULP implementations due to its relatively simple design, small area footprint, and high efficiency (energy/conversion). SAR ADC is a good fit for the resolution range required by the medical sensing applications. Due to the recent interest in the Level-Crossing ADCs for WIMDs, this chapter will also briefly cover the Level-Crossing ADCs and their inherent signal-adaptive behavior.

The rest of this chapter consists of 7 sections and is organized as follows. In Section 2, WIMDs are introduced. In Section 3, a summary of electrophysiological signal characteristics relevant to the current context is given. The next section introduces the SAR ADC. Section 5 provides the theoretical background for ADC power consumption. Section 6 offers a comprehensive review of the Reconfigurable and Adaptive SAR ADCs. The power consumptions of these ADC architectures are compared against a recording in the MIT-BIH Arrhythmia Database (Moody and Mark 2001) from the PhysioNet database (Goldberger 2000). Section 7 provides further discussion and application-specific uses of the reviewed ADCs. Section 8 concludes the chapter. The brief introduction of the related work on Level-Crossing ADCs is left to the Appendix 1. Matlab/Octave sources for various examples are given in the chapter are available in Appendix 2.

2. WEARABLE AND IMPLANTABLE MEDICAL DEVICES

The block diagram of WIMDs is shown in Figure 1. The WIMD has an electrophysiological signal monitoring and recording unit (MRU), and/or a stimulation unit (SU). The MRU consists of the following components:

1. **Sensing Electrodes:** The electrodes allow the conversion between the ionic currents in the body to the electronic currents in the WIMD circuits (Ha et al. 2014).
2. **Amplification and Signal Conditioning Circuits Also Known as Analog Front-End (AFE):** The AFE typically includes an instrumentation amplifier such as (Harrison and Charles 2003) with a bandpass filter and possibly a notch filter to filter the 50/60 Hz line noise.
3. **An Analog-to-Digital Converter (ADC):** The ADC converts the analog electrophysiological signal into the digital domain. The ADC in the MRU is typically a SAR ADC, a ΣΔ ADC, or a Level-Crossing ADC.
4. **A Digital Signal Processing Unit (DSP):** The DSP can be implemented as a microcontroller (for instance a Texas Instruments MSP430), a Digital Signal Processor (Bolomey et al. 2009), or as an ASIC (Application-Specific Integrated Circuit).

Figure 1. Block diagram of Wearable/Implantable Monitoring and Recording Unit (MRU), and Stimulation Unit (SU). A WIMD can include either or both of the MRU and the SU.

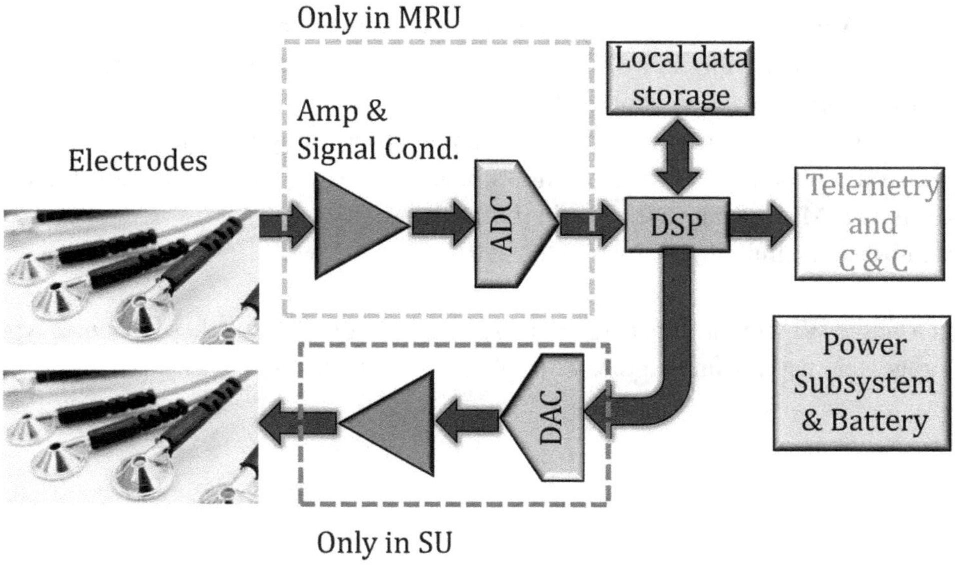

5. **Local Data Storage for Keeping the Recorded Data for a Short or a Long Term:** This can be magnetic storage for earlier wearable devices. Implantable devices and recent wearable devices use either SRAM, or flash-based storage.
6. **A Telemetry and Command and Control Unit:** This unit transmits the data recorded to an external device, and allows other entities (such as the tablet of a physician) to send commands and configuration changes to the IMD. The telemetry unit can be based on near-field communication (NFC) via an inductive link at frequencies mostly ranging from a few hundred kHz to a few tens of MHz. NFC-based telemetry is especially suitable if only intermittent communication is needed with pre-defined but not precise time intervals. For instance, in offline monitoring, a patient typically uses a wand every night to download the information recorded during the day. Usually, the bandwidth of the near-field communication especially for transcutaneous links is limited due to the attenuation by the tissue, distance and misalignment between the transmit and receive coils (Kiani et al. 2011). As the bandwidth requirement increases and low-power radios and protocols become available, far field links are getting more prevalent. In particular, wireless protocols such as Low-Energy Bluetooth, and ZigBee, as well as wide-band communication (Yuce et al 2009) are gaining attention. More recently, there are encouraging developments in using optical links for high data rate applications such as neural signal recording from large-scale cortical arrays (Liu et al 2014). Due to the bandwidth limitations, and the high energy cost of transmission, minimizing the data size before transmission is vital. Thus, electrophysiological data compression is also an important and active area of research (Kim et al. 2012).
7. **Power Subsystem:** The WIMDs are typically battery powered. As techniques for energy harvesting from the body for instance from the heartbeat (Tsai et al. 2013) and from the power in the RF band (Xia et al. 2014) matures, devices powered via energy harvesting are also gaining attention. IMDs can also be powered via transcutaneous inductive links. More frequently, inductive power is used to charge rechargeable batteries, wirelessly.

The SU consists of similar components as the MRU, but with different functionality:

1. Stimulation electrodes for stimulating the tissue.
2. A Digital-to-Analog Converter (DAC) to shape the stimulation current.
3. A digital signal processing unit (DSP) similar to the one in MRU. However, the DSP in SU is mostly used to provide stimulation patterns.
4. Local data storage for storing waveform, and stimulation configuration data.
5. A Command & Control Unit, which allows other entities to send commands and configuration changes to the IMD similar to the one in the MRU.
6. Power Subsystem similar to the one in the MRU.

Note that a single WIMD can consist of both the MRU and the SU, and these modules typically share the DSP, storage, telemetry, Command & Control, and power subsystem.

3. ELECTROPHYSIOLOGICAL SIGNALS

The following electrophysiological signals can be measured directly via various types of electrodes:

- Scalp and intracranial EEG (Electroencephalogram) for brain activity,
- Single- and Multi-unit monitoring for activity measured from individual neurons.
- MEG (Magnetoencephalography) for monitoring magnetic fields due to neuronal activity,
- ECG or EKG (Electrocardiogram) for heart monitoring,
- EMG (Electromyogram) for muscle movement,
- EOG (Electrooculogram) for eye movement, and
- GSR (Galvanic Skin Response)

Here, we briefly describe two types of electrophysiological signals as examples, namely the EEG, and the multi-unit action potentials.

3.1. Scalp and Intracranial EEG (Electroencephalogram)

Scalp EEG is a non-invasive method of monitoring electrical activity in the brain (i.e., brain waves) through electrodes attached to the scalp. Scalp EEG amplitude typically ranges between 0.5 and 20 μV. The brain waves are classified according to their frequency as

1. Delta waves: 0.5 - 4 Hz,
2. Theta waves: 4-8 Hz,
3. Alpha waves: 8-13 Hz, and
4. Beta waves: Above 13 Hz.

EEG has wide applications including epilepsy diagnosis, seizure monitoring, and sleep disorders. EEG is also widely used in the intensive care units (ICU) for patient monitoring. For both its clinical and research use, EEG measurements ultimately aim to identify existence, lack of, or characteristics of certain features such as rhythmicity, and signal amplitude in brain waves and how these features change in time. For instance, alpha rhythm emerges in the posterior regions of the brain in a relaxed state when eyes are closed. This can be used to establish a baseline for the relaxed state. The dominant frequency can also help identify the sleep stages. For instance, delta waves dominate the sleep stage known as slow-wave sleep.

In intracranial EEG (iEEG or Electrocorticography (ECoG)) arrays of electrodes are placed under the skull on or under the dura mater in the brain. There are up to 256 electrodes in a typical array. Multiple electrode arrays or strips can be used to get a wider coverage of the brain areas. ECoG signals typically have higher amplitude (up to 100 μV) and wider frequency band (up to 500 Hz) compared to scalp EEG.

3.2. Multi-Unit Monitoring

Multi-unit monitoring is the measurement of individual neuron activity (as action potentials) in the brain via microelectrode arrays such as the Blackrock Microsystems Utah array (Blackrock 2001). Utah Arrays have up to 96 electrodes per array. The neurons communicate via voltage spike trains, where each spike corresponds to an action potential. The timing of the action potentials encodes the neural information (O'Driscoll et al. 2011). Due to the neuron density, each electrode typically receives the signal of multiple neurons (up to 6 neurons is common). In order to encode the output of individual neurons, each spike should be associated with a single neuron through a process called spike sorting.

Multi-unit monitoring is used extensively in basic neuroscience research. Multi-unit recordings also have tremendous utility in Brain Computer/Machine Interfaces (BCIs/BMIs). BCI allows the intent of a tetraplegic patient to be expressed by translating neuronal activity in the brain to control signals. Then, these control signals are used to operate a device such as a computer cursor, or a robotic arm. Similarly, an amputee can use the BCI to control a prosthetic device.

Multi-unit monitoring usually consists of many channels (64 or more), where each channel is sampled at relatively high sampling frequencies at up to around 100 kS/S, with relatively low resolution (typically 8-10 bits) using WIMDs.

4. A BRIEF INTRODUCTION TO THE SAR ADC

4.1. SAR ADCs for WIMDs

ADC selection and design are critical for implementing a successful wearable/implantable system. Multiple ADC architectures including Successive Approximation Register ADCs (SAR ADCs), ΣΔ ADCs (Lu et al. 2015), and Level-Crossing ADCs (LC ADCs) (Li et al. 2014) are proposed for wearable, and/or implantable medical devices.

Ha et al. (2014) identified SAR ADCs and ΣΔ-ADCs as most suitable for low-power physiological measurement applications. Jonsson (2011) reported SAR ADC architecture as the most energy-efficient over a broad range of resolutions based on data from 1400 papers published between 1974 and 2010. In particular for low resolution and low sampling rate applications, where WIMD is a good fit, SAR ADCs can provide a low-power solution due to their simple op-amp free architecture (Sinangil et al. 2012). Yet, low-power SAR ADCs for higher resolution (Ha et al. 2014b) and high-sampling rate applications are also gaining attention (Lee 2015).

Recently, LC ADCs have become popular because of their very promising power efficiency. However, LC ADCs can suffer from slope overload, and they need their comparators to be active, continuously. At the system level, LC ADCs need special digital signal processing for non-uniform sampling. Due to space constraints, LC ADCs are only briefly covered in this chapter.

4.2. Operation of the SAR ADCs

The block diagram of a traditional SAR ADC is shown in Figure 2. Before starting the conversion, the S/H circuit samples and holds the analog input V_{in}. The Successive-Approximation Register first produces the code corresponding to the middle value in its range ($V_{ref}/2$), which consists of all zero digits except the MSB, as the starting point of the search. This digital value is converted to its analog counterpart voltage value by the Digital-To-Analog Converter (DAC) block. This voltage value is compared with the value in the S/H via the comparator. The comparator determines whether the code generated by the SAR is above or below the input value V_{in}. Based on the comparator output the SAR block generates a new output code with the goal of approximating the V_{in} value. As the conversion progresses, the output code produced by the SAR ADC gets closer to the input value. The operation completes after N iterations. If the input value V_{in} is within the input range of the ADC, the output code produced by the SAR ADC is guaranteed to be within 1 LSB of the input value V_{in} under an ideal setting. An example conversion for the SAR ADC is shown in Figure 3. In this example, the analog input V_{in} has the value of 0.65625 V (code value 21 for the 5-bit ADC). The SAR ADC starts comparison from 0.5 V ($V_{ref}/2$), and oscillates around the input value, until it converges to the input value in five cycles, determining one bit of the output code per cycle.

Due to the wide variety of parameters, architectures and application areas of ADCs, various ADC figure-of-merits (FoMs) are developed. The most widely used and accepted FoM for ADCs, which is also known as the Walden FOM (Walden, 1999) can be given as

$$FOM = \frac{P}{2^{ENOB} \times FS}\left[\frac{J}{conv}\right] \tag{1}$$

where P is the power consumption, and FS is the Nyquist sampling rate. ENOB is the effective number of bits and can be given as

$$ENOB = \frac{SNDR - 1.76}{6.02}\left[bits\right] \tag{2}$$

where SNDR is the *Signal-to-Noise-and-Distortion Ratio*. Another useful and simpler FoM for ADCs with approximately same ENOBs is

$$FOM = \frac{P}{Speed}[W/(S/s)] \tag{3}$$

Figure 2. SAR ADC block diagram

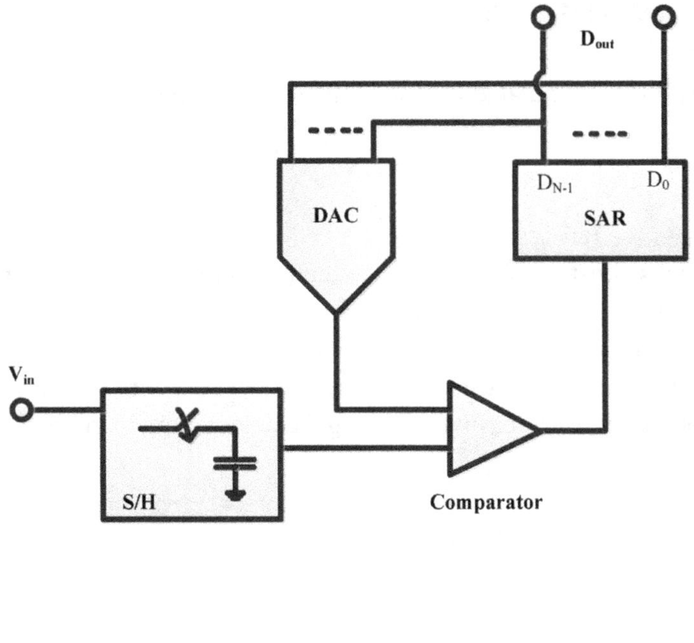

Figure 3. An example conversion for a 5-bit SAR ADC for the input V_{in}=10101

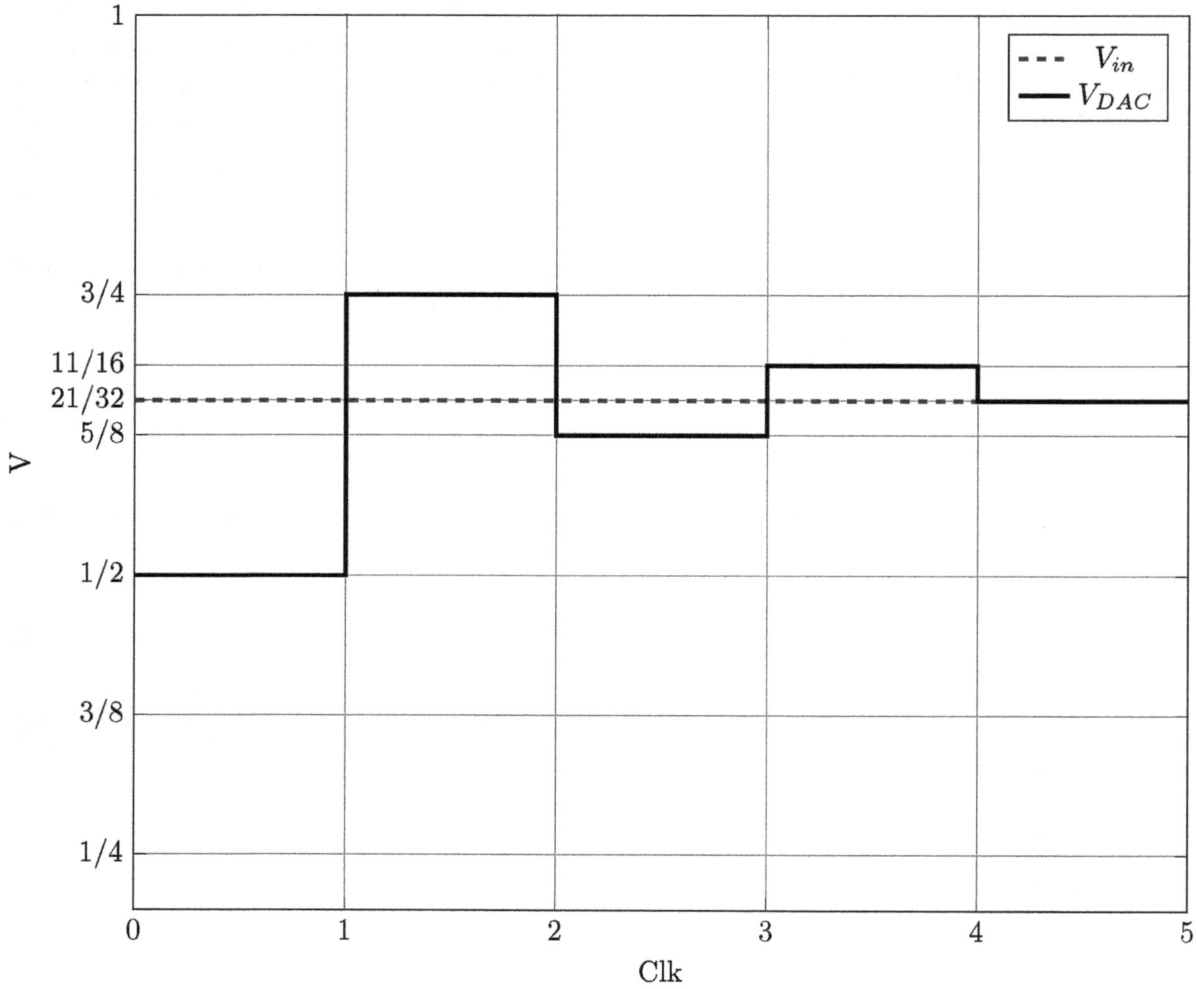

5. ADC POWER CONSUMPTION

Kenington and Astier (2000) formulated the theoretical limit for power consumption on ADCs as

$$P_{ADC_{\min}} = kTf_s 10^{(6N+1.76)/20}[W] \tag{4}$$

where k is the Boltzman's constant, T is the device temperature, f_s is the sampling frequency, and N is the resolution of the ADC. The theoretical minimum power consumption for ADCs at resolutions and sampling rates typical in wearable and implantable medical devices is shown in Figure 4. Accordingly, increasing the resolution of the ADC causes an exponential increase in the power consumption O'Driscol et al. (2011). On the other hand, increasing sampling frequency only increases the power consumption linearly.

Figure 4. Theoretical Minimum ADC Power for different sampling frequencies and resolutions common in WIMDs

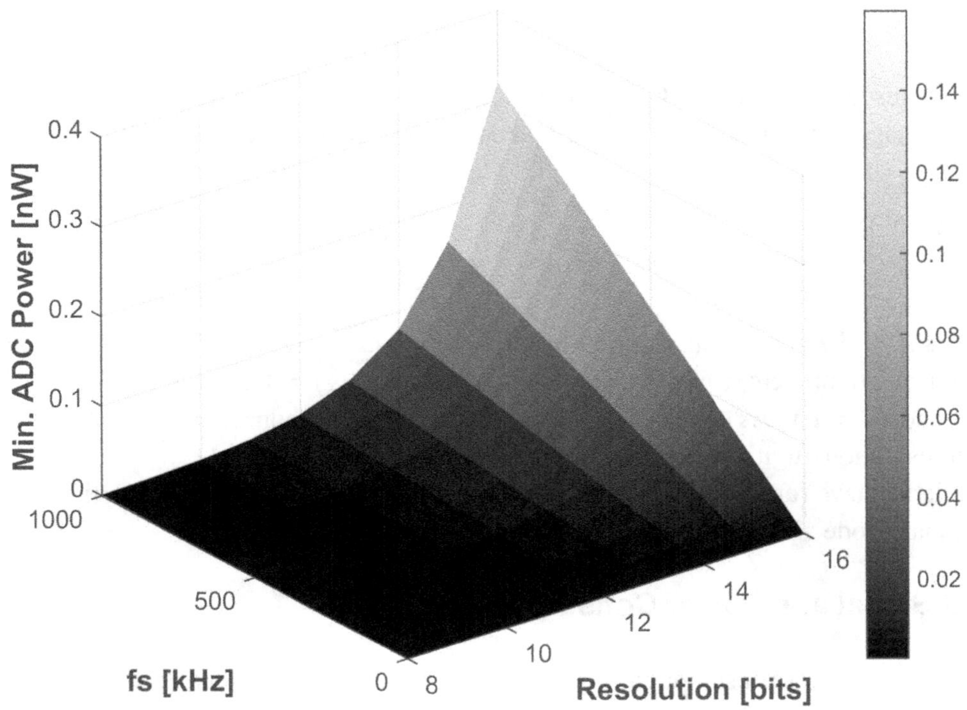

An N-bit traditional SAR ADC requires *N* approximation cycles (i.e., conversion following the sampling and holding of the analog input. During this process, the DAC generates *N* different reference values, and the input is compared to these *N* values. According to the comparison result, the SA Register is updated. For the DAC to generate the reference values, a capacitive array is charged and discharged. Accordingly, the power of the SAR ADC is formulated below. Note that, depending on the architecture, not all of this power is drawn directly from the power supply, but it can be received from the input instead.

$$P_{ADC} = P_{S/H} + P_{comp} + P_{DAC} + P_{SAR} \quad (5)$$

where $P_{S/H}$, P_{comp}, P_{DAC}, and P_{SAR} represents the power consumption for the Sample and Hold circuit, the comparator, the Digital-to-Analog Converter, and the Successive Approximation Register logic, respectively. For SAR ADCs, the power consumption is dominated by P_{DAC} (O'Driscoll et al. 2011). Thus, the rest of the discussion will be focusing on the DAC power, and more specifically to the power consumption of the DAC capacitive array, which is the power drawn from the voltage reference. It should be noted that many charge redistribution SAR ADCs combine the S/H with the DAC by using a dual-purpose capacitive array: (1) This array is used as an S/H capacitor, and (2) it is also used for generating the analog voltage corresponding to the SAR register value. The power consumption of this combined structure is very similar to the traditional structure as shown in (Saberi 2011).

In Equation (9) of (Saberi 2011), the power consumption of the DAC capacitive array is given. This equation is repeated below

$$P_{V_{ref}} \approx \frac{2^N f_{clk} C_0}{N+1} \times \left[\left(\frac{5}{6} - \left(\frac{1}{2}\right)^N - \frac{1}{3} \cdot \left(\frac{1}{2}\right)^{2N} \right) \cdot V_{ref}^2 - \frac{1}{2} V_{in}^2 - \left(\frac{1}{2}\right)^N \cdot V_{in} V_{ref} \right] \tag{6}$$

where N is the resolution of the ADC, f_{clk} is the ADC clock frequency, and C_0 is the unit capacitance. V_{in} is the input voltage corresponding to the digital code $D_1 D_2 \ldots D_N$ given as

$$V_{in} = \sum_{i=1}^{N} \left(D_i / 2^i\right) V_{ref} \tag{7}$$

Figure 5 shows the power consumption for different digital codes at different resolutions based on the above formulation. Here, it is assumed that sampling rate is f_s=10 kS/s, $f_{clk} = fs \cdot (N+1) = 110\ kHz$, $C_0 = 50\ fF$, and $V_{ref} = 1\ V$. As expected, the power consumption is reduced as the resolution is reduced. For a given resolution, digitizing small input values requires more energy.

The Matlab/Octave function to calculate Equation 6 and the Matlab/Octave function to convert between the binary code and input voltage are given in Appendix 2.

5.1. Input Signal and Power Consumption

For signal-specific power consumption it is informative to consider the input-dependent terms in Equation 6.

Figure 5. Power consumption of the DAC capacitive array for different digital codes at different resolutions

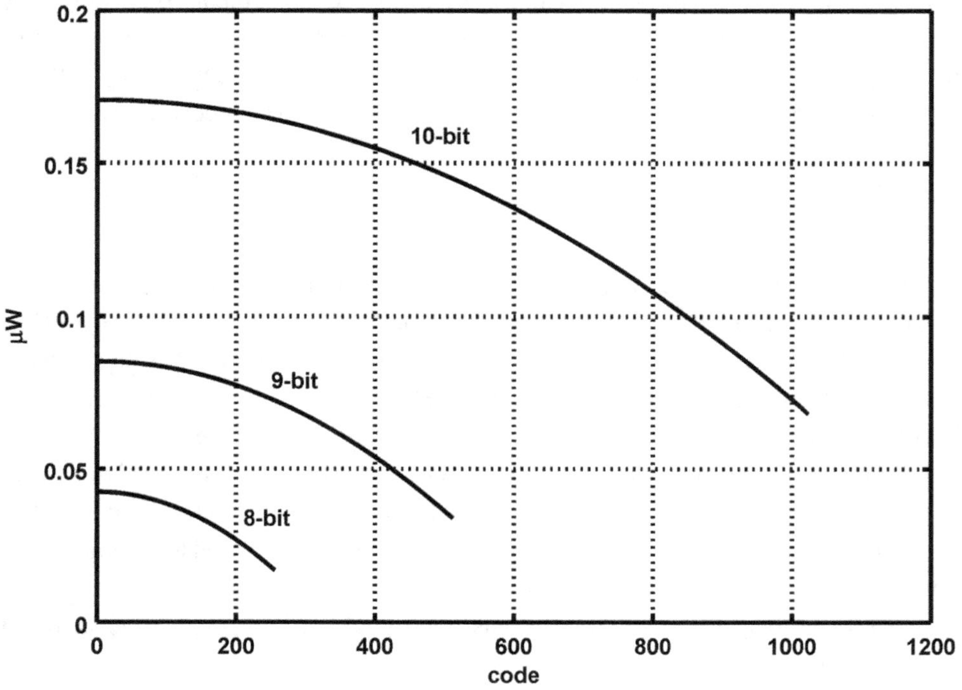

To achieve this goal let C define the constant in Equation 6 for a given ADC as

$$C = \frac{2^N f_{clk} C_0}{N+1} \tag{8}$$

Also, we can approximate the first term in Equation 6 as follows.

$$\left(\frac{5}{6} - \left(\frac{1}{2}\right)^N - \frac{1}{3} \cdot \left(\frac{1}{2}\right)^{2N}\right) \approx \frac{5}{6} \tag{9}$$

This approximation, which is achieved by neglecting the second and third terms, introduces an error of approximately 4% or less to this term for $N \geq 5$ (the remaining terms in the parenthesis adds up to 0.036 for $N = 5$). We can then simplify Equation 6 as

$$P_{V_{ref}} \approx C \times \left[\frac{5}{6} \cdot V_{ref}^2 - \frac{1}{2} \cdot V_{in}^2 - \left(\frac{1}{2}\right)^N \cdot V_{in} V_{ref}\right] \tag{10}$$

Without loss of generality, let us consider a single-ended input where the input voltage V_{in} has a range of $[0, V_{ref})$. Then, depending on the input value, $P_{V_{ref}}$ has a range defined as

$$\left(\frac{1}{3} - \left(\frac{1}{2}\right)^N\right) \cdot C \cdot V_{ref}^2 \leq P_{V_{ref}} \leq \frac{5}{6} \cdot C \cdot V_{ref}^2 \tag{11}$$

Thus, input signal which has lower average amplitude with respect to V_{ref} will cause higher power consumption. This leads to the conclusion that there is room for improvement for traditional SAR ADCs, which treat the whole amplitude range equally.

As a reference, the power consumption from the V_{ref} for a sine wave with amplitude $V_{pp} = V_{ref}/2$ is shown in Figure 6, for three different DC voltages at $V_{ref}/4, V_{ref}/2$ and $3/4V_{ref}$. For lower DC levels, the required energy for conversion increases. More specifically, changing the DC level from $3/4V_{ref}$ to $1/2V_{ref}$ increases the power consumption for sampling one period of the sine wave from 27.40 nJ to 40.21 nJ (a 47% increase).

5.2. Amplitude Variation in Electrophysiological Signals

In the discussion above, we have concluded that amplitude characteristics of the input signal have an impact on the power consumption of an ADC, and the SAR ADC tends to consume more power for signals with low amplitude. At this stage, it is informative to revisit the electrophysiological signals for their amplitude characteristics.

Typically, electrophysiological signals have *silent periods* between epochs of high activity. For instance, the EEG signal amplitude is higher in deep sleep with respect to the alert and focused state. In

Figure 6. Energy consumption of the DAC capacitive array for a 10 Hz sine wave input at different DC levels and sampled at 10 kS/s

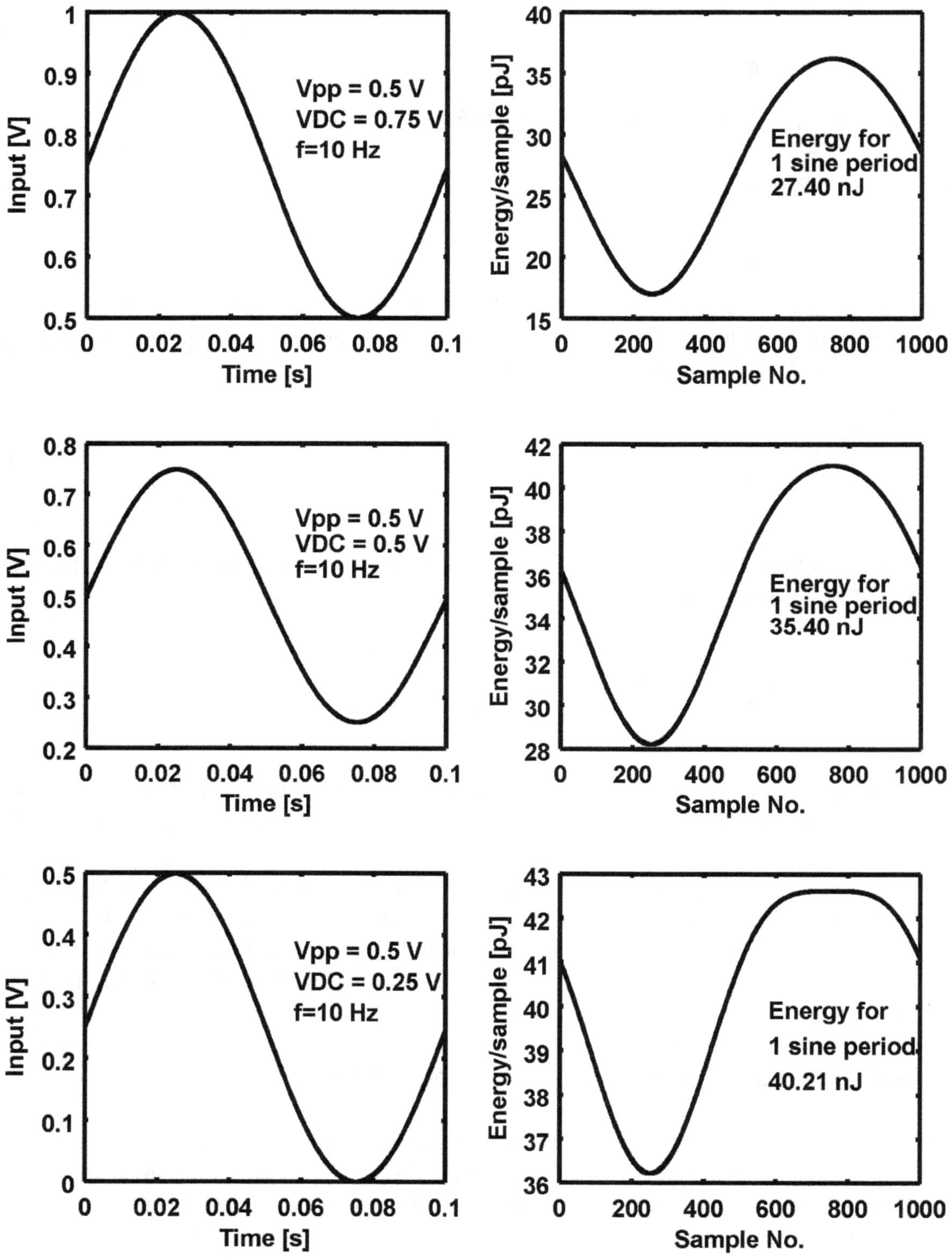

Figure 9. QRS complex in EKG

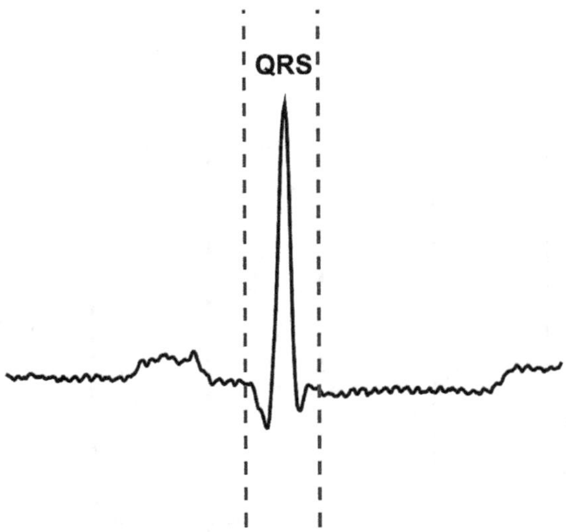

EEG-based seizure detection, the signal amplitude during a seizure is much higher than the rest of the signal for most of the seizure types. For multi-unit recordings, the signal amplitude is very low when there is no spike. For EKG (Electrocardiography), the QRS complex (the sequence of Q, R, and S waves representing the ventricular depolarization as shown in Figure 9) has higher amplitude than the rest of the signal.

It should also be noted that for all the electrophysiological measurements, artifacts due to various reasons (such as muscle artifacts or artifacts caused by the data acquisition circuits) can produce high amplitude signals.

In Figure 7, an EKG sample with 10 beats is shown as an example of amplitude variation in the electrophysiological signals. As can be seen from this figure, the average of the peak values is approximately 6 times higher than the average signal amplitude. This is one indicator that taking into amplitude variation can improve power consumption.

6. INPUT-AWARE SAR ADCS

Various SAR ADC architectures take advantage of the characteristics of the signals. Here, these SAR ADCs are classified into three broad categories:

1. Pre-configured SAR ADCs,
2. Reconfigurable SAR ADCs, and
3. Signal-adaptive SAR ADCs.

Pre-configured SAR ADCs target a specific application, and their design is optimized for the characteristics of that particular application. These pre-configured SAR ADCs represent a more traditional approach, namely the application specific custom design.

Figure 7. Average amplitude of the EKG signal, and average amplitude of the peak values. The data is normalized between 0 and 1 V

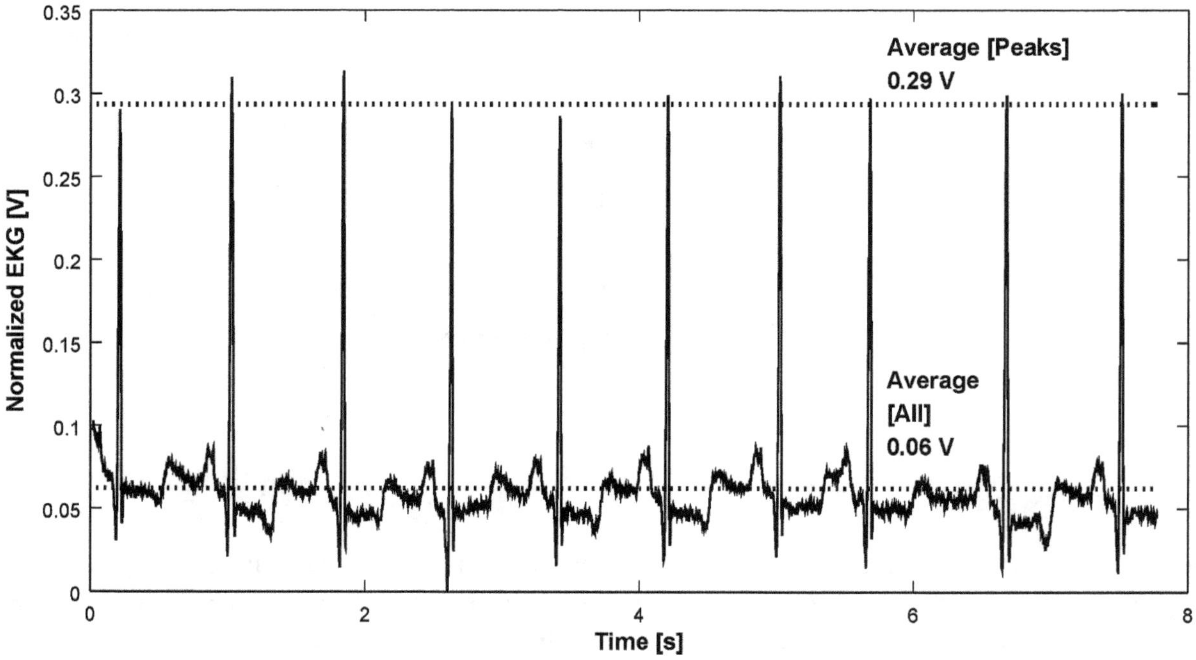

Reconfigurable SAR ADCs have parameters, which are updated depending on the changes in the input signal characteristics, or in the application characteristics. Here, the term reconfigurable SAR ADC is reserved for the case where the parameters are changed infrequently. In other words, the frequency of parameter updates for reconfigurable SAR ADC is much lower than the signal band that is being digitized.

The signal-adaptive SAR ADCs adapt their parameters as the input signal changes. The signal-adaptive SAR ADC parameters can be updated as frequently as possible, *i.e.*, it is possible to update the parameters for each sample or even at every clock cycle.

6.1. Pre-Configured SAR ADCs

To maximize the performance against power needed for a given application, many ADC designs for WIMDs optimize the ADC parameters for this single application. Yet, they have limited or no configurability to adapt to the changes in the application, or to the changes in the input signal. For instance, Do et al. (2012) proposed an ADC design approach targeting *maximal simplicity* for neural signal recording. Their ADC has only two modes targeting two of the common neural signal modalities: In the first mode, the ADC runs at 2.5 kS/s and it is used to measure local field potentials. In the second mode, the ADC runs at 25 kS/s and it is used to measure action potentials. Their design uses a single supply (1 V). The authors avoid any complex capacitor charge/discharge scheme to minimize the transistor count and switch activity. They achieved 9 bit (ENOB 8.56b) using an area of 0.04 mm^2 at 0.18 μm technology, while consuming 160 nW. They reported a FoM of 17 (18) for 2.5 kS/s (25 kS/s) operation.

6.2. Reconfigurable and Signal-Adaptive ADCs

In contrast to the minimalist approach above, and to satisfy the varying needs of WIMDs without sacrificing the performance or the battery life, many of today's state-of-the-art ultra-low power (ULP) ADC implementations offer reconfigurable and/or adaptive ADCs. These ADCs can be reconfigurable at a wide range of ADC parameters, as the application's needs change in time. Alternatively, these ADCs can also be adaptive to the instantaneous changes in the input: They can shut down unnecessary blocks, or scale the operation in time (i.e., sampling rate), or in accuracy (i.e., resolution). The reconfigurable and adaptive ADCs can adapt to changes in the input by solely following the input trend.

Reconfigurable and signal-adaptive ADCs can be classified based on the parameter(s) that is configurable, or the parameter(s) they adapt:

1. The resolution: As demonstrated in (O'Driscol et al. 2011),
2. The sampling rate: As demonstrated in (Artan et al. 2012), and
3. The power supply voltage: As demonstrated in (Yaul and Chandrakasan 2014).

In many ADC designs, more than one parameter is adapted concurrently to maximize the benefits. As an example to this multi-parameter adaptive approach, Yaul and Chandrakasan (2014) proposed an ADC, which adapts its resolution, sampling rate, and power supply.

6.2.1. Reconfigurable ADCs

Reconfigurable ADCs have been widely used in the context of wireless circuits for portable applications for instance (Rusu et al. 2006), (Stojcevski et al 2001), and (Razak et al. 2010), where for (Razak et al. 2010) both resolution and sampling rate are configurable for wireless signal processing. In the context of WIMDs, two designs are considered. In the first design ((O'Driscoll and Meng 2010) and (O'Driscoll et al. 2011)), a variable resolution SAR ADC for digitizing neural signals sensed by a microelectrode array was proposed. The second design (Tang et al. 2014) is a variable resolution SAR ADC, which was proposed for retinal prostheses.

O'Driscoll's design consists of 96 ADCs, one for each electrode of the microelectrode array, where the resolution of each ADC can be configured to have a 3- to 8-bit resolution. Each of their ADCs consumes 0.23 to 0.90 μW at 100 kS/S on a 0.1 μm process. O'Driscoll et al. (2011) observed that due to the variations in the relative distance of individual neurons to an electrode, the signal-to-noise ratio (SNR) of the recorded signal changes. Rather than designing the ADC to include the worst-case SNR for all electrodes, they opted to have reconfigurable ADCs, and periodically (every 12 hours) recalibrate the ADC resolution. Due to the nature of the device behavior in the tissue, the spike-sorting algorithm should be retrained every 12 hours to calibrate the device against cell growth and electrode movement. This retraining performance is used to decide how to update the resolution for each ADC.

Tang et al. (2014) proposed a 5- to 9-bit configurable SAR ADC for retinal prostheses, which is powered from a 0.9 V supply voltage, and occupies 0.15 mm^2 on a 0.18 μm process. Their ADC can reach to 100 kS/s using 1.33 μW of power. The resolution of this ADC can be programmed by changing the number of clock cycles generated by an asynchronous clock generator (ACG). The number of clock cycles can be configured using an on-chip delay line. At the application level for the retinal prosthesis, the resolution of the ADC is configured depending on the illumination level of each pixel of the retinal prosthesis, such that if a pixel has high illumination, its resolution is reduced.

The energy and power consumption of the DAC capacitive array sampling a 10-second portion of the EKG signal given above is shown in Figure 8 for different resolutions. As expected, the power consumption approximately halves for 1-bit reduction in resolution. It should also be noted that, the energy to sample the QRS complex; in particular the peak value is much smaller than the energy required to sample the rest of the signal.

Figure 8. Energy consumption and average power of the DAC capacitive array when sampling EKG signal at different resolutions

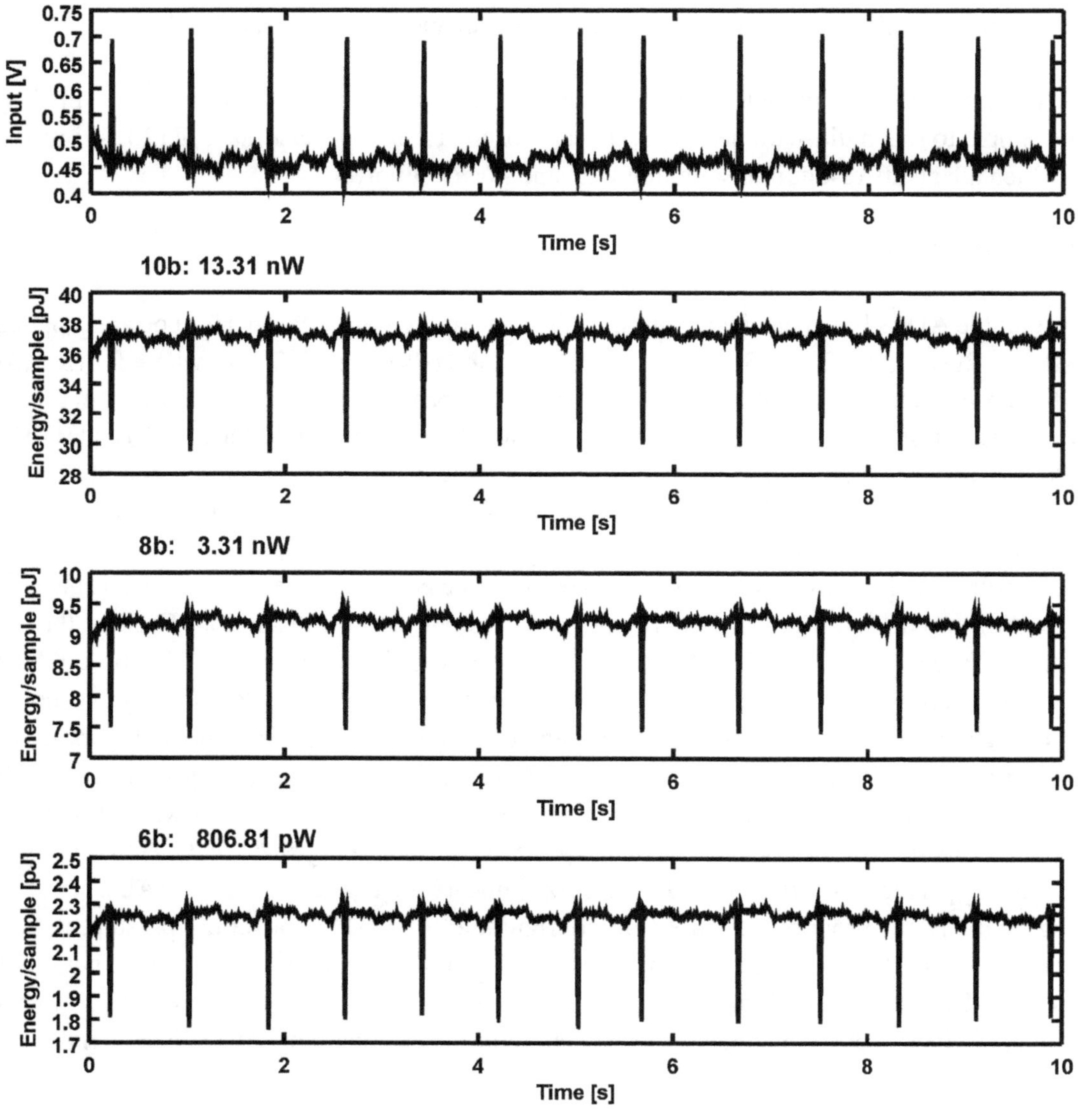

6.2.2. Signal-Adaptive ADCs

Another type of adaptive ADCs, which is referred as signal-adaptive ADC in this chapter, takes advantage of input signal-specific properties. To elaborate on this definition, the following example will be used. Consider ADC design for sampling the Electrocardiogram (ECG) signal. A normal ECG signal has low amplitude regions interspersed with periodic high amplitude and high speed activities. Most notably, the periodic QRS complex, especially at the *R*-peak (*i.e.,* the heart beat) (Figure 9), has typically higher amplitude than the overall average. Furthermore, in the QRS complex, the signal changes rapidly. In other words, in this region,

1. the input dynamic range requirement of the ADC is much wider, and
2. the upper cutoff frequency of the signal bandwidth is much higher.

Furthermore, in most cases, the features of clinical or research interest in an ECG appears around the QRS complex. As such, the rest of the signal is less critical. A traditional design of such an ADC should have a specification that can handle this amplitude and bandwidth range and end up with oversampling most of the input, and either should have very high resolution, or a large LSB. Due to oversampling and high resolution such an ADC will not be optimum for power consumption.

Now, let us consider a *signal-adaptive ADC* for the same example, for ECG digitization. In this case, the digital signal processing system following the ADC can have a simple modification to provide a function that approximately identifies when the signal of interest should appear and raise a flag to indicate this to the ADC. The signal-adaptive ADC can treat this signal as a simple selector between a low-power (low accuracy mode), and a high-power (high accuracy mode). When not in the vicinity of the heart beat and potentially when the signal has low amplitude and low activity, (i.e., when the flag is not raised), the ADC can reduce its resolution, sampling rate, power supply voltage (or input dynamic range), or any other parameter to save power. On the other hand, when the flag is raised the ADC can operate at full power mode, and digitize the signal of interest (i.e. the heart beat) at full sampling rate, full resolution, etc.

6.2.3. Input Tracking

Although, input tracking ADCs (Sepehrian et al. 2011) are not SAR ADCs, studying input tracking ADCs help better understand the power trade-offs in signal-adaptive SAR ADCs. Thus, in this section, we will first introduce the input tracking ADCs.

Traditional SAR ADCs starts the Successive Approximation operation by first deciding whether the MSB is 1 or 0. Once the MSB is determined, the successive approximation proceeds to determine the next significant bit. This operation continues until the LSB is decided, which concludes the approximation as all the bits have been determined. This allows quick convergence to the input signal in a predetermined time (*N* iterations for an N-bit ADC) without any *a priori* information or assumption about the signal or the signal statistics. However, most of the signals, including the electrophysiological signals are continuous in nature with a bounded *slope* most of the time (Chiang et al. 2013).

Intuitively, if, for a given application, errors with low probability are tolerable, or can be corrected by other means with some performance penalty, this predictability of the signal can be used to allow the successive approximation to converge faster eliminating many unnecessary operations. This basic principle is applied to architecture and circuit design of many ADCs including and beyond SAR ADCs (for instance Level Crossing ADCs, and tracking ADCs).

Traditional Tracking ADCs (Sepehrian et al. 2011) rely mainly on this principle, such that each sample is predicted to be equal to the previous sample. If the input value is the same as the previous sample, the conversion completes in a single iteration. If the input value is different from the previous sample, the new value is approximated by linear increments (or decrements). In this traditional tracking ADC, an up/down counter is used instead of the SAR logic. If the new sample is greater than the previous sample, the counter, which holds the value of the previous sample, is incremented by one. The counter value is applied to the DAC to generate the new value for the next comparison. The counter will be incremented until the value in the counter is the same as the input value. Similarly, if the sample is greater than the previous sample, than the counter is decremented until the new input value is equal to the counter value. In the worst-case this leads to a conversion time $O(2^N)$, when the previous sample is at 0, and the input value is at 2^N-1 (or vice versa).

6.2.4. Tracking SAR ADCs

The worst-case conversion time due to the linear search approach in Tracking ADCs, limit their adoption. More recently, architectures combining the adaptive approach of the Tracking ADCs with the relatively rapid convergence of the SAR ADCs are proposed.

For predicting the value of the next sample, (Van Rethy et al. 2013) proposed a method based on the values of previous samples. In their approach, Van Rethy et al. define n^{th} *order approximations*, where $n \geq 0$. For 0th order prediction, the input is predicted to be the same as the previous sample. In general, for an nth order prediction, the input is predicted to be the nth order extrapolation of the previous n signals (e.g. linear for the first order, quadratic for the second order, etc.). Once, the prediction is made, they use a linear search from the predicted value to identify the new sample. The *sub-range* this linear search will be carried on is selected such that the decision is made in $N + O(1)$ comparisons. Thus, the number of comparisons for their linear search algorithms is in par with the successive approximation. Alternatively, Van Rethy et al. also proposed to have a *restricted binary search* algorithm, which runs binary search but only in a limited range (i.e., limited number of bits) around the predicted value. For ECG data, the linear search offers power savings at low resolutions but proved to be not scalable to higher resolution beyond 12-bit. The restricted binary search on the other hand performed well for a wider range of resolutions (Van Retty et al. 2013). Yet both linear and restricted binary search approaches provide the best result at 8-bit resolution. They reported to have 40-50% improvement in energy consumption for 6-10 bit resolution, and 25-40% for higher resolutions.

In (Chiang et al. 2013), the authors proposed the Adaptive Tracking ADC. The Adaptive Tracking ADC uses a partial binary search instead of the linear search in the traditional Tracking ADCs. This way the convergence can still be fast as in the SAR ADC, yet, the conversion can start from the vicinity of the previous sample, thus avoiding most of the unnecessary conversions. Chiang et al. (2013) also pointed out an important difference between the continuous analog signal, and the discrete code such that an LSB change in the analog input signal can change all the bits in a binary coded number (e.g. an LSB increase in the analog input corresponding to the code *0111*, will result in the code *1000*. The latter code has no digits equal to the former code, although they differ in only one LSB in the analog signal). Then, the authors propose an efficient SAR logic circuit and algorithm to converge rapidly to the input while minimizing number of comparisons, and energy consumed. Chiang et al. (2013) reported their 12-bit adaptive SAR ADC can reach to 25 kS/s sampling rate for a power consumption of 289 nW at 0.18 μm using a 1 V power supply.

Yaul and Chandrakasan (2014) adopted the tracking approach, however rather than using an incremental approach, the authors used a successive approximation. Unlike traditional SAR ADCs, however, Yaul and Chandrakasan's successive approximation algorithm called "LSB-First Successive Approximation" starts with the LSB, and goes up to the MSB. This way, the number of steps required for successive approximation is reduced thus the ADC power is also reduced as long as the signal variation stays within certain limits. Depending on the signal activity, their approach can reduce power significantly.

In (Huang et al. 2012) and (Lie et al. 2014b), bypass window based methods are proposed. In the bypass window based methods, few initial iterations of the successive approximation can be skipped, when the input signal is in a pre-defined window, i.e., in the bypass window.

Avila et al. (2014) proposed a generalized digital assistance method for adaptive ADCs. The authors observed that the maximum change in between two consecutive samples is a function of the maximum slope of a sine wave at the frequency equal to the bandwidth of the input signal. The authors formulated the maximum change, $\Delta v_{in_{\max}}$ in an input signal between two samples as

$$\Delta v_{in_{\max}} = \left\lceil \frac{2^N \pi BW}{FS} \right\rceil \tag{12}$$

where BW is the bandwidth of the preceding stage. This result is derived based on the maximum slope of a sine wave, which has a frequency of FS/2 frequency and peak-to-peak amplitude equal to the input dynamic range of the ADC.

It is beneficial to represent the slope of the input signal in terms of the LSB of the ADC per sample time (i.e., $1/f_s$) with the assumption that the sampling time is fixed. Since it is usually necessary to precede the ADC with an amplification stage for low-resolution ADCs, using the amplified slope rather than the actual slope of the signal can be more informative.

For instance, consider a 12-bit ADC with an input dynamic range of 1 V. This ADC has an LSB of $1/2^{12} = 244.14\ \mu V$. If this ADC is used to sample a signal with a slope of $100\ \mu V$, which is 10× amplified before it is applied to the ADC, then the slope, and amplified slope are *0.41* LSB, and *41* LSB, respectively.

Now if the ideal case, where the amplified slope is bounded by L LSBs, is considered and if the slope has a discrete uniform distribution between 0 and L LSBs, the expected number of iterations for a Tracking ADC is $L/2$ iterations/conversion. To have a smaller number of iterations in this tracking ADC with respect to a traditional SAR ADC, we should have $L/2 < N/2 \rightarrow L < N$. Following from the previous example, for the 12-bit Tracking ADC to have smaller number of iterations compared to the traditional SAR ADC, the input slope should be bounded by *1.2* LSB, and the amplified input slope should be bounded by *12* LSB.

One metric of comparison for power or energy efficiency of an adaptive ADC is the number of comparisons needed (Van Rethy et al. 2013). However, this is a first order metric that does not take into account the variation in the cost of individual comparisons (e.g. DAC power consumption is different for different comparisons based on the reference input to the comparator). This is mainly due to the non-linear relation between the capacitor values representing the DAC bits as illustrated in the paragraphs above.

6.2.5. Adaptive Sampling Rate

In the above discussion, more *significant* samples, i.e., samples that correspond to the relevant portions of the signal (e.g., spikes) are sampled at higher resolution or in general resolution is traded-off for lower power. In this section, we look into proposals that adapt sampling rate depending on the significance of a particular region in the input and sample that region at a higher rate.

Artan et al. (2012) observed that for multi-unit monitoring for applications such as BCI, the accurate sampling of the spikes is vital, but the rest of the signal is also important to determine baseline such as noise levels. Thus, Artan et al. proposed to have an adaptive sampling approach where the full sampling rate is used for sampling spikes, and a lower sampling rate is used for sampling the rest of the input signal. The authors proposed a closed-loop approach using the spike detection circuit readily available in multi-unit monitoring systems to identify the starting point of the spikes. The reported approach saves up to 62% power without affecting the accuracy of the spike detection.

(Zheng et al. 2014) has a similar approach, where the main goal is to better sample the peaks of spikes, which is crucial for accurate spike sorting. Zheng et al. (2014) used two clocks, a coarse clock (16 kHz) for identifying the spike peaks, and a fine clock (64 kHz) for sampling the rest of the signal. Their 9-bit ADC consumes $3\ \mu W$ from a 3.3 V power supply on 0.35 μm technology.

The proposals related to adapting the sampling rate are not limited to multi-unit monitoring. In (Zaare and Maymandi-Nejad 2013), the sampling rate is adapted based on the signal rate for ECG applications. Similar to the two approaches above, (Zaare and Maymandi-Nejad 2013) observed that the ECG can be divided into low and high-activity regions (around QRS complex), and use this observation to switch between two different sampling rates. The authors have shown that their 10-bit adaptive SAR ADC consumes 54 nW, at a sampling rate of 250 S/s on 0.18 μm technology using a 1 V power supply.

To evaluate the adaptive sampling approach, the EKG example given above is used. Here, two different sampling rates are applied: the fast sampling rate for the QRS complex and the slow sampling rate for the rest of the signal. The original EKG signal is sampled at $f_s = 360$ Hz. The adaptive sampling approach is tested at three different sampling rate pairs as shown in Table 1. The total energy consumption of the DAC capacitive array for the total duration of this EKG (over half an hour) for each of these cases is also given in Table 1.

As seen in Table 1, the adaptive sampling can save up to 66% energy while still sampling the QRS complex at full sampling rate. Additionally, the total energy used when sampling at $[f_s, f_s/2]$ is close to

Table 1. Sampling rates for the adaptive sampling experiments

Fast Sampling Rate	Slow Sampling Rate	Energy [μJ]	Energy Saving: Both Sampled at Fast Rate *vs.* Adaptive
f_s		27.09	N/A
$f_s/2$		13.55	N/A
$f_s/4$		6.78	N/A
f_s	$f_s/2$	15.06	44%
f_s	$f_s/4$	9.04	66%
$f_s/2$	$f_s/4$	7.53	44%

the energy when sampling the entire signal at $f_s/2$. More specifically, the energy consumption is only 11% higher than the energy consumption at $f_s/2$, while sampling the QRS at f_s. This is because the QRS complex takes only a fraction of the entire EKG signal duration.

Table 2 provides performance parameters of the key preconfigured, reconfigurable, and signal-adaptive ADCs covered in this chapter. These ADCs provide resolutions ranging from 3 to 12 bits, and total power consumption ranging from 8.7 nW to 13 μW. The work of Yaul and Chandrakasan (2014) stands out as the proposal with the widest range of power options.

7. APPLICATIONS AND DISCUSSIONS

The long-term (12 hours) calibration/reconfiguration interval of O'Driscoll et al. (2011) is very fitting to their application for BCI (Brain-Computer Interfaces). The BCI application would not gain much power savings from dynamically changing the resolution, rapidly. In other applications including multi-unit monitoring outside BCI, dynamical adaptation of the ADC parameters rapidly (e.g. for every sample) can reduce power further while preserving accuracy. The two settings although similar do have important distinctions depending on the target application. For instance, for Brain-Computer Interface (BCI) applications based on multi-cellular recordings, the key point is to identify an action potential (i.e. a spike) is fired, and to identify which neuron fired (i.e. spike sorting). In this setting, the variation on the background noise levels and distortion in spike waveforms appear on the scale of days. In contrast, in EEG, there are changes in average value of the signal depending on the activity (e.g. sleep spindles, alpha waves, seizure waveforms, and interictal spikes) that can gain from dynamic adaptation of the resolution of the signal much more rapidly. Furthermore, even for BCI, some adaptive approach can be useful to reduce number of samples (For instance see (Artan et al. (2012) and Zheng et al. (2014)).

If an application relies heavily on the morphology of the signal epochs (e.g. spikes for BCI), and these morphologies drift in time (e.g. due to biofouling), a periodic reconfiguration is necessary to maximize the SNR. In such a case, using Reconfigurable ADC architectures is crucial. If such an architecture requires extensive retraining and associated data storage and processing, such parameters should also be considered when determining overall system power.

Table 2. Comparison of preconfigured, reconfigurable, and signal-adaptive ADCs

Reference	V_{dd} [V]	Resolution [bits]	Area [mm²]	Power [μW]	Technology [μm]
(Do et al. 2012)	1	9	0.04	0.16	0.18
(O'Driscoll et al 2010)	1.2, 2.5	3-8	0.07	0.23-0.9	0.13
(Tang et al. 2014)	0.9	5-9	0.15	1.33	0.18
(Chiang et al. (2013)	1	12		0.289	0.18
(Yaul and Chandrakasan 2014)	0.5-1	10	0.12	0.008-13	0.18
(Huang et al. 2012)	0.6	10	0.082	1.04	0.18
(Zheng et al. 2014)	3.3	9		3	0.35
(Zaare and Maymandi-Nejad 2013)	1	10		0.054-0.570	0.18

For other applications, there is a constant reliance on instantaneous signal values for processing (e.g. BCI) that the electrophysiological signal has a repetitive behavior that is of primary interest. For these applications, switching between two levels of accuracy can be a better trade-off between simplicity and power consumption. These two levels can be based on the resolution, or sampling rate of the ADC.

On the other hand, for applications that monitor for *rare events* such as detecting epileptic seizures, there could be multiple levels of accuracy. Here, the design should be able to scale down rather quickly if there is no event of interest (e.g. no seizures), but the system should still be responsive based on the alert status at any given moment. Most importantly, the system should be able to ramp up its accuracy in case that the rare event is imminent. Such systems require circuits to determine which accuracy level to use at any given time regardless of number of accuracy levels. These circuits should be designed carefully to consume negligible power to avoid negating the benefits of the multi-level design (Artan et al. 2012).

8. CONCLUSION

In this chapter, power consumption of the Successive Approximation Register (SAR) ADCs targeting wearable and implantable devices is discussed. It is shown that for digitizing electrophysiological signals, there is room for improvement to optimize the power consumption of SAR ADCs. Signal-aware SAR ADCs can be made to consume very low-power, especially when the needs for the target application is taken into account. In particular, reconfigurable, and signal-adaptive SAR ADCs are reviewed and evaluated for their power consumption. Significant power reduction is shown when the ADC parameters such as resolution and sampling rate is modulated based on the requirements of the target application, and these results are reinforced with simulations against electrophysiological data. More specifically, it is shown that even by using a pair of power/accuracy settings, up to 67% power saving can be achieved for monitoring the QRS complex in EKG.

Level-Crossing ADCs are briefly covered (See Appendix 1) for their inherent signal-specific behavior. As the future work, the level-crossing ADCs will be investigated further for optimizing their power consumption for different target applications.

Here number of comparisons is used as a metric to evaluate the suitability of adapting an ADC parameter based on the signal characteristics. It is further pointed out that not each comparison consumes equal energy but the consumption amount is dependent on the actual values that are compared. These metrics, however do not address the accuracy of the evaluated methods. Another important metric is the percentage of wrong predictions for a given dataset. This number determines how often an algorithm fails to converge to an input, and need to take additional measures (i.e., more iterations, or starting over the search from the beginning). A related parameter is the amount of extra work done (or energy spent) for wrong predictions. In this chapter, the cases covered have a high success rate for prediction so that the energy spent for wrong predictions is not significant. It should be noted that this can be achieved by any of the approaches above provided that a conservative set of parameters is used. The case, where parameters are selected for highly-speculative operation can definitely benefit power consumption if designed well. However, such speculative design is beyond the scope of this chapter, and is left as future work. The focus of the chapter is on adapting the ADC parameters to the signal to reduce power consumption without specifying what algorithm to use to make the decision in real-time to change the parameters. The counterpart work, which evaluates algorithmic aspects that leads to making the adaption decisions in real-time based on the target application is beyond the scope of this chapter, however will also be covered in our future work.

REFERENCES

Artan, N. S., Xu, X., Shi, W., & Chao, H. J. (2012). Optimizing Analog-To-Digital Converters for Sampling Extracellular Potentials. In *Proceedings of* the *34th Annual International Conference of the IEEE Engineering in Medicine and Biology Society (EMBC 2012), 1663-1666*. doi:10.1109/EMBC.2012.6346266

Avila, D., Alvarez, E., & Abusleme, A. (2014). Digital assistance for energy reduction in ADCs using a simple signal prediction algorithm. In *Proceedings of IEEE 5th Latin American Symposium on Circuits and Systems (LASCAS)*, 1-3. doi:10.1109/LASCAS.2014.6820264

Blackrock. (2001). *Blackrock Microsystems Utah Array*. Retrieved August 31, 2015, from http://www.blackrockmicro.com/content.aspx?id=78

Bolomey, L., Meurville, E., & Ryser, P. (2009). Implantable ultra-low power DSP-based system for a miniature chemico-rheological biosensor. *Procedia Chemistry*, *1*(1), 1235–1238. doi:10.1016/j.proche.2009.07.308

Chiang, K. C., Artan, N. S., & Chao, H. J. (2013). A Signal-Specific Approach for Reducing SAR-ADC Power Consumption. In *Proceedings of* the *IEEE Biomedical Circuits and Systems Conference (BioCAS), 278-281*. doi:10.1109/BioCAS.2013.6679693

Danova, T. (2015). *The Wearables Report: Growth trends, consumer attitudes, and why smartwatches will dominate. Business Insider*. Retrieved August 31, 2015, from http://www.businessinsider.com/the-wearable-computing-market-report-2014-10

Do, A. T., Lam, C. K., Tan, Y. S., Yeo, K. S., Jia, H. C., Zou, X., & Je, M. et al. (2012). A 160 nW 25 kS/s 9-bit SAR ADC for neural signal recording applications. In *Proceedings IEEE 10th International New Circuits and Systems Conference (NEWCAS)*, (pp. 525-528). doi:10.1109/NEWCAS.2012.6329072

EntroMedics. (2015). *EntroMedics VBLOC Maestro® Rechargeable System*. Retrieved August 30, 2015, from http://www.enteromedics.com/usa/vbloc_maestro_system.asp

Fitbit. (2015). *FitBit Activity Monitors*. Retrieved August 30, 2015, from http://www.fitbit.com

Goldberger, A. L., Amaral, L. A. N., Glass, L., Hausdorff, J. M., Ivanov, P. Ch., Mark, R. G., & Stanley, H. E. et al. (2000). PhysioBank, PhysioToolkit, and PhysioNet: Components of a New Research Resource for Complex Physiologic Signals. *Circulation*, *101*(23), e215–e220. doi:10.1161/01.CIR.101.23.e215 PMID:10851218

Ha, H., Lee, S.-K., Kim, B., Park, H.-J., & Sim, J.-Y. (2014). A 0.5-V, 1.47- μW 40-kS/s 13-bit SAR ADC With Capacitor Error Compensation. *IEEE Transactions on Circuits and Wystems. II, Express Briefs*, *61*(11), 840–844. doi:10.1109/TCSII.2014.2350378

Ha, S., Kim, C., Chi, Y. M., Akinin, A., Maier, C., Ueno, A., & Cauwenberghs, G. (2014). Integrated Circuits and Electrode Interfaces for Noninvasive Physiological Monitoring. *IEEE Transactions on Bio-Medical Engineering*, *61*(5), 1522–1537. doi:10.1109/TBME.2014.2308552 PMID:24759282

Harrison, R. R., & Charles, C. (2003). A low-power low-noise CMOS amplifier for neural recording applications. *IEEE Journal of Solid-State Circuits*, *38*(6), 958–965. doi:10.1109/JSSC.2003.811979

Huang, G.-Y., Chang, S.-J., Liu, C.-C., & Lin, Y.-Z. (2012). A 1-μW 10-bit 200-kS/s SAR ADC With a Bypass Window for Biomedical Applications. *IEEE Journal of Solid-State Circuits*, *47*(11), 2783–2795. doi:10.1109/JSSC.2012.2217635

Inose, H., Aoki, T., & Watanabe, K. (1966). Asynchronous delta-modulation system. *Electronics Letters*, *2*(3), 95–96. doi:10.1049/el:19660077

Jonsson, B. E. (2011). An empirical approach to finding energy efficient ADC architectures. In *Proceedings 2011 International Workshop on ADC Modelling, Testing and Data Converter Analysis and Design and IEEE ADC Forum.*

Juniper Research. (2014). *Smart Wearables Market to Generate $53bn Hardware Revenues by 2019.* Retrieved August 25, 2015, from http://www.juniperresearch.com/press/press-releases/smart-wearables-market-to-generate-$53bn-hardware

Kenington, P. B., & Astier, L. (2000). Power consumption of A/D converters for software radio applications. *IEEE Transactions on Vehicular Technology*, *49*(2), 643–650. doi:10.1109/25.832996

Kiani, M., Jow, U. M., & Ghovanloo, M. (2011). Design and Optimization of a 3-Coil Inductive Link for Efficient Wireless Power Transmission. *IEEE Transactions on Biomedical Circuits and Systems*, *5*(6), 579–591. doi:10.1109/TBCAS.2011.2158431 PMID:21922034

Kim, T. H., Artan, N. S., Viventi, J., & Chao, H. J. (2012). Spatiotemporal compression for efficient storage and transmission of high-resolution electrocorticography data. In *Proceedings of Engineering in Medicine and Biology Society* (pp. 1012–1015). EMBC.

Lee, B. G. (2015). Power and Bandwidth Scalable 10-b 30-MS/s SAR ADC. *IEEE Transactions on Very Large Scale Integration (VLSI) Systems*, *23*(6), 1103–1110.

Li, Y., Mansano, A. L., Yuan, Y., Zhao, D., & Serdijn, W. A. (2014). An ECG Recording Front-End With Continuous-Time Level-Crossing Sampling. *IEEE Transactions on Biomedical Circuits and Systems.*, *8*(5), 626–635. doi:10.1109/TBCAS.2014.2359183 PMID:25330494

Li, Y., Zhao, D., & Serdijn, W. A. (2013). A 0.8V 8-bit low-power asynchronous level-crossing ADC with programmable comparison windows. In *Proceedings of IEEE Biomedical Circuits and Systems Conference (BioCAS)*, (pp. 138-141). doi:10.1109/BioCAS.2013.6679658

Lima Silva, V. M., Lisboa de Souza, A. A., & Cavalcanti Catunda, S. Y. (2014). Flexible A/D converter architecture targeting sparse signals. In *Proceedings, IEEE International Instrumentation and Measurement Technology Conference (I2MTC)*, (pp. 1496-1500). doi:10.1109/I2MTC.2014.6860995

Liu, T., Bihr, U., Becker, J., Anders, J., & Ortmanns, M. (2014). In vivo verification of a 100 Mbps transcutaneous optical telemetric link. In *Proceedings of IEEE Biomedical Circuits and Systems Conference (BioCAS)*, (pp. 580-583). doi:10.1109/BioCAS.2014.6981792

Liu, Y. P., Yuan, C., & Hung, Y. L. Y. (2014). A capacitor constructed bypass window switching scheme for energy-efficient SAR ADC. In *Proceedings of IEEE International Symposium on Circuits and Systems (ISCAS)*, (pp. 1352-1355). doi:10.1109/ISCAS.2014.6865394

Lu, L., Yan, G., Zhao, K., & Xu, F. (2015). An Implantable Telemetry Platform System With ASIC for in Vivo Monitoring of Gastrointestinal Physiological Information. *IEEE Sensors Journal*, *15*(6), 3524–3534. doi:10.1109/JSEN.2015.2392105

Medtronic. (2015a). *Medtronic Activa® RC Neurostimulator.* Retrieved on August 22, 2015, from http://www.medtronic.eu/your-health/essential-tremor/device/dbs-therapy/ActivaRCNeurostimulator/index.htm

Medtronic. (2015b). *Medtronic MiniMed 530G with Enlite*. Retrieved on August 22, 2015, from http://www.medtronicdiabetes.com/home

Moody, G. B., & Mark, R. G. (2001). The impact of the MIT-BIH Arrhythmia Database. *IEEE Eng in Med and Biol*, *20*(3), 45–50. doi:10.1109/51.932724 PMID:11446209

O'Driscoll, S., & Meng, T. H. (2010). Adaptive signal acquisition and wireless power transfer for an implantable prosthesis processor. In *Proceedings of IEEE International Symposium on Circuits and Systems (ISCAS)*, (pp. 3589-3592). doi:10.1109/ISCAS.2010.5537805

O'Driscoll, S., Shenoy, K., & Meng, T. (2011). Adaptive Resolution ADC Array for an Implantable Neural Sensor. *IEEE Transactions on Biomedical Circuits and Systems*, *5*(2), 120–130. doi:10.1109/TBCAS.2011.2145418 PMID:23851200

Ravanshad, N., Rezaee-Dehsorkh, H., Lotfi, R., & Lia, Y. (2014). A Level-Crossing Based QRS-Detection Algorithm for Wearable ECG Sensors. *IEEE Journal of Biomedical and Health Informatics*, *18*(1), 183–192. doi:10.1109/JBHI.2013.2274809 PMID:24403416

Razak, Z., Erdogan, A., & Arslan, T. (2010). ASIC Design of an Adaptive Control Unit for Reconfigurable Analog-to-Digital Converters. In *Proceedings IEEE Computer Society Annual Symposium on VLSI (ISVLSI)*, (pp. 179-184). doi:10.1109/ISVLSI.2010.79

Rusu, A., Gonzalez, D. R. L., & Ismail, M. (2006). Reconfigurable ADCs enable smart radios for 4G wireless connectivity. *IEEE Circuits and Devices Magazine*, *22*(3), 6–11. doi:10.1109/MCD.2006.1657844

Saberi, M., Lotfi, R., Mafinezhad, K., & Serdijn, W. A. (2011). Analysis of Power Consumption and Linearity in Capacitive Digital-to-Analog Converters Used in Successive Approximation ADCs. *IEEE Transactions on Circuits and Systems. I, Regular Papers*, *58*(8), 1736–1748. doi:10.1109/TCSI.2011.2107214

Second Sight. (2015). *Second Sight Argus II Retinal Prosthesis System*. Retrieved on August 31, 2015, from http://www.2-sight.com/g-the-argus-ii-prosthesis-system-pf-en.html

Sepehrian, H., Saberi, M., & Lotfi, R. (2011). A signal-specific successive-approximation analog-to-digital converter.*IEEE International Symposium on Circuits and Systems (ISCAS)*, (pp. 1624-1627).

Sinangil, M. E., Yip, M., Qazi, M., Rithe, R., Kwong, J., & Chandrakasan, A. P. (2012). Design of Low-Voltage Digital Building Blocks and ADCs for Energy-Efficient Systems. *IEEE Transactions on Circuits and Wystems. II, Express Briefs*, *59*(9), 533–537. doi:10.1109/TCSII.2012.2208675

St. Jude Medical. (2015). *St Jude Medical Nanostim Leadless Pacing*. Retrieved on August 29, 2015, from http://sjm.com/leadlesspacing.aspx

Stojcevski, A., Veljanovski, R., Singh, J., Faulkner, M., & Zayegh, A. (2003). A Real-Time Reconfigurable Pipelined Architecture with Advanced Power Management for UTRA-FDD. In *Proceedings of the IEEE Personal Indoor and Mobile Radio Communications (PIMRC)*, (pp. 64-68). doi:10.1109/PIMRC.2003.1264233

Strickland, E. (2013). Teeny Tiny Pacemaker Fits Inside the Heart. *IEEE Spectrum*.

Tang, H., Sun, Z. C., Chew, K. W. R., & Siek, L. (2014). A 1.33 µW 8.02-ENOB 100 kS/s Successive Approximation ADC With Supply Reduction Technique for Implantable Retinal Prosthesis. *IEEE Transactions on Biomedical Circuits and Systems*, *8*(6), 844–856. doi:10.1109/TBCAS.2014.2300186 PMID:25608284

Tang, W., Osman, A., Kim, D., Goldstein, B., Huang, C., Martini, B., & Culurciello, E. et al. (2013). Continuous Time Level Crossing Sampling ADC for Bio-Potential Recording Systems. *IEEE Transactions on Circuits and Systems. I, Regular Papers*, *60*(6), 1407–1418. doi:10.1109/TCSI.2012.2220464 PMID:24163640

Trakimas, M., & Sonkusale, S. R. (2011). An Adaptive Resolution Asynchronous ADC Architecture for Data Compression in Energy Constrained Sensing Applications. *IEEE Transaction on Circuits and Systems I:Regular Papers*, *58*(5), 921–934.

Tsai, J., Wang, J., & Su, Y. (2013). Piezoelectric rubber films for human physiological monitoring and energy harvesting. In *Proceedings of IEEE 26th International Conference on Micro Electro Mechanical Systems (MEMS)*, (pp. 841-844). doi:10.1109/MEMSYS.2013.6474374

Van Rethy, J., De Smedt, M., Verhelst, M., & Gielen, G. (2013). Predictive sensing in analog-to-digital converters for biomedical applications. In *Proceedings of the International Symposium on Signals, Circuits and Systems (ISSCS)*, (pp. 11-12). doi:10.1109/ISSCS.2013.6651263

Walden, R. H. (1999). Analog-to-digital converter survey and analysis. *IEEE Journal on Selected Areas in Communications*, *17*(4), 539–550. doi:10.1109/49.761034

Xia, L., Cheng, J., Glover, N. E., & Chiang, P. (2014). 0.56 V, –20 dBm RF-Powered, Multi-Node Wireless Body Area Network System-on-a-Chip With Harvesting-Efficiency Tracking Loop. *IEEE Journal of Solid-State Circuits*, *49*(6), 1345–1355. doi:10.1109/JSSC.2014.2305074

Yaul, F. M., & Chandrakasan, A. P. (2014). A 10 bit SAR ADC With Data-Dependent Energy Reduction Using LSB-First Successive Approximation. *IEEE Journal of Solid-State Circuits*, *49*(12), 2825–2834. doi:10.1109/JSSC.2014.2352304

Yip, M., & Chandrakasan, A.P. (2013). A Resolution-Reconfigurable 5-to-10-Bit 0.4-to-1 V Power Scalable SAR ADC for Sensor Applications. *IEEE Journal of Solid-State Circuits, 48*(6), 1453-1464.

Yuce, M. R., Keong, H. C., & Chae, M. S. (2009). Wideband Communication for Implantable and Wearable Systems. *IEEE Transactions on Microwave Theory and Techniques*, *57*(10), 2597–2604. doi:10.1109/TMTT.2009.2029958

Zaare, M., & Maymandi-Nejad, M. (2013). A new power efficient wireless non-uniform adaptive-sampling ECG recording system. In *Proceedings 21st Iranian Conference onElectrical Engineering (ICEE)*. doi:10.1109/IranianCEE.2013.6599768

Zhang, X., & Lian, Y. (2014). A 300-mV 220-nW Event-Driven ADC With Real-Time QRS Detection for Wearable ECG Sensors. *IEEE Transactions on Biomedical Circuits and Systems*, *8*(6), 834–843. doi:10.1109/TBCAS.2013.2296942 PMID:25608283

Zheng, L., Leene, L. B., Liu, Y., & Constandinou, T. G. (2014). An adaptive 16/64 kHz, 9-bit SAR ADC with peak-aligned sampling for neural spike recording. In *Proceedings of IEEE International Symposium on Circuits and Systems (ISCAS)*. doi:10.1109/ISCAS.2014.6865652

KEY TERMS AND DEFINITIONS

Input-Aware Circuit: Any analog or mixed signal circuit that is designed to have a wide range of system parameters that can be tailored to the characteristics of an input signal. This tailoring can be in the form of reconfiguration (See *reconfigurable circuit*), or adaptation (See *signal-adaptive circuit*)

Level-Crossing ADC: An analog-to-digital converter architecture with non-uniform sampling, where a sample is taken only when and every time the analog input signal reaches one of a set of predetermined thresholds.

Reconfigurable Circuit: An analog or mixed-signal circuit, whose system parameters can be updated in the field usually as a result of the changes in the characteristics of the input signal. These updates are not expected to be very often, *i.e.*, the frequency of parameter updates for reconfigurable circuits is much lower than the bandwidth of this circuit.

Signal-Adaptive Circuit: An analog or mixed-signal circuit that monitors the characteristics of its input signal in real-time, and change its system parameters on the fly to optimize performance for the current input.

Successive Approximation ADC (SAR ADC): An analog-to-digital converter architecture, where the code corresponding to the analog input signal is identified by starting with an initial guess, and progressively converging on the input value by trial-and-error. The approximation is usually achieved by a binary search starting from the middle of the possible input range, and covering the entire input range. However, non-binary search and/or search covering a subset of the input range are also used in SAR ADCs.

Tracking ADC: Like a SAR ADC, tracking ADC is an analog-to-digital converter architecture, where the code corresponding to the analog input signal is identified by starting with an initial guess, and progressively converging on the input value by trial-and-error. The key distinguishing property of the tracking ADC is that the initial guess is a function of the recent samples, and typically is equal to the most recent sample, or the average of the few recent samples. It relies on the observation that most of the real-world signals are continuous and does not abruptly change their values drastically.

Wearable/Implantable Medical Devices (WIMDs): A device either worn on the body as part of a garment, or as a standalone gadget, or surgically placed within the body to help diagnose, monitor, or treat a medical condition.

APPENDIX 1: LEVEL-CROSSING ADC

Level-crossing ADCs (LC ADCs) rely on level-crossing principles introduced decades ago (Inose et al. 1966), yet they recently gain attention due to their applicability in ultra-low power systems.

A traditional ADC samples the input at a fixed sampling rate without considering the temporal variation in the actual signal. As an acute example, an ADC sampling a constant input will still take many samples that have the same exact value or values slightly different than the constant input value depending on the noise. For systems with tight power constraints such as WIMDs, the traditional approach is not optimum. Unlike traditional uniform sampling ADCs, LC ADCs only samples the input if the input passes any one of the amplitude thresholds. Thus, the sampling rate of the LC ADC is not fixed, and its sampling is not uniform. Instead, the sampling rate is determined by the speed of the instantaneous variations in the input signal. These ADCs do not need or assume any knowledge about the signal they are sampling beyond very basic statistics about the signal.

Considering the constant input example above, if the difference between the constant input and either the threshold above or below this value is higher than the instantaneous noise, the LC ADC will not take any new samples. In Figure 10, we show an analog signal and compare how it is sampled using the LC and uniform sampling approaches. In this example, the input range is assumed to be 0 to 1 V, and the uniform sampling is done every 2 ms, i.e., at a sampling rate of 500 Hz. As such during the 12 ms shown in the figure, the uniform sampling results in 6 samples (at times 2 ms, 4 ms, 6 ms, …, 12 ms). The LC sampling, on the other hand, is set to generate a sample only if the input crosses the voltage threshold values $0.2c$, where $c = 1 \ldots 5$. For the 12 ms duration in this example, only 4 samples are produced using LC sampling at voltage threshold values. It can be easily observed that in the example, the signal has a small slope until around 8 ms. During this time, LC has taken only one sample, whereas the uniform sampling has 4 samples. However, after 8 ms, the signal has a larger slope and climbs rapidly. In this region, the LC has one additional sample compared to the uniform sampling. As far as the amount of information carried with respect to the power consumption, LC shows two advantages. First, LC ADC

Figure 10. Sampling with a level crossing ADC vs. uniform sampling

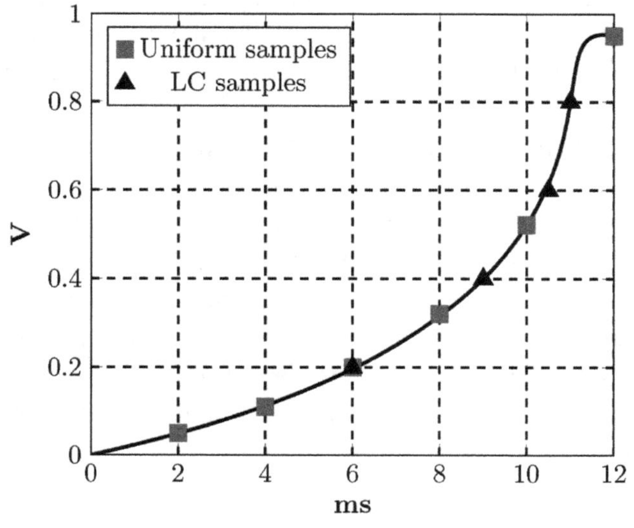

Table 3. Comparison of Level Crossing ADCs

Reference	V_{dd} [V]	Resolution [bits]	Area [mm²]	Power [μW]	Technology [μm]
(Trakimas and Sonkusale 2011)	0.7	8	0.96	25	0.18
(Li et al. 2013)	0.8	8	0.50	0.84	0.18
(Tang et al. 2013)	3.3	5	2.25	125.4	0.5
(Ravanshad et al. 2014)	1.2	7	0.136	0.62	0.13
(Zhang and Lian 2014)	0.3	5	0.36	0.22	0.13

does not get unnecessary samples for slow changing signals (as in the first 8 ms in the example), which saves power. Secondly, it gets as many samples as needed for rapidly changing signals (after 8 ms mark), so that it provides equal or better accuracy compared to the uniform sampling.

For LC ADC design, there is a trade-off between the step size (i.e., difference between two *levels*), and the energy consumption. Observe that majority of the circuit of an LC ADC is in sleep mode until a level-crossing event occurs. For a given signal, if there are fewer levels, it is expected that number of level-crossing events are also smaller compared to the case with more levels. Thus, if number of levels is small, it is expected that the ADC consumes less power. On the other hand, to distinguish between two signal amplitudes, there should be at least one level crossing between these two signal amplitudes. Thus, having more levels increases the resolution of the ADC.

Yaul and Chandrakasan (2014) pointed out three major issues with the LC ADCs, namely slope overload, a continuously on and power consuming comparator, and the DSP requirements for processing the asynchronous samples.

Table 3 shows specifications of various recent LCADC proposals.

APPENDIX 2: MATLAB/OCTAVE SOURCES

Algorithm 1. Matlab/Octave Function to calculate the power drawn from the V_{ref} for a sine wave as shown in Figure 6.

```
function [Ps, Ener] = PVref_plot_sine_for_print (N, Vref, fclk, C0)
  Ctotal = C0*2^N;
  d=0:359/1000:359-1/1000;
  size(d);
  idx0=round((sind(d)+1)* (2^(N-1) - 1)/2 + 2^(N-1));idx0 = idx0';
  idx1=round((sind(d)+1)* (2^(N-1) - 1)/2 + 2^(N-2));idx1 = idx1';
  idx2=round((sind(d)+1)* (2^(N-1) - 1)/2);idx2 = idx2';
  codes0=de2bi(idx0, N, 2, "left-msb");
  codes1=de2bi(idx1, N, 2, "left-msb");
  codes2=de2bi(idx2, N, 2, "left-msb");
  for i=0:size(idx1)-1
    Vin0(i+1) = code_to_v(codes0(i+1,:),N,Vref);
```

```
    Vin1(i+1) = code_to_v(codes1(i+1,:),N,Vref);
    Vin2(i+1) = code_to_v(codes2(i+1,:),N,Vref);
  end
  % Power
    Ps0  = PVref(Vin0, N, Vref, fclk, C0);
    Ps1  = PVref(Vin1, N, Vref, fclk, C0);
    Ps2  = PVref(Vin2, N, Vref, fclk, C0);
  % Energy
    Ener0 = Ps0*(N+1)/fclk;
    Ener1 = Ps1*(N+1)/fclk;
    Ener2 = Ps2*(N+1)/fclk;
    Ener = [Ener0; Ener1; Ener2];
    Ps = [Ps0; Ps1; Ps2];
endfunction
```

Algorithm 2. Matlab/Octave Function to calculate the power drawn from the V_{ref} *as in Equation 6.*

```
function [P] = PVref (Vin, N, Vref, fclk, C0)
  Coeff = (2^N * fclk * C0)/(N+1);
  Item1  = (5/6 - (1/2)^N - 1/3 *(1/2)^(2*N))*Vref^2;
  Item2  =  -1/2.*Vin.^2 - (1/2)^N.*Vin.*Vref;
  P = Coeff * (Item1 + Item2);
endfunction
```

Algorithm 3. Matlab/Octave Function to convert the binary code to voltage value.

```
function [Vin] = code_to_v (code, N, Vref)
  idx=1:N;
  powtwo=pow2(idx);
  Vin = sum(code./powtwo).*Vref;
endfunction
```

Algorithm 4. Matlab/Octave Function to plot the QRS complex example in Figure 9. *The data is from the MIT-BIH Arrhythmia Database.*

```
function  plot_heart_QRS ()
  load('mat/100m.mat');
  plot(val(1,12800:13100),'k', 'linewidth', 2);
  line([130 130],[800 1300], 'linewidth',2, 'linestyle', '--');
  line([170 170],[800 1300], 'linewidth',2, 'linestyle', '--');
  text(136,1250, 'QRS', 'fontsize', 16, 'fontweight', 'bold')
  axis off
 endfunction
```

Chapter 9
The Design of Ultra Low Power RF CMOS LNA in Nanometer Technology

Kavyashree P.
UTL Technologies Ltd., India

Siva S. Yellampalli
UTL Technologies Ltd., India

ABSTRACT

In this chapter, an ultra low power CMOS Common Gate LNA (CGLNA) with a Capacitive Cross-Coupled (CCC) g_m boosting scheme is designed and analysed. The technique described has been employed in literature to reduce the Noise Figure (NF) and power dissipation. In this work we have extended the concept for low voltage operation along with improving NF and also for significant reduction in current consumption. A g_m boosted CCC-CGLNA is implemented in 90nm CMOS technology. It has a gain of 9.9dB and a noise figure of 0.87dB at 2.4GHz ISM band and consumes less power (0.5mw) from 0.6V supply voltage. The designed g_m boosted CCC-CGLNA is suitable for low power application in CMOS technologies.

INTRODUCTION

The IEEE 802.15.4 standard was specially designed to cater for the fundamental lower network layers of wireless personal area network (WPAN) which focuses on low-cost, low-speed ubiquitous communication between devices (Oh & Lee, 2005). The emphasis is on very low cost communication of nearby devices with low power consumption and little to no underlying infrastructure. The concept of IEEE 802.15.4 standard is to provide communications over distances up to about 10 meters and with maximum transfer data rates of 250 kbps. Low power consumption has been the centre of attention for this technology. In the context of mobile wireless applications, lower power consumption can lead to longer battery life or in another word longer time over which a mobile device can be used without having to recharge. The IEEE 802.15.4 frequency bands align with the license free radio bands that are available around the globe. Of the bands available, the 2.4 GHz band is the most widely used in view of the fact

DOI: 10.4018/978-1-5225-0190-9.ch009

Figure 1. RF receiver

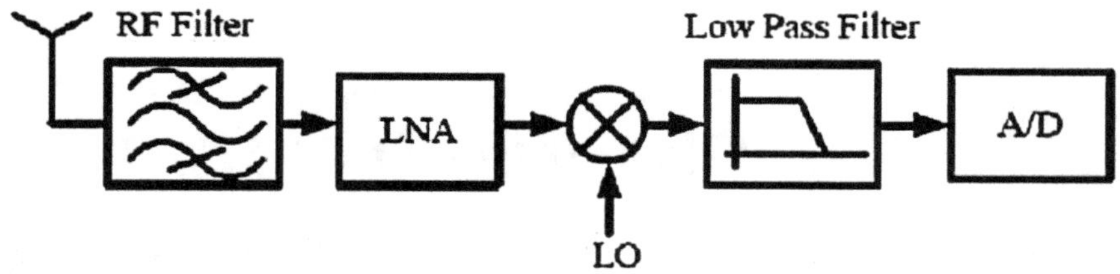

that it is available globally and this brings many economies of scale. Figure 1 shows the generic radio frequency (RF) mobile receiver.

The generic RF receiver consists of antenna, low noise amplifier (LNA), mixer, local oscillator, low pass filter and a data converter. From the Figure 1 it is seen that LNA is the first active amplification block in the receiving path. Being the first block of the receiver, the LNA plays a crucial role in amplifying the received signal while adding little noise to it. In addition, the input of the LNA needs to be matched to the output of the filter following the antenna to prevent the incoming signal from reflecting back and forth between the LNA and the antenna. While the LNA is a relatively simple design compared to other RF components in a cellular receiver chain, the performance tradeoffs challenge the LNA design engineer. LNA design typically involves making choices between directly competing performance parameters such as: noise, gain, linearity and power consumption. In the IEEE 802.15.4 standard, the noise figure (NF) and linearity requirement can be relaxed in order to achieve other important parameters such as gain and power consumption.

In this chapter we develop a thorough understanding of low-power LNA design and introduce to new ultra low power LNA design for IEEE 802.15.4 standard. The LNA is optimized for low noise figure and low power consumption.

1. TOPOLOGIES OF CMOS LNA

In Low noise amplification, two topologies are widely adopted: common source amplifier with inductive degeneration and common gate amplifier (Lee, 1998).

1.1 Common Source LNA (CSLNA)

The variable resistive input impedance can be achieved by adding an inductance in series with the source of the MOSFET to degenerate the common source amplifier as shown in Figure 2 and its small signal model is shown in Figure 3.

1.1.1 Input Impedance of CSLNA

It is instructive to calculate the input impedance looking into gate of MOSFET when its source is degenerated by inductor Ls. The input impedance of CSLNA is

$$Z_{in} = \frac{V_{in}}{I_{in}} = SL_s + \frac{1}{SC_{gs}} + \left(\frac{g_m}{C_{gs}}\right)L_s \qquad (1)$$

Figure 2. Common source low noise amplifier (CSLNA)
(Lee, 1998)

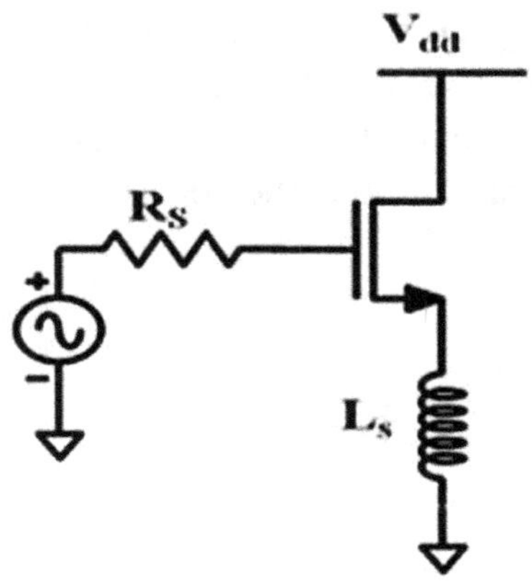

Figure 3. Small signal model of inductive source degenerated CSLNA
(Lee, 1998)

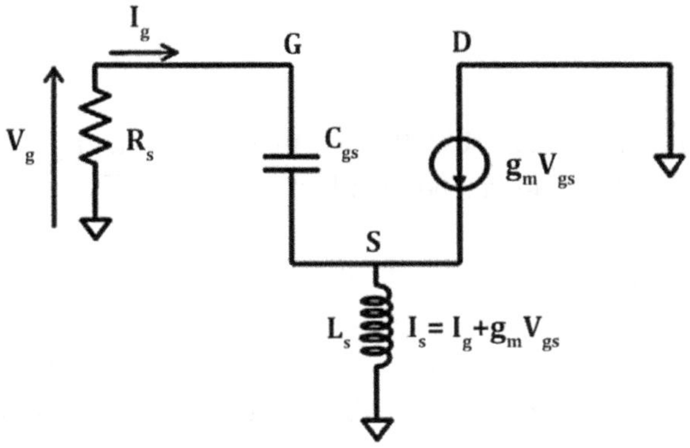

Note that there is a resistive component in the input impedance. To get a resistive input impedance of 50Ω at the operating frequency, the inductive component L_s resonate with capacitive component C_{gs} and cancel each other now $(g_m/C_{gs})\ L_s$ is set equal to 50Ω.

$$j\omega_0 L_s + \frac{1}{j\omega_0 C_{gs}} = 0 \tag{2}$$

$$Z_{in} = \left(\frac{g_m}{C_{gs}}\right) L_s = \omega_T L_s \tag{3}$$

One of the most attractive advantages of this topology is that the resistive component of the input impedance is noiseless because it is synthesized using an inductor, unlike other techniques where a noisy resistor is added in the signal path to provide the 50Ω termination resistance (Lee, 1998). This explains the low noise performance and popularity of the inductively degenerated common source LNA.

As the input matching circuit is a pure series RLC resonant circuit. At resonance, the voltage across C_{gs} is enhanced by Q times (Lee, 1998), where Q is the quality factor of the input matching RLC circuit. In other words, the Q enhancement mechanism provides a free gain of Q for both the input signal and the noise from the source resistance R_s. The added gain from the input matching circuit helps to suppress channel thermal noise, which accounts for the superior noise performance of the common-source LNA with inductive degeneration. Furthermore, the effective transconductance is also enhanced by a factor of Q as

$$G_m = g_m Q = \frac{g_m}{2\omega_0 C_{gs} R_s} \tag{4}$$

Where $Q = \frac{1}{\omega C_{gs} R_s}$

If the input is matched to R_s we have

$$G_m = \frac{1}{2R_s}\left(\frac{\omega_T}{\omega_0}\right) \tag{5}$$

Usually in an RF system R_s is equal to 50Ω. It is worth nothing that the effective transconductance G_m is only related to the ratio of ω_T to ω_0 and is independent on MOSFET small signal transconductance g_m.

1.1.2. Noise Figure of CSLNA

A major advantage of the common-source amplifier with inductive degeneration is that the resistive input impedance is noiseless, unlike other topologies where a noisy resistor is added in the signal path to create a 50Ω terminating impedance. Figure 4 depicts a small-signal circuit for noise analysis of CSLNA with inductive degeneration (Lee, 1998). The noise factor of this topology is

$$F = \frac{i_{n,out}^2}{i_{RS}^2} = 1 + \frac{\gamma}{\alpha}\left(\frac{\omega_0}{\omega_T}\right) \tag{6}$$

$$i_{RS}^2 = \frac{4kT\Delta f}{R_s} \tag{7}$$

$$i_d^2 = 4kT\gamma g_{do}\Delta f \tag{8}$$

$$i_g^2 = 4kT\delta g_g \Delta f \tag{9}$$

$$Q = \frac{1}{\omega_0 C_{gs} R_s} \tag{10}$$

Figure 4. Small signal model of inductive source degenerated CSLNA for noise derivation (Lee, 1998)

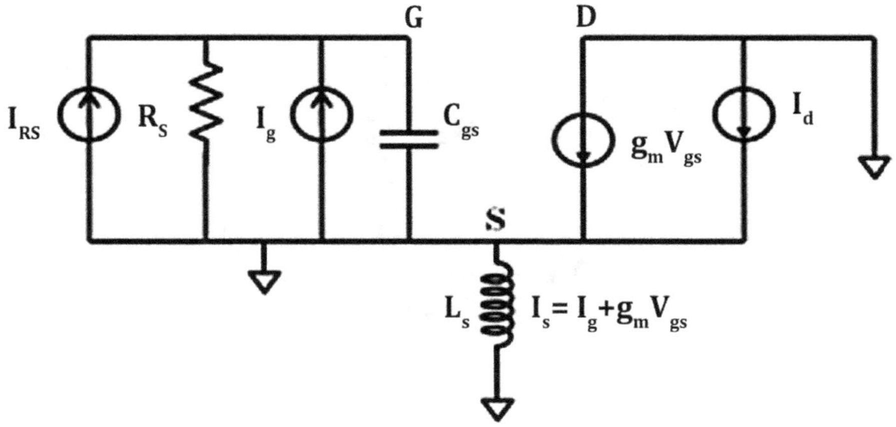

The total noise current at the MOSFET drain terminal is

$$i_{n,out}^2 = i_{RS}^2 + i_d^2 + i_g^2 \quad (11)$$

$$F = \frac{i_{n,out}^2}{i_{RS}^2} = 1 + \frac{\gamma}{\alpha}\left(\frac{\omega_0}{\omega_T}\right) \quad (12)$$

Where α, γ, and δ are bias dependent parameters (Lee, 1998) and ω_0 and ω_T are the operating and unity current gain frequencies.

1.2 Common Gate LNA (CGLNA)

The circuit topology of a conventional CGLNA and its small signal model of CGLNA are shown in Figure 5 and Figure 6 respectively. The CGLNA is well known for wideband applications (Tran, Boon, Do, & Yeo, 2012). It is relevant to consider the input impedance and voltage gain of a CGLNA.

1.2.1 Input Impedance and Voltage Gain

The input impedance of a CGLNA is given by

$$Z_{in} = \frac{V_{in}}{I_{in}} = R_s + \frac{1}{1/r_{ds} + g_m} + \frac{R_L}{1 + g_m r_{ds}} \quad (13)$$

Note that when $R_s = \frac{1}{g_m}$ then $Z_{in} = R_s$ (14)

Where $R_s = \frac{1}{g_m}$ (15)

Figure 5. Common gate low noise amplifier (CGLNA)
(Tran, Boon, Do, & Yeo, 2012)

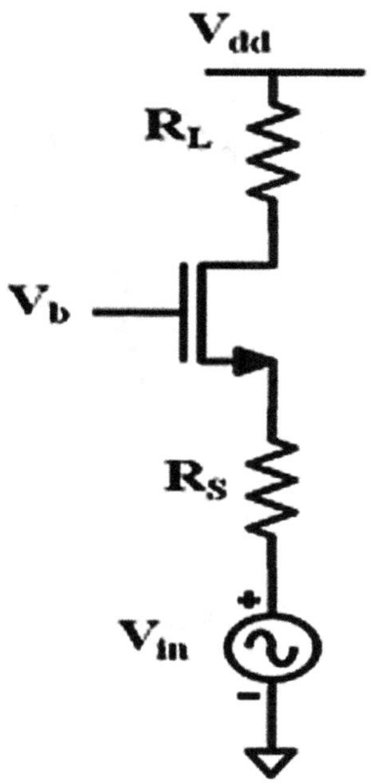

Figure 6. Small signal model of conventional CGLNA
(Tran, Boon, Do, & Yeo, 2012)

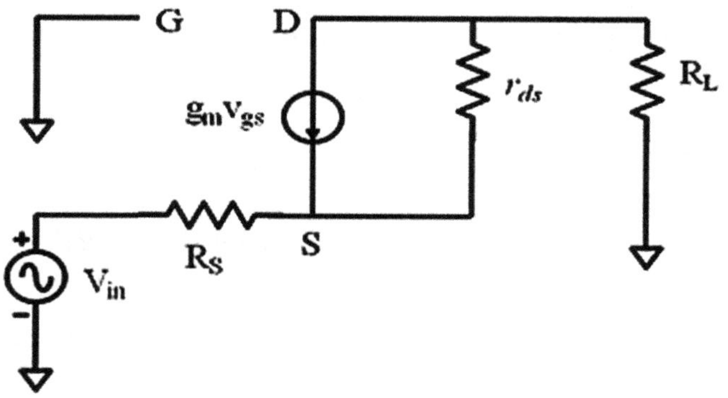

The voltage gain of a CGLNA is given by

$$A = \frac{V_{out}}{V_{in}} = \frac{\frac{1}{r_{ds}} + g_m}{\frac{1}{R_L} + \frac{1}{r_{ds}}} \tag{16}$$

Note that when $r_{ds} \to \infty$ then $A = g_m R_L$

1.2.2 Noise Figure

The Figure 7 shows the small signal model of CGLNA for noise analysis, the effect of gate noise is omitted since its contribution to the total noise factor is negligible compared to other noise sources. The three main noise sources that are considered in this noise analysis are I_d, V_n, and V_{RL} which are the transistor drain current noise, the input voltage noise due R_s and the voltage thermal noise due to R_L respectively.

Firstly, the output noise due to V_n is analysed by the small signal model circuit is given in Figure 7

$$i_{n1}^2 = \left(\frac{V_{in}}{R_{in}}\right)^2 = \frac{4kTR_s\Delta f}{\left(R_s + \frac{1}{g_m + 1/r_{ds}} + \frac{R_L}{1+g_m r_{ds}}\right)^2} \tag{18}$$

Secondly, the output noise due to I_d is given by the small signal model circuit in Figure 7

$$i_{n2}^2 = \left(\frac{V_2}{R_L}\right)^2 = \left(\frac{V_1}{R_s}\right)^2 = \frac{i_d^2}{\left[1+g_m R_s + (R_s + R_L)\frac{1}{r_{ds}}\right]^2} = \frac{4kT\gamma g_{do}\Delta f}{\left[1+g_m R_s + (R_s + R_L)\frac{1}{r_{ds}}\right]^2} \tag{19}$$

Thirdly, the output noise due to V_{RL} according to small signal model circuit in Figure 7

$$i_{n3}^2 = \left(\frac{V_{RL}}{R_s + R_L + (1+g_m R_s)r_{ds}}\right)^2 = \frac{4kTR_L\Delta f}{\left[R_s + R_L + (1+g_m R_s)r_{ds}\right]^2} \tag{20}$$

Figure 7. Small Signal model of CGLNA for noise derivation (Tran, Boon, Do, & Yeo, 2012)

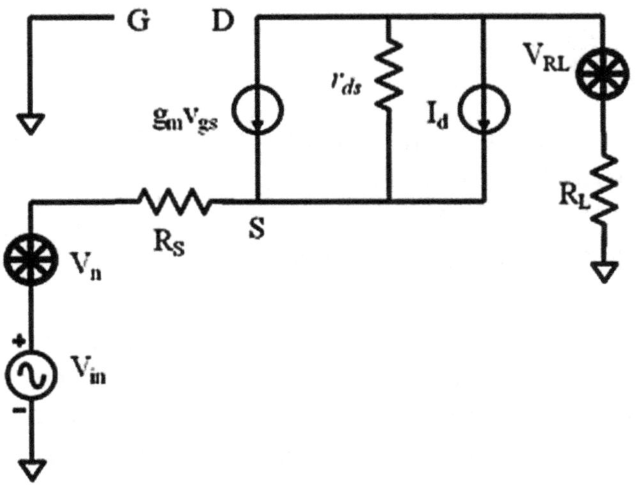

From Equations (18), (19) and (20), the total noise factor is

$$F = 1 + \frac{i_{n2}^2}{i_{n1}^2} + \frac{i_{n3}^2}{i_{n1}^2} \tag{21}$$

$$F = 1 + \frac{4kT\gamma g_{do}\Delta f}{4kTR_s\Delta f}\left(\frac{R_s + \dfrac{1}{1/r_{ds} + g_m} + \dfrac{R_L}{1 + g_m r_{ds}}}{1 + g_m R_s + (R_s + R_L)\dfrac{1}{r_{ds}}}\right)^2 + \frac{4kTR_L\Delta f}{4kTR_s\Delta f}\left(\frac{R_s + \dfrac{1}{1/r_{ds} + g_m} + \dfrac{R_L}{1 + g_m r_{ds}}}{R_s + R_L + (g_m R_s + 1)r_{ds}}\right)^2 \tag{22}$$

$$F = 1 + \frac{\gamma g_m}{\alpha R_s}\left(\frac{R_s + \dfrac{1}{1/r_{ds} + g_m} + \dfrac{R_L}{1 + g_m r_{ds}}}{1 + g_m R_s + (R_s + R_L)\dfrac{1}{r_{ds}}}\right)^2 + \frac{R_L}{R_s}\left(\frac{R_s + \dfrac{1}{1/r_{ds} + g_m} + \dfrac{R_L}{1 + g_m r_{ds}}}{R_L + R_s + (g_m R_s + 1)r_{ds}}\right)^2 \tag{23}$$

Where γ is the excess channel noise co-efficient and α is the ratio between g_m and the zero bias drain conductance g_{do}. The second term in Equation (23) represents the channel thermal noise of transistor M_1 and the last term in Equation (23) (Tran, Boon, Do, & Yeo, 2012) represents the noise contribution due to the load R_L.

If $r_{ds} \to \infty$ then Equation (23) (Tran, Boon, Do, & Yeo, 2012) becomes

$$F = 1 + \frac{\gamma}{\alpha}\frac{1}{g_m R_s} \tag{24}$$

When the input is matched, that is $R_s = 1/g_m$ then NF becomes

$$F = 1 + \frac{\gamma}{\alpha} \tag{25}$$

The conventional CGLNA suffers a relatively high NF. Noise reduction techniques are used to improve the NF of the CGLNA (Tran, Boon, Do, & Yeo, 2012)

2. COMPARISON OF CSLNA AND CGLNA

There are basically two topologies in LNA-design, the common source amplifier and the common gate amplifier (Sobhy et al., 2011). They can be single ended or differential, come in cascaded versions, and sometimes they are combined in different ways (Sobhy et al., 2011). This section presents and compares the two basic topologies of LNA in a neat and structured way. Figure 8 shows general scheme of the two basic topologies.

Figure 8. a) The common source LNA (CSLNA) b) the common gate LNA (CGLNA) (Sobhy et al., 2011)

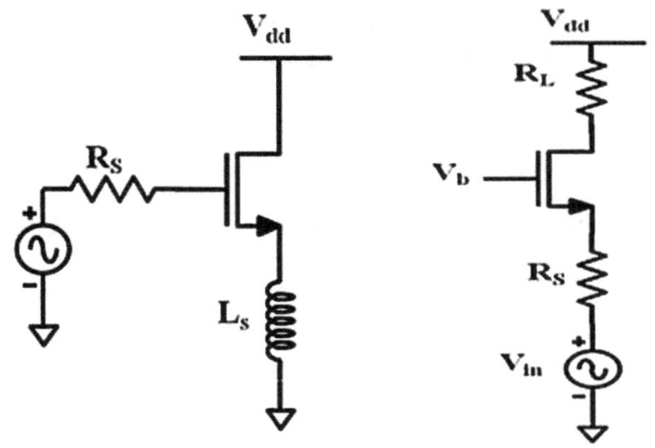

2.1 Input Matching

For a common source LNA a usual way to obtain the input matching is with inductive degradation. One inductor is placed in series with the source and one in series with the gate. The input impedance of the common source LNA is given by Equation (26).

$$Z_{in,CSLNA} = \frac{V_{in}}{I_{in}} = SL_s + \frac{1}{SC_{gs}} + \left(\frac{g_m}{C_{gs}}\right)L_s \quad (26)$$

The values for L_s are chosen such that at a desired frequency the inductor will resonate away C_{gs}, while the last term in the Equation creates ohmic impedance. Since there is no physical ohmic resistance in the input matching network, the input match is noiseless (Sobhy et al., 2011). For the common gate LNA with resistive source degradation, the input impedance looking into the source-terminal is given by Equation (27).

$$Z_{in,CGLNA} = \frac{V_{in}}{I_{in}} = \frac{1}{g_m} \quad (27)$$

This provides the possibility of an ohmic input matching only dependent of the transconductance. A fundamental difference between the input matching networks is that CSLNA is a series resonant circuit while CGLNA is a parallel one, the associated quality factors are given in Equation (28) and (29).

$$Q_{CS-LNA} = \frac{1}{2\omega_0 C_{gs} R_s} > 1 \quad (28)$$

$$Q_{CG-LNA} = \frac{\omega_0 C_{gs} R_s}{2} < 1 \quad (29)$$

It is known that the sensitivity of Z_{in} to parasitic components is proportional to the quality factor of the matching network (Sobhy et al., 2011). Hence, CG-LNA with its lower Q parallel resonant network is more robust against typical production process, voltage, and temperature variations.

2.2. Noise

The expressions for the noise factor of the Common source LNA and the Common gate LNA is given by

$$F_{CSLNA} = 1 + \frac{\gamma}{\alpha}\left(\frac{\omega_0}{\omega_T}\right) \tag{30}$$

$$F_{CGLNA} = 1 + \frac{\gamma}{\alpha} \tag{31}$$

Where α is a device parameter depending on the channel length, γ is a device parameter depending on V_{ds} and the channel length, and ω_T is the maximum operating frequency of the transistor, here defined as g_m/C_{gs} (Do, Boon, Yeo, & Cabuk, 2008). If one compares the expressions one sees that the noise factor for the common gate is independent of frequency while the noise factor of the common source is a linear function of the ratio between operating frequency ω_0 and ω_T. This feature has made the common gate topology popular in ultra-wide band LNAs. Equation (31) assumes no gate induced noise since the gate node is shortened to AC-ground. It is easy to see that one can lower the nose factor in the common source by having a large ω_T. To achieve this one would need a large g_m and a small C_{gs}. This is possible with a large drain current.

2.3 Gain

The expressions for gain for the two basic topologies derived in (Do, Boon, Yeo, & Cabuk, 2008) are:

$$G_{m,CSLNA} = \frac{1}{2R_s}\left(\frac{\omega_T}{\omega_0}\right) \tag{32}$$

$$G_{m,CGLNA} = \frac{1}{2R_s} \tag{33}$$

The value of ω_T/ω_0 typically lies in the range 5~10 depending on the operating frequency and process details. Therefore, common source LNA provides higher gain than its conventional common gate LNA.

2.4 Reverse Isolation and Stability

Reverse isolation S_{12} represents amplification in the reverse direction. For an amplifier, it is hoped that this reverse transmission is minimized. In a direction conversion receiver the large signal local oscillator (LO) can be coupled to the output of LNA. With finite reverse isolation it can be transmitted back to input of LNA. Being reflected back from antenna, this leakage from the LO is mixed with the LO signal itself, generating a DC offset voltage. Known as LO self mixing, this is a serious problem in DR.

Reverse isolation is also related to stability, as can be seen from stern stability factor K.

$$K = \frac{1 + |S_{11}S_{22} - S_{12}S_{21}|^2 - |S_{11}|^2 - |S_{22}|^2}{2|S_{12}S_{21}|} \tag{34}$$

The amplifier is stable if K>1. So, lower S_{12} or better reverse isolation means better stability. In a conventional CSLNA, C_{gd} provides a feed forward path between input and output that degrades reverse isolation and stability. In contrast, since the Miller effect on C_{gd} does not exist in CGLNA, it exhibits better reverse isolation. Also, cascoding is not necessary in a conventional CGLNA so there is no added noise from cascode transistors.

2.5 Power Consumption

The common source is sensitive to variations of inductance and capacitance in the input matching network. To avoid this problem one can design low-Q input matching network by utilizing a constant g_m optimization approach. This solution usually consumes a high DC current. Simulations of different common source and common gate topologies show that the common source LNA consumes more power than the common gate LNA (Tran, Boon, Do, & Yeo, 2011). From the point view of power consumption, CGLNA is more efficient than CSLNA. A few more points are discussed in the comparison between the topologies in (Tran, Boon, Do, & Yeo, 2011). The results are summarized in Table 1.

3. IEEE 802.15.4 SPECIFICATIONS

The IEEE 802.15.4 standard covers three frequency bands, i.e., 868 MHz, 915 MHz, and 2.4 GHz. The popular 2.4 GHz industrial, scientific, and medical (ISM) band was chosen for this work because it is an unlicensed band and is accepted worldwide. Furthermore, 2.4 GHz is a high enough frequency (compared to 868 MHz, and 915 MHz band) to permit integration of high quality on-chip inductors with reasonable size (Cook, 2007). Receiver's specifications (Balankuntty, Shih-An, Yiping, & Kinget, 2010) that are

Table 1. Comparison between common source and common gate LNA

	Common Source LNA	Common Gate LNA
Noise Factor	+	-
Gain	+	-
Power Consumption	-	+
Input Matching	-	+
Sensitivity	-	+
Reverse Isolation	-	+

Table 2. IEEE 802.15.4 Receiver requirements

Receiver's Specification	Requirement
Bandwidth	2.4GHz
NF	< 5dB
IIP3	> -17dB

important to the LNA design is shown in Table 2. The IEEE 802.15.4 standard has a system bandwidth spanning 83.5 MHz. This 83.5 MHz bandwidth is divided into 16 channels each with a bandwidth of 2 MHz, and spaced by 5 MHz.

Since we aim to optimize our LNA design in two directions, one is for low NF and low power and the other one is for high gain. We need to formulate two sets of LNA specification requirement for our designs. The work in (Balankuntty, Shih-An, Yiping, & Kinget, 2010) and (Trung-Kien et al., 2006) introduced two different receiver systems for the 2.4 GHz ISM band. One used active mixer and the other used passive mixer to realize the front-end. Based on the performances of all the blocks right after the LNA in (Balankuntty, Shih-An, Yiping, & Kinget, 2010) and (Trung-Kien et al., 2006), we formulate the required specifications for our LNAs. For the system using active mixer at its front-end, we set the LNA's targeted gain to be 10 dB. If passive mixer is used, higher LNA's gain is required. Our targeted gain for this case is 21 dB. The other LNA's specifications are derived in Table 3 respectively S_{11}, S_{22} and S_{12} represent the input matching, output matching and reverse isolation of an LNA.

4. ULTRA-LOW POWER CMOS LNA WITH CAPACITIVE CROSS-COUPLING TECHNIQUE

The wireless communication industry is currently experiencing tremendous growth, there is an increasing demand for portable wireless devices. These portable devices require low power dissipation to maximize battery lifetime. In this chapter, the power consumption can be reduce by capacitive cross coupling tech-

Table 3. IEEE 802.15.4 LNA requirements

LNA's Specification	LNA's Requirements
Frequency	2.4 GHz
Gain	> 9 dB
NF	< 5 dB
IIP3	> -15 dB
S_{11}	< -10dB
S_{12}	< -17 dB
S_{22}	<-10 dB
Power	< 2mW

nique and operating the circuit at low supply voltage. Some low power applications, such as wireless medical telemetry, require the portable devices to operate at a low supply voltage with a small battery. The negative feedback g_m boosted CGLNA which is presented in this chapter is designed for very low supply voltage such as 0.6 V. It is a differential stage non-cascode LNA. The NF and gain problems in the conventional CGLNA are improved by employing the g_m boosted capacitive cross-coupling (CCC) technique.

4.1 General G_M Boosting Technique

Since noise figure is the most important characteristic of an LNA, the noise calculations of the CG-LNA stage are briefly revisited. It can be shown that the noise factor is

$$F_{CGLNA} = 1 + \frac{\gamma}{\alpha}\frac{1}{g_m R_s} \tag{35}$$

As the Equation (35) implies, increasing g_m reduces the noise factor. In RF systems, however, input matching requires that $g_m = 1/R_s$, which results in $F = 1+\gamma/\alpha$. From another point of view, it is the input-matching requirement that prevents increasing g_m to lower the noise factor, and this in turn sets a lower bound on the noise factor. Note that in CGLNA, impedance matching may be traded against noise figure. More specifically, if g_m can be increased to decrease the noise figure (Zhuo et al., 2005).

The discussion above illuminates the tight link between noise figure and input matching in CGLNA. If the condition of input matching can be separated from noise performance, the possibility of improving noise figure arises. The scheme suggested in (Zhuo et al., 2005) is based on an important observation, two g_m values are used in calculating the noise factor of the common-gate amplifier. One is the effective transconductance looking into the source terminal; denote it as G_m. The other is the intrinsic transconductance of the amplifying device, which is related to drain current channel noise i_d and is denoted as g_m. In general, G_m is not necessarily equal to g_m. However, in a conventional CGLNA in which the gate terminal is shorted to ac ground, $G_m = g_m$. A design challenge in improving CGLNA is to modify the topology so that $G_m \neq g_m$. specifically, to make $G_m \neq g_m$ a coupling mechanism is introduced between the gate and source terminals (Zhuo et al., 2005).

Figure 9. depicts the topology of the studied g_m boosted CGLNA. In this structure, an inverting amplification is introduced between the source and gate nodes of the MOSFET. Consequently, the effective transconductance looking into the source terminal is boosted from $G_m = g_m$ to $G_m = (1+A)\, g_m$. where A is the gain from source to gate. Most important, the resulting noise factor becomes

$$F_{CGLNAg_m-boosted} = 1 + \frac{\gamma}{\alpha}\frac{1}{G_m R_s} = 1 + \frac{\gamma}{\alpha(1+A) g_m R_s} = 1 + \frac{\gamma}{\alpha(1+A)} \tag{36}$$

It is clear that the noise factor of the structure shown in Figure 9 is reduced by the factor (1+A).

Figure 9. Basic CGLNA with g_m-boosting (W. Zhuo, 2005)

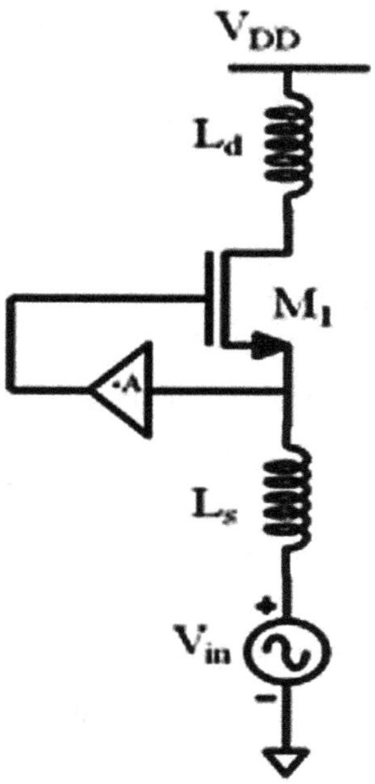

4.2 Capacitor Cross Coupled CGLNA

One possible way to achieve passive inverting amplification in the g_m boosted CGLNA is shown in Figure 10 wherein a differential topology allows the differential active devices to be capacitor cross-coupled (Woo, Kim, Lee, & Laskar, 2012).

Figure 10. Capacitor cross coupled differential CGLNA (Woo, Kim, Lee, & Laskar, 2012)

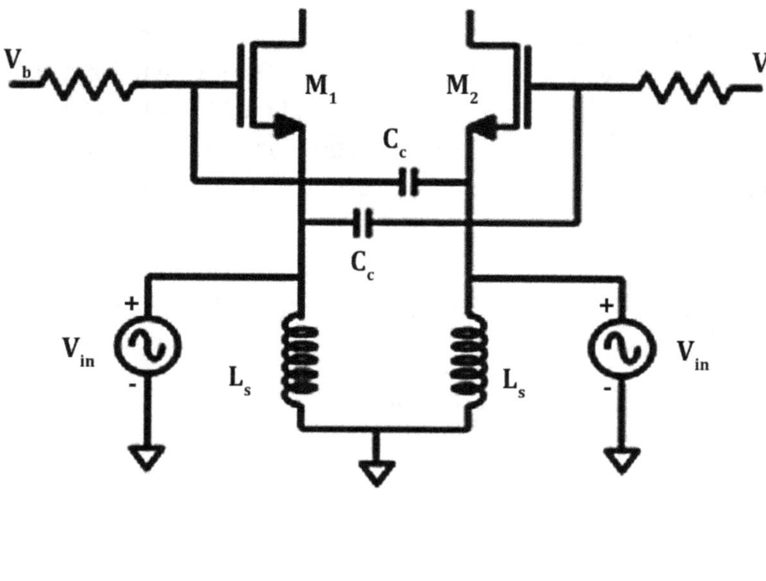

The inverting amplification value A, for the topology of Figure 10 is approximately given by the capacitor voltage division ratio.

$$A = \frac{C_c}{C_c + C_{gs}} \tag{37}$$

The effective transconductance is

$$G_m = (1 + A)g_m \tag{38}$$

$$G_m = \left(\frac{C_{gs} + 2C_c}{C_{gs} + C_c} \right) g_m \tag{39}$$

The noise factor is given by

$$F_{CGLNA,CCC} = 1 + \frac{\gamma}{\alpha} \left(\frac{C_{gs} + C_c}{C_{gs} + 2C_c} \right) \tag{40}$$

Thus, $C_c >> C_{gs}$ results in $A \approx 1$and $G_m \approx 2g_m$

$$F_{CGLNA,CCC} \approx 1 + \frac{\gamma}{2\alpha} \tag{41}$$

Comparing to the conventional CGLNA, the noise factor (F) of the g_m boosted CCC-CGLNA is reduced by half and the effective transconductance is increased with reduction in power consumption by same factor.

4.3 Design and Simulation Results

The circuit of Figure 11 is simulated in *SPECTRE-RF* at 2.4 GHz. The simulations are performed using a 0.6V supply, 90nm RF CMOS technology. The transistor W/L ratios of the designed g_m boosted CCC-CGLNA is derived in Equation (45). The coupling capacitors, employed in the design, are implemented using a *MIMCAP* device supported by *CADENCE* design systems. The biasing resistors are implemented using poly resistors. The capacitors (C_C) are used to cross couple M_1 and M_2 transistors. The simulated values of on-chip inductors L_s and L_d are calculated using the Equations (43) and (44) are 8.9 and 7.3 nH.

Figure 11. Schematic of Capacitor cross coupled g_m boosted CGLNA

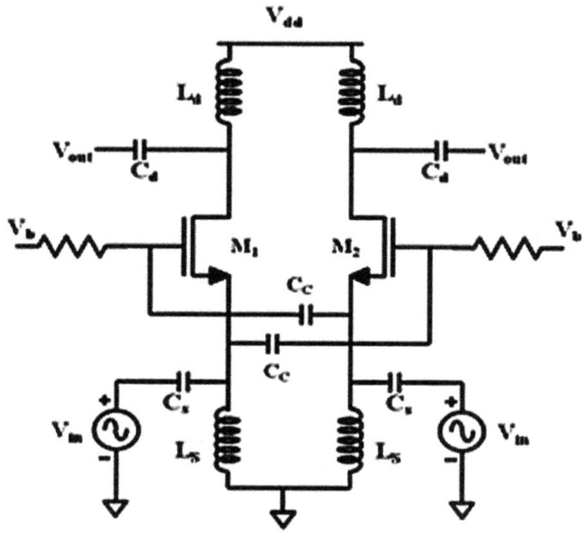

The value of source inductance L_s can be estimated from

$$f_o = \frac{1}{2\pi\sqrt{L_1.C_{gs}}} \tag{42}$$

$$L_s = \frac{\left(\frac{1}{2\pi f_o}\right)^2}{C_{gs}} = \frac{\left(\frac{1}{2\pi 2.4E^9}\right)^2}{0.48E^{-12}} = 8.9nH \tag{43}$$

Drain Inductance L_d can be obtained as

$$L_d = \frac{\left(\frac{1}{2\pi f_o}\right)}{C_o} = \frac{\left(\frac{1}{2\pi 2.4E^9}\right)}{0.6E^{-12}} = 7.3nH \tag{44}$$

The W/L ratio is calculated by

$$W = \frac{g_m^2 L_{\min}}{2K_n I_D} \tag{45}$$

$$W = \frac{20ms^2.90nm}{2.4.54E^{-5}.12mA} = 33\mu m \tag{46}$$

4.3.1. Power Gain

To compensate noise contribution of subsequent stages in the receiver chain, it is desirable to have a large gain for the low noise amplifier. The simulated g_m boosted capacitor cross coupled common gate low noise amplifier exhibits a gain of 9.9 dB at 2.4 GHz as can be seen from the plot of the power gain shown in Figure 12.

4.3.2 Noise Figure (NF)

The noise figure variation with frequency is shown in Figure 13. The simulated noise figure of the g_m boosted CCC-CGLNA are 0.87 dB at 2.4 GHz. This is an excellent NF value for CGLNA topology. The CCC-CGLNA thus can achieve a very low noise figure compared to the conventional CGLNA.

Figure 12. Simulated Gain of CGLNA

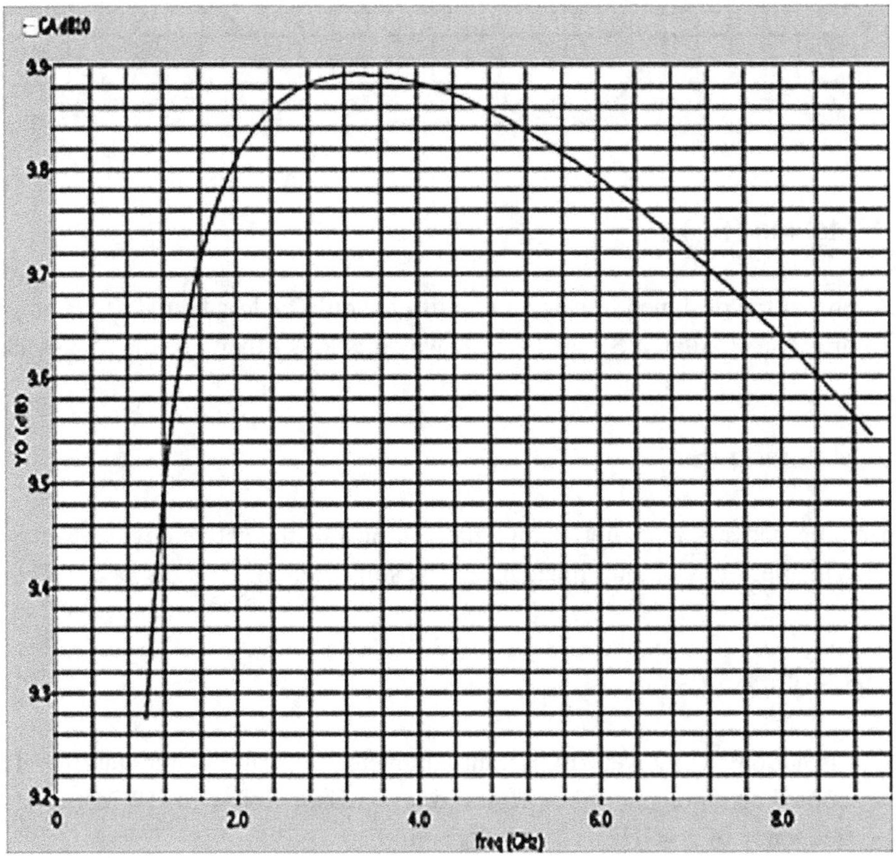

Figure 13. Simulated Noise Figure of CGLNA

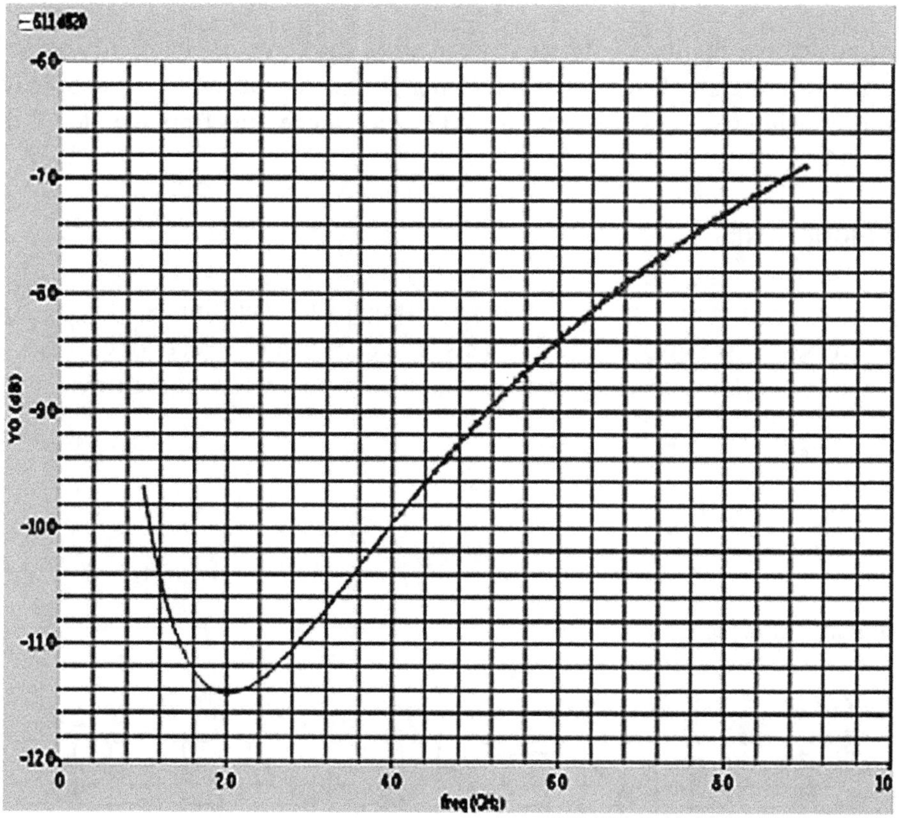

4.3.3 Input Matching (S_{11})

The input matching of the designed low noise amplifier should be less than –10 dB while maintaining lowest noise figure. The simulated S_{11} is -11.7 dB which is lower than -10 dB at 2.4GHz as shown in Figure 14.

4.3.4 Output Matching (S_{22})

The output matching network does not change the DC bias of the active device. Since the low noise amplifiers are having very low output impedance. The simulated S_{22} is -9 dB at 2.4GHz is as shown in Figure 15.

4.3.5 Reverse Isolation (S_{12})

The input output isolation (S_{12}) is very important parameter to ensure better stability. The g_m boosted capacitor cross coupled common gate LNA achieved reverse isolation with -19.7dB has been simulated at the operating frequency of 2.4GHz as shown in Figure 16.

Figure 14. Simulated Input Matching of CGLNA

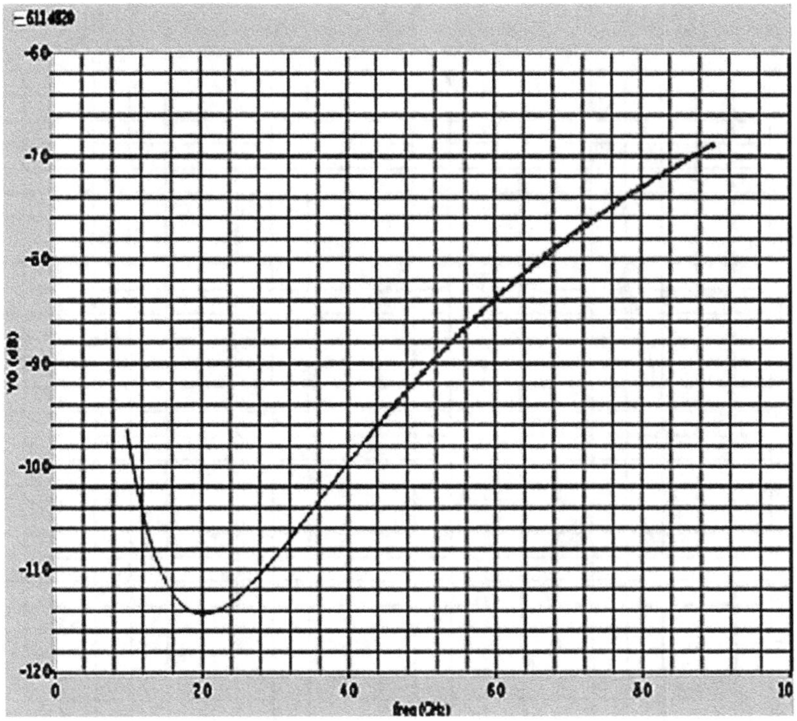

Figure 15. Simulated Output Matching of CGLNA

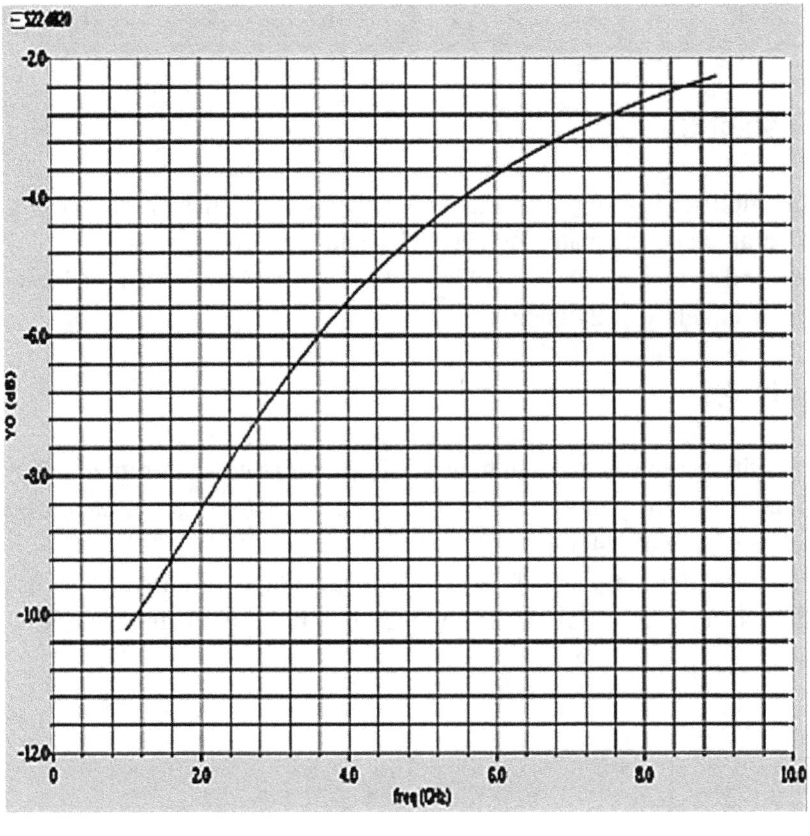

Figure 16. Simulated Reverse Isolation of CGLNA

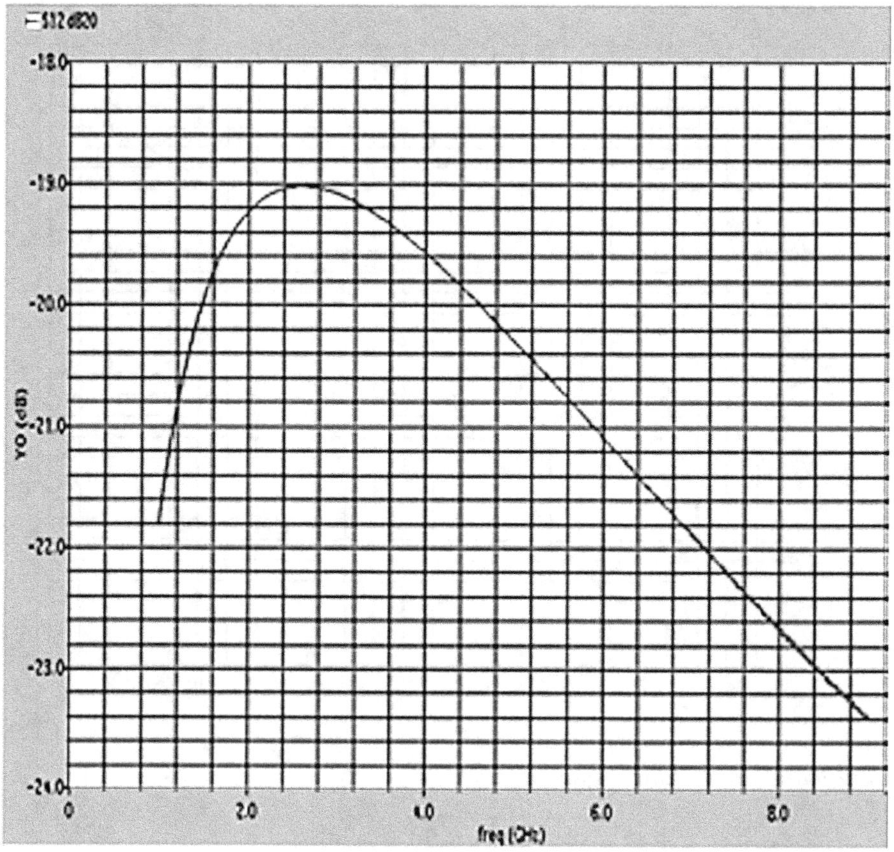

4.3.6 Stability Factor (KF)

The stability of an amplifier is a very important consideration in a design of an LNA and can be determined from the *S* parameters, the matching networks, and the terminations.

The plot of the stability factor is shown in Figure 17. The CCC-CGLNA is stable with stability factor greater than 1 at the frequency of 2.4GHz.

4.3.7 Linearity (IIP3)

The amplifier's linearity was measured using the input referred third-order intercept point (IIP3). Figure 18, Shows the simulated linearity plots for the CCC-CGLNA. This amplifier achieved the target IIP3 of -1.017 dBm at the frequency of 2.4GHz.

The Table 4 Illustrates the summary of the simulation results for the LNA designs. The performance of the designed g_m boosted CCC-CGLNA is compared with the performance of the recently reported low noise amplifiers.

Figure 17. Simulated stability factor of CGLNA

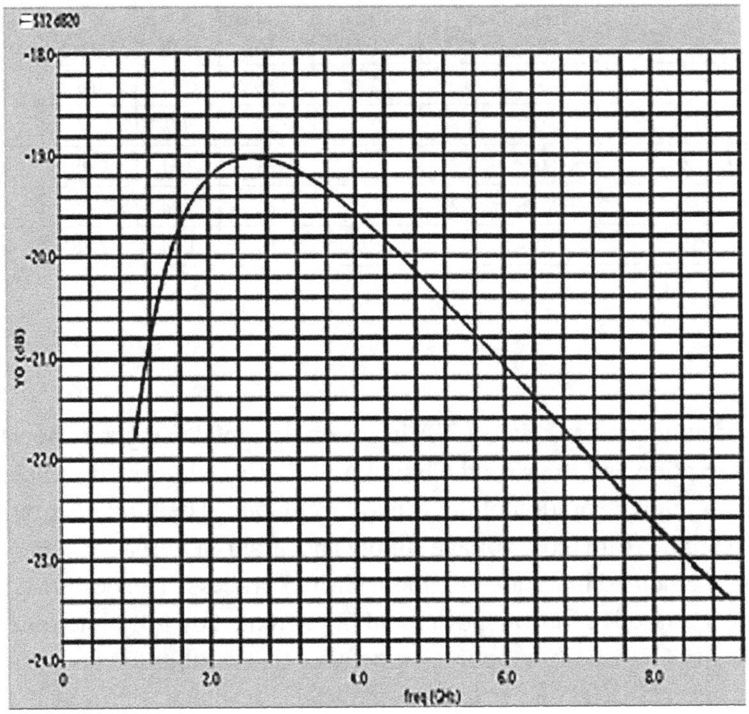

Figure 18. Simulated Linearity of CGLNA

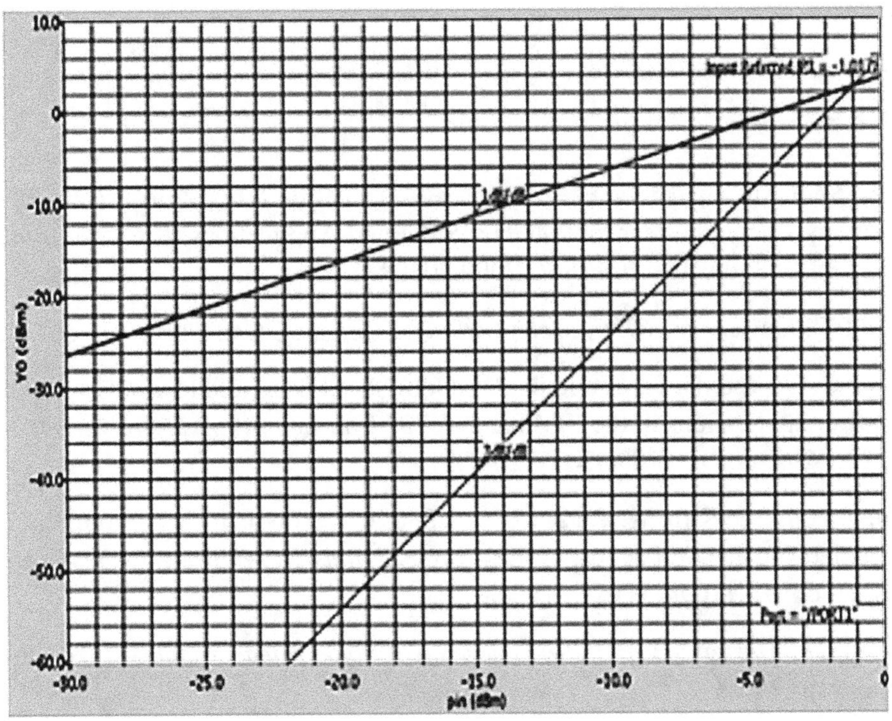

Table 4. Comparison of the low noise amplifier designs

Reference	Tech [nm]	V_{DD} [V]	Freq [GHz]	Gain [dB]	NF [dB]	P_{DC} [mW]	S11 [dB]	S22 [dB]	IIP3 [dBm]
This Work	90	0.6	2.4	9.9	0.87	0.5	-11	-9	-1.017
(Ye, Horng, & Wu, 2013)	180	1.8	2.2	10	1.9	7.4	-10	-16	-35
(Wang,Zhang,Fan, & Liu,2014)	180	1.8	2.68	23	2.61	0.98	-10	-10	-2.7
(Lin, Wang, & Lee, 2014)	90	1.2	1.9	12	3.24	9.96	-8.7	-10	-1.8
(Han, Jung, & Kim, 2015)	130	1.2	0.9	18	1.94	2.88	-10	-10	9.06

CONCLUSION

In this chapter, an ultra low power CMOS LNA has been designed using a g_m boosting technique to reduce the power consumption and improve the NF of a CGLNA is described. A capacitor cross-coupled LNA is presented as an implementation of the general technique. The fully integrated differential CCC-CGLNA consumes 0.83mA from 0.6V voltage supply and attains a measured NF of 0.875dB at 2.4GHz the amplifier were implemented in a standard 90nm CMOS process. The designed technique makes the common gate topology attractive for low-power and high-frequency fully integrated designs.

A general g_m boosted CCC-CGLNA topology is detailed that exhibits lower noise figure and consumes less power than conventional CGLNA. It also preserves the advantages of insensitivity to parasitic input capacitance, higher reverse isolation and better stability, which makes it attractive for emerging high frequency applications. Implementation of the g_m boosted CCC-CGLNA scheme is also discussed and simulations shown good agreement between analysis and simulated results. The g_m boosted topology provides new opportunities for circuit innovation implementations using other coupling techniques are under development.

REFERENCES

Balankuntty, A., Shih-An, Y., Yiping, F., & Kinget, P. R. (2010). A 0.6 V Zero-IF/Low-IF Receiver with Integrated Fractional-N Synthesizer for 2.4GHz ISM-Band Applications. *IEEE Journal of Solid-State Circuits*, *45*(3), 538–553. doi:10.1109/JSSC.2009.2039827

Cook, B. W. (2007). *Low energy RF transceiver design*. Berkeley, CA: University of California.

Do, A. V., Boon, C. C., Yeo, K. S., & Cabuk, A. (2008). A subthreshold low-noise amplifier optimized for ultra-low-power applications in the ISM band. *IEEE Transactions on Microwave Theory and Techniques*, *56*(2), 286–292. doi:10.1109/TMTT.2007.913366

Han, G. H., Jung, H. D., & Kim, W. T. (2015). A 2.88mW + 9.06dBm IIP3 Common-Gate LNA with Dual Cross Coupled Capacitive Feedback. *IEEE Transactions on Microwave Theory and Techniques*, *63*(3), 1019–1025. doi:10.1109/TMTT.2014.2377036

Lee, T. H. (1998). *The Design of CMOS Radio-Frequency Integrated Circuits* (1st ed.). New York, NY: Cambridge University Press.

Lin, S. Y., Wang, C. C., & Lee, J. H. (2014). A 9.96mW 3.24dB NF 1.9GHz Wideband Low Noise Amplifier Using 90nm CMOS Technology. In *Proceedings of the IEEE Radio and Wireless Symposium*, (vol. 14, pp. 208-210). IEEE.

Oh, N., & Lee, S. (2005). Building a 2.4-GHz radio transceiver using IEEE 802.15.4. *IEEE Circuits Devices Magazine.*, *21*(6), 43–51. doi:10.1109/MCD.2005.1578587

Sobhy, E. A., Hemy, A. A., Hoyos, S., Entesari, K., & Sanchez-Sinencio, E. (2011). A 2.8mW Sub 2-dB Noise Figure Inductorless Wideband CMOS LNA Employing Multiple Feedback. *IEEE Transactions on Microwave Theory and Techniques*, *59*(12), 3154–3161. doi:10.1109/TMTT.2011.2169081

Tran, T. T. N., Boon, C. C., Do, M. A., & Yeo, K. S. (2012). *Ultra-low power Low noise amplifier designs for 2.4GHz ISM band applications*. (Doctoral dissertation). Nanyang Technological University, Singapore.

Tran, T. T. N., Boon, C. C., Do, M. A., & Yeo, K. S. (2011). Ultra-low power sub-mA series input resonance differential common gate LNA. *IET Electronic Letters*, *47*(12), 703-704.

Trung-Kien, N., Krizhanovskii, V., Jeongseon, L., Seok-Kyun, H., Sang-Gug, L., Nae-Soo, K., & Cheol-Sig, P. (2006). A Low Power RF Direct-Conversion Receiver/Transmitter for 2.4GHz Band IEEE 802.15.4 Standard in 0.18um CMOS Technology. *IEEE Transactions on Microwave Theory and Techniques*, *54*(12), 4062–4071. doi:10.1109/TMTT.2006.885556

Wang, M., Zhang, M., Fan, X., & Liu, Y. (2014). A Power Reduction Technique for Wideband Common Gate Low Noise Amplifiers. In *Proceedings of the IEEE 57th International Midwest Symposium on Circuits and Systems,* (vol. 10, pp. 969-972). IEEE.

Woo, S., Kim, W., Lee, C., & Laskar, J. (2012). A Wideband low-power CMOS LNA with positive-negative feedback for noise, gain, and linearity optimization. *IEEE Transactions on Microwave Theory and Techniques*, *60*(10), 3169–3178. doi:10.1109/TMTT.2012.2211379

Ye, R., Horng, T., & Wu, J. (2013). Two CMOS Dual Feedback Common Gate Low-Noise Amplifiers with wideband Input and Noise Matching. *IEEE Transactions on Microwave Theory and Techniques*, *61*(10), 3690–3699. doi:10.1109/TMTT.2013.2277997

Zhuo, W., Shekhar, S., Embabi, S. H. K., Gyvez, J. P. D., Allstot, D. J., & Sanchez-Sinencio, E. (2005). A capacitor cross-coupled common-gate low-noise amplifier. *IEEE Transactions on Circuits and Wystems. II, Express Briefs*, *52*(12), 875–879. doi:10.1109/TCSII.2005.853966

Chapter 10
Low Power Arithmetic Circuit Design for Multimedia Applications

Senthil C. Pari
Multimedia University, Malaysia

ABSTRACT

The objective of this chapter is to describe the various designed arithmetic circuit for an application of multimedia circuit that can be used in a high-performance or mobile microprocessor with a particular set of optimisation criteria. The aim of this chapter is to describe the design method of binary arithmetic especially using by CMOS and Pass Transistor Logic technique. The pass transistor techniques are reduced the noise margin for small circuit, which can be explained in this chapter. This chapter further describe the types of arithmetic and its techniques. The technique design principle procedure should make the following decisions: circuit family (complementary static CMOS, pass-transistor, or Shannon Theorem based); type of arithmetic to be used. The decisions on the designed logic level significantly affect the propagation delay, area and power dissipation.

1. INTRODUCTION

Reductions in power dissipation and improvements in speed require optimisation at all levels of the logic design. The proper circuit style and methodology are important factors to be considered for low power design. High-speed computation is now the expected norm for the average user. Likewise, another significant change in user attitudes is the desire to access computation from any location, without physically connecting to a wired network. According to Swaroop Ghosh, Kaushik Roy (2008), the requirement of portability thus places severe restrictions on size, weight, and power. The power is particularly important since absolute battery technology only provides 20 W-hrs of energy for each pound of weight. Improvements in battery technology are being made, but it is unlikely that a dramatic solution to the power problem is forthcoming. It is projected that only a 30% improvement in battery performance will

DOI: 10.4018/978-1-5225-0190-9.ch010

be obtained over the next 5 years. Different circuit design techniques must be considered in order to maximise efficiency in terms of speed and power dissipation.

The increasing demand for low-power **V**ery **L**arge **S**cale **I**ntegration (VLSI) can be addressed at different design levels, including the architectural level, circuit layout, and the process technology level. At the circuit level, there is considerable potential for power savings by means of proper choice of a logic style for implementing combinational circuits. Marios Psilogeorgopoulos (1999) determine, all the important parameters governing power dissipation, switching capacitance, transition activity and short-circuit currents are strongly influenced by the chosen logic style. The low-power logic styles reported in the literature have mainly focused on particular logic cells, namely, full adder, full subtractor, multiplier, divider, XOR, NAND, OR and AND gates. In this chapter, low power logic circuit investigations are extended to a much wider set of logic gates and to arbitrary combinational circuits. Actual logic gates designs are implemented into multiplier and divider circuit.

The power dissipation has become a critical design metric for an increasingly large number of VLSI circuits. Mohab Anis (2002), determine, the exploding market for portable electronic appliances fuels the demand for complex integrated systems that can be powered by lightweight batteries with long times between recharges. The goal of extending the battery lifespan of portable electronics can be achieved by reducing the energy expended per arithmetic operation, but low-power consumption need not necessarily imply low energy. A circuit can consume very low power by clocking at an extremely low frequency to execute an arithmetic operation, but it may then take a very long time to complete the operation. One of the objectives of the present work is to investigate the effect of reduced supply voltage on the optimal energy efficiency of the full adders designed with different logic styles based on Ultra Deep Sub Micron (UDSM) process technology. Several circuit design techniques are compared in terms of speed and power dissipation. All kind of circuits is to be implemented and design technique can be chosen depending on the applications. The designed circuit are given different performance, which become important and disallow the formulation of universal rules for optimal logic styles.

1.1 Need for Low Power

The demand of low power microelectronics can be traced to the invention of the transistor in 1947. According to Gunok Jung, (2005), the elimination of the crushing need for several watts of heater power and several hundred volts of anode voltage in vacuum tubes in exchange for transistor operation in the tens of mill watts range was a breakthrough of unparalleled importance in electronics. The motivation for low power electronics comes from two reasonably distinct sources:

1. The most recent need is for ever-increasing packing density in order to further enhance the speed of high performance systems. This, in turn, imposes severe restrictions on power dissipation density.
2. The broadest need is for conservation of power in desktop systems, where a competitive lifecycle cost-to-performance ratio demands low power operation to reduce power supply and cooling costs.

Marios Psilogeorgopoulos, (1999) found that high speed system and low power operation classes need to encompass a substantial majority of current applications of electronic equipment. Low power electronics has become mainstream in the effort to achieve Giant Scale Integration (GSI). As digital computer arithmetic is an aspect of logic design with the objective of developing appropriate algorithms in order to achieve an efficient utilisation of the available hardware. The hardware can only perform a relatively simple and primitive set of Boolean operations. The hardware design is measured for speed, power and chip area. Abid. Z, (2008), described, the efficiency of a design gives a strong link between the algorithms and the technology used for its implementation. This chapter presents the Shannon theorem approach for the design of an adder circuit for use in an arithmetic logic unit (ALU) circuit.

1.2 Basics of Low Power Circuit and Logic Design

A Design of primitive adder cell should give low power, which can obey its basic design method. An integrated low power methodology requires optimization at all design abstraction layers as mentioned below.

- **System**: Partitioning, and Power reduction
- **Algorithm**: Complexity, Concurrency, Regularity
- **Architecture**: Parallelism, Pipelining, Redundancy, Data Encoding
- **Circuit Logic**: Logic Styles, Energy Recovery, and Transistor Sizing
- **Technology**: Threshold Reduction, Multi threshold Devices.

The design cell dynamic power varies as V^2_{DD}. According to dynamic power dissipation (Equation 1), the reducing the supply voltage reduces power dissipation. Also selective frequency reduction technique can be used to reduce dynamic power. According to Chip-Hong Chang, (2005), Multi threshold voltage can be used to reduce leakage power at system level. Transistor resizing can be used to speed-up circuit and reduce power. The Sleep transistors can be used effectively to reduce standby power. Parallelism and pipelining in system architecture can reduce power significantly. Clock disabling, power-down of selected logic blocks, adiabatic computing, software redesign to lower power dissipation are the other techniques commonly used for low power design.

1.2.1 Logic Design for Low Power

The logic design selections between static versus dynamic topologies, conventional CMOS versus pass-transistor logic styles and synchronous versus asynchronous timing styles have to be made during the design of a circuit. In static CMOS circuits, the component of power due to short circuit current is about the 10% of the total power consumption. However, in dynamic circuits there no short circuit current problem, since there is no any direct dc path from supply voltage to ground. Only in domino-logic circuits there is such a path, in order to reduce sharing, hence there is a small amount of short-circuit power dissipation. The above mentioned circuit power dissipation can be reduced swing to lower power by pass transistor logic technique. (e.g., reduced bit-line swing in memory). The pass transistor output model is shown in Figure 1. The pass transistor logic dynamic power can be calculated by transistor switching activity, which can be given in Equation 1.

$$P = C_L * V_{DD} * (V_{DD} - V_t) \tag{1}$$

where,

P = dynamic power,

Figure 1. Pass transistor logic

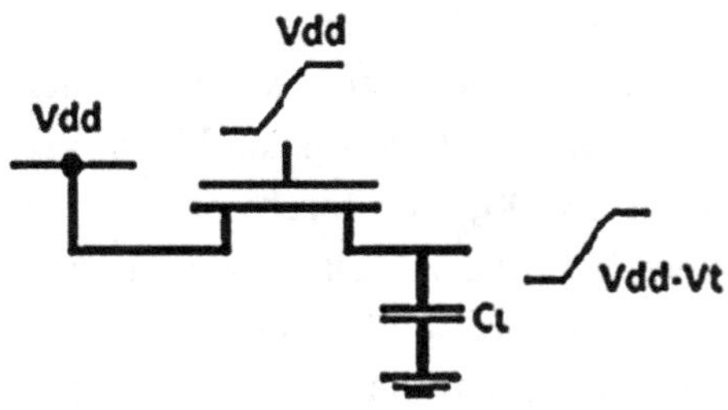

C_L= Load capacitance,
V_{DD} = Supply voltage and
V_t= threshold voltage.

During standard design of logic circuit, the standard logic optimization for low power, technology parameters such as supply voltage are fixed, which is illustrated in Figure 2. A Path equalization with buffer insertion is one of the techniques, which ensures that signal propagation from inputs to outputs of a logic network follows paths of similar length to overcome glitches. When paths are equalized, most gates have aligned transitions at their inputs, thereby minimizing spurious switching activity/glitches.

Most of the designers are design a circuits by AOI method, which denotes as re mapping method. Remapping of AND/OR circuits has a power minimization techniques include local transformations. The high-activity node of remapping circuit, which marked with removed and replaced by new mapping onto an and OR gate. Static/Leakage power, originates from substrate currents and subthreshold leakages. PSwitching was predominant for technologies 1μm and above. However, for deep-submicron

Figure 2.

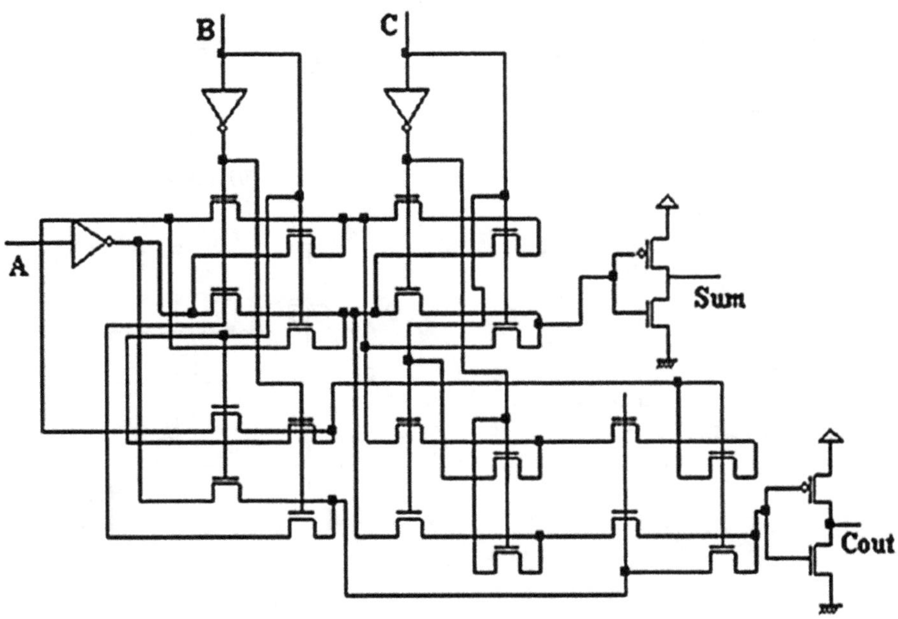

processes below 180nm, PLeakage becomes dominant factor. Leakage power is a major concern in recent technologies, as it impacts battery lifetime. CMOS technology has been extremely power-efficient when transistors are not switching or in stand-by mode, and system designers expect low leakage from CMOS chips. To meet leakage power constraints, multiple-threshold and variable threshold circuit techniques are often used. In multiple-threshold CMOS, the process provides two different threshold transistors. Low-threshold are employed on speed-critical sub-circuits and there are fast and leaky. High-threshold transistors are slower but exhibit low sub-threshold leakage, and they are employed in noncritical/slow paths of the chip. As more transistors become timing-critical multiple-threshold techniques tend to lose effectiveness.

Variable-threshold circuits can be solved the quiescent/static leakage problem, but they require control circuits that modulate substrate voltage in stand-by, which is shown in Figure 3. Gunok Jung, (2005), describes the fast and accurate body-bias control with control circuit is quite challenging, and requires carefully designed closed-loop control. When the circuit is in standby mode, the bulk/body of both PMOS and NMOS are biased by third supply voltage to increase the V_t of the MOSFET. However the designed circuit transistor are keeping constant threshold voltage during normal operation which is switched back to reduce the V_t.

The desired designed circuit has short circuit power dissipation. The proper choice and design method are minimize the power dissipation. The pass transistor circuit design are affected by short circuit power dissipation, which is caused by the short circuit currents that arise when pairs of PMOS/NMOS transistors are conducting simultaneously. In static CMOS circuits, short-circuit path exists for direct current flow from V_{DD} to ground, when $V_{Tn}<V_{in}< V_{DD}-|V_{Tp}|$, which is shown in Figure 4. The other way to reduce short circuit power is to keep the input and output rise/fall times the same. The short circuit

Figure 3.

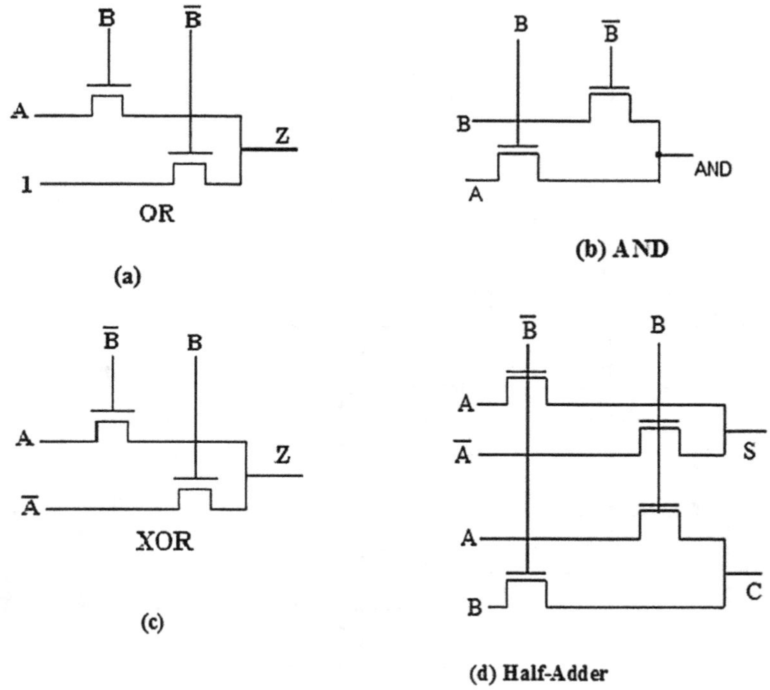

possessing Short-circuit power is caused by the short circuit currents that arise when pairs of PMOS/NMOS transistors are conducting simultaneously. In static CMOS circuits, short-circuit path exists for direct current flow from VDD to ground, when threshold voltage less than input voltage. If $V_{DD}<V_{tn} + |V_{tp}|$, then short-circuit power can be eliminated. If the load capacitance is very large, the output fall time is larger than the input rise time. The drain-source voltage of the PMOS transistor is 0.Hence the short-circuit power will be 0.

1.3 Non-Clocked Pass Gate Logics

According to short circuit power dissipation and static power dissipation are delayed the output and increased the power dissipation in the circuit. The pass transistor logic is trade off above mentioned flaws. The pass transistor logic are designed without clocking input, which is known as non-clocked pass transistor circuit. The Non pass transistor logic, the transistor switching activity is an eliminating the glitches. This can be done by controlling the widths and lengths the delays of each pass transistor. Compared with CMOS node, switching capacitance of a node in the PTL would be less due to the smaller size of the transistors in PTL implementation. So, the PTL circuits consume less power, and their lower propagation delay path tends to result in higher capacitance than other styles. The extra capacitance is associated with driving large PFETs as well as NFETs. Transistor sizing is a key issue for achieving performance targets in non-clocked circuits. Chip-Hong Chang, (2005), found that power is minimised when circuitry utilises the smallest possible devices, reducing the capacitive load of the preceding stages. The delay of a non-clocked logic path has two predominant components: (i) the intrinsic delay and (ii) RC delay of the output signal driving fan out interconnect. The propagation delay is constant due to output logic is given as energy (E) x Total capacitance (C), total delay can be simply represented as a function of:

1. Input rise time,
2. Output load, and
3. An intrinsic gate delay dependent only on processing parameters and operating conditions.

Figure 4.

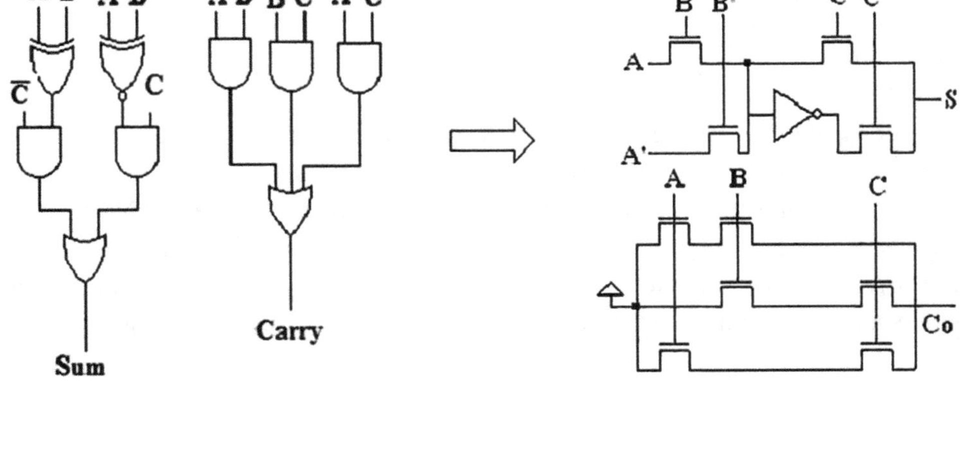

The non-clocked circuits are quite reliable. Nodes that determine the value of successive stages are never left to float. The careful device size selection allows PFET/NFET device ratio tuning to achieve desired switching points and unity gain points for specific noise concerns. Since control devices receive full gate voltage even after their capacitive load has been charged, the device remains inverted, and current is readily available if needed to overcome short circuit paths or leakage mechanisms caused by minor defects. The pass gate logic has earned the reputation of being a risky design style with good reason. There are a number of issues raised during the pass transistor operation that are listed in below:

1. **Limited Fan-In Capability**: for a pass gate with fixed transistor size, the amount of current to be discharged to ground through the structure must be limited in order to achieve acceptable down levels with reasonable delay, such as power must be in nW and delay pico seconds. In some configurations, a pass gate circuit may be expected to provide a level to multiple output nodes. Fartash Vasefi, (2005) determined, the cumulative gate capacitance of 'n' output buffers on node C_{out} is allowed the current must be discharged to ground.
2. **Excessive Fan Out**: Chip-Hong Chang, (2005), the output inverter provides a logic signal to subsequent pass gates that must be sized to provide sufficient high and low levels to the entire path it serves. If fan-out is excessive, the inverting buffer can be overwhelmed. In Figure. 5, node A (differential node) is tied to *n* logic paths; the inverting buffer in the differential node must deliver timely outputs with acceptable levels when coupled to additional paths, represented by the swing restoration point at output node of Figure 5.
3. **Noise:** Unlike static combinational CMOS, noise on signal lines caused by interconnects coupling can be propagated through a pass gate circuit via device gates or sources, differential node or swing restoration node as shown in Figure 6. According to John P. Uyemura (1999), this consideration doubles the opportunity for noise-driven disturbances of the output data. In CPL, the full adder circuit can be switched momentarily to its 'ON' state if swing restoration node B (Figure 5) couples up to 1's threshold voltage above, or if the differential node couples down to a voltage below 1's threshold voltage. If input C_{in} is high, the restoration appears on the input of the restoration inverter temporarily and disturbs the output. Noise in static logic cannot be disregarded, due to following reasons.
   a. Rail –to-rail voltage
   b. Poor up or down supply level, which can create voltage differentials on inputs
   c. Violation of decode exclusivity
   d. Full high level restoration
   e. Body effect

Pass transistor logic often uses fewer transistors, runs faster, and requires less power than the same function implemented with the same transistors in fully complementary CMOS logic. So, the non-clocked pass logics are better replacement of small scale application circuit compared with CMOS circuit. This book chapter has discussed six types of non-clocked pass transistor full adder logic. The six types of pass gates are developed by using multiplexing control input techniques (MCIT). The non-clocked pass transistor adders may be used in digital media and other new microprocessor applications, because they

Figure 5.

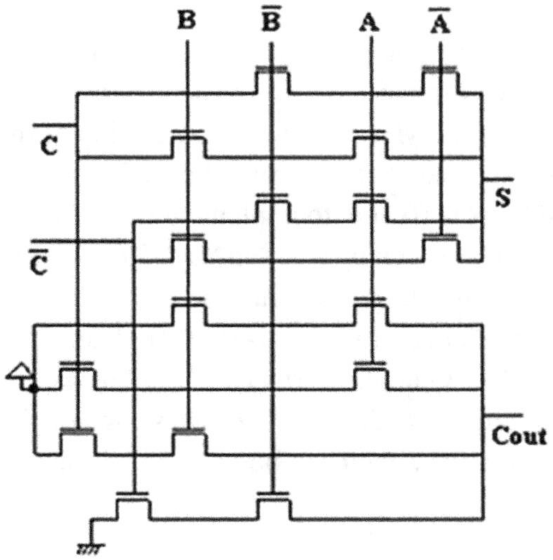

Figure 6.

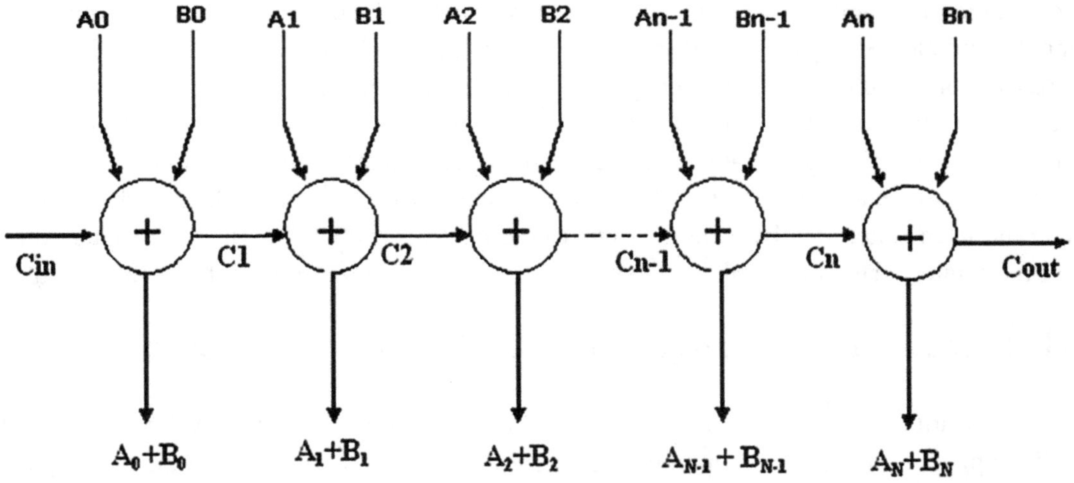

are fast-performing and execute the bits without any signal loss. The non-clocked logic adder circuits are present everywhere at once due to following considerations:

- Low power consumption
- Straightforward delay rule timing
- Inherent reliability and noise immunity
- Process variation and defect tolerance
- Migratability into successive technology
- Deterministic diagnostic capability

1.3.1 Complementary Pass Transistor Logic (CPL)

The full adder circuit designed by using complementary pass transistor logic (CPL) has swing restoration ability. Its dual-rail structure uses 34 transistors. The basic difference between the pass-transistor logic and the complementary CMOS logic styles is that the source side of the pass logic transistor network is connected to some input signals instead of the power lines. The advantage is that one pass-transistor network (either PMOS or NMOS) is sufficient to implement the logic function, which results in smaller number of transistors and input loads, especially when an NMOS network is used. However, pass-transistor logic has an inherent threshold voltage drop problem. The CPL full adder circuit is shown in Figure. 5. The output is a weak logic "1" when logic "1" is passed through an NMOS and a weak logic "0" when logic "0" is passed through a PMOS. Output inverters are used to ensure drivability. The strength of the CPL circuit is high speed operation due to fewer delay stages and small tree devices' +NFET transconductance. The weaknesses of the CPL circuit are body effect, limited logical depth and high leakage power when not cross coupled.

1.3.2 Swing Restored Pass Gate Logic (SRPG)

A number of pass gate based logic families use differential pass gate trees, similar to the CPL structure, to execute a required logic function. They, however, vary in their load circuit structure to improve CPL's power delay product. Swing restored Pass Gate Logic replaces CPL's cross–coupled PFET devices with four cross–coupled latches, as shown in Figure. 6. Kuo-Hsing Cheng, (2006), found that the SRPG logic has the ability to implement any Boolean logic function in a tree, which is restored to full swing by the latch. Besides load current, a trade-off exists between the performance of the transition and the maximum load that each stage may drive. The latch's NFET devices must be sized small enough to achieve and maintain the latch state. Ideally, once the logic tree accomplishes the flipping of the latch, the single latch NFET connected to ground takes over the business of producing strong high and low voltage levels.

1.3.3 Differential Cascade Voltage Swing Pass Transistor Logic (DCVSPG)

The DCVSPG full adder circuit is shown in Figure. 7. The performance of DCVSPG logic can be extended by implementing pass gate topology. The DCVSL is a differential style of logic that requires both true and complementary signals to be routed to gates. The two complementary NFET switching trees are connected to cross-coupled PFET transistors. Massimo, (2002), describes, depending on the differential inputs, one of the outputs is pulled down by the corresponding NFET network. The cross-coupled PFET transistors then latch the differential output. Since, the inputs drive only the NFET transistors of the switching trees, the input capacitance is typically two or three times smaller than that of the conventional CMOS logic. True and complement versions of logical inputs A, B and C are used as inputs to the DCVSG circuit. If A, B and C inputs are high, the NFET devices are 'ON'. In DCVSPG, both the NFET and PFET contribute to pull up performance, and both true and complement outputs are actively driven to their logical value. The PFET device sizing sensitivity problem in conventional DCVS is also eliminated, so that improperly sized PFET does not affect functionality. Ruiz. G.A. (2004), determined, the DCVS logic with the pass gate is a means of extending the performance benefits associated with DCVSL into pass gate topologies during operation of DCVSL pass gate logic, the charge is drawn from the power supply V_{DD}, then transported to the node capacitance and returned to the ground terminal, resulting in an irreversible energy conversion from electric energy to heat. As a result, when a node capacitance is charged (or discharged), it leads to an energy dissipation of $\frac{1}{2}CV_{DD}^2$. Thus reducing the energy dissipation has been equated to reducing the switching activity. The energy consumption is Depending on the differential inputs, one of the outputs is pulled down by the corresponding NFET network. The cross-coupled PFET transistors then latch the differential output.

1.3.4 Energy Economised Pass Transistor Logic (EEPL)

The EEPL full adder circuit is shown in Figure. 8. EEPL reduces power consumption and delay by interrupting the feedback of the latches and forming the load circuit in the structure, which allows a reduction in the width of the NFET devices. The device width reduction further contributes to the power reduction. The circuit action simultaneously provides regenerative positive feedback and shorter delay compared to CPL circuits. EEPL will be a valuable logic element in low power applications where performance is still essential.

1.3.5 Push-Pull Pass transistor Logic (PPPL)

PPPL achieves a performance advantage using complementary logic trees, and cross coupling to pull up (or) pull down devices, which recover full rail voltage. The Push-Pull Pass transistor Logic (PPPL) features a new innovation of recovering full V_{DD} levels while still exploiting the speed of differential pass gate trees, as shown in Figure 9. The complementary pass and control variables are connected to the drain and gate of the pass-transistors in theoretical logic network to implement the logic functions. The level restoration of PPPL, the drain of the PMOS is connected to the output of the n-channel network to push the degraded 'high' signal to the supply voltage (V_{DD}) while the drain of the NMOS is connected to that of the p-channel network to pull down degraded 'low' signal to the ground (Vss) with their gates cross-coupled together. Consequently, these cross-coupled transistors behave as pull-up and pull-down devices, and provide the driving capability since above two transistors turn on simultaneously during restoration., the power dissipation of the could be calculate from leakage current and V_{out}. The output of the PPPL full adder circuit is taken by both buffering devices PFET and NFET. The sum, its comple-

Figure 7.

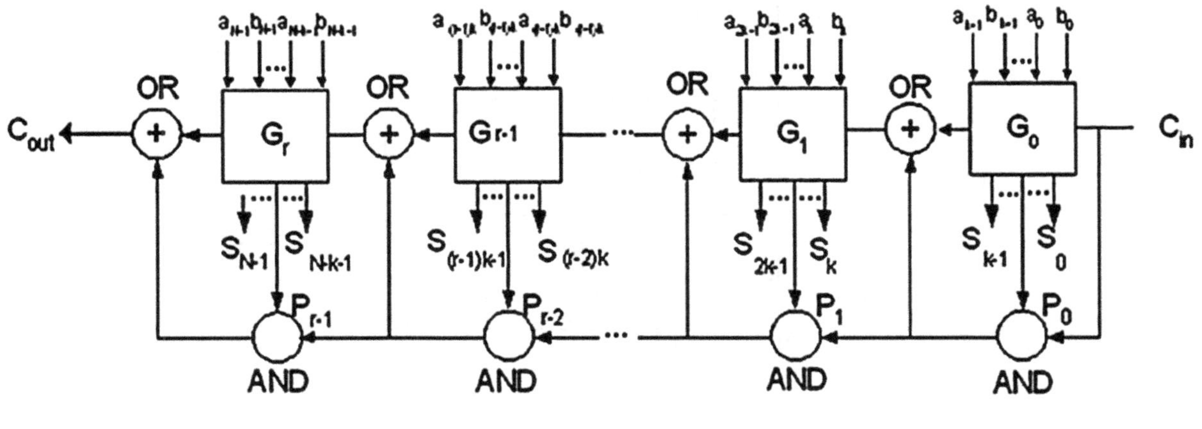

Figure 8.

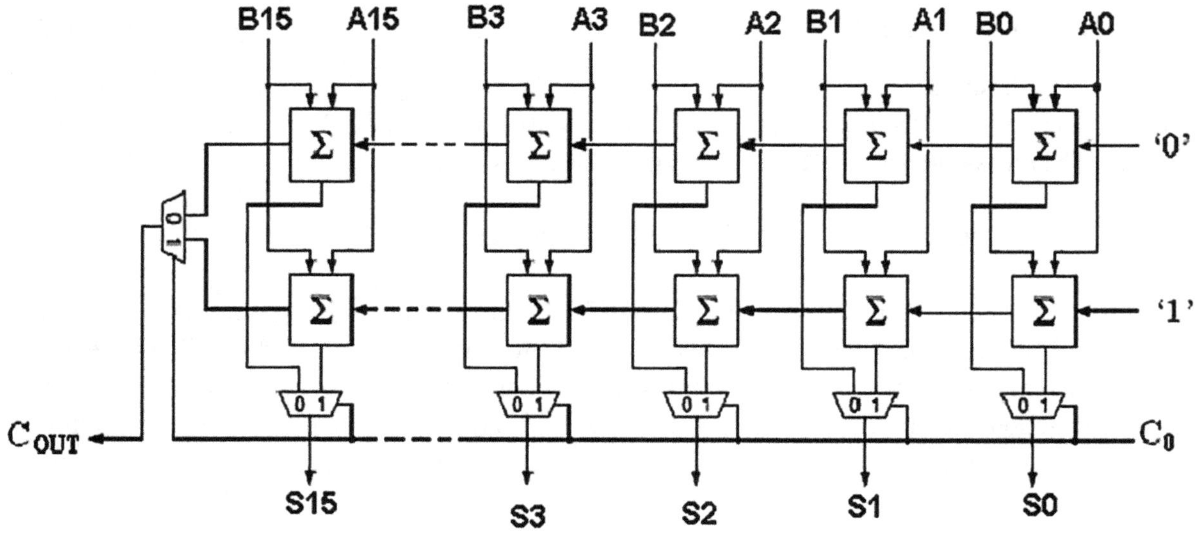

ment, carry and its complement struggle to acquire polarity output voltage. Due to swing restoration, the loading effect is created. The loading effect is minimised by using buffers of PFET and NFET in the output node for getting full rail voltage of output signal polarity.

1.3.6 Single-Ended Pass Gate Logic (SEPG)

The SEPG full adder circuit is shown in Figure 10. The SEPG logic design proposes reduction in size, power dissipation and propagation delay. The logic uses a simple and flexible set of logic designs to assign the input and can be used bidirectionally. These inputs could be connected to either supply if

Figure 9.

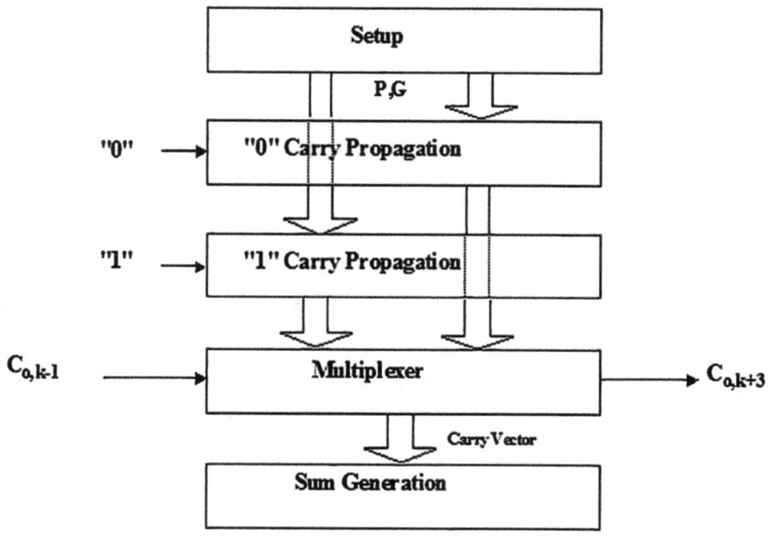

needed for a particular logic voltage level (logical '1'). The SEPG cell has to be reduced power dissipation when reduce packaging costs, reduce chip and system cooling. SEPG concentrates on synthesising the function of full logic blocks rather than individual logic functions. The specialised SEPG inverter is essentially a simple inverter with half-latched "keeper" PFET devices that bring the inverter input the rest of the way up from $V_{DD} - V_t$ to V_{DD}.SEPG logic addresses the issues associated with implementing pass gate logic in an automated design methodology. Thetraditional design methodologies to execute functions of limited scope in pass gates, which realise minimal advantages over static logic. SEPG logic maintains logic flexibility in order to support general logic function solutions that do not necessarily resemble the standard Boolean solution.

1.4 Multiplexing Control Input Techniques (MCIT)

Multiplexing Control Input Techniques refer to a CMOS-type logic family that possesses certain advantages. It is part of the larger family of CMOS with transmission gates, or CMOSTG. It is common to use this logic family for multiplexers and latches. CPL uses pass transistors to select between possible inverted output values of the logic. The output node drives an inverter to generate the non-inverted output signal. Since inverted and non-inverted inputs are needed to drive the gates of the pass-transistors, the complement of the logic also needs to exist to select between the possible non-inverted output values and then to drive an inverter to generate an inverted version of the output. The pass transistor logic gates are designed by using a reduction of Boolean identities and multiplexing control input techniques (MCIT). The two inputs (control input and actual input) of the pass transistor circuit can pass through the source and gate of the transistor terminals. The multiplexing control input technique (MCIT) could assume any input as a control signal. For the logical AND, the multiplexing technique assumes that input B is the control signal (see Figure 11) that is fed through the gate terminals. The actual input A is fed through the source terminal of the pass transistors. The design concept of the full adder circuit first transistor occupies inputs A and B, and the second transistor gets inputs B and B-complement, as shown in Figure 11(b). The output is generated from the AND gate after reduction by Boolean identities. In a similar way, the OR gate Boolean identities are simplified, assuming B as a control signal for the MUX circuit and A as the input variable to the OR gate. The AND, OR, XOR, Half adder, and multiplexer circuits are designed by using the Multiplexing Control Input Technique. In this book chapter, logic circuits are designed using pass transistors. The pass transistor logic gates are designed using reduction of Boolean

Figure 10.

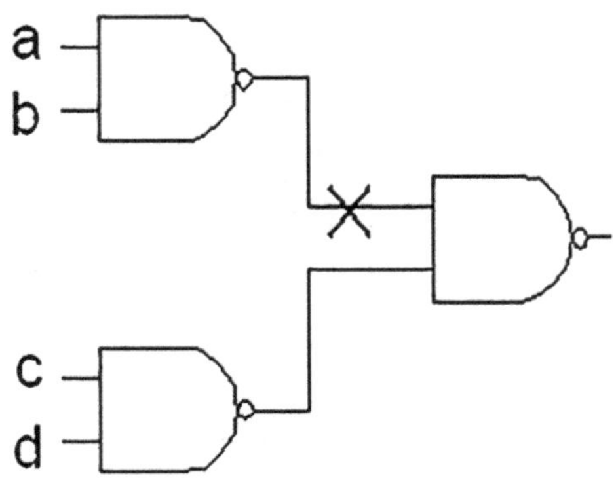

identities and multiplexing control input techniques (MCIT). According to C Senthilpari, (2008), the two inputs (control input and actual input) are supplied to the pass transistor circuit through the source and gate of the transistor terminals. According to Boolean identities, the inputs are activated.

1.4.1 OR Gate

The OR gate Boolean identities are simplified and shown in Figure 11 (a). In the design, it is assumed that B is a control signal for the Multiplexing method circuit and A is the input variable. The OR gate is designed by using the following equation:

$Z = B' (AB + AB') + B (A'B + AB)$

$= B' (A) + B (B)$

$= B' (A) + B (1)$ (1)

1.4.2 AND Gate

The AND gate pass transistor design concept is shown in Figure 11(b). With B is a control signal for the Multiplexing technique and A is the input variable. The two input AND gate is designed by using the Equation (2).

Figure 11.

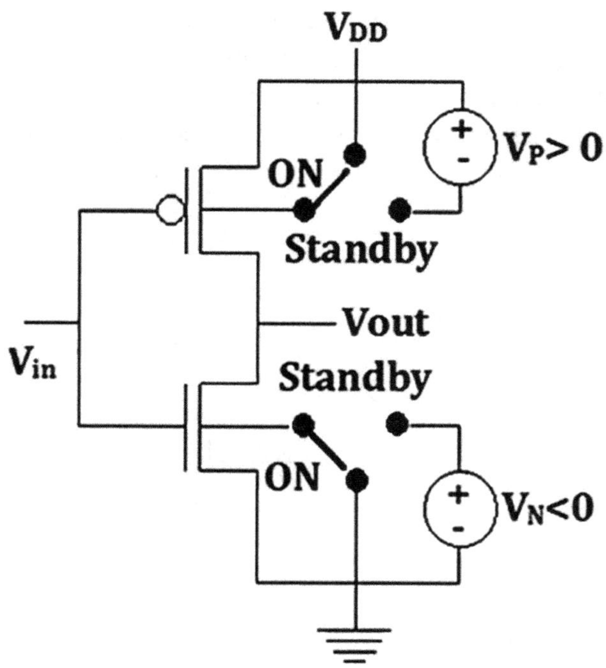

Z = B' (AB + A'B) + B (AB' + AB)

= B' (B) + B (A)

= *B' (0) + B (A)* (2)

1.4.3 XOR Gate

The control input B is fed through the gate of the MOSFET and input A and its complement are fed through source inputs of the logic circuit, as shown in Figure 12(c). A total of 6 transistors are used in the design of the XOR circuit. The Equation (3) needed for the implementation of the XOR gate is as follows:

Z = B' (AB + AB') + B (A'B + A'B')

= B' (A) + B (a')

= *B'A + B A'* (3)

The half adder circuit is designed by arranging the 2 input AND gates and 2 input XOR gates, as shown in Figure 12(d).

The multiplexing control input technique generates the pass-transistor functions. The 1-bit adder circuit, for the 'n' input variable function uses 'n-1' as controls and only one as input. Due to complementary data inputs, the gate is always working with a control data signal, as in the multiplexer. Since all gates use the same topology, only the inputs are permuted; therefore, the bit flows through the n-1 controls (C Senthilpari, (2011)). The transition of bits is very fast and there is no swing. The NMOS tree of the pass transistor logic is still static because the output defining nodes are always tied to V_{DD} or GND through a low resistance path. Due to this condition, the propagation delay is reduced tremendously. In the designed ripple carry n-bit adder, the stage *1* uses three inputs, which is to be implementing a 1-bit addition, i.e., two input data bits (A_i, B_i) and a carry input (C_{i-1}) from the previous stage. According to Shenn-Fu Hsiao, (2002), the speed of this adder depends on the carry propagation time through its stages to a large extent. Table 1 shows the truth table of the full adder carry-out and its

Figure 12.

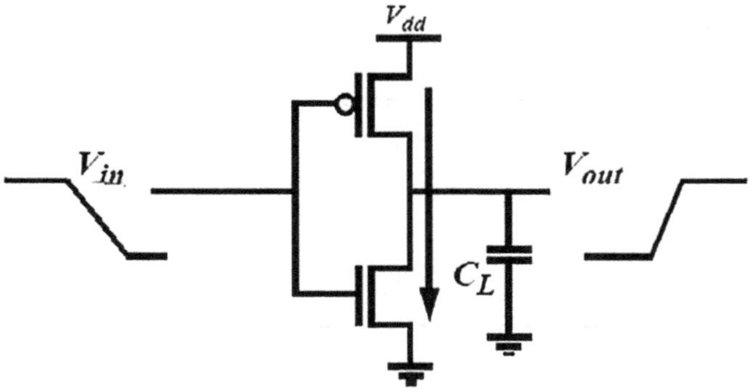

complement. When $A_i=B_i=0$ or $A_i=B_i=1$, the carryout is generated at the ith stage. The ith stage will propagate incoming carry C_{i-1} and ($\overline{C_{i-1}}$) to the next higher stage. P_i indicates an output of the stage, as given in table 2. The carry-out and its complement can be expressed as follows:

$$C_i = A_iB_i + P_iC_{i-1} = G_i + P_iC_{i-1} \tag{4}$$

$$\overline{C_i} = \overline{A_i}\,\overline{B_i} + P_i\overline{C_{i-1}} = N_i + P_i\overline{C_{i-1}} \tag{5}$$

where $G_i = A_i B_i$ and $N_i = \overline{A_i}\,\overline{B}$. Note that $G_iP_i=0$, $N_i P_i=0$, and $N_i G_i=0$; that is, these signals are mutually exclusive. Figure 5 shows an efficient implementation in CPL logic of the basic cells that enable the C_{OUT} and $\overline{C_{OUT}}$ to be obtained. The cell in Figure 13 is more suitable since it is not necessary to complement the input data (A_i, B_i). Moreover, the noise margin problem presented by $\overline{sum}$ a $\overline{carry}$ cells that can be eliminated by output restoring inverters. This problem is due to the fact that the high level is lower than the supply voltage level and close to the threshold voltage of the NMOS pass-transistors (P_i). These inverters also provide electrical insulation from the cell and increase its fan-out. To ensure full swing when a high level is transmitted via pass transistors, the level restoring inverter shown

Table 1. Average energy consumption per instruction of different adders

Feature Size	Variable	CPL	DCVS	SRPG	EEPL	PPPL	SEPG
0.60μm	Operating frequency MHz	496.524	473.397	481.023	611.097	531.914	461.893
	Energy/Instruction (nJ)	1.7623	1.9205	1.7402	1.4932	1.4261	1.8890
0.35μm	Operating frequency MHz	773.694	762.078	773.694	409.920	921.658	672.947
	Energy/Instruction (pJ)	778	797	778	667	629	715
0.25μm	Operating frequency MHz	920.217	814.796	993.048	760.340	991.86	844.594
	Energy/Instruction (pJ)	368.7	368.17	368.775	269.25	287.48	348.08
0.18μm	Operating frequency MHz	1895.95	1872.90	1895.95	1553.90	1734.90	1789.86
	Energy/Instruction (pJ)	186.96	184.51	186.96	142.56	159.18	186.96

Table 2. Full – Adder carryout

A_i	B_i	C_i	$\overline{C_i}$	P_i
0	0	0	1	0
0	1	C_{i-1}	$\overline{C_{i-1}}$	1
1	0	C_{i-1}	$\overline{C_{i-1}}$	1
1	1	1	0	0

in Figure 12 is required. This inverter has a weak fed back PMOS transistor, and the N- and P-transistors should be adjusted to balance the rising and falling output time.

1.5 Shannon Based (Proposed) Adder Design

The Shannon expansion theorem was developed by using the idea of Boolean functions, which can be reduced to the Boolean identity:

$$F = x.Fx + x'.Fx' \tag{6}$$

where F is any function and F_x and $F_{x'}$ are positive and complement Shannon cofactors of F, respectively. A positive Shannon cofactor of function F with respect to variable x is defined as that function with all x's set to 1. A negative Shannon cofactor is the same but sets all x's to 0. The Shannon expansion theorem is an important idea in Boolean algebra, which paves the way for binary decision diagrams and many other techniques relevant for digital circuits. According to Claude Elwood Shannon, the equation is simply stated as the expansion of a function:

$$f(X_1, X_2, ..., X_n) = X_1.f(1, X_2, ...X_n) + X_1^{'}.F(0, X_2, ..X_n) \tag{7}$$

followed by the Shannon expansion for two variables and noting that the expansion can be continued for any number of variables.

Figure 13.

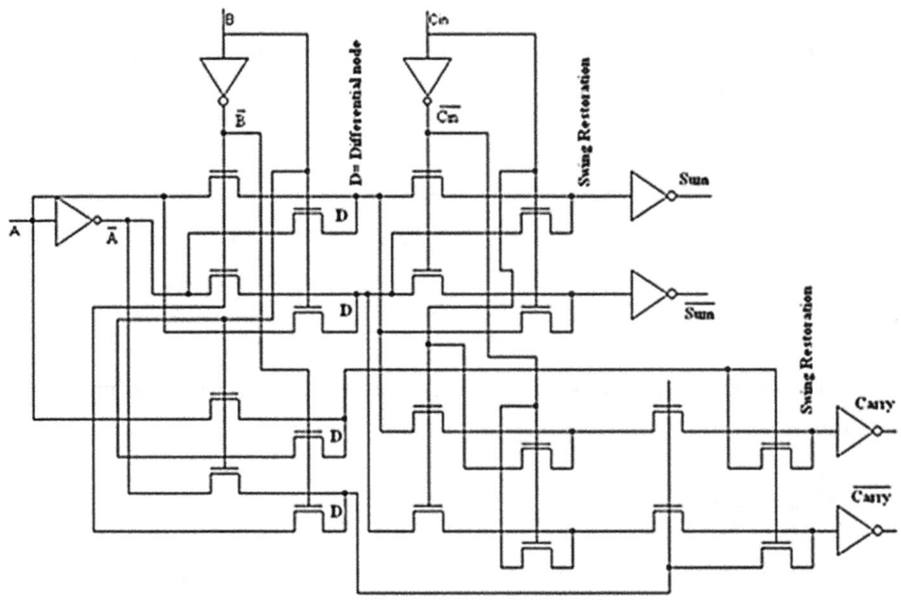

1.5.1 Example of Shannon Theorem

According to Claude Elwood Shannon (1950), the three inputs can be rewritten using Boolean algebra. Let the function f be expressed as:

$$f = xyz + xy'z + x'y'z + x'yz + x'y'z' \tag{8}$$

This function Equation (7) may be rewritten in terms of any two variables, namely, a variable and its complement:

$$f = x'g'_x + xg_x \tag{9}$$

where g'_x generates functions of y' and z' and g_x generates functions of y and z. This means that the logical 1 values are separated from the logical '0' values.

To make the final expression of Equation (7) equivalent to composition of functions, simply apply the distributive theorem to the function of x:

$$f = x'(y'z + yz + y'z') + x(yz + y'z) \tag{10}$$

In Shannon's expansion, the term x is very significant but problems can arise in simple equations (Claude Elwood Shannon (1950)). If there is no x variable in one or more of the functions, the problems can lead to confusion. Dealing with x variable in one or more of the functions can be simple, but the solution is not always intuitive. Claude Elwood Shannon (1950) found in Boolean algebra, can AND any literal or term with 1 and still achieve the same truth value, which is shown in below equation.

$$f = yz + xyz' + x'y'z \tag{11}$$

If desire to expand the variable x and simply do not have enough information in the first term to accomplish this task. If AND the literal with 1, or, in this case, $(x' + x)$:

$$f = yz(x' + x) + xyz' + x'y'z, \tag{12}$$

Then expand:

$$f = x'yz + xyz + xyz' + x'y'z \tag{13}$$

This function now contains the variable about which can expand the expression, so now there should be no problem doing so:

$$f = x'(yz + y'z) + x(yz + yz') \tag{14}$$

The expansion is now complete. Of course, it can perform Shannon's expansion about any desired variable, as long as function can provide for that variable in the expression without changing the truth value of the expression. The Shannon function can perform multiple expansions of a single function (e.g., about *x*, then about *y*) or many variables at once (e.g., about *xy*). The result is a functionally equivalent expression for the variables involved.

1.5.2 Expanding and Minimising

The logic gates are designed by basic design logic functions that are

- The Shannon function can expand a given SOP function.
- The Shannon function can minimise a given SOP function.

The expanding and minimisation of Shannon functions can be performed by using a logic device called a multiplexer. Multiplexers take *n* select inputs and 2^n data inputs and give one output. The Boolean function can expand any number of variables as select inputs and their respective composed (complement and true) functions as the corresponding data inputs.

The Shannon Theorem states that any logic expression can be expanded into two terms, the first term setting a variable to 1, another variable set to 0 and multiplying by the inverse. By repeating Shannon's Theorem for each variable in the expression, the fullest reduction can be achieved. The method lends itself to automation, which is particularly useful in multiplexer and pass transistor circuit design. According to C. Senthipari, (2008), Shannon's Theorem is stated in a generalised form is given in equation. A function of many variables, $f(a_0, a_1, a_2, ..., a_i, ..., a_n)$ can be written as the sum of two terms, one with a particular variable (say a_i) set to 0, and one with it set to 1.

$$f_{(a0,\,a1,\,a2,\,...,\,ai,\,...,\,an)} = a_i'\, f_{(a0,\,a1,\,a2,\,...,\,0,\,...,\,an)} + a_i f_{(a0,\,a1,\,a2,\,...,\,1,\,...,\,an)} \tag{15}$$

Shannon's Theorem is applied to the logical function using n-1 variables as control inputs and three data lines set to a logical '1'. These source inputs are connected to the V_{DD} lines for logical "1" and connected to ground for logical '0'. The remaining n^{th} variable is connected from the data input to the source input. The control signals flow vertically, and the data flow horizontally. N-type transistors are placed at the intersections to satisfy the expanded Shannon function. According to Boolean identities, pairs of transistors are connected in one input, which is making as 1 transistor (A.A=A) the opposite pair such as (A.A'=0) which is cancel each other. The Shannon expression's output depends upon the pass logic. If expression have '0' logic, the connection input is given by 0 and by '1' for the connection input '1'.

1.5.3 Architecture of the Proposed Adder Cell

The proposed full adder circuit combines the MCIT for the sum operation and the Shannon Theorem for the carry operation; the sum and carry circuits are designed based on standard full adder sum and carry equations. An input B and its complement are used as the control signals of the sum circuit, as shown in Figure 12. The two-input XOR gate is developed using the multiplexer method. The output node of the two-input multiplexer circuit is the differential node. According to Equation (16), the sum circuits need three inputs. In order to avoid an increasing number of transistors due to the addition of a third input, the XOR gate output is fed through the NOT gate from the differential node to C and $\overline{C}$

inputs of the full adder, thereby reducing the number of transistors to six in the sum circuit. The C and $\overline{C}$ output node is called the differential node of the circuit. The differential node output is a summing output.

The carry circuit is designed using Shannon complementary pass transistor logic and uses only the inputs A, B and C. The circuit was designed using the fundamental Shannon equation given in (26). The source inputs are connected to logical '1', yielding an always 'on' condition for the transistor. The actual inputs AB, BC and CA are connected in parallel to give the carry output Cout = AB + BC + CA. There are no complementary inputs used in the full adder carry circuit. The circuit works according to carry Equation (15). In this circuit, all of the pass inputs are connected to the V_{DD} line so that the pass gates are always ON. The proposed full adder circuit (Figure 12 and 13) was designed using the Shannon theorem. The full adder sum and carry circuits are designed based on standard adder circuit equations. An input B and its complement are used as the control signal of the sum circuit, as shown in Figure24. The two-input XOR gate is developed using the multiplexer method. The output node of the two-input multiplexer circuit is the differential node. In order to avoid increasing the number of transistors due to the addition of a third input, the CPL XOR gate is multiplied with C's complement input, and the XNOR gate is multiplied with input C, thereby reducing the number of transistors in the sum circuit.

Compared with the Mixed-Shannon case, the XOR gate arrangement avoids the critical path delay but results in a larger number of transistors. The C and $\overline{C}$ output node is called the differential node of the circuit. The differential node output is a summing output, as given by Equation (16). The two complementary ($\overline{C}$ and $\overline{B}$) inputs are used in the full adder carry circuit to balance the circuit and to avoid a floating wire. The circuit works according to the carry equation. In this circuit, all of the pass inputs are connected to the V_{DD} line so that the pass gates are always ON. The control input terminals are connected to the function inputs.

The proposed circuit are simulated by various feature size and results are shown in Table 3. The proposed Shannon based adder, Mixed-Shannon adder, CPL adder and CMOS adder circuits are simu-

Table 3. 1-Bit proposed Shannon based, CPL and CMOS Adder Cells' power dissipation, propagation delay and area

Adder Type	Supply Voltage	0.5V	0.7V	1.0V	1.2V	2V	2.5V	3.5V
Shannon	Power µW	0.042	0.371	0.442	0.569	1.984	12.04	24.88
	Delay ps	2	4	6	13	23	29	54
	Area µm²	55x7	22x10	23x11	28x14	37x26	62x32	84x52
Mixed Shannon	Power µW	0.309	0.437	0.513	0.658	2.529	14.8	31.13
	Delay ps	6	12	18	22	25	49	70
	Area µm²	16x7	16x9	17x10	20x9	19x12	52x30	82x48
CPL	Power µW	0.817	2.247	3.249	4.907	9.936	14.782	29.69
	Delay ps	14	23	31	81	87	205	324
	Area µm²	43x8	19x10	20x12	24x14	49x28	62x35	98x56
CMOS	Power µW	1.49	3.76	7.06	10.69	14.04	25.34	32.10
	Delay ps	654	854	971	995	1000	1401	1602
	Area µm²	25x7	53x10	67x14	74x11	93x26	117x32	186x52

lated for various feature sizes from 50nm to 0.35μm using corresponding supply voltages. The proposed Shannon adders give better performance in terms of power dissipation, propagation delay, and area than CPL and CMOS full adder circuits, due to regular arrangement of the transistor in a tree structure and a shorter critical path, as seen in Table 2.

The proposed 1-bit adder cell consumes less power, gives lower delay and less PDP than the various proposed 1-bit adder cells due to regular arrangement of the transistor tree structure, shorter critical path and the multiplexing method of design.

1.6 Types of Adders

Wu-An Kuo, (2006), found that an addition is a fundamental operation for any digital system. Fast and accurate operation of a digital system is greatly influenced by the performance of the resident adders. The adders are very important components in digital systems because of their extensive use in basic digital operations such as subtraction, multiplication and division. Hence, improving the performance of the digital adder would greatly advance the execution of binary operations. According to Yun Kim, (2001), the performance of a digital circuit is gauged by analysing its power dissipation, layout area and operating speed. In this section, the implementation technique of several types of adders and their characteristics and performance will be discussed. These are:

- Ripple carry adder, or carry propagate adder
- Carry look-ahead adder
- Carry skip adder
- Carry select adder
- Carry save adder

1.6.1 Ripple Carry Adder

Adiseno (2003) studied, Aparallel adder performs addition on all bits simultaneously, so it is faster than serial adders. The simplest parallel adder is a ripple carry adder. An n-bit ripple carry adder is constructed by cascading 'n' full adders, as shown in Figure 14. The carry output of each full adder (FA) is connected to the carry input of the full adder of the next higher position. The amount of hardware is proportional to *n*. In designing a full adder for a fast ripple carry adder, it is required to minimise the delay from the carry-in, Ci, to the carry-out, C_i+1. A full adder can be realised with logic gates such as AND gates, OR gates and XOR gates, as seen in Figure 14. Since the speed of adders is important for the speed of the entire computer, full adders are usually realised with more sophisticated transistor circuits using MOSFETs such that a carry Ci can propagate quickly to higher bit positions through pass transistors. A carry propagates through transmission gate T. When $A_i = B_i = 0$,Tbecomes non-conductive, and NMOSFETs become conductive, then the carryout, C_{i+1} becomes 0 because the carry-out terminal C_{i+1} is connected to the ground. When $A_i = B_i = 1$, T becomes non-conductive and PMOSFETs become conductive. Then the carry-out, C_{i+1}, becomes 1 because the carry-out terminal C_{i+1} is connected to the power supply V_{DD}. When $A_i = 0$ and $B_i = 1$, or $A_i = 1$ and $B_i = 0$, T becomes conductive, and the carry-out terminal is connected to neither V_{DD} nor the ground, so a carry-in C_i is sent to C_{i+1} as a carry-out.

Figure 14.

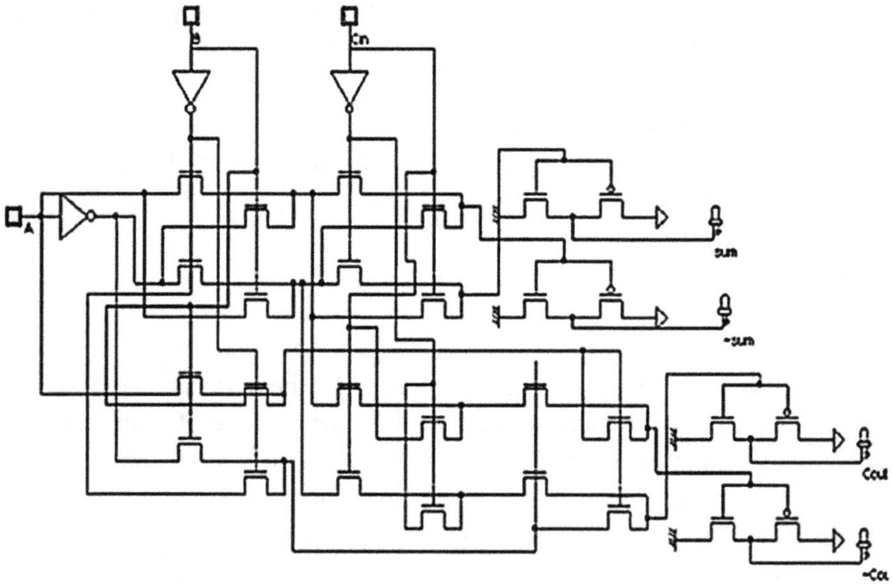

Vesterbacka. M,(1999), found that, A ripple carry adder with Manchester carry chain is referred as a Manchester adder. This idea combines very well with the carry skip technique mentioned below in Section 1.6.3. Oscar (2004), the full adder cannot send a carry over many positions in a ripple carry adder. For speedup, one must insert an inverter after every few positions to send a high output power over many higher bit positions. In the simultaneous addition in all n-bit positions, a carry propagates n positions in the worst case, but, on average, it propagates only about log (2n) positions. The average computation time of a ripple carry adder can be reduced by detecting carry completion because it need not always wait for the worst delay. According to Victor (2007), an adder with such a mechanism is called a carry completion detection adder and is useful for asynchronous systems. When a full adder is realised with ordinary logic gates, for example, a NOR gate only, the total number of logic gates in the ripple carry adder is not minimised, even if the number of logic gates in each FA is minimised. But the number of logic gates in the ripple carry adder can be reduced by using modules that have more than one input for a carry-in and more than one output for a carry-out. The number of NOR gates of such a module is minimised by the integer-programming logic design method, and it has been found that there are 13 minimal modules.

Different types of ripple carry adders can be realised by cascading such minimal modules. Some of these adders have carry propagation times shorter than that of the ripple carry adder realised with FAs having NOR gates. Also, there is a Ripple Carry Adder (RCA) that has a minimum number of NOR gates when this adder is constructed by cascading the three consecutive minimal modules, where the module for the least significant bit position is slightly modified and one NOR gate is added to convert the carry out in multiple lines into the carry out in a single line. It is the total number of NOR gates in this ripple carry adder is minimum for any value of 'n'. Also, there is a ripple adder such that the number of connections, instead of the number of logic gates, is minimised.

1.6.2 Carry Look-Ahead Adder (CLA)

A *carry look-ahead adder* is a type of adder used in Vander Zanden's digital logic. The carry look-ahead adder depends on two things:

- Calculating, for each digit position, whether that position is going to propagate a carry if one comes in from the right.
- Combining these calculated values allows us to deduce quickly whether, for each group of digits, that group is going to propagate a carry that comes in from the right.

Tadahiro Kuroda (1995), describes, a carry look-ahead logic uses the concepts of *generating* and *propagating* carries. It is most natural to think of generating and propagating in terms of binary addition. The addition of two 1-digit inputs *A* and *B* is said to *generate* if the addition will always carry, regardless of whether there is an input carry (equivalently, regardless of whether any less significant digits in the sum carry).

In the case of binary addition, $A + B$ will be generated if both *A* and *B* are 1. If $G(A,B)$ represents the binary predicate, then it will generate *A* AND B as given below:

$$G\,(A.B) = A.B \tag{16}$$

The addition of two 1-digit inputs *A* and *B* is said to *propagate* if the addition will carry whenever there is an input carry (equivalently, when the next less significant digit in the sum carries). In the case of binary addition, $A + B$ propagates if and only if at least one either *A* or *B* is 1. If we write $P(A,B)$ to represent the binary predicate that is true if and only if $A + B$ propagates, we have:

$$P\,(A,\,B) = A+B \tag{17}$$

Sometimes a slightly different definition of *propagate* is used. By this definition, $A + B$ is said to propagate if the addition will carry whenever there is an input carry but will not carry if there is no input carry. It found that generate and propagate bits, which are used by the carry look-ahead logic, it does not depends on the definition of chain structure in carry look ahead adder. In the case of binary addition, this definition is expressed by:

$$P'\,(A,\,B) = A \oplus B \tag{18}$$

For binary arithmetic, OR is faster than XOR and takes fewer transistors to implement. However, for a multi-level carry look-ahead adder, it is simpler to use $P'\,(A,\,B)$. Bisdounis. (1998), Using these concepts of generates and propagates, there will be carry propagation when either the addition generates *or* the next less significant bit carry out. According to Boolean algebra, with C_{ini} as the carry bit of digit *i*, and P_i and G_i as the propagate and generate bits of digit *i* respectively, the carry input for i+1 bit ($C_{in\,(i+1)}$) is given as:

$$Cin_{(i+1)} = G_i + (P_i.Cin_i) \tag{19}$$

Implementation of CLA for each added binary bit in the carry look-ahead logic will determine whether the bit pair will generate or propagate the carry. This allows the circuit to "pre-process" the two numbers (x_0, y_0) being added to determine the carry ahead of time. When the actual addition is performed, there is no delay from waiting for the ripple carry effect or from the time it takes for the carry from the first full adder to be passed down to the last full adder.

1.6.3 Carry Skip Adder (CAS) Technique

A full adder circuit is capable of handling the addition of two 1-bit 2's complement numbers. It is possible to create a logical circuit using several full adders to add multiple bit numbers. It adds additional group carry bypass paths to its ripple path and carries can bypass the ripple path when the group propagate signal is high. Figure 14 shows the basic structure of an n-bit carry-skip adder with k-bits per group. The critical signal is generated at LSB, ripples through the remaining k-1 bits in the least significant group, bypasses the order of n (k–2) groups, and ripples through the k-1 bits in the most significant group. The 1-bit adder circuits are designed using the multiplexing control input technique (MCIT) for different non-clocked pass gate logics.

Working Principle of CSA Adder

The non-clocked pass transistor circuits are evaluated in terms of power dissipation (P_D), propagation delay, PDP and area for different submicron feature sizes. The 16-bit CSA adder's results are compared with other reported results. Each stage of the adder has to wait until the previous stage has calculated and propagated its carry output signal. Obviously, the longest delay results for operands like A = 0b0000000000000000, B = 0b1111111111111111 or A=0b0010101010101010l and B=0b1010101010101010. Therefore, the total delay of a carry skip adder (CAS) is proportional to the number of bits. Faster adders are often required for bit widths of 16 or greater. The parameters analysed, such as power dissipation, propagation delay, PDP and area are dependent on the maximum voltage of any pass gate structure. This pass gate structure uses single NFETs to perform logic, and there is no problem due to threshold voltage loss or slow logic '1' transfers if one use the maximum supply voltage. These concerns are addressed by adding a static inverter at the output node. The design of the inverter can be an important consideration for this type of logic circuit. The power output is measured from the layout simulation of the adder circuit. The power delivered in the load is calculated by multiplying the value of the quiescent current and the output voltage of the respective node. In a realistic circuit, however, a small amount of quiescent current I_{DQ} flows across the reverse drain-bulk regions. The quiescent DC power delivered in transistors of the proposed adder circuit is calculated by using the standard formula. According to the output voltage swing, design style and transition of logic '0' (low) to logic '1' (high), the output voltage and power delivered in the load are different for different design logics. The supply voltage of the designed cell is an important tool for measuring the output voltage.

1.6.4 Carry Select Adder

The ripple carry adder's worst delay can be reduced by partitioning the adder into two blocks; one for higher bit positions and the other for lower bit position. For the block with the higher bit position, calculate two sums in parallel, one assuming a carry input of '0' from the block for lower bit positions and the other assuming a carry input of '1'. After that, select the correct sum based on the actual carry output from the lower bit position block. The delay is approximately halved because the calculations in these two blocks are carried out concurrently, as the adder is portioned into two blocks of the same size. The carry select adder is shown in Figure 15. The delay can be reduced even more by portioning the adder into more blocks. One can reduce the amount of hardware by calculating two candidate sums using only one adder in each block. Applying the principle used to develop the carry select adder to each block, one can realise a two-level carry select adder. Henrik (2006), found that the recursive application of the principle yields a multi-level carry select adder. A conditional sum adder can be regarded as the extreme case.

1.6.5 Carry Select Adder (CSA) Technique

The common way to design the 16-bit carry-select adder builds it as a cascade form of 1-bit full-adder, a 3-bit carry-select block, a 4-bit carry-select adder and a 6-bit carry select adder. The ripple-carry adder is one in which each adder bit has to wait for the arrival of its carry-input signal before the actual addition can start. In total, the carry propagation time through an n-bit adder block is reduced from O(n) to the number of stages times the delay of the multiplexers. Naturally, using 'n' blocks of 1-bit carry-select adders would incur a complexity of 'n' multiplexers, again resulting in O(n) delay. Therefore, a partition with slowly increasing block-size is chosen. In the Carry Select Adder (CSA), the first (least-significant) block consists of a simple full adder, followed by a 3-bit carry-select block, and finally a 4-bit carry-select block. While the delay of the standard ripple-carry adder with n-bits is O(n), the delay through the carry-select adder behaves as $0\sqrt{n}$ at a hardware cost of O (3n).

A carry-select adder is divided into sectors, each of which, except for the least significant, performs two additions in parallel, one assuming a carry-in of zero, the other a carry-in of one. Figure 16 shows the block diagram of the proposed 16 bit adder circuit using the CSA technique. Weste, (1993), described that the circuit is mainly divided into sectors of lengths 1, 2, 3, 4, and 6, proceeding from least-significant

Figure 15.

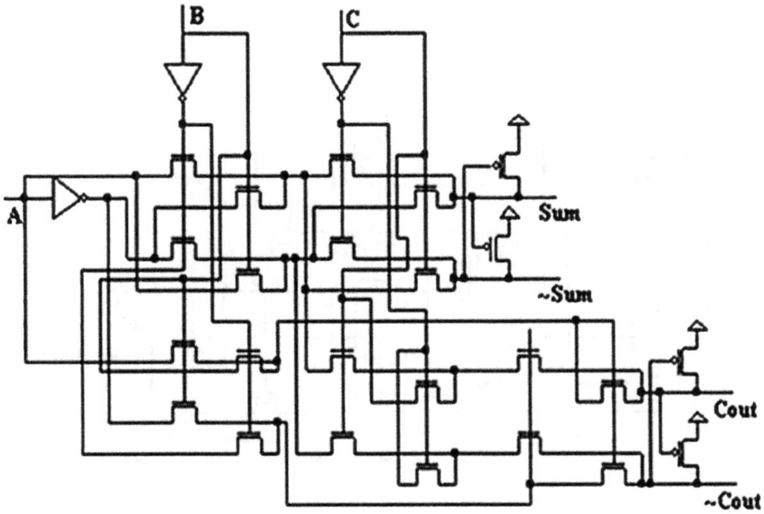

(LSB) to most-significant bit (MSB). The CPL full adder circuit is designed by using the multiplexing control input technique. The pass transistor function can be generated by using the K-map. When the full adder Boolean expression result =1, the pass transistor function is represented by the input variables, and when the full adder Boolean expression result =0, the pass transistor function is represented by the complement of the input variables. The write expression to include the product terms of the control signals ANDed with the resultant of the minimised sum-of-products expression. Due to complementary data inputs, the gate always works on a control data signal, as the multiplexer.

The source input acts as an input of the signal. The 4-bit sector of Figure 16 illustrates the general principle. Within the sector, there are two 4-bit ripple-carry adders receiving the same data inputs but different carry-ins. The upper adder has a carry-in of zero, and the lower adder has a carry-in of one. According to Gustavo (1998), the actual carry-in from the preceding sector selects one of the two adders. If the carry-in is zero, the sum and carry-out of the upper adder are selected. If the carry-in is one, the sum and carry-out of the lower adder are selected. Logically, the result is not much different than if a single ripple-carry adder were used. The difference, of course, is in performance. Instead of having to ripple through four full adders, the carry now only has to pass through a single multiplexer.

The proposed 16-bit adder is designed using the carry select adder technique for different non-clocked pass gate families. The proposed adder circuit results are compared with other existing adder circuit results. Ashmila (2005), found that the 1-bit adder circuit has 34 transistors. The inputs A_i and B_i are the ith bits of the input data, and C_{i-1} is a carry input of the ith stage. The outputs of the sum and carry are obtained using the computing method, as given in Table 4.

- **G (A.B)** (generate) occurs when a C_0 is internally generated within the adder (occurs independent of C_i).
- **P (A+B)** (propagate) indicates that C_i is propagated(passed) to C_0.
- **P'(A XOR B)** (propagate) is used in some adders for the P term since it can be reused to generate the sum term.

Figure 16.

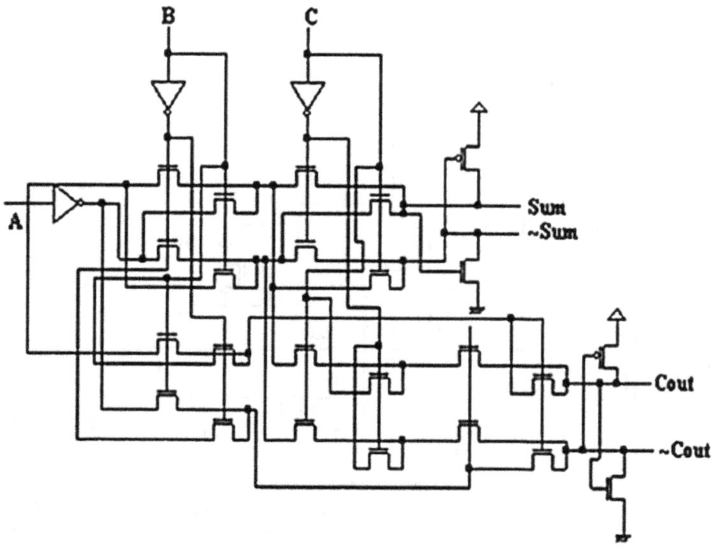

Table 4. Truth table of CSA Adder

A	B	Ci	G(A.B)	P(A+B)	P'(A⊕B)	Sum	Cout	Carry status
0	0	0	0	0	0	0	0	Delete
0	0	1	0	0	0	1	0	Delete
0	1	0	0	1	1	1	0	Propagate
0	1	1	0	1	1	0	1	Propagate
1	0	0	0	1	1	1	0	Propagate
1	0	1	0	1	1	0	1	Propagate
1	1	0	1	1	0	0	1	Generate
0	0	0	0	0	0	0	0	Delete

- **D** (A. B) (delete) ensures that a carry bit will be deleted at C_0.

The Boolean expressions for S and $C_0$are:

Sum = A'B'C_i+ A'BC_i+ AB'C_i+ ABC_i= A **XOR**B **XOR**C. (20)

- Addition/Subtraction: S and C_0 can be written in terms of G and P' as C_0 (G, P') = G + P'Ci (or P in this case). S (G, P') = P' XOR Ci, where G and P' are independent of Ci. (Also, C_0 and S can be expressed in terms of delete (D)).

According to Mudassir, (2005), $C_i = G_i + P_i C_{i-1}$ $S_i = P_i \oplus C_{i-1}$ where the Ci and Si are the sum and carry of the net output of the ith adder circuit, and $P_i = A_i \oplus B_i$ and $G_i = A_i B_i$ are carry propagate and carry generate signals of the full adder circuit.

Guyot. A, (1987), found that the carry select adder (CSA), the problem of carry propagation delay is overcome by dividing the data word into two parts and completing the addition of the most significant (MS) and least significant (LS) parts simultaneously and by taking both possible values of the input carry and evaluating the result for both possibilities. According to Mitra Mirhassani, (2008), The CSA adder speeds up the addition operation, but it still has area and power consumption disadvantages since the MS adder uses dual carry ripple adders (CRAs) and the correct, or true, sum from each of the MS adders is selected by using a multiplexer. This CSA speed up technique attempts to reduce the additional area of the second MS adder in the CSA technique, and the use of dynamic logic has resulted in some area and operating speed savings. Different CSAs with more area-efficient structures or with an improved critical path delay have been proposed. These CSAs can be roughly classified according to the type of select circuit in the CSA, whether based on carry selection, whose main objective is high speed, or on adder selection. The total power of the proposed adder circuit is calculated from the layout simulation result using the standard formula V_{DS} x I_{DQ}, where I_{DQ} is the operating current of the NMOS tree transistor. The propagation delay is calculated using the standard formula, namely, the difference between the rise and fall time of one full square wave. The proposed 16-bit CSA adder circuit is compared with non-clocked pass transistor logic; the CPL is found to be mostly dominant over other existing pass transistor logics (Figure 16).

1.6.5 Carry Save Adder

When several numbers are sequentially added up, it is not necessary to propagate the carry during each and every addition. The carries generated during an addition may be saved as partial carries and added to the next operand during the next addition. We can accelerate each addition by postponing the carry propagation. This leads to the concept of carry save addition. The carry save additions add up numbers by a series (multiple-operand addition) followed by a carry propagate (carry propagate) addition. Morris Mano (1994), found that the adder for carry save addition is known as a carry save adder, while the adders mentioned previously are called carry propagate adders. Morris Mano (1994), described a carry save adder sums up a partial sum and a partial carry from the previous stage as well as an operand and produces a new partial sum and partial carry. An n-bit carry save adder consists of just n full adders without interconnections among them. The operation of the carry save adder is shown in Figure 17.

The 16-bit carry save adder (CSA) circuit is designed by using the carry save technique for various non-clocked logic families, such as SRPG, DCVSPG, EEPL, PPPL, CPL and SEPG. The 16-bit CSA non-clocked pass transistor adder's simulated results are given in Table 5, and the results clearly show the dominance of CPL logic over other non-clocked logic. Lower power dissipation occurs for a 16-bit adder circuit designed using CPL logic for a 50nm feature size. The power consumption, delay and PDP for any logic design decrease as one scales down the feature size. The other non-clocked logic, SRPG, is comparable to CPL logic. One interesting observation is that when the feature size is higher (120nm), the reduction is less than smaller feature size due to the shrinkage effect of the semiconductor. In the same manner, there is less percentage reduction in propagation delay with smaller feature sizes. The analyses are extended to calculate the values of energy per instruction (EPI), Latency and throughput. The 16-bit ultra submicron non-clocked pass gate adders are analysed using the BSIM 4 advanced model, and simulated results are given in Table 5.

Figure 17.

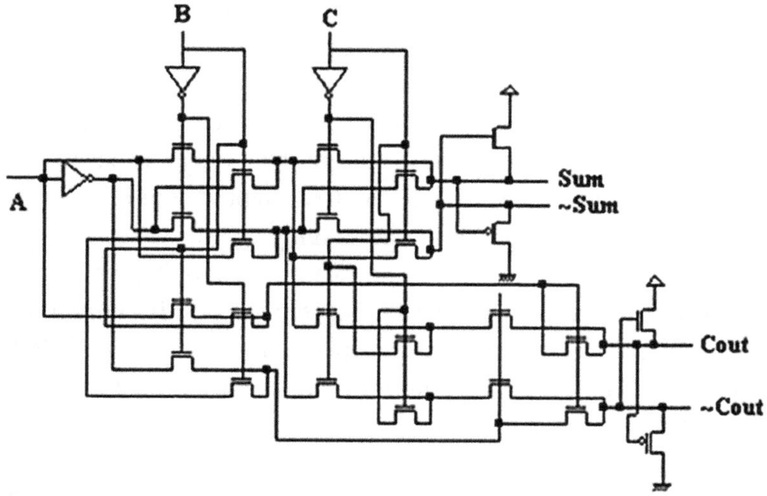

Table 5. Simulation result for power dissipation, delay, PDP, area, number of transistors, EPI, throughput and latency for different pass transistor logics

Feature Size	Variable	CPL	DCVS	SRPG	EEPL	PPPL	SEPG
120nm (1.2V)	P.Diss (P_D) W	59 x10^{-6}	62 x10^{-6}	59.22 x10^{-6}	1.14 x10^{-3}	0.684 x10^{-3}	0.4326 x10^{-3}
	Delay (t_d) ns	0.36761	0.37344	0.36761	2.4395	0.3446	0.4074
	PDP(P_D x t_d)	2.1688x10^{-14}	2.3153x10^{-14}	2.1769x10^{-14}	2.7810x10^{-12}	2.537x10^{-13}	1.7624x10^{-13}
	Area μm^2	19346	12000	11072	15000	10217	15750
	No. of Transistor	1120	902	1120	902	984	970
	EPI (pico/Joules)	68.258	60.285	31.243	50.99	51.851	47.460
	Throughput (Gbits)	2.03	2.00	0.389	2.03	1.878	1.962
	Latency (nanosec)	7.881	7.975	41.032	7.881	8.5184	8.153
90nm (1V)	P.Diss (P_D) W	27 x10^{-6}	32x10^{-6}	54.9 x10^{-6}	0.87 x10^{-3}	0.445 x10^{-3}	0.208 x10^{-3}
	Delay (t_d) ns	0.3622	0.36875	0.3622	1.6364	0.3510	0.3842
	PDP(P_D x t_d)	9.7794 x10^{-15}	11.8x10^{-15}	19.884x10^{-15}	1.4236 x10^{-12}	1.561x10^{-13}	7.991 x10^{-14}
	Area μm^2	9500	5500	10778	6513	7014	7014
	No. of Transistor	1120	902	1120	902	984	970
	EPI (pico/Joules)	38.476	37.225	19.905	30.476	34.556	29.508
	Throughput (Gbits)	2.052	2.025	0.567	2.050	1.963	2.016
	Latency (nanosec)	7.795	7.9	28.182	7.795	8.147	7.913
70nm (0.7V)	P.Diss (P_D) W	3.12 x10^{-6}	10x10^{-6}	31.69x10^{-6}	0.25 x10^{-3}	0.123x10^{-3}	0.1440 x10^{-3}
	Delay (t_d) ns	0.29777	0.30613	0.29777	0.34354	0.2439	0.2844
	PDP(P_D x t_d)	9.2904x10^{-16}	3.0613x10^{-15}	9.4363x10^{-15}	8.5885x10^{-14}	2.999x10^{-14}	4.0953x10^{-14}
	Area μm^2	9140	4810	7228	6734	6734	6734
	No. of Transistor	1120	902	1120	902	984	970
	EPI (pico/Joules)	21.874	18.574	13.119	15.574	17.239	11.574
	Throughput (Gbits)	2.364	2.316	2.134	2.343	2.275	2.132
	Latency (nanosec)	6.764	6.907	7.496	6.827	7.030	7.502
50nm (0.5V)	P.Diss (P_D) W	0.113 x10^{-6}	0.9 x10^{-6}	0.54 x10^{-6}	1.195x x10^{-6}	1.32 x10^{-6}	1.32 x10^{-6}
	Delay (t_d) ns	0.17297	0.17985	0.18013	0.32152	0.1636	0.1740
	PDP(P_D x t_d)	1.9545x10^{-17}	1.6186x10^{-16}	9.727 x10^{-17}	3.8421x10^{-16}	2.159x10^{-16}	2.2968x10^{-16}
	Area μm^2	4238	2408	4356	3096	3440	3440
	No. of Transistor	1120	902	1120	902	984	970
	EPI (pico/Joules)	11.665	9.962	5.339	7.341	8.925	6.908
	Throughput (Gbits)	3.356	3.280	2.239	3.277	3.134	3.240
	Latency (nanosec)	4.767	4.877	7.144	4.882	5.104	4.937

According to Ed. Grochowski (2006), the designed adder cells have better performance in transferring data in the hardware system, of the order of picoseconds/joules. The throughput is an essential factor for transferring data to computational tasks. The CPL circuit gives higher EPI compared to other non-clocked logic families, which is shown in Table5.

2. CONCLUSION

The chapter presented the design and layout simulation results of non-clocked pass transistor CSA adders. The auxiliary components of OR, AND, XOR and Half-adder circuits were designed by using the Multiplexing Control Input Technique (MCIT). The 1-bit adder circuits were designed by using non-clocked pass transistor logic, Mixed-Shannon and Shannon theorem. The CPL adder circuit dominates other non-clocked pass transistors in terms of power dissipation, propagation delay and speed. In a similar way, the Shannon theorem based adder cells are designed using the Shannon theorem. The Mixed-Shannon circuit used 12 transistors, and the Shannon theorem based adder circuit used 16 transistors, for the design of a cell. The Mixed-Shannon 1-bit adder circuit gives more power dissipation such as 0.309μWatt for feature size 50nm and corresponding voltage 0.5V., higher propagation delay, and lower area than the full Shannon 1-bit adder circuit. Even though the number of transistors is increased in the full Shannon 1-bit adder, the power dissipation and propagation delay are lower than in the Mixed-Shannon 1-bit adder circuit due to a shorter critical path in the full Shannon adder circuit. The proposed Shannon based adder cell dominates over Mixed-Shannon adder and non-clocked pass transistor logic adders in terms of power dissipation 42nW, propagation delay2ps, number of transistors 16, area (55x7)μm^2, throughput (4.21Gbps), latency (4.767ns) and EPI (11.665pico/joules) of the circuits. So, this proposed adder can be used in the Multimedia devices, such as personal digital assistants (PDA), cellular phones, game consoles, and digital cameras require real time audio and video signals processor. The Signal processor circuit are using arithmetic as a core component of real time application circuit.

REFERENCES

Abid, Z., El-Razouk, H., & El-Dib, D. A. (2008). Low power multipliers based on new hybrid full adders. *Microelectronics Journal, 39*(12), 1509–1515. doi:10.1016/j.mejo.2008.04.002

Adiseno, H. M., & Olsson, H. (2003). *A 1.8-V Wide-Band CMOS LNA for Multiband Multistandard Front-End Receiver.* Radio Electronics-LECS/IMIT, Royal Institute of Technology (KTH). Retrieved from Sweden.www.imec.be/esscirc/essderc-esscirc-2003/papers/all/292.pdf

Alioto, M., & Palumbo, G. (2002). Analysis and Comparison on Full Adder Block in Submicron Technology. *IEEE Transaction on Very Large Scale Integration (VLSI) Systems, 10*(6), 806–823.

Anis, Mahmoud, & Elmasry. (2002). *Dynamic and Leakage Power Reduction in MTCMOS Circuits Using an Automated Efficient Gate Clustering Technique.* DAC 2002, New Orleans, LA.

Ashmila, E. M., Dlay, S. S., & Hinton, O. R. (2005). Adder methodology and design using probabilistic multiple carry estimates. *IEE Proc.-Comput. Digit. Tech., 152*(6), 697-703. doi:10.1049/ip-cdt:20045185

Bisdounis, L., Isdounis, L., Gouvetas, & Koufopavlou. (1998). A comparative study of CMOS circuit design styles for low-power high-speed VLSI circuits. *Int. J. Electronics, 84*(6), 599-613.

Chang, Gu, & Zhang. (2005). A Review of 0.18-μm Full Adder Performances for Tree Structured Arithmetic Circuits. *IEEE Transaction on Very Large Scale Integration (VLSI) Systems, 13*(6), 686–695.

Cheng, K.-H., Cheng, S.-W., & Lee, W.-S. (2006). 64-bit pipeline carry lookahead adder using all-n-transistor TSPC logics. *Journal of Circuits, Systems, and Computers, 15*(1), 13–27. doi:10.1142/S0218126606002915

Eriksson, H., Larsson-Edefors, P., & Eckerbert, D. (2006). Toward Architecture-Based Test-Vector Generation for Timing Verification of Fast Parallel Multipliers. *IEEE Transaction on Very Large Scale Integration (VLSI) Systems, 14*(4), 370–379.

Ghosh, S., & Roy, K. (2008). Exploring high-speed low-power hybrid arithmetic units at scaled supply and adaptive clock-stretching. *ASPDAC, 2008*, 635–640.

Grochowski & Annavaram. (Eds.). (2006). Energy per Instruction Trends in Intel Microprocessors. Technology Intel Magazine, 1-8.

Gustafsson, O., Dempster, A. G., & Wanhammar, L. (2004). Multiplier blocks using carry-save adders. *IEEE Proceeding, ISCAS*. doi:10.1109/ISCAS.2004.1329311

Guyot, A., Hochet, A. B., & Muller, M. J. (1987). A way to build efficient carry-skip adders. *IEEE Transactions on Computers, C-36*(4), pp1144–pp1151. doi:10.1109/TC.1987.1676855

Hsiao, S.-F., Yeh, J.-S., & Chen, D.-Y. (2002). High-performance Multiplexer-based Logic Synthesis Using Pass-transistor Logic. *VLSI Design, 15*(1), 417–426. doi:10.1080/1065514021000054736

Jung, G., Sundarajan, V., & Sobelman, G. E. (2005). A robust self- Resetting CMOS 32-bit parallel adder. Department of Electrical and Computer Engineering, University of Minnesota.

Kim, Y., Song, B.-S., Grosspietsch, J., & Gillig, S. F. (2001). A carry-free 54b x 54b Multiplier Using Equivalent Bit Conversion Algorithm. *IEEE Journal of Solid-State Circuits, 36*(10), 1538–1545. doi:10.1109/4.953482

Kuo, Hwang, & Wu. (2006). A Power-Driven Multiplication Instruction-Set Design Method for ASIPs. *IEEE Transaction on Very Large Scale Integration (VLSI) Systems, 14*(1), 81–85.

Kuroda & Sakurai. (1995). Overview of low-Power ULSI techniques. *ICICE Transaction of Electronics, E78-C*(4), 334-344.

Mano, M. (1994). *Computer system architecture* (3rd ed.). Prentice Hall.

Mirhassani, M., Ahmadi, M., & Jullien, G. A. (2008). Low-Power Mixed-Signal CVNS-Based 64-Bit Adder for Media Signal Processing. *IEEE Transactions on Very Large Scale Integration (VLSI) Systems, 16*(9), 1141–1150.

Mudassir, R., & Abid, Z. (2005). New parallel multipliers based on low power adders. 2005 IEEE CCECE/CCGEI, Saskatoon.

Navarro-Botello, V., Montiel-Nelson, J. A., & Nooshabadi, S. (2007). High performance low power CMOS dynamic logic for arithmetic circuits. *Microelectronics Journal*, *38*(4-5), 482–488. doi:10.1016/j.mejo.2007.03.018

Psilogeorgopoulos, Munteanu, Chuang, Ivey, & Seed. (1999). *Contemporary Techniques for Lower Power Circuit Design PREST Deliverable D2.1*. The Department of Electronic and Electrical Engineering The University of Sheffield Mappin Street Sheffield S1 3JD.

Ruiz, G. A. (1998). Evaluation of three 32-bit CMOS adders in DCVS logic for self-timed circuits. *IEEE Journal of Solid-State Circuits*, *33*(4), 604–613. doi:10.1109/4.663566

Ruiz, G. A., & Granda, M. (2004). An area-efficient static CMOS carry-select adder based on a compact carry look-ahead unit. *Microelectronics Journal*, *35*(12), 939–944. doi:10.1016/j.mejo.2004.09.002

Senthilpari, C., Mohamad, Z. I., & Kavitha, S. (2011). Proposed low power, high speed adder-based 65-nm Square root circuit, Vol41(2) pp 445-451. *Microelectronics Journal*.

Senthilpari, C., Singh, A. K., & Diwakar, K. (2008). Design of a low-power, high performance, 8× 8bit multiplier using a Shannon-based adder cell. *Microelectronics Journal*, *39*(5), 812–821. doi:10.1016/j.mejo.2007.12.016

Shannon, C. E. (1950c). A Symmetrical Notation for Numbers. *The American Mathematical Monthly*, *57*(2), 90–93. doi:10.2307/2304993

Uyemura. (1999). *CMOS logic circuit design*. Kluwer Academic Publishers.

Vasefi & Abid. (2005). *Low Power N-bit Adders and Multiplier using Lowest-Number of transistor 1 bit adders*. Saskatoon: Department of Electrical and Computer Engineering IEEE CCECE/CCGEI.

Vesterbacka, M. (1999). A 14-transistor CMOS full adder with full voltage swing nodes. In *Proc. IEEE Workshop Signal Processing Systems*, (pp. 713–722). doi:10.1109/SIPS.1999.822379

Weste, N. H. E., & Eshraghian, K. (1993). *Principles of CMOS VLSI Design: A systems perspective* (2nd ed.). Addison-Wesley Publishing Company.

Chapter 11
Case Study:
System on a Chip for Electric Stimulation

Martha Salome Lopez
Universidad de Monterrey, Mexico

ABSTRACT

Integrated circuits have been predominantly designed and developed by large firms and manufacturers; nowadays, any electronic engineer should be able to develop specific and innovative low-power designs using available open cores. This chapter presents the design process for a specific chip, beginning with a definition of its function, design considerations, power analysis, performance optimization, and chip optimization. The hardware and software for this circuit were developed for low-power implementation: it includes a processor, memory blocks, ports, buses, and a proposed application program, so it can be used as a starting point for other low-power very-large-scale integration (VLSI) circuits. The chip uses frequency synthesis and configuration parameters to deliver electric signals on a variety of waveforms and patterns. This design can be used in many research fields and application areas, where experiments or portable devices need low-power, programmable, and configurable electric signal generators.

INTRODUCTION

This system on a chip (SoC), developed by the author, is based on open cores and uses DDFS (direct digital frequency synthesis) algorithms. It is intended to operate as a standalone module or to interconnect with other devices or processors. It has been presented in medicine and biology forums (Lopez et al., 2011), and it has been used in scientific instruments (Lopez et al., 2013).

This SoC uses an OR1200 processor (Open RISC 1200), program memory, data memory, external address bus, data bus, communication port, interrupt controller, clock, reset, and general purpose input/output (GPIO) ports. All of its hardware cores are open source. The application program is proposed and optimized for minimum power and memory access. At the end of this chapter, an analysis of possible applications is presented, where this SoC can be used with little or no modifications.

A processor based SoC was selected over a gate-level design for several reasons: a different application program can be developed and uploaded with no modifications on the hardware design, the modular

DOI: 10.4018/978-1-5225-0190-9.ch011

design allow upgrading the processors or extending the memory easily, additional cores and functionality can be added thanks to the modular design, and it still is a specific application design while keeping the openness of a configurable and programmable system.

Throughout the entire design process, a low-power goal was pursued, both from a hardware and software perspective, as this integrated circuit is intended for embedded and portable systems.

The developed SoC has an area of 4.03 mm^2, 189 pins, an operating frequency of 180 MHz, and a power consumption of 0.52 watts when fabricated on 180 nm technologies. The design flow is explained, showing calculations for every design stage. The flow covers the synthesis process, area optimization, power calculations, performance optimization, IO ring definition, and place and route procedure.

As for the application program, the implemented methodology that is used to generate and deliver electric signals is presented, comparing a memory-intensive methodology versus a computation-intensive methodology. A selection of signal parameters and patterns is defined in order to ensure that a wide variety of applications is covered by this design. Waveforms (square, triangle, sine, and saw tooth) can be simultaneously delivered in separate output channels, or they can also be superimposed in one output channel.

A simple user interface is developed, so anyone can use this system without needing to understand the integrated circuit theory or the frequency synthesis methodology. The chip is developed featuring a modular design, so cores can be updated or upgraded using new releases. Similarly, application areas not currently covered can be achieved with small modifications to this design.

This system can be used as an example to develop other systems, and also can be used in other applications such as particle manipulation, pollution detection, water and food quality, cell detection, nano-manipulation, and biomedical analysis. Generally speaking, any application that needs electric signals on different waveforms, frequencies and patterns, can use this system.

Background

This chapter reviews the design flow for an application-specific integrated circuit that can be easily customized for other applications. Design parameters such as power, area, and performance are presented through several optimization rounds (Lopez et al., 2011). Open cores are used throughout the entire design, so no licensed circuits are needed. The design tradeoffs are presented, and a practical case with its actual numbers and results are exposed. This integrated circuit is used as an example on how to design your own SoC or a scientific instrument (Lopez et al., 2013) using open cores (OpenCores, 2014) and CAD tools (Jupiter, 2006) to achieve a low-power system. This chapter presents the definition of this system, as well as its design flow and implementation. Previous related works have been published, but they lack one or more of the following characteristics: SoC and labs on chip that are not programmable (Gascoyne & Vykonkal, 2004); circuits that are oriented to deliver a limited range of frequencies and only sinusoidal waveforms (Manaresi et al., 2003); so-called general purpose systems that are limited in purpose and configurability, varying only in terms of some signal parameters and serving a specific purpose (Tierney, Rader, & Gold., 1971); other systems that do not provide different waveforms nor signal superposition capabilities (Choi et al., 2006); and, finally, systems oriented to one specific application, with no possibilities of being modified (Chang, Lee & Chiu,. 2008) (Ibrahim et al., 2009). Applications that use a similar system can be found in many areas, such as nano-manipulation (Villamejane et al, 2010), isolation of malaria (Gascoyne et al, 2002), separation of malignant cells (Alazaam et al, 2008), high voltage stimulation for manipulation (Yuk K. et al, 2007), and micro-fluidics dielectrophoresis (Li, H.

et al, 2005), among others. The system presented here has the requisite characteristics and functionality to provide electric signals to stimulate a variety of experiments by selecting a combination of patterns and mixing waveforms of different frequencies, within a wide frequency range. The objective of this chapter is to illustrate the process of designing an integrated circuit, while understanding and applying the tradeoffs between power consumption, circuit area, and performance.

SYSTEM ON CHIP DESCRIPTION AND COMPONENTS

The implemented system can be briefly described by the functions it performs. It is intended to take configuration parameters as input, so as to deliver different waveforms in several patterns: a single signal in one output channel, different signals in separate output channels, or a single signal in one output channel containing superimposed frequencies. Once the methodology is understood, it is easily modified to generate other signal patterns and waveforms. The SoC is able to:

- Synthesize and deliver four different waveforms.
- Provide different operation modes: single output signal, dual separate signals, and dual superimposed signals.
- Allow user-defined operation parameters: waveform type, frequency, and phase.
- Allow user-defined numbers of data samples for DDFS on every run.
- Deliver single or superimposed signals, either of the same or different waveforms.
- Perform recurrent tests for the continuous delivery of signals.
- Use preloaded data tables for four waveforms: square, triangle, sine, and saw tooth.

The system is designed to ease integration into existing devices, so it provides input/output Wishbone buses for data and instructions. Also, the system can be application independent by using ROM-based Bios that loads different application software at each reboot. Existing devices with a different communication bus, like AMBA, can be connected to this system using a converting bridge without changing the current design.

The memory system consists of a ROM containing the boot-loader, which uploads the application software at the beginning of the operation, and Harvard architecture of RAM to contain the program and data tables. The OR1200 processor is a good option from the available open source cores, and its corresponding instruction set is appropriate for this design, in terms of performance, circuit area and power consumption. There are other processors available as open source, such as the OR1000 and the OR2000. The OR1000 is a previous and smaller version of open risc architectures, and the OR2000 has an extended instruction set which does not add significant improvement for this application and needs more circuit area. On the final stages of this work a new processor was released, the OR1200HP, which is hyper pipelined, and can lead to higher operating frequencies but does not have a significant reduction on power consumption.

On-chip communication is achieved using the Wishbone bus, which allows for smooth integration of the components inside the system, as shown in Figure 1.

The hardware and software architectures are low-power oriented; they use minimum-sized cores and memories, and implement a frequency synthesis methodology that minimizes execution times and memory access.

Figure 1. Block diagram of the system

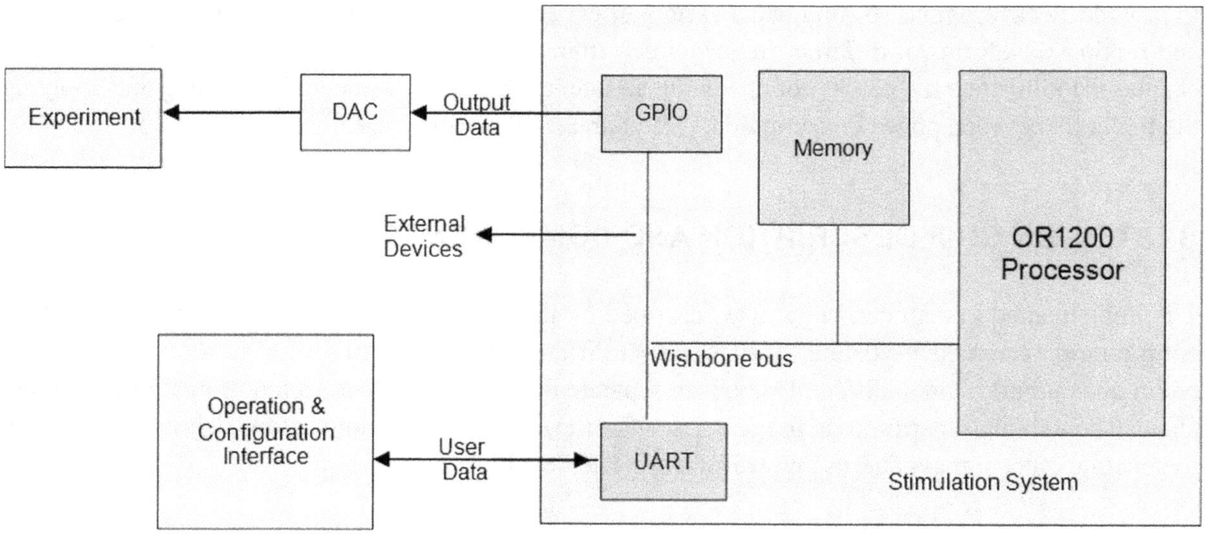

The user interface, connected via a standard serial port, allows for system configuration. The user can also select the mode of operation (waveform type, frequency, superposition), and visualize the delivered data.

DESIGN CONSIDERATIONS

The mix of single-frequency signal generation with signal superposition and system configuration capabilities enables control over a wide range of signal parameters and patterns for electric stimulation tests. With a few modifications, other waveforms and patterns can be delivered.

Execution times are minimized for maximum output frequency; this depends solely on processor performance. Moreover, the proposed application algorithm and program memory space is minimized and does not increase, regardless of operation mode.

Data memory is defined for minimum area and power: data samples for sine, triangle, and saw tooth waveforms are stored in tables and used to process temporary tables for selected output frequency. Output and buffer tables are calculated after a time (period) match operation. Any periodic waveform can be delivered if it is stored in memory.

Data pre-processing and table preparation reduces power consumption and computation instructions during signal generation. For signal updating, multiple simultaneous writes to the output port are made, so no loss in output frequency occurs when two or more signals are being delivered simultaneously. This port partitioning scheme is particularly convenient for this application.

In summary, the system combines efficient, low-power oriented hardware and software resources: minimum signal generation source code, no computation during frequency synthesis, and minimum memory access times. Power aware decisions are made along the whole design process; either they are hardware related or software related. The more significant decisions have to do with memory size and execution times for the application program. Decisions about core optimization had no significant impact on power consumption.

This SoC features bus-based hardware architecture; it uses a selected set of open source cores and the Wishbone on-chip communication bus. The OR1200 processor is selected due to its previously demonstrated implementations in field-programmable gate arrays (FPGAs) and its open instruction set. The on-chip memory array may contain the application software developed specifically for stimulating fluidic devices or a boot-loader to upload different applications. Chip data memory contains a complete cycle of the three base waveforms in 256 samples of 8-bit data each. A universal asynchronous receiver/transmitter (UART) port is used to configure operation, to program sequence tests, to control operation, and to visualize data during execution. The four 8-bit GPIO ports deliver processed data points for output signals according to the selected configuration (sine, triangle, or saw tooth) and to the presenting signal (dual or superimposed frequencies).

The main design challenges where to define a memory size suitable for this application but big enough for possible additions that can fit other applications; to decide between memory intensive and computation intensive DDFS methodologies, since their resource needs are completely opposite; and to solve the pin dominance of the resulting chip by eliminating the less necessary functions.

Figure 2 shows the interconnection between in-chip blocks. It is demonstrated that processor and Bus Interface Units connect directly to Wishbone instruction and data buses, while memories and peripherals connect via wrappers. Table 1 details the function of each primary block in the SoC.

Figure 2. System on chip block diagram

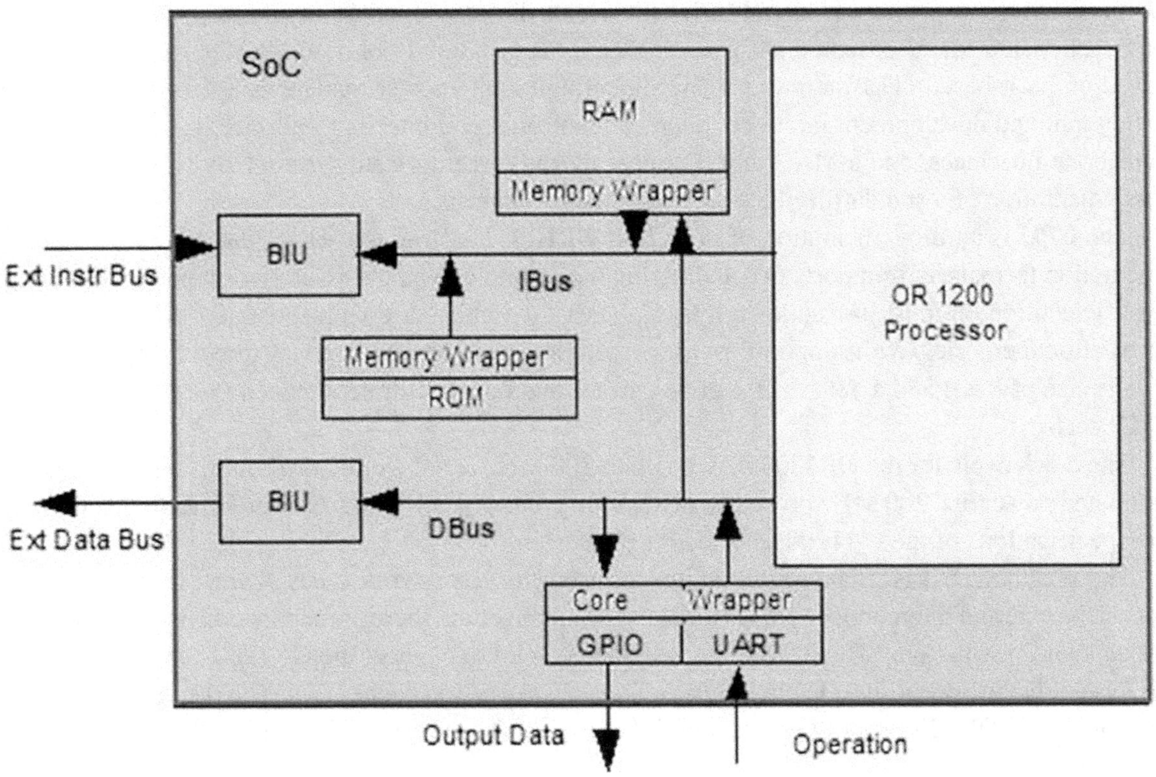

Table 1. Functional description for primary system blocks

Block	Description
CPU (OR1200)	RISC CPU, Harvard architecture, cache memory for data and instructions, operates at 250 MHz max using 180 nm standard cell TSMC technology.
Wishbone Bus	On-chip bus for cache, main memories, and interface peripherals.
GPIO	Grouped in four I/O 8-bit ports, from open source cores; used as inputs for configuration and operation, as outputs for data.
Clock and Reset	Receives clock from crystal oscillator, generates clock and reset signals for system operation, base clock for processor blocks, and ¼ base clock for Wishbone bus.
UART	Serial port controller provides connection for external configuration and operation device. Open core source is used.
RAM and ROM	Memory blocks built with Artisan memory generators (Standard library SRAM Generator from Artisan, 2004) (Standard library 0.13um - 0.25 um ROM Generator from Artisan, 2004) for Verilog, vclef, and gdsII views.
Interrupt controller	Exceptions handler from open cores included in OR1200 architecture.

The OR1200 is a RISC, Harvard architecture processor with basic digital signal processor (DSP) capabilities. As an open source, customizable core, it is not optimized for power or size. This particular implementation for the processor is based on the open source files of the OR1200.

The Open RISC 1200 is a synthesizable CPU core; it is a configurable open source Verilog implementation of the Open RISC 1000 architecture. The OR1200 is intended for use in a variety of embedded applications. Some open source software, such as Linux, has been ported over to the OR1200 platform. The GNU tool chain, including GCC, has also been ported to the architecture to aid in software development.

This processor specifies a Central CPU/DSP block, direct mapped data cache, direct mapped instruction cache, data MMU based on hash-based DTLB (Translation Look-aside Buffer), instruction MMU based on hash-based ITLB, a power management unit and a power management interface, tick timer, debug unit and development interface, interrupt controller and interrupt interface, instruction and data Wishbone interfaces, and a MAC unit. Peripherals and a memory subsystem may be added using the implementation of a standardized 32-bit Wishbone bus interface.

The CPU is an implementation of the 32-bit ORBIS32 instruction set architecture. It features five instruction formats and supports two addressing modes. It has a single-issue five-stage pipeline, single-cycle execution on most instructions. It uses a Harvard architecture with separate MMUs for data and instruction memories, with support for virtual memory, a hash-based one-way direct mapped TLB with a page size of 8 KB and a default size of 64 entries, and one-way direct mapped D-Cache and I-Cache, 8 KB each.

The clock cycle for the OR1200 may go up to 200 MHz at a 0.18 μm, 6ML fabrication process. The estimated power of a 200 MHz processor in a .18 μm process is <1 W at full throttle. Table 2 summarizes the specifications of the OR1200 selected for the design.

The Wishbone bus is a standard communication bus for open source cores. It serves as the in-chip bus and as the external interconnection. As for Harvard architecture, there are separated buses and memories for data and instructions. The instruction interface is used to connect the OR1200 core to the memory subsystem for the purpose of fetching instructions or instruction cache lines. The data interface is used to connect the OR1200 core to external peripherals and the memory subsystem to read and write data or data cache lines.

Table 2. Specifications of the OR1200

Concept	Specification
Architecture	32-bit scalar RISC, Harvard
Pipeline	5-stage, 32-bit integer instructions
DSP	Basic capabilities
Caches	Separate instruction and data, one-way direct mapped. Configurable to 1 KB, 2 KB, 4 KB, 8 KB
Virtual memory	64-entry hash-based one-way direct mapped TLB for data and instruction MMU
Speed	Worst case: 150 dhrystone 2.1 MIPS at 150 MHz (typical corner 250 MHz) Best case for 180 nm implementation: 300 dhrystone, 2.1 MIPS at 300 MHz
Size	Default configuration about 40 K ASIC gates, 1 M transistors
RTL status	Not optimized for speed or area
Instruction set	Instruction unit handles only the ORBIS32 instruction class. ORFPX32/64 and ORVDX 64 instruction classes are not supported
Configurable	Major characteristics can be set by the user (see Core HW configuration table)
Communication bus	Wishbone, internal and for SoC interconnection
GPRs	General purpose register file is implemented as two synchronous dual-port memories, 32 words, 32 bits per word
Exceptions	Transparent to user software, the same mechanism to handle all types of exceptions; control is transferred to an exception handler. See Exceptions table
Interrupt controller	Direct enabled Int0 and Int1, Masked Int [31:2]
Tick timer	Clocked by RISC clock, restart able, mask interrupt, max count 2^32
Power management	Dynamically activated modes: slow and idle, doze, sleep, and clock gating
Debug unit	Basic debugging; no watch points, break points, or program flow control registers
Clock and reset	Core has several clock inputs: clk_cpu, clk_dc, clk_ic, clk_dmmu, clk_immu, clk_tt; all clocks must be in phase and as low skew as possible. Reset signal *rst* resets all flip flops when asserted high; when not asserted, reset exception start
Wishbone interface	Rev. B compliant, 32-bit bus width may connect to external peripherals and external memory subsystem

Instruction memory stores the application program for system configuration, system operation, frequency synthesis, and output data delivery. The data memory stores data samples for waveforms (look-up tables containing sine, triangle, and saw tooth data), temporary tables for processed data, and buffer output tables for the final processed data. Possible implementations are explored: storing application software in on-chip ROM increases circuit area and restrains functionality; and loading application software to RAM using a boot-loader reduces memory area and enables the modification and debugging of application software in development boards before being loaded into the chip.

A similar concept is applied for waveform data tables: data can be stored in on-chip ROM with a fixed data width and fixed samples per waveform cycle; this increases the in-chip data memory use and reduces the flexibility to change the data size and samples per cycle. Otherwise, the sample data can be modified outside the chip, as well as the number of data samples per cycle, and these are then uploaded to RAM using the boot-loader. This shows the trade-off between a fully customized design versus a configurable, programmable circuit design. In both cases, more memory implies more power consumption.

A UART port has been selected over a USB port for system configuration and operation, since it uses less power and circuit area; also, speed is not important during configuration and operation. For extended interfacing capabilities, Wishbone data and instruction buses have been added; they can be used for external memory access and to connect this system to external peripheral systems, drivers, or bridges.

Parallel ports have been used as output channels: a 32 bit (4×8 bit) GPIO port is used to deliver output data. Standard data is 8-bit wide, so up to four channels are available. Output data are delivered in one of these forms, depending on the operation mode: in mode 1, 8-bit data samples containing one single frequency are delivered; in mode 2, 16-bit data samples containing two separate frequencies are delivered in two 8-bit channels, 8-bit each; and in mode 3, 8-bit data samples containing two superimposed frequencies are delivered in one 8-bit channel. Table 3 lists the processor cores and the peripherals for the SoC.

SYSTEM ON CHIP DESIGN FLOW

This section presents a description of the design methodology, the computer-aided design (CAD) tools used in the chip design flow, and the procedures and results on the clusters involved in the design: synthesis, timing, memory block generation, place and route of the system cores, power consumption analysis, IO ring design, and integration at the chip level.

Open source cores have been used to integrate a modular architecture over an open source bus. The design flow has been developed in the following priority order when a compromise is needed: minimum power consumption, maximum performance, and minimum chip area. Other design parameters, relevant to its physical design, are also taken into account: power grid design, IO pin selection, IO ring design, block placement, pin routing, and layout. As a reference, a tutorial on design flow is available (ASIC Design Flow Tutorial, 2010). Table 4 shows the tools used for each of the main design tasks. Tasks have been performed sequentially in the first design steps; in the later design steps, they were performed in

Table 3. Processor and peripheral cores

Core	Block Description	Core	Block Description
or1200_alu	Arithmetic and Logic Unit	or1200_lsu	Load and Storage Unit
or1200_cfgr	Configuration Registers	or1200_mult_mac	Multiply and MAC Unit
or1200_ctrl	Control Unit	or1200_operandmuxes	Operand Muxes
or1200_dc_top	Data Cache	or1200_pic	Programmable Interrupt Controller
or1200_dmmu_top	Data Memory Management Unit	or1200_pm	Power Management Unit
or1200_du	Debug Unit	or1200_rf	Register File
or1200_except	Exceptions Unit	or1200_sb	Store Buffer
or1200_freeze	Freeze Unit	or1200_sprs	Special Purpose Registers
or1200_genpc	General Program Counter	or1200_tt	Tick Timer
or1200_gpio	General Purpose Input Output	or1200_uart	Universal Asynchronous Rec/Trans
or1200_ic_top	Instruction Cache	or1200_wb_biu	Wishbone Bus Interface Unit
or1200_if	Instruction Fetch	or1200_wbmux	Wishbone Mux
or1200_immu_top	Instruction Memory Management Unit		

parallel and iteratively to work on the tradeoffs of the design: power consumption, chip area, and system performance. Note that for each licensed design tool, there is a license-free equivalent, generally available from specialized Websites (OpenCores, 2014)

Figure 3 illustrates the design flow. All four design parameters have to be measured and calculated each round. Each new round involves slight design changes that may affect the result in one or more variables. Generally speaking, if one of the parameters improves on a round, some of the others may get worse. It is suggested that different persons or teams work separately on each parameter, so a tight struggle between improvements is generated after each round, and each team has to convince the others about the improving strategy to be implemented in the next round.

Circuit Synthesis and Timing

The circuit synthesis process consists of converting the hardware description language (HDL) of each core into circuits, and to interconnect the resulting circuits. Timing is the analysis performed on those circuits to determine the system's response time (i.e., the time it takes to generate an output or a resulting data). As an analogy, when you build a circuit in a prototyping board, you know there are different implementations, and each one will take different times to produce the outputs, depending on the circuits

Table 4. CAD tools used in design flow

Design Aspect	Tasks	Tools
Timing	Synthesis process at the block level and processor level. Set initial time constraints at the block level. Iteratively push timing constraints at the block and chip level. Deliver block and global timing analysis.	Synopsys, Design Compiler (DC) (Design Compiler User Guide, 2009)
Place and Route	Initial area estimations at the block level. Deliver an initial floor plan for routing. Placement and routing at the block level. Global routing. Placement optimizations.	Synopsys, Design Compiler (DC) Synopsys, Integrated Circuit Compiler (ICC) (Library Data Preparation for IC Compiler, 2010)
Memories	Generate logical and physical views for RAM and ROM blocks. Generate different aspect ratio implementations for placement optimization. Create memory wrappers for integration.	Artisan, Memory Generators (Ibrahim et al., 2009)(Standard library SRAM Generator from Artisan, 2004) Synopsys, Design Compiler (DC)
Power	Initial power estimations at the block level. Iterative Power Grid Design.	Synopsys, Design Compiler Synopsys, PrimeTime PX
IO Ring	Iterative IO ring design.	Synopsys, Integrated Circuit Compiler (ICC) (Library Data Preparation for IC Compiler, 2010)
Clock	Clock tree synthesis at the chip level.	Synopsys, Integrated Circuit Compiler (ICC) (Library Data Preparation for IC Compiler, 2010)
Integration	Integrate Power grid to floor plan. Integrate IO ring to floor plan. Integrate memory blocks at the chip level. Integrate clock tree at the chip level.	Synopsys, Integrated Circuit Compiler (ICC) (Library Data Preparation for IC Compiler, 2010) JupiterXT, Synopsys (Jupiter, 2006)
Verification	Functional verification at the block and processor level.	ModelSim (Design Compiler User Guide, 2009)

Figure 3. Chip design flow for four areas: timing, power, IO pins, area

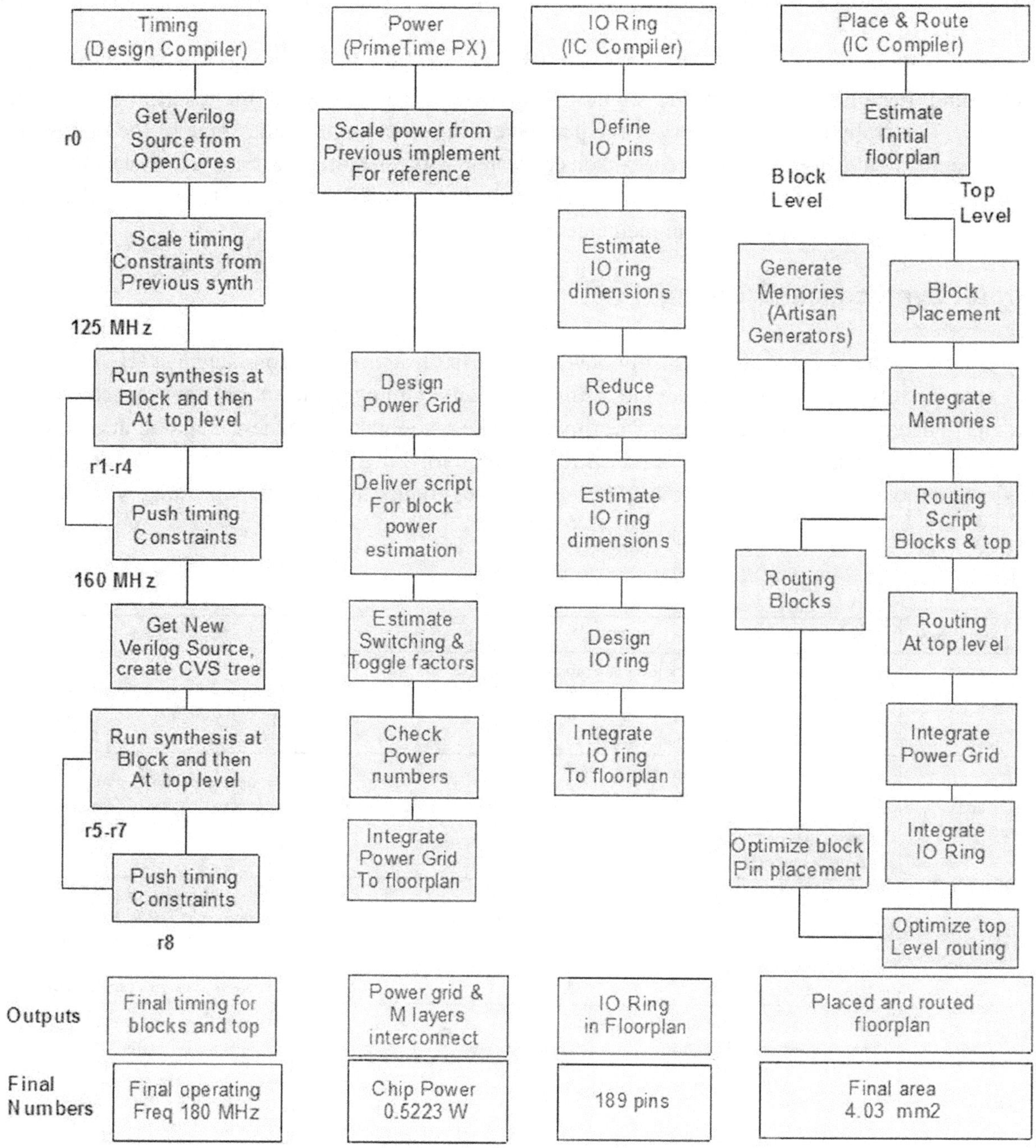

you use and the longest data paths from inputs to outputs. During the synthesis process of this design, the local and global timing has been optimized and the initial area and power estimations have been completed. Eight rounds of synthesis have been performed until timing convergence and closure have been achieved. Synthesis rounds to obtain the minimum slack time, area estimations, and power estimations have been performed with the Design Compiler (Design Compiler User Guide, 2009) from Synopsys.

Selected processor cores from Open Cores have been used for bus compatibility and minimum-edition source codes. The available versions of open source cores are not optimized for power, performance, or area, so optimization was done. Furthermore, a simulation for functionality, at the processor level, has been completed before starting the synthesis process.

The tasks performed and the obtained slack times from the synthesis rounds, as summarized in Table 5, were:

Round 0: Estimate the initial timing constraints for each core based on gate count, estimated size, and worst case data path.
Rounds 1 to 4: Run the synthesis at the block level, pushing timing constraints; consider adjacent blocks in the data path to push constraints for the required arrival times and arrival times.
Rounds 5 and 6: Run the synthesis at the top level. Identify critical paths. Push timing constraints on critical paths. Use a new version of the open core source files; go back to round 5.
Rounds 7 and 8: Update the CVS tree (source version control) to match the last rounds. Round 8 is set to be the final round for pushing the block and global timing constraints.

The global path delay at the top-level synthesis has been calculated by:

Global path delay= Arrival time + pass-through + (cycle time – required arrival time) + interconnect delay,

where the first three terms depend on both block level synthesis and top-level synthesis, and the last term depends on floor plan and time of flight for metal 3 and metal 4.

The final results show that the minimum clock cycle is the requested 4 ns plus the worst slack time achieved of –0.978 ns, which leads to a minimum required clock cycle of 4.978 ns, equivalent to a processor frequency of 200.8 MHz. Figure 4 shows the occurrence distribution of the single- and double-cycle paths in the architecture, as these are the ones in which the worst slack times were observed. Both power and timing calculations are made by executing benchmark programs.

Table 5. Slack time results from the synthesis rounds

Round	Slack	Action taken	Improvement
1	–7.9 ns	Push timing constraints at the block level	
2	–6.4 ns	Push timing constraints at the block level	1.5 ns
3	–4.5 ns	Consider pin info in constraints	1.9 ns
4	–4.3 ns	Push synthesis effort	0.2 ns
5v1	–2.3 ns	Push synthesis effort	2 ns
5v2	–1.2 ns	Identify multi-cycle paths	1.1 ns
5v3	–0.75 ns	Reduce multi-cycle paths	0.45 ns
6	–1.0 ns	Change several source files	–0.25 ns
7	–0.82 ns	Change several source files	0.18 ns
8	–1.1 ns	Iterate place and route	–0.28 ns
Final	–0.978 ns	Consider interconnect delay	0.12 ns

Figure 4. Occurrence distribution of slack times

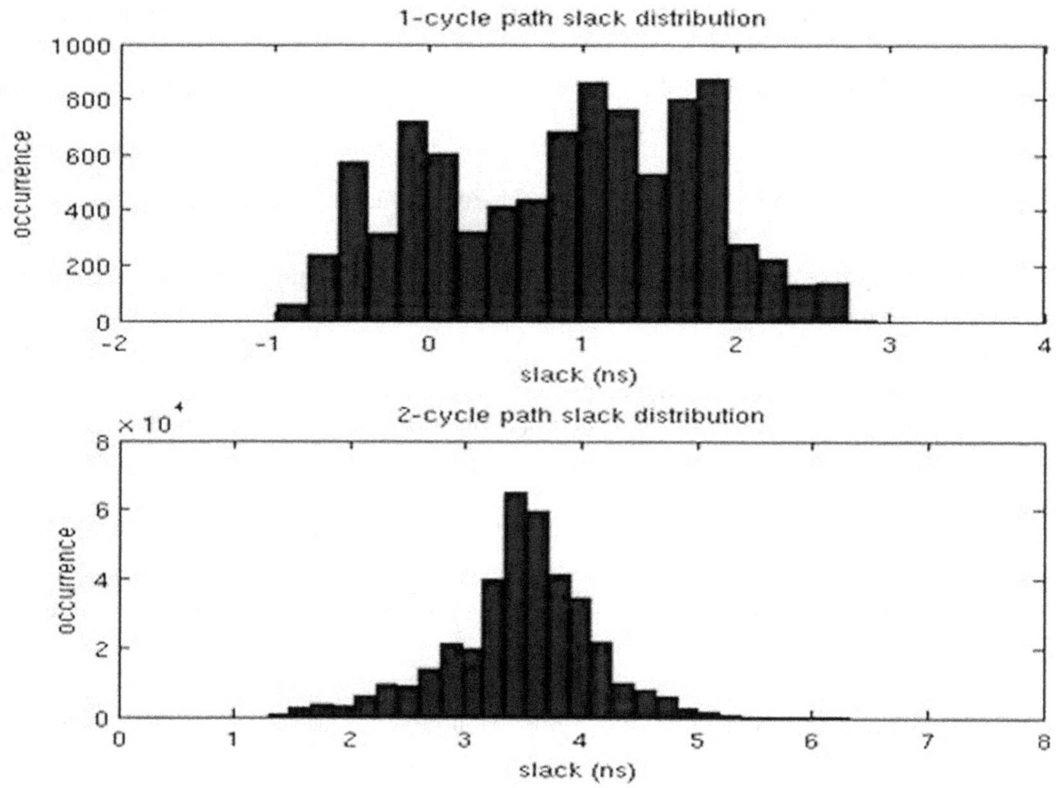

Memory Blocks

Memories for this system (instruction cache, data cache, in-chip RAM, and in-chip ROM) have been generated using a memory generator tool from Artisan (Ibrahim et al., 2009), which takes as input an abstract description of the memory blocks and produces several memory formats suitable for various tools and purposes. Using a memory generator instead of synthesizing a memory can optimize speed, density, and power. It can also control the memory block aspect ratio for efficient floor planning, deliver timing and power models to integrate into other design tools, and enable configurable word-write mask and redundancy options. Different sizes of memory blocks (2 KB, 4 KB, and 8 KB) were generated to facilitate block placement and routing.

A set of views can be generated: PostScript data sheet, ASCII data table, GDSII layout file, LVS netlist, Synopsys model, Primetime models, TLF models, VCLEF footprint, Verilog model, and VHDL model. The relative footprint shows how the aspect ratio of memory changes as the words, bits, and Mux parameters are varied. The instance and power ring are included in the footprint.

RAM architecture, timing specifications, and physical characteristics:

- Synchronous random access memory is triggered by the CLK rising edge.
- Pins: CLK, CEN, WEN, OEN, A [m-1:0], D [n-1:0], output Q [n-1:0].

- Memory blocks are cut in symmetrical sides to easily clock distribution and layout.
- Dual-port memories provide dual ports for all input and output signals.
- Power rings can be generated around the SRAM; size them properly. Size depends on the chip-level power distribution; the number, width, and placement of supply wire connections to the power rings; and the current consumption. Recommendation: supply the current evenly from the edge of the instance where the pins are located.
- Top metal layer: metal1 to metal4 are used in the design and are blocked for routing. Layers above m4 can be routed over the memory.
- I/O pins are located along the bottom edge of the memory block on any of the metal layers, and they are large enough to accommodate a predetermined on-grid width wire connection.
- Verification: the views produced by the generator can be verified with standard tools.

ROM architecture, timing specifications, and physical characteristics:

- Synchronous read-only memory is triggered by the CLK rising edge.
- Pins: CLK, CEN, A [m-1:0], Q [n-1:0]. If CEN is high, then memory is on standby mode and Q has the last data; if CEN is low, then memory is in read mode and Q has data from address A.
- Memory blocks include row and column decoders, clock generator, memory array, and amplifiers/IO buffers for the outputs.
- Power rings: multiple, evenly-spaced connections have been used from core V_{dd} and V_{ss} to the rings around the instance on the side where the I/O pins are located.
- I/O pins are located along the bottom edge of the memory block on any of the metal layers.
- ROM code file: an Artisan format ROM code file must be provided for each generated instance.
- Format: code file contains only 0s and 1s.
- The line number in the file is equivalent to (address-1).
- Each character of a line corresponds to the bits of a word. The character in column 1 of a line is the most significant bit.
- The address goes from m to 0; bits and columns go from n to 0.
- This file is needed for behavioral or physical views such as Verilog, VHDL, Testscan, Sunrise, GDSII, and LVS Netlist.

Place and Route (P&R)

P&R, the process of placing each individual block within the top-level design, has a major impact on chip area. It must use the area optimized netlist for each block, while efficiently using the interconnected area for routing. Several iterations for P&R have been made, especially since changes in synthesis, pin placement, and the block aspect ratio impact placement and routing results. The Integrated Circuit Compiler (ICC) from Synopsys has been used to route signals within blocks to place blocks within layout, to route signals between blocks, and to optimize block and pin placement for optimal routing.

A preliminary floor plan has been delivered for power grid design. Block placement reruns have been performed with different aspect ratios for each block, until better area utilization has been achieved. About 30% of space is left between blocks for interconnected routing, clock tree, and power grid.

Power Analysis

To estimate and analyze the power consumption for this system, a sequence of iterative tasks has been performed: power for each block has been estimated using CAD tools (Design Compiler User Guide, 2009); activity factors in power calculations have been included; Switching Power, Cell Internal Power, and Cell Leakage power have been calculated for each block; a power grid has been designed; and a power grid to the routed layout has been incorporated.

Power estimations for individual blocks are initially made from block size and gate count. Activity factors have been integrated using a code benchmark. Later, a preliminary floor plan from ICC was used as the base for the power grid.

Vertical power lines go on M5, horizontal on M6. No power ring was added to ease integration with power in M3 and M4, and to power pins in the IO ring. The power grid includes V_{dd}, clk, rst, and V_{ss} lines, with spacing and widths as follows: 5λ, 2λ, 2λ, 2λ, 2λ, 2λ, 5λ, and 2λ, respectively. The total spacing between two V_{dd} will be 22λ, which should be a multiple of power grid spacing in M3 and M4 for interconnection.

Power estimation for each block has remained similar, regardless the changes made in the source code that occurs during synthesis and integration. The power grid at M5/M6 considers block placement and individual block consumption, while the power grid inside blocks on M3/M4 is placed automatically. A verification procedure, which is run to obtain activity factors by block from simulation, instead of from estimations, has also been completed.

Dependencies for Power IR drop analysis, which is the voltage drop due to the resistance of interconnects in the power network are illustrated as follows:

- Additional code lines at the top level are needed for the power grid.
- The IO ring needs to be hooked up to the top-level design.
- All blocks should place their pins only in M3 and M4.
- The IO pads have to be hooked up to the power rail to acquire information about the external source of the power rails.
- Modules should be power routed in M4 and hooked up to the power grid via M4 and M5.
- Decoupling capacitor filler cells must be inserted in empty spaces – within individual modules and at the full-chip level between modules.
- Design rule check should be executed after hooking up power routes.
- Design decisions should be made for each individual block; these include determining where to put the power pins to minimize the route to the power grid.

Results from the final run of power estimation are shown in Table 6, where the weighed value is calculated as follows:

Total power = Net switching power + Cell internal power

Weighted value = (Block activity × Total power) + Cell leakage power.

Table 6. Power consumption values (Watts)

Block	Block Activity	Net Switching Power	Cell Internal Power	Cell Leakage Power	Total Power	Weighted Value
alu	0.4	4.8500E−03	6.1020E−03	1.0290E−07	1.1000E−02	4.380903E−03
cfgr	0.02	2.4420E−04	3.0160E−04	1.0400E−08	5.4580E−04	1.092640E−05
ctrl	0.4	1.4490E−03	3.0220E−03	7.0550E−08	4.4710E−03	1.788471E−03
dc_top	0.5	3.0410E−03	5.9600E−02	1.0080E−05	6.2600E−02	3.133058E−02
dmmu_top	0.5	1.2410E−03	9.6630E−03	6.0270E−06	1.0900E−02	5.458027E−03
du	0.01	1.7200E−02	1.9800E−02	6.5830E−07	3.7100E−02	3.706583E−04
except	0.4	3.1590E−03	1.1200E−02	1.9230E−07	1.4300E−02	5.743792E−03
freeze	0.4	2.5380E−05	6.8790E−05	1.9090E−09	9.4170E−05	3.766991E−05
genpc	0.5	3.0450E−03	5.7010E−03	1.2740E−07	8.7470E−03	4.373127E−03
gpio	0.5	7.4720E−04	2.3390E−03	1.9830E−07	3.0860E−03	1.543298E−03
ic_top	0.5	1.3880E−03	5.7500E−02	1.0040E−05	5.8900E−02	2.945404E−02
if	0.5	7.3330E−04	2.2270E−03	3.9420E−08	2.9600E−03	1.480189E−03
immu_top	0.5	2.3510E−03	1.0600E−02	6.0780E−06	1.3000E−02	6.481578E−03
iwb_biu	0.1	7.1960E−04	1.4030E−03	8.6690E−08	2.1220E−03	2.123467E−04
lsu	0.6	5.7150E−03	4.5360E−03	9.4160E−08	1.0300E−02	6.150694E−03
mult_mac	0.1	6.9460E−03	2.8100E−02	5.1140E−07	3.5000E−02	3.505111E−03
operandmux	0.4	2.2710E−03	2.6400E−03	6.4490E−08	4.9110E−03	1.964464E−03
pic	0.01	5.3930E−04	1.1050E−03	3.0240E−08	1.6450E−03	1.647324E−05
pm	0.01	2.3360E−04	4.0410E−04	9.5910E−09	6.3770E−04	6.386591E−06
rf	0.5	2.9220E−03	2.6000E−02	7.3280E−07	2.9000E−02	1.446173E−02
sb	0.2	5.6610E−04	7.5770E−04	1.5890E−08	1.3240E−03	2.647759E−04
sprs	0.02	5.5530E−03	5.2600E−03	1.3510E−07	1.0800E−02	2.163951E−04
tt	0.01	9.8770E−04	1.7000E−03	4.6980E−08	2.6880E−03	2.692398E−05
uart	0.5	4.8040E−04	2.8240E−03	2.8800E−07	3.3050E−03	1.652488E−03
wb_biu	0.1	5.0040E−04	9.5720E−04	4.9520E−08	1.4580E−03	1.458095E−04
wbmux	0.4	2.3900E−03	3.6520E−03	7.5520E−08	6.0420E−03	2.416876E−03
Total		6.9298E−02	2.67463E−01	3.57668E−05	3.36936E−01	1.234937E−01

These results were obtained using a benchmark code during the design stages. When the system was implemented the application was run for a specific program, the power consumption was 0.52 W due to memory access and GPIO writes.

Input–Output Ring

The input–output (IO) pad ring on a chip acts as a communication link between the chip's core and the outside world. The IO pad ring is a collection of open metal areas usually located at the periphery of the chip. When a chip is being packaged, a mechanical wire bonder connects the open metal surface of an IO pad with the corresponding package pin.

The circuit functions of an IO pad ring are:

- **ESD Protection:** The IO pad ring has diode protection circuitry that protects the gates connected to the pads from any external electrostatic discharge.
- **Buffer on the Output Signal:** Digital output pads have buffers to enable the driving of external capacitances, in the order of 30 pF.
- **Buffer on the Input Signal:** Digital input pads can have buffers to isolate the external input from the signals inside the chip. A digital input pad can also have noise-tolerant functionality, which removes any noise that might have coupled to the external input. A Schmitt trigger circuit is used to perform this function. The circuit generates a cleaner signal, mitigating any effect on the circuit's performance. However, Schmitt trigger circuits are slow and power hungry.
- **Mixed Voltage Interface:** An IO pad ring usually provides a mixed voltage interface. The external IO pads usually run at a higher voltage, while the core's chip runs at a lower voltage to minimize power. The IO pad ring contains level shifter circuits that perform this function.

Due to limitations in the resolution of the mechanical wire-bonding tool, a minimum open metal surface area and a minimum pad pitch need has been maintained. Maintaining a minimum pad pitch resulted in a limitation of the number of IO pins possible. Table 7 shows the final IO pins for the SoC.

The design of the IO ring for this system has been based on the pins needed for this specific application; it has been subsequently integrated to the complete and routed layout. These pads are ready-made and available in TSMC's digital IO pad library. These pads have been piled together in a rectangle to create the pad ring. The open metal area surface of the pad has a 50 μm length. The pad length itself is 70 μm. To maintain sufficient spacing between two metal area surfaces – where the wire bonder would attach the bonding wires – a spacer of 10 μm is inserted between each pad. This increased the pad pitch to 80 μm.

Table 7. SoC IO pin list

Signal	From Block	Signal	From Block
clk_i	RISC 250 MHz	iwb_we_o	Wishbone
rst_i	RISC rst	iwb_sel_o	Wishbone
clmode_i	RISC clock control	GPIO (31:0)	GPIO
pic_ints_i (3:0)	PPIC interrupts	pm_cpustall_i	Power
iwb_clk_i	Wishbone	pm_dc_gate_o	Power
iwb_rst_i	Wishbone	pm_ic_gate_o	Power
iwb_ack_i	Wishbone	pm_dmmu_gate_o	Power
iwb_err_i	Wishbone	pm_immu_gate_o	Power
iwb_rty_i	Wishbone	pm_tt_gate_o	Power
iwb_dat_i (31:0)	Wishbone	pm_wakeup_o	Power
iwb_cyc_o	Wishbone	pm_cpu_gate_o	Power
iwb_adr_o (31:0)	Wishbone	pm_1volt_o	Power
iwb_stb_o	Wishbone	pm_clk_sd_o (3:0)	Power

The input pads were chosen without Schmitt trigger functionality (PDIDGZ) because of power and performance. Only general purpose IO pads (PRU08SDGZ) had Schmitt trigger circuitry inside of them. They also had control-enabled IO configuration functionality. The output pads were chosen based on the current driving capability required from the pad. The average current is estimated for the pad, assuming a 30pF load and a 25% rise time at 62.5 MHz. Based on the average current calculation for some of the pads, the average current requirement is 4.8 mA. An output pad (PDO16CGZ) with a current capability of 16 mA is therefore chosen to safely meet the current requirements. Figure 5 shows a close up of the physical IO ring.

To determine whether the area is pad limited or core chip limited, a very conservative area estimate was calculated by adding the block areas; this value was multiplied by two to account for wire routing

Figure 5. Close-up view of the IO pad ring; pad pitch is 80 μm

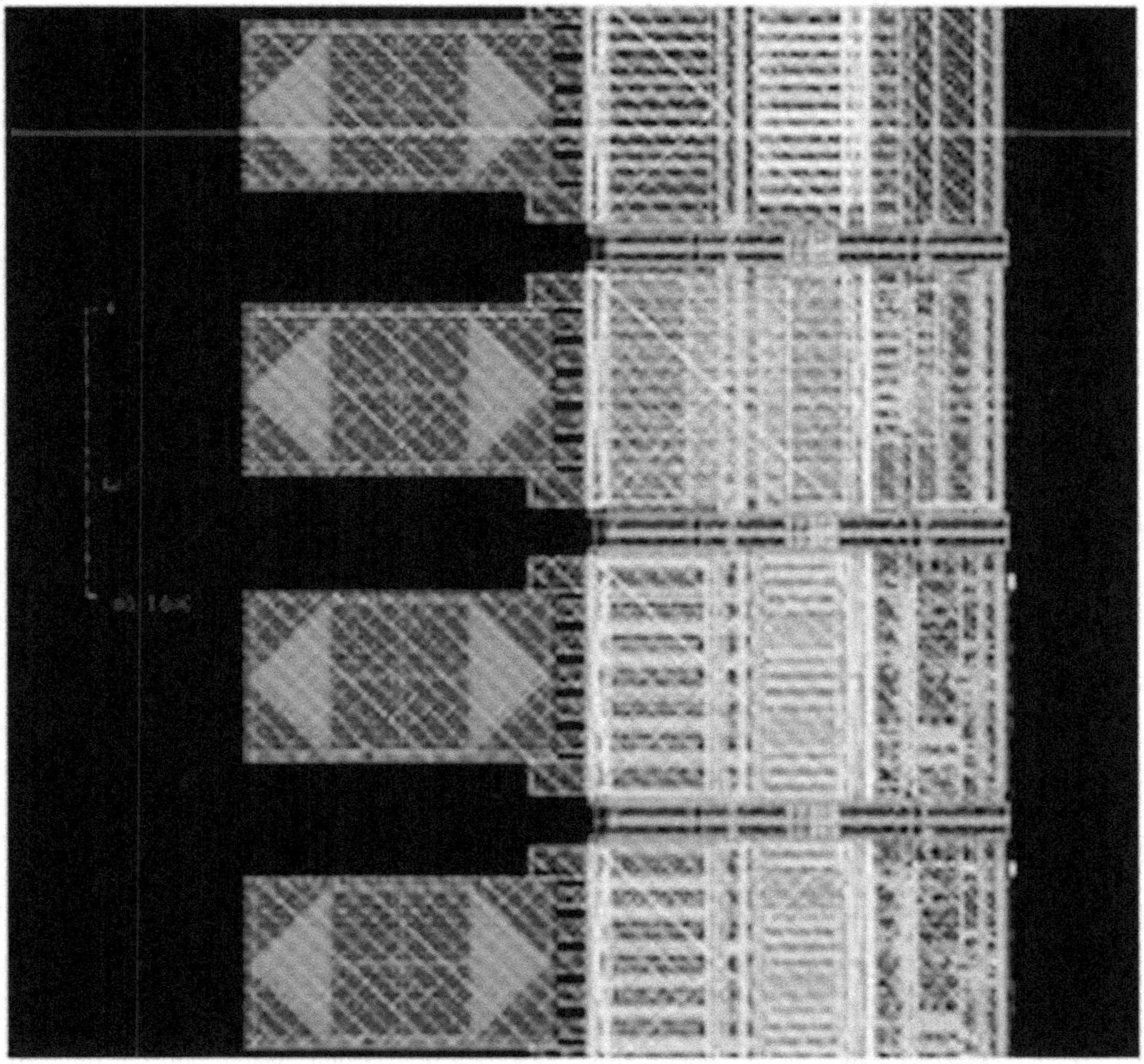

overheads. The area, assuming a square, came out to be 0.583976 mm^2 (0.7 mm × 0.7 mm), which denoted a severely pad-limited die size.

For the ($V_{dd} - V_{ss}$) pair pin calculation, there are two power supply voltages on the chip: the first supply is 3.3 V, which is used for IO signals from the external world to the IO pads; the other voltage is 1.8 V, which is the V_{dd} for the core chip. The IO pad ring converts the voltage levels.

An adequate amount of ($V_{dd} - V_{ss}$) pins is needed to allow for a sufficient current source and sink-in capability. The greater the number of IO pads, the larger the number of ($V_{dd} - V_{ss}$) pins. For this design, a 3.3 V ($V_{dd} - V_{ss}$) pair for every 8 IO pins was chosen, while two 1.8 V ($V_{dd} - V_{ss}$) pairs were placed on each side of the die.

There are more 3.3 V ($V_{dd} - V_{ss}$) pairs due to the amount of power dissipation of the IO pads needed to drive the external 30 pF buffers. The core chip itself requires less power, so fewer IO pins were dedicated to it.

Clock Tree Synthesis (CTS)

CTS is a separate design process that consists of building a balanced buffer tree from the clock input pin to all clock sinks in the design blocks. Clock design includes clock generation, clock regeneration, and clock distribution. Tree design, leafs, sinks, and location have been set according to this chip's needs. The input files required for the clock design are top-level DEF files and top-level Verilog net lists.

To perform CTS, SOC Encounter from Cadence was used. When beginning with a placed net list, the flow involves performing CTS; completing global routing and in-block routing; specifying operation conditions and extraction mode; delaying calculation mode; performing RC extraction; delaying calculation after RC extraction; and engaging in a final timing analysis.

The clock tree design is critical for system synchronization: if the clock does not arrive on time to each block, depending on its location within the data path, the instruction flow becomes erroneous. Clock distribution and clock pin placement are design placement dependent, so each new P&R run requires a new clock tree design. Figure 6 shows a sample run of clock tree generation:

Integration

Integration work is an inter-dependent task, since results from power, timing, placement, and routing affect other results; as such, iterations have been performed until satisfactory results have been reached in all areas. The primary designs of the power grid and clock tree have been completed based on the first preliminary floor plan delivered.

As part of the integration process, different versions of definition files (one for verification, one for timing, and one for P&R) have been merged into one common file to check for and eliminate inconsistencies. When cores were not completely compatible with the bus-based architecture, wrappers were needed, particularly since the cores are generic open source code, and subsequent customization is thus needed; this is especially the case for hardware integration and in-chip communication.

A Wishbone-compliant wrapper has been added for an 8 K SRAM for the memory module, a 4×8 bit GPIO core, and a standard UART. When synthesizing at the top level, some signal and bus inconsistencies arise: missing pins, incompatible bus widths, and unreferenced instantiations.

Most Verilog sources have undergone editing, and the current control version is maintained for code consistency since it is critical for integration. When integrating at the chip level, the hardware hierarchy

Figure 6. Clock tree synthesis results, at the chip level

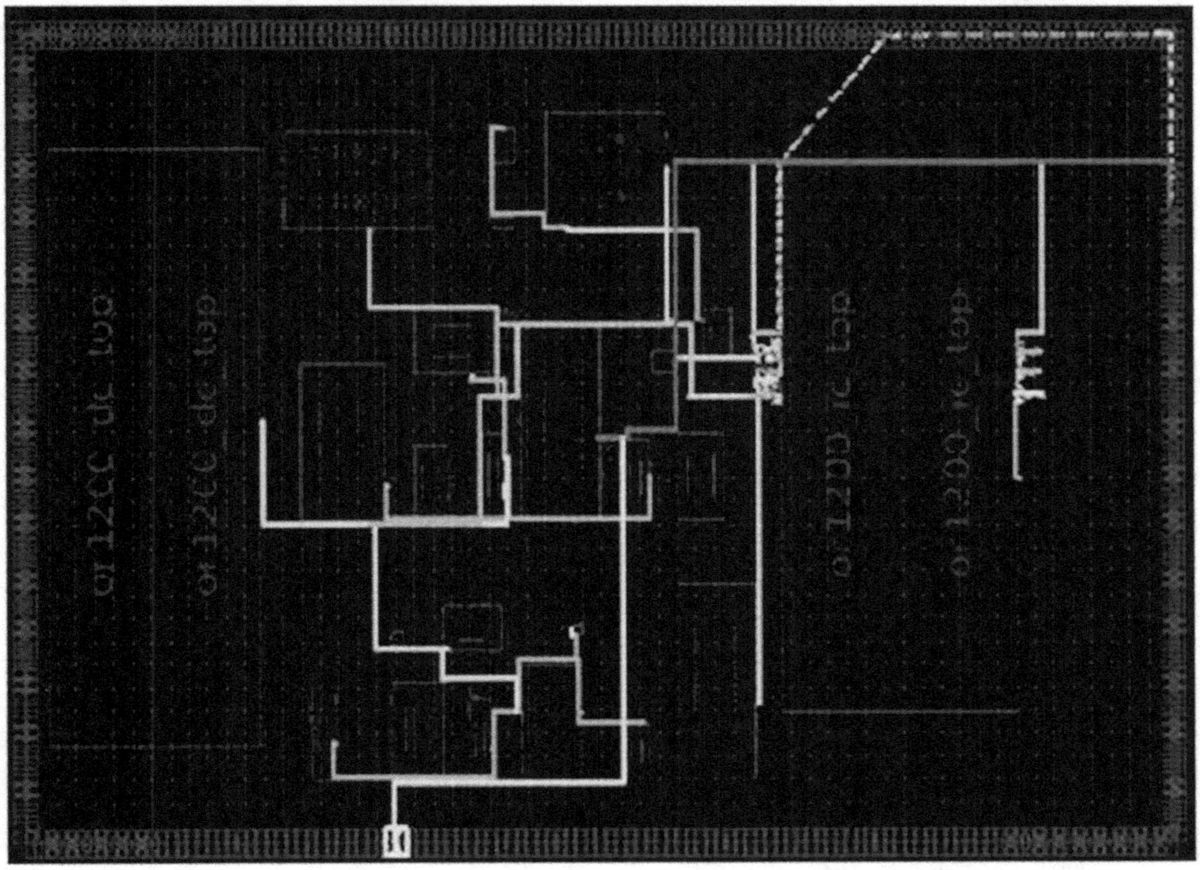

has been redefined. The block level is the open source for each individual block; the CPU level is the unit built by interconnecting the individual blocks; the OR1200 level is the processor built with CPU, memories, debug, and system units; and the SoC level is the system built with an OR1200 processor, SRMA, ROM, GPIO, UART, power grid, and clock tree. Tables 8 and 9 present the details of the tasks performed and the results obtained during the last two rounds of integration (7th and 8th rounds).

DDFS METHODOLOGY

DDFS has been historically achieved using several analog and digital approaches. The actual devices that deliver the output signals for a wide frequency range can be found in wireless communications, but they are application-specific and do not allow for the use of additional functions.

This approach integrates frequency synthesis with multi-waveform generation, multi-waveform superposition, and operation configurability, so implementation can be customized for multiple applications and processes. In this way, a small, low-power implementation is achieved. The original digital frequency synthesis schemes calculate sine values on the fly (computation-intensive scheme), or they access pre-stored values from memory (memory-intensive scheme). The algorithm developed for this

Table 8. Tasks and results, 7th round of synthesis and integration

Cluster	Tasks/Problems/Results
P&R	The preliminary floor plan has been delivered for a power grid design. To finalize it, it needs a resized IMMU cache, 30% interleaving space for routing, and clock and power lines. Perform block placement reruns with different aspect ratios for each block, until better area utilization is achieved.
Timing	Worst slack time at –800 ps over a 4 ns period. That leads to a frequency of 208 MHz.
Top-level synthesis	Synthesis script and netlist for top-level OR1200: ready. Synthesized or1200_top.v, net list: generated. To do: resynthesize to be consistent with updated definition files due to inconsistencies in source files.
Integration	Different versions (one for verification, one for timing, and one for P&R of defines.v file) should merge into one common file. Example inconsistencies: "OR1200_ARTISAN_SSP" has been commented out, `define OR1200_ASIC is commented out in one version, FPU-related macros to be removed.
Verification Verification for power	Verification is complete with the new Verilog files, redo when defines.v file is common to all clusters. Use SAIF from Synopsys (1st option for tool compatibility) or VCD from Model-Sim to write out the activity factor for a given timing window of simulation. SAIF: forward_saif file required for Model-Sim to generate a backward_saif, the required output file. VCD: To write out VCD: Read DC synthesized net list into Model-Sim. Write out a VCD from Model-Sim from the designed test bench using the time window for which power numbers are going to be generated; zip VCD: "zip -r DESIGN.vcd.zip DESIGN.vcd". To convert between formats: Invoke Synopsys DC and use "vcd2saif" command.
Source	Wrappers are written for SRMA, GPIO, and UART. These blocks, along with OR1200, will build the SoC top level. UART and GPIO connect to the data Wishbone of the processor. New pins are created for GPIO wrapper: *aux_i, ext_pad_i, ext_pad_o, ext_padoe_o, clk_pad_i.* For UART: *int_o, stx_pad_o, srx_pad_i, rts_pad_o, cts_pad_i, dtr_pad_o, dsr_pad_i, ri_pad_i, dcd_pad_i.* SRAM: Wishbone-compliant wrapped 8 K SRAM memory module. No syntax errors. Needs functional verification, run Design Compiler to get the gate-level Verilog.

Table 9. Tasks and results, 8th round of synthesis and integration

Cluster	Tasks/Problems/Results
P&R	Rerun the global P&R scripts, since the floor plan has changed significantly. All of the objects are placed. Rebuild each using synopsys/icc/2010 and create the or1200_top library.
Power	As the Verilog source files change, new SDC files for the new net lists are needed for each block. After generating the sdc files for each block, check for consistency among the gate.v net list and new sdc file. Upload new files in the corresponding folder, as existing scripts take them from there. Results from the last-run power analysis are shown in a table below. Needs from the Global P&R for Power IR drop analysis (due to the resistance of the interconnects in power network, there is a voltage drop): **1.** Add code lines for the power grid. **2.** IOs still not hooked up in the milky way. **3.** Check all blocks; place pins in M3 and M4. **4.** Hook up the IO pads to the power rail to get information about the external source of the power rails. **5.** Modules should be power routed in M4 and hooked up to the power grid via M4–M5. **6.** Insert decoupling capacitor filler cells in empty spaces (within individual modules and at the full-chip level between modules). **7.** Design rule check should be executed after hooking up the power routes.
Timing	Final run: squeeze out as much extra timing as possible. For each pass-through in a critical path, a log has been used to show the average path slack, the average time for the pass-through, and the number of times that the pass-through appears in a critical path. Locate the pass-through routes in each block that occupy a good amount of time in a critical path that allow you to tighten, while also locating a path that provides you with enough slack that you can relax. Due to changes in the definition files between r6 and r7, the timing became slightly worse, now at 200 MHz. Some pins in the debug unit and control unit are causing the problem. These pins pop up only in this release, probably due to changes in the define file.
Integration	Design decision: where and how to put the power grid on blocks.
Source	Define files at the processor's top level are modified to meet the requirements of the cache memory blocks (`define OR1200_ASIC and `define OR1200_ARTISAN_SSP 0), register file block (type of register file RAM: `define OR1200_RFRAM_GENERIC), and Wishbone bus (OR1200_CLKDIV_4_SUPPORTED) (this will allow us to use 50 MHz for the external Wishbone bus). The power management unit (`define OR1200_PM_IMPLEMENTED) and eliminated references to the floating process unit are not used since they are not implemented.

system integrates the advantages of the memory-intensive and computation-intensive approaches into one synthesized methodology, while keeping a low implementation area, low-power, and high-performance design. A comparison between the computation-intensive and memory-intensive schemes is shown in Figure 7.

A computation-free algorithm loads data from the output buffer table and stores it in one or more output ports, depending on the operation mode. External conditioning circuitry, including a DAC, a current-to-voltage converter, and a voltage amplifier, converts the sampled data into the finally delivered analog signal.

APPLICATION PROGRAM AND USER INTERFACE

Application software, preloaded in the chip, is developed to perform three main tasks: obtain configuration and operation parameters from the user, process data to obtain selected waveforms, and execute frequency synthesis to deliver processed signals.

Waveform data samples are stored in the data RAM to reduce ROM needs. For the three stated waveforms, 3×256 bytes of data space is needed. As these values are stored into RAM along with the uploading program code, new data tables with different waveforms can be included in the application's source files, and the software architecture remains unchanged.

Configuration and operation parameters enable the user to select the type of waveform desired, the operation mode to be executed, and the time period to deliver the outputs. Using the operation mode, users select one of three of the following parameters: one output signal with one single frequency, two separate output signals with two different frequencies, or one output signal with two superimposed frequencies. To complete the frequency synthesis process, the output data for delivery takes data from the processed output table and sends it to the output port, using 8 bits for modes 1 and 3, or 16 bits

Figure 7. Memory-intensive versus computation-intensive DDFS methodologies

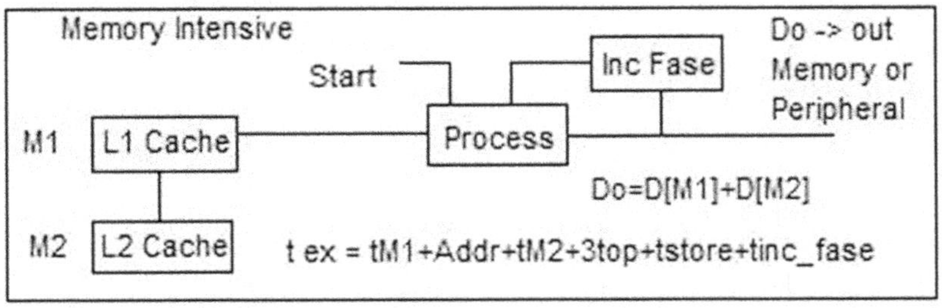

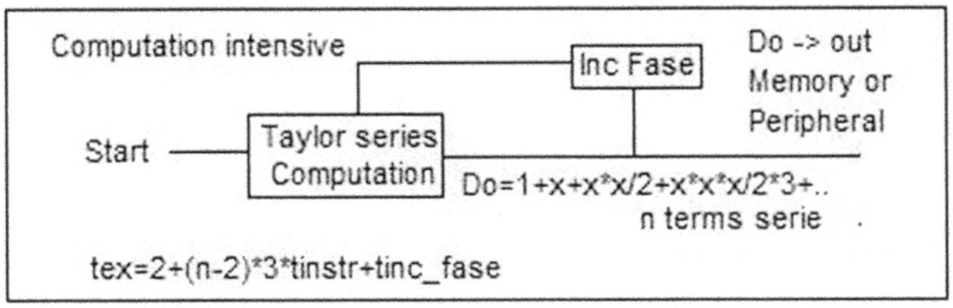

for operation mode 2. Figure 8 shows the application software flow. The application program can be described by its main functions, listed in Table 10.

A simple user interface defines the experiment environment by selecting several operation parameters: select the operation mode between three output options (one single frequency, two separated frequencies, and two superimposed frequencies); set the frequencies (base and superimposed) for the experiment; select the number of data samples desired for each frequency; select the exposure time for the test; and start the operation when ready for the experiment.

A UART port has been selected for system configuration and operation. It is selected over a USB port since it requires a smaller circuit area and speed is not important during configuration and operation setting. For extended interfacing capabilities, Wishbone data and instruction buses have been added; they can be used for external memory access and to connect this system to external peripheral systems, drivers, or bridges. The user interface functionality has been summarized in Table 11, according to the system configuration and operation needs.

Figure 8. Application software flow

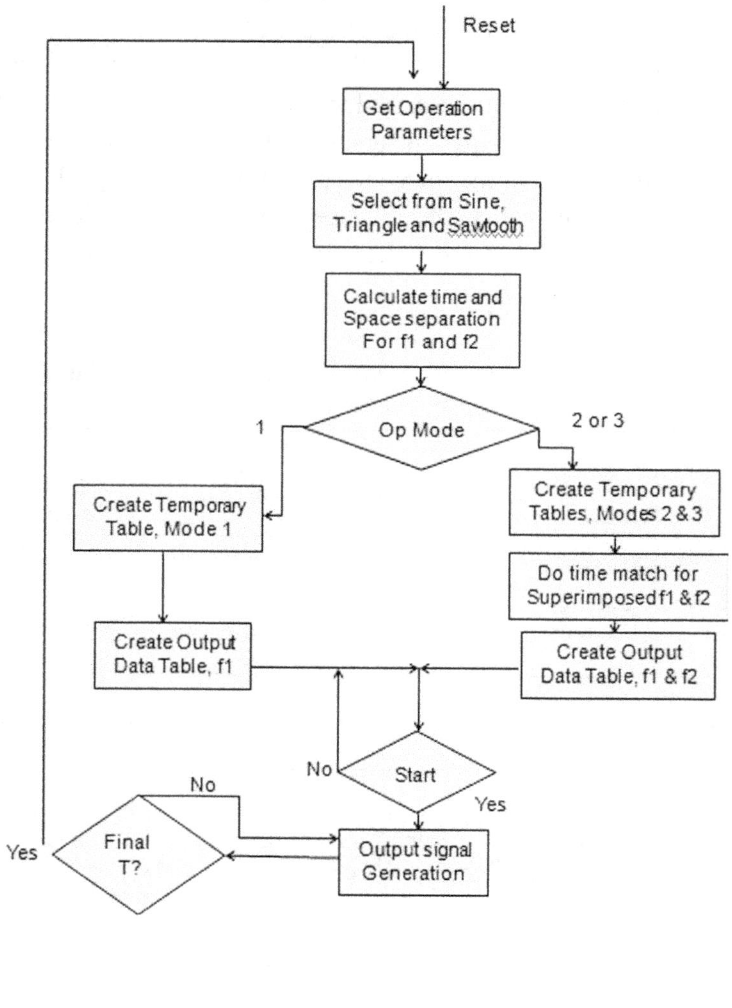

Table 10. Application program function list and description

Function	Description
Store waveform data tables	Stores 256 8-bit values for each of the three waveforms: sine, triangle, and saw tooth.
Get operation parameters	Get operations parameters via the user interface. Displays parameter list and obtains the input values.
Validate input values	Check for frequency multiplicity in operation mode 2; check for 2-multiple number of data samples; check for the frequency out of range.
Check frequency range	Separate operations into low and high frequencies at 400 Hz. Low frequencies use up to 256 data samples per waveform cycle; high frequencies use up to 32.
Calculate data separation	Desired output frequency and number of voltage steps determine how many data points will be extracted from the original sine table in order to construct the desired output signal. Two data separation parameters are needed for operation modes 2 and 3.
Calculate base time	According to the desired frequency and the number of voltage steps, there is a base time that indicates whether the time between data points are sent to the output port.
Create Table-Mode1	Extract data from the original sine table needed to construct one output signal, at one single frequency: each 8-bit data from the original sine table is stored as the 8 least significant bits of the 32-bit output port.
Create Table-Mode2	Extract data from the original sine table needed to construct two output signals, at two separated frequencies: if two 8-bit output ports can be stored at the same time with 32-bit data, two data points from the original sine tables must be concatenated before stored in the buffer memory table.
Create Table-Mode3	Extract data from the original sine table needed to construct one output signal, at two superimposed frequencies: data points for the different frequencies should be added to achieve superimposition.
Time match	When in superposition mode, perform a time-match between data samples for frequencies f_1 and f_2, since the period and time between samples are different. See Appendix B2.
Clock set	Set the clock for system and parallel port control from the board's main oscillator.
Configure port	Enable and configure parallel port A as output, as an 8-bit set, output driving current.
Offline monitor	When in offline mode, data in the final output table are displayed to monitor the frequency superposition methodology.
Output signal generation, high frequencies	Continuous, uninterrupted loop for loading the data from buffer memory and storing it on output ports. For low power, no memory other than buffer is read; no instructions other than those for signal generation are executed.
Output signal generation, low frequencies	Loading data from buffer memory and storing it on output ports uses a counter to create wait cycles and to extend the time between samples. In this mode of slow output frequencies, the restriction of no computation during synthesis is not necessary.
Generate multiple sequences	When selected on a user interface, the cyclic delivery of same or different signal patterns is performed for a specific and individual period of time. Multiple tests can be done with sequenced stimulation patterns.
Start/stop external interrupt	The Start/Stop button is enabled as an external interrupt, at startup for configuration and various parameters, and during execution to stop signal generation.

Table 11. User interface: primary functions

Function	Description
Load	Load application software from available selections.
Select	Define the waveform to be used for the next experiment.
Configure	Select the waveform, samples per cycle, output frequencies, operation mode, and exposure time for signal generation.
Operate	Start signal generation, disabling other functionalities to maximize output frequency. Maintain continuous operation until the exposure time finishes or a stop request is received.
Monitor	Displays feedback information from the operation, and data are sent to the output port. This function can be disabled to eliminate the execution time for monitoring and to maximize the output frequency.

IMPLEMENTATION

To perform reliable experiments, the user begins by defining the test parameters from a previously known base used in manual experiments, such as the frequency value known to be effective for a particular manipulation experiment on a specific type of particle. From there, the user can modify parameters such as waveform, frequency, exposure time, or a sequence of patterns. These parameters can be changed one at a time, or as a set, for each test run. Once a set of parameters is found to be effective for a specific manipulation experiment, that test can be precisely repeated with no manual intervention.

During the SoC design stages, customized software and hardware architectures were developed leading to an open, configurable, modular, and scalable circuit design.

The implementation of this integrated circuit opens possibilities in areas such as implantable devices, portable labs, point of care, and so on. It is a flexible stimulation system that delivers a variety of signal waveforms within a wide frequency range and superimposes equal or different waveforms in one signal, while allowing users to select operation parameters such as signal type, output frequencies, the number of samples per cycle, and time of exposure.

This modular design can be used as a standalone stimulation instrument in micro-fluidic devices, or it can be integrated into other systems; it is a small, low-power, high-performance IC design with interfacing instruction and data buses. The 189-pin IC runs on a 190 MHz clock, consuming around half a Watt during operation, and it is tiny enough to be considered for portable or implantable devices. Tables 12 and 13 summarize the final parameters for this design, using TSMC 180 nm technology.

Table 12. SoC design parameters

Parameter	Value	Parameter	Value
Area	2.44×2.44 mm^2	Data memory	8 KB
Power	0.523 W	ROM	256 bytes
Clock cycle	1/(190 MHz)	System clock	190 MHz
IO pins	189 pins	Wishbone bus clock	47.5 MHz
Instruction memory	8 KB	GPIO operating frequency	5.94 MHz

Table 13. Execution times by function group

Function	Clock Cycles	Execution Time, ns
Parameter setup, 8 parameters, 4 cycles each	4×8	168
Data processing for temporary tables	$2\ (4 \times 32)$	1347
Data processing for output table	$5\ (32)\ (f_2/f_1)$	3368
Output update	4	21

SAMPLE APPLICATIONS

The SoC presented in this chapter can be used in several applications, so the user can begin with this design and develop a new one by customizing it.

A general overview of related applications shows the usability of this work:

- Water pollution analysis, for human consumption or industrial processes.
- Food contamination analysis, to look for bacteria or particles of non desired substances.
- Testing prosthetic devices, either in developing stages or before implanting them.
- Medical lab tests – faster and in situ tests can be performed on drop-size fluid samples searching for bacteria or inert particles.
- Any research experiment that requires an electric signal as part of the setting.
- As electric signals applied over micro-samples of fluids generate electric fields inside the sample, dielectrophoresis is widely used as a particle-manipulation technique.

Table 14 presents the actual works where this system could be used. The main advantages of this system over those in use are that it is easily programmable by users with no background on electronics, the interface allows users to set a different set of experiment parameters, the interface runs on every computer with no need for a specific software, new waveforms or patterns can be easily integrated into the application program, and the frequency range for the delivered signals is not limited to one application. This means that by using the system proposed here, those applications can improve operation, repeat or change experimental runs faster, and change parameters easily to obtain and explore different results. The system presented here is considered low-power compared to others shown on Table 14 because they are designed based on functionality not on power consumption, and although some of them are close to 2 Watts (compared to 0.52 Watts for our system), they present no documentation on the design process targeting low power.

Table 14. Referenced works on electric stimulation

Reference	System Specifications	Application
Gascoyne and Vykonkal (2004)	Sine only, 4 channel, f=200 KHz, not configurable	General purpose sample handling
Manaresi et al. (2003)	Sine only, 2 channel, 4-9 Volts, specific for manipulating eukaryotic cells, not configurable	Individual cell manipulation and detection
Tierney, Rader, and Gold (1971)	Sine only, 2 channel, 40-100 Volts, 1 and 2 KHz only, not configurable	Audio and electro-acoustics
Choi et al. (2006)	Specific for a particular electrode structure only, not configurable	Flow-through micro-electroporation
Chang, Lee and Chiu (2008)	Specific for a particular type and size of beads, not configurable	Micro-bead manipulation
J Wu (2007)	2.2 Volts, up to 1 KHz only, not configurable	For ACEK experiments
Watkins et al. (2009)	SOI CMOS exciter chip, not configurable	An electrical microcytometer
T. Ch. Shih et al. (2007)	Sine only, 8 Volts, 1 MHz, not configurable	Multi sorting of proteins and DNA
Villemejane et al. (2010)	High voltage outputs but low frequencies only, not configurable	Living cell nanomanipulation

CONCLUSION

A small and low-power integrated circuit can be achieved for specific applications using available open cores and following a simplified design flow. The case presented here shows that a variety of electric signals and patterns were found to be useful in a variety of applications, where complex electric stimulation and varying signal parameters are needed. This system is an integrated, flexible, and configurable solution that covers a wide variety of setups and applications; it has configuration options that are able to define specific and automated test procedures. The developed SoC has an area of 4.03 mm^2, 189 pins, an operating frequency of 180 MHz, and a power consumption of 0.52 watts when fabricated on 180 nm technologies. The final numbers of the three design parameters are considered good for the 8 optimization rounds and the natural limitations of open source components. It can be inferred from this work that the design presented here can be used as it is or modified, to fit other applications needs. This work also shows, step by step, how to design an integrated circuit for a specific application using open source components and tools only.

REFERENCES

Aalazaam, A., Roman, D., Nerguitzian, V., Stiharu, I., Bhat, R., Yasmeen, A., & Mosutafa, A. (2008). Real time continuous dielectrophoretic separation of malignant cells. *IEEE Conference on Microsystems and Nanoelectronics Research*. doi:10.1109/MNRC.2008.4683374

ASIC Design Flow Tutorial. (2010). Nano-Electronics and Computing Research Center, San Francisco State University.

Chang, F., Lee, Y., & Chiu, Ch. (2008). Multiple Electrodes Arrayed Dielectrophoretic Chip with Application on Micro-Bead Manipulation. *IEEE Proceedings of the Nano/Micro Engineered and Molecular Systems*. IEEE.

Choi, Y. S., Kim, Y. J., Im, M., Kim, B. G., Yun, K. S., & Yoon, E. (2006). Three Dimensional Electrode Structure Controlled by Dielectrophoresis for Flow-Through Micro Electroporation System. *IEEE International Conference on Micro Electro Mechanical Systems*. doi:10.1109/MEMSYS.2006.1627837

Design Compiler User Guide. (2009). Synopsys, Version C-2009.06.

Gascoyne, P., Mahidol, C., Ruchirawat, M., Satayavivad, J., Watcharasit, P., & Becker, F. (2002). Microsample preparation by dielectrophoresis: isolation of malaria. *Lab on Chip Magazine, 2*, 70-75. DOI: 10.1039/b1109990c

Gascoyne, P., & Vykonkal, J. (2004). Dielectrophoresis-based sample handling in general-purpose programmable diagnostic instrument. *IEEE Proceedings, 92*, 22-42. DOI: 10.1109/JPROC.2003.820535

Ibrahim, M., Elsayed, F., Ghallab, Y., & Badawy, W. (2009). An Electric Field Array Microsystem for Lab-on-Chip and Biomedical Analysis. *IEEE Conference on Biomedical Circuits and Systems*. doi:10.1109/BIOCAS.2009.5372076

Jupiter XT. (2006). *Synopsys, Top-Down Hierarchical Flow, User Guide, Version Y-2006*. Author.

La Rosa, C., Tilley, P., Fox, J., & Kaler, K. (2008). Microfluidic Device for Dielectrophoresis Manipulation and Electrodisruption of Respiratory Pathogen Bordetella pertussis. *IEEE Transactions on Biomedical Engineering*, 2426-2432. DOI: 10.1109/TBME.2008923148

Li, H., Yanan, Z., & Akin, D. (2005). Characterization and modeling of a microfluidic dielectrophoresis filter for biological species. *IEEE Journal of Microelectromechanical Systems, 14*(1), 103–112. doi:10.1109/JMEMS.2004.839124

Library Data Preparation for IC Compiler. (2010). *User Guide, Synopsys, Version D-2010.03*. Author.

Lopez, M. S., Gerstlauer, A., Avila, A., & Martinez, S. O. (2011). A programmable and configurable multi-port System-on-Chip for stimulating electrokinetically-driven microfluidic devices. In *Proceedings of the Annual International Conference of the IEEE Engineering in Medicine and Biology Society, EMBS* (pp. 8361-8364). doi:10.1109/IEMBS.2011.6092062

Lopez, M. S., Moncada, H., Perez, V., Lapizco, B., & Martinez, S. (2013). An electric stimulation system for electrokinetic particle manipulation in microfluidic devices. *Review of Scientific Instruments, 84*(3). DOI: 10.1063/1.4793559

Manaresi, N., Romani, A., Medoro, G., Altomare, L., Leonardi, A., Tartagni, M., & Guerrieri, R. (2003). A CMOS Chip for Individual Cell Manipulation and Detection. *IEEE Journal of Solid-State Circuits, 38*(12), 2297–2305. doi:10.1109/JSSC.2003.819171

OpenCores. (2014). *OpenCores*. Retrieved in 2014, http://opencores.org/

Shih, T. C., Chu, K,. & Liu, C. (2007). A Programmable Biochip for the Applications of Adaptive Multisorting Using Dielectrophoresis Array. *IEEE Journal of Microelectromechanical Systems, 16*(4), 816-825. DOI 10.1109/JMEMS.2007.897558

Standard library 0.13um - 0.25 um ROM Generator from Artisan: User Manual. (2004). *revision ug_2004q3v1*.

Standard library SRAM Generator from Artisan. (2004). User Manual (revision ug_2004q1v0). Author.

Tierney, J., Rader, C., & Gold, B. (1971). A digital frequency synthesizer. *IEEE Transactions on Audio and Electroacoustics, 19*(1), 48–57. doi:10.1109/TAU.1971.1162151

Villemejane, J., Mottet, G., Francais, O., Pioufle, B., Woytasik, M., & Dufour, E. (2010). Nanomanipulation of Living Cells on a Chip Using Electric Field. *IEEE International Symposium in Electronic Design, Test and Application*. DOI doi:10.1109/DELTA.2010.26

Watkins, N., Vankatesan, B., Toner, M., Rodriguez, W., & Bashir, R. (2009). A Robust Electrical Microcytometer with 3-Dimensional Hydrofocusing. *La bon a Chip, 9*(22), 3177-3184. DOI .10.1039/b91224a

Wu, J. (2008). Interactions of Electrical Fields and Fluids: Laboratory on a Chip Applications, IET. *Journal of Nanobiotechnology, 2*(1), 14–27. doi:10.1049/iet-nbt:20070023 PMID:18298196

Yuk, K., McConaghy, C., Gascoyne, P., Schwartz, J. A., Vykoukal, J., & Andrews, C. (2007). A High-Voltage SOI CMOS Exciter Chip for a Programmable Fluidic Processor System Current, K.W.; Biomedical Circuits and Systems. *IEEE Transactions on Biomedical Circuits and Systems*, *1*(2), 105–115. doi:10.1109/TBCAS.2007.908110 PMID:23851665

KEY TERMS AND DEFINITIONS

ASIC: Application-specific integrated circuit – a circuit designed specifically for certain applications that, after fabrication, cannot be modified for use with a different purpose.

Clock Tree: The set of lines or branches that conforms the net to supply the clock signal to all the cores or sub-circuits in an integrated circuit.

Dielectrophoresis: The effect produced by electric signals when applied to fluid samples by generating an electric field that is capable of moving the particles suspended in it.

Electric Signal Testing: The process of needing and using electric signals to test a system or device to determine whether it produces the expected result, behavior, or output.

Frequency Synthesis: The process of generating cyclic signals or waveforms of a certain frequency, using certain methodologies, techniques, or algorithm – either with analog or digital resources.

IO Ring: The connections and structures that conform the input and output signals in a chip between the internal circuitry and the external pin of the integrated circuit.

Open Cores: Source code in the hardware description language that represents a circuit for a specific purpose or functionality; it is available for free use.

Signal Generation: The process (analog or digital) that produces and delivers signals – either electric, electromagnetic, or of any sort.

System on Chip: Integrated circuit that includes all the blocks or modules needed to perform a complete function by itself, independent and comprehensive enough to be considered a whole and functional system.

Wishbone Bus: A standard bus for interconnecting open cores, and that can be adapted to connect to other standard buses.

APPENDIX

Code for the Application Program

```
//
// Version date: Feb the 3rd, 2014
// Details: separates frequency ranges in low (<400 Hz) and high (>400Hz)
// Delivers superimposed frequencies in any mix of available waveforms

#include <stdlib.h>
#include <stdio.h>
#include <math.h>
#include "hw_memmap.h"
#include "hw_types.h"
#include "sysctl.h"
#include "hw_sysctl.h"
#include "gpio.h"
#include "hw_gpio.h"


//#define PORT_DATA(GPIO_PIN_0 |GPIO_PIN_1 | GPIO_PIN_2 | GPIO_PIN_3 | //
GPIO_PIN_4 | GPIO_PIN_5 | GPIO_PIN_6 | GPIO_PIN_7)
#ifdef DEBUG
void
__error__(char *pcFilename, unsigned long ulLine)
{
}
#endif

#define Pi 3.14159265358979323846264338327

static unsigned int sinedatint[256] =
{ 127, 130, 133, 136, 139, 142, 145, 148, 151, 154, 157, 160, 163, 166, 169,
172,  175, 178, 181, 184, 186, 189, 192, 194, 197, 200, 202, 205, 207, 209,
212, 214,  216, 218, 221, 223, 225, 227, 229, 230, 232, 234, 235, 237, 239,
240, 241, 243,  244, 245, 246, 247, 248, 249, 250, 250, 251, 252, 252, 253,
253, 253, 253, 253,   254, 253, 253, 253, 253, 253, 252, 252, 251, 250, 250,
249, 248, 247, 246, 245,  244, 243, 241, 240, 239, 237, 235, 234, 232, 230,
229, 227, 225, 223, 221, 218,  216, 214, 212, 209, 207, 205, 202, 200, 197,
194, 192, 189, 186, 184, 181, 178,  175, 172, 169, 166, 163, 160, 157, 154,
151, 148, 145, 142, 139, 136, 133, 130,  127, 123, 120, 117, 114, 111, 108,
105, 102, 99, 96, 93, 90, 87, 84, 81, 78, 75, 72, 69, 67, 64, 61, 59, 56, 53,
51, 48, 46, 44, 41, 39, 37, 35, 32, 30, 28, 26, 24,  23, 21, 19, 18, 16, 14,
```

```
13, 12, 10, 9, 8, 7, 6, 5, 4, 3, 3, 2, 1, 1, 0, 0, 0, 0, 0,
0, 0, 0, 0, 0, 0, 1, 1, 2, 3, 3, 4, 5, 6, 7, 8, 9, 10, 12, 13, 14, 16, 18, 19,
21, 23, 24, 26, 28, 30, 32, 35, 37, 39, 41, 44, 46, 48, 51, 53, 56, 59, 61,
64, 67, 69, 72, 75,78, 81, 84, 87, 90, 93, 96, 99, 102, 105, 108, 111, 114,
117, 120, 123 };

static unsigned int toothsawdat[256] =
{0, 1, 2, 3, 4, 5, 6, 7, 8, 9, 10, 11, 12, 13, 14, 15, 16, 17, 18, 19, 20, 21,
22, 23, 24, 25, 26, 27, 28, 29, 30, 31, 32, 33, 34, 35, 36, 37, 38, 39, 40,
41, 42, 43, 44, 45, 46, 47, 48, 49, 50, 51, 52, 53, 54, 55, 56, 57, 58, 59,
60, 61, 62, 63, 64, 65, 66, 67, 68, 69, 70, 71, 72, 73, 74, 75, 76, 77, 78,
79, 80, 81, 82, 83, 84, 85, 86, 87, 88, 89, 90, 91, 92, 93, 94, 95, 96, 97,
98, 99, 100, 101, 102, 103, 104, 105, 106, 107, 108, 109, 110, 111, 112, 113,
114, 115, 116, 117, 118, 119, 120, 121, 122, 123, 124, 125, 126, 127, 128,
129, 130, 131, 132, 133, 134, 135, 136, 137, 138, 139, 140, 142, 142, 143,
144, 145, 146, 147, 148, 149, 150, 151, 152, 153, 154, 155, 156, 157, 158,
159, 160, 161, 162, 163, 164, 165, 166, 167, 168, 169, 170, 171, 172, 173,
174, 175, 176, 177, 178, 179, 180, 181, 182, 183, 184, 185, 186, 187, 188,
189, 190, 191, 192, 193, 194, 195, 196, 197, 198, 199, 200, 201, 202, 203,
204, 205, 206, 207, 208, 209, 210, 211, 212, 213, 214, 215, 216, 217, 218,
219, 220, 221, 222, 223, 224, 225, 226, 227, 228, 229, 230, 231, 232, 233,
234, 235, 236, 237, 238, 239, 240, 241, 242, 243, 244, 245, 246, 247, 248,
249, 250, 252, 252, 253, 254, 255 };

static unsigned int triangdat[256] =
{0, 2, 4, 6, 8, 10, 12, 14, 16, 18, 20, 22, 24, 26, 28, 30, 32, 34, 36, 38,
40, 42, 44, 46, 48, 50, 52, 54, 56, 58, 60, 62, 64, 66, 68, 70, 72, 74, 76,
78, 80, 82, 84, 86, 88, 90, 92, 94, 96, 98, 100, 102, 104, 106, 108, 110, 112,
114, 116, 118, 120, 122, 124, 126, 128, 130, 132, 134, 136, 138, 140, 142,
144, 146, 148, 150, 152, 154, 156, 158, 160, 162, 164, 166, 168, 170, 172,
174, 176, 178, 180, 182, 184, 186, 188, 190, 192, 194, 196, 198, 200, 202,
204, 206, 208, 210, 212, 214, 216, 218, 220, 222, 224, 226, 228, 230, 232,
234, 236, 238, 240, 242, 244, 246, 248, 250, 252, 254, 255, 254, 252, 250,
248, 246, 244, 242, 240, 238, 236, 234, 232, 230, 228, 226, 224, 222, 220,
218, 216, 214, 212, 210, 208, 206, 204, 202, 200, 198, 196, 194, 192, 190,
188, 186, 184, 182, 180, 178, 176, 174, 172, 170, 168, 166, 164, 162, 160,
158, 156, 154, 152, 150, 148, 146, 144, 142, 140, 138, 136, 134, 132, 130,
128, 126, 124, 122, 120, 118, 116, 114, 112, 110, 108, 106, 104, 102, 100, 98,
96, 94, 92, 90, 88, 86, 84, 82, 80, 78, 76, 74, 72, 70, 68, 66, 64, 62, 60,
58, 56, 54, 52, 50, 48, 46, 44, 42, 40, 38, 36, 34, 32, 30, 28, 26, 24, 22,
20, 18, 16, 14, 12, 10, 8, 6, 4, 2 };

unsigned int timeindex1int[256],timeindex2int[256],timeindexbuffint[256];
```

```
unsigned int TempTable1int[256],TempTable2int[256];// Temporary tables, //
modes 2 & 3, scale 0 to 255
unsigned int BuffTable[4096];// output table, all modes
double trunc(double arg);
float tbs,tbs1,tbs2;// time between samples, para senal 1 y 2
int dat_samples_buff1,dat_samples_temp1,dat_samples_temp2;//data samples
int opmode;//operation modes: 1 single signal, 2 superposition, 3 separate //
signals
int signaltype,signaltype2;//signal type: 1 sine, 2 saw-tooth, 3 triangle
float freq1, freq2;//frecuency for output signals 1 & 2
int N, n1, n2;//samples per waveform cycle

void GetOperParam()
{
        //printf ("Operation mode: 1, 2 or 3:\n  ");
        //        scanf ("%d", &opmode);
        opmode=2;
                //printf ("Selected Operation mode  = %d\n", opmode);
                //printf ("Signal type, 1 sine, 2 tooth, 3 triang:\n  ");
                //scanf ("%d", &signaltype);

        signaltype2=2;
        signaltype=2;

        //printf ("Selected signal type  = %d\n", signaltype);
                //printf ("Output single/low frequency in KiloHertz:\n ");
                //scanf ("%f", &freq1);
        freq1=500;
                //printf ("Samples per cycle: \n");
                //scanf ("%d", &n1);
        //if (freq1>399) {
                //n1=240000/freq1;         //write to port, high frequencies
                //n2=16;
        //}
        //if (freq1<400) {
                //n1=64;
                //n2=64;//write to port, low frecuencies
        //}

        n1=64;
        if (n1<9) n1=16;
                if ((n1>8) & (n1< 17)) n1=16;
                if ((n1>16) & (n1< 33)) n1=32;
```

```
                if ((n1>32) & (n1< 65)) n1=64;
                if ((n1>64) & (n1< 129)) n1=128;
                if (n1>128) n1=256;
                //printf("   %d\n", n1);
                if (opmode>1)
                {
                //printf ("Output high frequency2 in KiloHertz:\n ");
                //scanf ("%f", &freq2);
                        freq2=5000;
                //printf ("Samples per cycle 2: \n");
                //scanf ("%d", &n2);
                n2=64;
                if (n2<9) n2=8;
                if ((n2>8) & (n2< 17)) n2=16;
                if ((n2>16) & (n2< 33)) n2=32;
                if ((n2>32) & (n2< 65)) n2=64;
                if ((n2>64) & (n2< 129)) n2=128;
                if (n2>128) n2=256;
                //printf("   %d\n", n2);
                //printf ("Leaving function GetOperParam\n");
                }

}

void SineDisplay(int N)
{
                unsigned int iter;
//                printf ("Entering function SineDisplay  ORIGINAL SINE
TABLE\n");
                for (iter = 0; iter < N; iter++)
                {
        //                      printf("[%d] ", iter);
                //        printf("   %d\n", sinedatint[iter]);
                }
}

void BuffTableGen(int N, int n)
{
        unsigned int dsepi;
         double dsepd;
         unsigned int iter;
         unsigned int i;
         unsigned int j;
```

```
//printf ("Entering function BuffTableGen mode 1\n");
         i=0;
         dsepd=N/n;
         dsepi=dsepd;
         j=dsepi;
        // printf("data separation int, modo1   = %d\n ", dsepi);

         for (iter = 0; iter < N; iter=iter+dsepi)
                 {
        //                 /*printf("Data number     = %d\n ", iter);*/
                         if (signaltype == 1) BuffTable[i]=sinedatint[iter];
                         if (signaltype == 2) BuffTable[i]=toothsawdat[iter];
                         if (signaltype == 3) BuffTable[i]=triangdat[iter];
                 //        /*printf("Dato original   = %.6ef\n ",
sinedat[iter]);*/
                 //        printf("%d\n ", BuffTable[i]);
                         dat_samples_buff1=i;
                         i=i+1;
                 }
        //         printf("Amount of data samples in buffer table = %d\n ", //
dat_samples_buff1);
                 //printf ("Leaving function BuffTableGen, mode 1\n");
}

void TempTable1Calc(int N, int n1)//prepare temp table, signal 1, modes 2&3
 {
 unsigned int dsepi;
 double dsepd;
 unsigned int iter;
 unsigned int i;
 float t;
//         printf ("Entering function TempTable1Calc, modes 2 & 3\n");
 i=0;
 dsepd=N/n1;
 dsepi=dsepd;
 if ((dsepd-dsepi)>0.495)
         {dsepi++;//if separation es > .495 round up to next integer
        // printf ("separation table 1%d \n", dsepi);
         }
 //printf("data separation in TEMPORARY TABLE 1   = %d\n ", dsepi);
 //printf ("it   time         out data\n");
 for (iter = 0; iter < N+1; iter=iter+dsepi)
        {
```

```
                t=tbs1*i;
        //        printf("[%d] >", iter);
                //printf("[%d]  ", i);
                //printf("%.3ef ", t);
                if (signaltype==1) TempTable1int[i]=sinedatint[iter];
                if (signaltype==2) TempTable1int[i]=toothsawdat[iter];
                if (signaltype==3) TempTable1int[i]=triangdat[iter];
                timeindex1int[i]=100000*t;
                //printf("t=%d ", timeindex1int[i]);
                //printf("%d    ", TempTable1int[i]);
                //printf("%x\n ", TempTable1int[i]);
                dat_samples_temp1=i;
                i=i+1;
        }
        //printf("Data samples in temporary table1  = %d\n ", (dat_samples_
temp1+1));
        //printf ("Leaving function TempTable1Calc, modes 2 & 3\n");
}

void TempTable2Calc(int N, int n2)//prepare temp table signal 2, modes 2&3
 {
 unsigned int dsepi;
 double dsepd;
 unsigned int iter;
 unsigned int i;
 float t;
        //printf ("Entering function TempTable2Calc, modes 2 & 3\n");
 i=0;
 dsepd=N/n2;
 //printf(" %.2ef ", dsepd);
 dsepi=dsepd;
 if ((dsepd-dsepi)>0.495)
         {dsepi++;
         // printf ("separation table 2%d \n", dsepi);
         }
 //printf("data separation in TEMPORARY TABLE 2   = %d\n ", dsepi);
 //printf ("it   time         out data\n");
 for (iter = 0; iter < N; iter=iter+dsepi)
        {
                t=tbs2*i;
        //        printf("[%d] >", iter);
                //printf("[%d]  ", i);
                //printf("%.3ef ", t);
```

```
                if (signaltype2==1) TempTable2int[i]=sinedatint[iter];
                if (signaltype2==2) TempTable2int[i]=toothsawdat[iter];
                if (signaltype2==3) TempTable2int[i]=triangdat[iter];
                timeindex2int[i]=100000*t;
                //printf("t=%d ", timeindex2int[i]);
                //printf("%d ", TempTable2int[i]);
                //printf("%x\n ", TempTable2int[i]);
                dat_samples_temp2=i;
                i=i+1;
        }
}


void BuffTableSuperposition()////prepare output table mode 2
 {
 unsigned int i,j,k,l,m,dato1,dato2;
 float t,tbsmin,tmax;
 unsigned int tint,auxt1,auxt2;

        //printf ("Entering function BuffTableSuperposition, mode 2\n");
        if (tbs1<tbs2)
                tbsmin=tbs1;
        else
                tbsmin=tbs2;

        if        (freq1<freq2)
                tmax=1/freq1;
        else
                tmax=1/freq2;

 t=0; i=0; j=0; k=0;
 dato1=TempTable1int[i];
 dato2=TempTable2int[i];
 BuffTable[k]=(dato1+dato2)/2;
        //printf("%d ", k);
         //printf(" %.2ef ", t);
         //printf(" %x ", data2);
        //printf("+ %x ", data1);
        //printf("= %d ", BuffTable[k]);
   //printf("= %x\n ", BuffTable[k]);

 do{
         t=t+tbsmin;
```

```
        tint=t*100000+1;
        auxt1=timeindex1int[i+1];
        auxt2=timeindex2int[j+1];
        m=0;
        if (tint<auxt1)
                l=0;
        else{
                m=1;
                i=i+1;
                        //printf("new data table111, index %d \n", i);
                dato1=TempTable1int[i];
                }
        //printf("m value after checking table1%d \n", m);

        if (tint<auxt2)
                l=0;
        else{
                m=2;
                j=j+1;
                if (j==n2) j=0;//returns to beginning of temp table, low
frequency
                dato2=TempTable2int[j];
                }

        if(m>0)
                {
               k++;
                if (opmode==2) BuffTable[k]=(dato1+dato2)/2;
                if (opmode==3) BuffTable[k]=dato1+dato2*256;
                timeindexbuffint[k]=tint;
//               printf("[%d] ", k);
 //               printf("t= %d ", tint);
        //        printf(" %x ", dato2);
               //printf("+ %x ", dato1);
               //printf("= %d ", BuffTable[k]);
               //printf("= %x\n ", BuffTable[k]);
                }
        }
       while (t<tmax);
       dat_samples_buff1=k;
       //printf("Amount of data samples in BUFFER TABLE = %d\n ",k);
  }
```

```
void tbsCalc(float freq, float n)//calculates time between samples
 {
//         printf ("Entering function tbsCalc\n");
                 tbs=1/(freq*n);
 }


void WriteToOutLow()//load data from buffer table, write to output port
{
         unsigned int i;
         unsigned int j;
    unsigned int k;
         //float t;
//                 printf ("Entering function WriteToOut\n");
//         printf("Time running between samples buffer1  = %.2ef\n", tbs1);
         //printf ("it   time        out data\n");
         // delayed cycle for slow signal generation
         k=11000/freq1; // k in inverse proportion of desired frequency
    for (;;)
         for (i = 0; i < dat_samples_buff1+1; i=i+1)
                  {
                           for (j = 0; j < k; j=j+1) {} // generates time be-
tween //samples for slow frequencies running k wait cycles between output
updates
                  //        printf("[%d] ", i);
                  //        printf(" %d ", timeindexbuffint[i]);
                  //        printf(" %x\n ", BuffTable[i]);
                           GPIOPinWrite(GPIO_PORTA_BASE,GPIO_PIN_0 |GPIO_PIN_1 |
GPIO_PIN_2 | GPIO_PIN_3 | GPIO_PIN_4 | GPIO_PIN_5 | GPIO_PIN_6 | GPIO_
PIN_7,BuffTable[i]);

                  }
//                         printf ("Leaving function WriteToOut\n");
}
void WriteToOutHigh()//load data from buffer table, write to output port
{
         unsigned int i;
         //unsigned int j;

//         printf("Time running between samples buffer1  = %.2ef\n", tbs1);
         //printf ("it   time        out data\n");
```

```
    for (;;)
         for (i = 0; i < dat_samples_buff1+1; i=i+1)
                   {
                    //for (j = 0; j < tbs1*12; j=j+1)//con 1e+5 is time between
//samples
             // t is accumulated time in cycle, last values is period T of //
waveform
                   //t=tbs1*i;
                   //        printf("[%d] ", i);
                   //        printf(" %d ", timeindexbuffint[i]);
                   //        printf(" %x\n ", BuffTable[i]);
                           GPIOPinWrite(GPIO_PORTA_BASE,GPIO_PIN_0 |GPIO_PIN_1 |
GPIO_PIN_2 | GPIO_PIN_3 | GPIO_PIN_4 | GPIO_PIN_5 | GPIO_PIN_6 | GPIO_
PIN_7,BuffTable[i]);

                   }
//                         printf ("Leaving function WriteToOut\n");
}

//***************************************************************************

int main(void)
{
    //
    // If running on Rev A2 silicon, turn the LDO voltage up to 2.75V.  //This
is a workaround to allow the PLL to operate reliably.
    //
    if(DEVICE_IS_REVA2)
    {
        SysCtlLDOSet(SYSCTL_LDO_2_75V);
    }


    //
    // Set the clocking to run directly from the crystal.
    // Default values assume an external crystal of 6MHz. See Luminary
    // driverlib documentation for other values.
    // Clock source can be:
    //              'SYSCTL_USE_OSC | SYSCTL_OSC_MAIN' - use the xtal with-
out PLL
    //              'SYSCTL_USE_PLL | SYSCTL_OSC_MAIN' - use the xtal with
PLL
    // If using the PLL, the oscillator runs at 200MHz and then you select
```

```
    // a division of this frequency to clock the core. Otherwise the        //
divider just
    //directly divides the XTAL frequency.
    // Use a 6MHz external XTAL directly with no division
    //printf ("Entering  sysctlclokset \n");
    SysCtlClockSet(SYSCTL_SYSDIV_1 | SYSCTL_USE_OSC | SYSCTL_OSC_MAIN | SYS-
CTL_XTAL_6MHZ);

    // Use the XTAL directly to clock the PLL with division by 4
    //SysCtlClockSet(SYSCTL_SYSDIV_4 | SYSCTL_USE_PLL | SYSCTL_OSC_MAIN |
    // SYSCTL_XTAL_6MHZ);

   // printf ("Entering sysctlperipheralenable \n");
    SysCtlPeripheralEnable(SYSCTL_PERIPH_GPIOA);

    //printf ("Entering define gpio as output \n");
    GPIOPinTypeGPIOOutput(GPIO_PORTA_BASE,GPIO_PIN_0 |GPIO_PIN_1 | GPIO_PIN_2
| GPIO_PIN_3 | GPIO_PIN_4 | GPIO_PIN_5 | GPIO_PIN_6 | GPIO_PIN_7);

    //printf ("Entering write to port \n");
    GPIOPinWrite(GPIO_PORTA_BASE,GPIO_PIN_0 |GPIO_PIN_1 | GPIO_PIN_2 | GPIO_
PIN_3 | GPIO_PIN_4 | GPIO_PIN_5 | GPIO_PIN_6 | GPIO_PIN_7,0x55);

    // printf ("Leaving write to port \n");

    // Use the XTAL directly to clock the PLL with division by 4
    // SysCtlClockSet(SYSCTL_SYSDIV_4 | SYSCTL_USE_PLL | SYSCTL_OSC_MAIN |
    // SYSCTL_XTAL_6MHZ);


    N=256;
            GetOperParam();//Get operation parameters
        //SineDisplay(N);//Display data samples for sine waveform
        if (opmode==1)// operation mode = 1?
        {
                tbsCalc(freq1,n1);//calculate separation between samples
                tbs1=tbs;
                BuffTableGen(N,n1);//generate buffer table extracting samples
        }
        if (opmode>1)
        {// operation mode= 2 or 3?
                tbsCalc(freq1,n1);//calculate separation between samples,
signal 1
```

```
                tbs1=tbs;
                TempTable1Calc(N,n1);//generate temp table for signal 1
                tbsCalc(freq2,n2);//calculate separation between samples,
signal 2
                tbs2=tbs;
                TempTable2Calc(N,n2);        // generate temp table for sig-
nal 2
                BuffTableSuperposition();// generate buffer table modes 2 & 3
            }
            if (freq1>399) WriteToOutHigh(); // write to output port, high //
frequencies
            if (freq1<400) WriteToOutLow(); //write to output port, low //
frequencies

        while(1);
```

Chapter 12
Low Power Design of High Speed Communication System Using IO Standard Technique over 28 nm VLSI Chip

Bhagwan Das
Unuversiti Tun Hussein Onn Malaysia, Malaysia

Mohammad Faiz Liew Abdullah
Universiti Tun Hussein Onn Malaysia, Malaysia

ABSTRACT

The low power design of Very Large Scale Integration (VLSI) system is one of the hot topic in research. In this chapter, the low power design for VLSI based high-speed communication is realized over 28 nm VLSI chip packed in UltraScale Field Programming Gate Array (FPGA) using proposed technique. The high-speed communication system is taken as case study for the low power design of VLSI system. Similarly, various VLSI design system can be realized to achieve the low power VLSI system design goal. High-speed communication systems provide the smooth operation for global internet traffic and requires high power devices and components.IO standard is powerful interface tool that provides low power consumption using the fast signal termination by mean of electrical characteristics. In result for this work, more than 96% power reduction is achieved for VLSI based high-speed communication system, when operated at 500 GHZ, 900 GHz, 10 THz and 17 THz carrier frequencies using the High-Speed Unterminated Logic IO Standard. The power analysis is performed using XPA analyzer in Xilinx suite.

INTRODUCTION

The low power design for communication systems are in real demand for green information and communication technologies (ICT). It is observed that more than 35% of total power consumption in the world is utilized by ICT devices (Cetinkaya & Akan, 2015) (Yigit, Gungor, Tuna, Rangoussi, & Fadel, 2014). In this concern, saving of ICT devices power is now becoming a social responsibility. In this chapter,

DOI: 10.4018/978-1-5225-0190-9.ch012

authors demonstrated the low-power design for VLSI based high-speed communication system using proposed IO Standard technique. The low power design of VLSI based high communication system is verified by implementing the designed system on 28 nm chip size packed in UltraScale FPGA device. The variation in proposed IO Standard technique varies the chip power and other chip parameters such as; temperature and other power consumption like; IO Power, logic power, clock power and etc. to produce the low power design for the VLSI based designed system. Furthermore, the designed VLSI based high speed communication system represents a springboard for developing future generation networks using proposed technique. The implementation of these system on UltraScale FPGA that deliver the affluence design with several low power features.

The chapter is organized by discussing the background of previous and currently used techniques to reduce the power consumption for VLSI design system. The basic terminologies related to the research work are also discussed to support the features of proposed technique. The methodology to develop the low power design for VLSI based high speed communication system is also discussed along with power. Finally, results and conclusion are discussed to define the output characteristics of low power VLSI based high speed communication system. This chapter will help the students, researcher and academicians for designing and implementing the low power VLSI based designed system on VLSI chip using UltraScale devices. The audience will enjoy the experience of interesting feature of low power design for VLSI based high speed communication system using proposed technique. Using proposed technique, several real time systems can be developed as a low power design.

Background

High-speed communication systems are the set of interconnected components that provide services to the users e.g. information services for near and far reach and other in (Mendez et al., 2014). In high-speed communication system, the thermal excoriation in components and other power challenges indulge extra cost (Whitfield & Doyel, 1984) (Cloetens, 2014) for the system. There are several techniques involved in developing low power design for the components (Wenger, 2014). One convenient and easy solution for all above problems is to design low power VLSI design using hardware descriptive language (HDL) or Verilog HDL (VHDL) (Verhelst et al., 2015). Formerly, the primary concerns of Very Large Scale Integration (VLSI) designers was to improve the area, cost and reliability of design; power consumption was secondary like other parameters (Mahoob & Mehra, 2015). Nevertheless, power consumption issue is dealt as primary issue to be solved for different systems. The numerous studies have contributed in this regard (Hinton, Jalali, & Matin, 2015). The power reduction can be accomplished through several available techniques e.g. clock gating (Sahni, Rawat, Pandey, & Ahmad, 2015), voltage scaling (Li , Xie, Wang, Nazarian, & Pedram, 2015) (Chandrakasan, Sheng, & Brodersen, 1992), variable frequency (Chandrakasan & Brodersen, 2012). Moreover, variable power supply (Gonzalez, Gordon, & Horowitz, 1997), multiplied threshold (Ghosh & Ledwich, 2012), chip reduction and packaging size (Tummala, Rymaszewski, & Klopfenstein, 2012), design improving techniques, power management techniques (Seo, Lee, Lee, & Kang, 2015) are also used for low power design. In (Yadav & Pandey, 2015) designed the low power half adder cell using threshold based CMOS design. (Xu, Gu, Wu, & Chang, 2015) developed the 140 GHz transmitter using CMOS for data rate of 2.5 Gbps. The energy efficient design of optical transmitter and laser driver is also designed by (Abdullah etal., (2015), (Abdullah etal., (2015).

In this chapter, the authors validated the low power design using simulation and experimental results. The low power design for VLSI based system is achieved using IO Standard technique that varies the chip power during realization over UltraScale FPGA. In the next section, the basic terminologies are discussed. The basic terminologies related to this work are categorized in VLSI power consumption and its type, need of low power design in VLSI system and basic terms related to proposed technique.

VLSI POWER CONSUMPTION AND ITS TYPES

VLSI power consumption is described as power converted from the source into heat dissipation. There are mainly two types of VLSI power dissipation; one is static power and other is dynamic power (Shang, Kaviani, & Bathala, 2002). Static power dissipation produces the leakage current for the transistor when input is applied without any dependency on clock frequency or switching. Dynamic power produces the dissipation through the transistors switching and dependent on clock frequency and switching (Horowitz, Indermaur, & Gonzalez, 1994). When the VLSI design is implemented, the power further is classified in design power, device power and static power utilized in terms of clock, IO, signal and leakage power for the designed system (Shang et al., 2002). VLSI power is described in Equation (1) and (2) (Yadav & Pandey, 2015);

$$P_{Total} = P_S + P_{SC} + P_L \tag{1}$$

PTotal is total average power, *Ps* is switching power, *Psc* is short circuit power, *PL* is leakage power of transistor. These are further designed as in Equation (2)- (4) (Bellaouar and Elmasry, 2012) ;

$$P_S = V^2 f_c \tag{2}$$

$$P_{SC} = VI_{SC} \tag{3}$$

$$P_l = VI_L \tag{4}$$

where, V is applied voltage, Isc is short circuit current, *IL* is leakage current of transistor and *fc* is clock frequency (Yadav & Pandey, 2015). VLSI power consumption of the electronic devices is also depending upon the operating frequency and the voltage applied to the device as describes in Equation (5) (Yadav & Pandey, 2015);

$$P_s = CV_{Vdd}^2 * f \tag{5}$$

In the next section, the need of low power consumption in VLSI based design system is discussed.

Need of Low Power Consumption in VLSI Design Systems

VLSI based design system can be achieved for different applications such as; telecommunication, automation and in etc. These VLSI design systems dissipate large amount of power even programmed on chip (Bacchus, Nguyen, & Benedict, 2014). This large heat dissipation increases the junction temperature, density of device and clock frequencies of the chip. To solve these problems, larger heat sinks are mounted on chip (Bacchus et al., 2014). Even installing heat sink is not solution, the VLSI design must be configured in such a way that when it is practically realized must consumes less power.

In particular case study of high speed communication systems design using VLSI design require low consumption. However many reported that only designing the VLSI based high speed communication system may not reduce the power consumption as much (Bellaouar & Elmasry, 2012).

To cater this, some additionally techniques must be used with the low power design of VLSI based high speed communication system. Recently, the interest of low power VLSI design have been shifted from low power IC design technique to other technique that provides low power at the time realization. For example, many IC's now a days come in already compact form of 16- 28 nm chip size. The VLSI design using these nm chips size and additional technique utilization can reduce up to 40% power consumption (Santarini, 2013). Figure 1 describes the power consumption status for VLSI design in 28 nm chip size packed in UltraScale FPGA.

It is illustrated that VLSI design using UltraScale consumes very less power for realization compare to other methods (Santarini, 2013). Instead of working on VLSI IC design systems the design system should be realized in UltraScale FPGA like devices will produce the low power design VLSI based systems and the power consumption issues can be resolved (Santarini, 2013). In the next section basic terms related to this work are discussed.

Basic Terminologies Related to Research Work

The advantage of low power design using nm chip size in UltraScale FPGA are already discussed. It is analyzed that VLSI IC design, alone itself is it able to produce low power during realization. Here some of the basic terms related to proposed technique are explained.

Figure 1. On chip power consumption in UltraScale FPGA power consumption
Hussein et al. (2011)

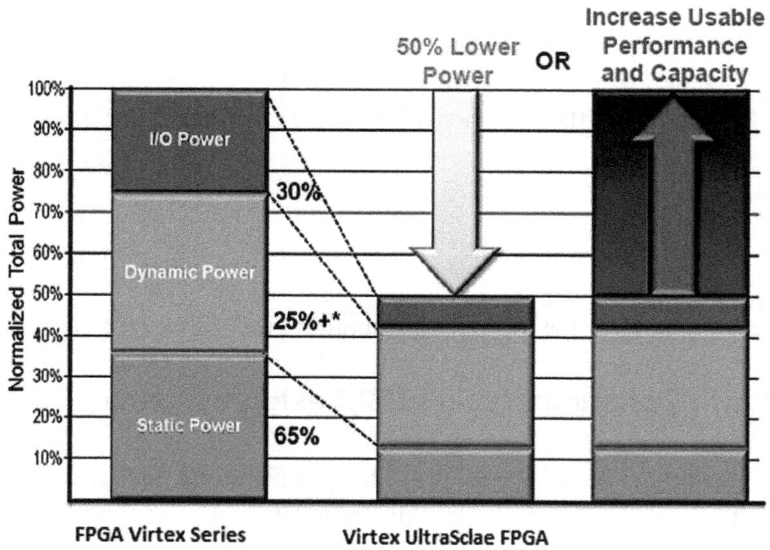

Field Programmable Gate Arrays (FPGA) is an example of programmable logic can perform any number of function, provide the reuse of the devices at any time and can be tested in live circuit (Trimberger, 2012). FPGA is 2D array of configurable logic blocks (CLBs) that implements the combinational logic or sequential logic (Trimberger, 2012). Figure 2 shows the internal architecture of FPGA. FPGA have mainly three blocks; logic blocks, routing and Input/ Output (I/O) Block (Maxfield, 2004). Advanced FPGA may also include the On-chip Memory, Embedded Processor, Clock Management and High-Speed Transceiver inside FPGA (Tessier, Pocek, & DeHon, 2015).

There are several manufactures that are producing the System on Chip (SoC) development board. Among these manufacturers, Xilinx is world class leader in producing high performance FPGA devices using 16-28 nm chip size (Santarini, 2013). Today, the UltraScale FPGA are packed with 16-28 nm chip size that consumes very less power (Santarini, 2013). Due to address of the massive I/O, memory, bandwidth requirements of next generation applications while efficiently routing and processing the data brought on-chip (Leibson & Mehta, 2013). The UltraScale architecture provides fascinating choice for designing the high-speed and high performance VLSI design system at low power consumption (Dorsey, 2010). The main purpose for which this devices is selected for this research work is that it has wide verity of IO Standards that re used to reduce the chip power and enables the high performance for VLSI designed at low power consumption. In the next section, the terms related to proposed IO Standard technique are discussed.

IO Standard

IO standard is an enhanced development in 28 nm chip packed in UltraScale (Santarini, 2013). IO Standard specifies the signal termination using electrical characteristics (Santarini, 2013). It provides the many control features such as low power consumption, high performance design, temperature efficient and etc. for VLSI based design systems. Before applying the IO standard technique for VLSI based systems

Figure 2. Internal architecture of FPGA
Maxfield (2004)

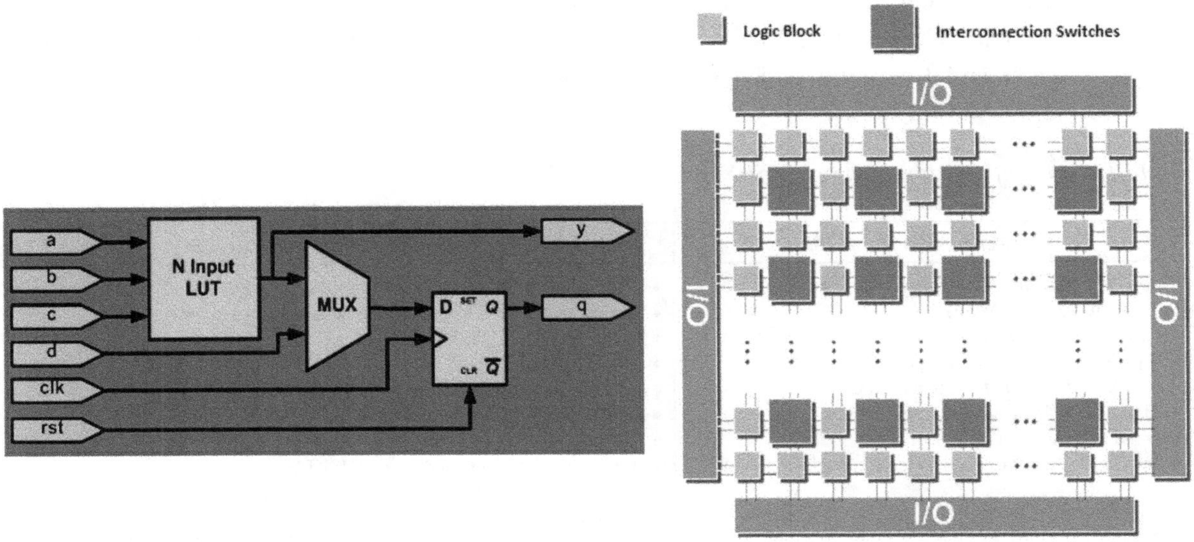

some basic features of IO Standards are discussed. In UltraScale FPGA an I/O tile consists of I/O buffers, input/output logics and input output programmable delays (Santarini, 2013). Each IOB contains both input and output logic and IO drivers. These drivers can be configured to various I/O standards. Each IOB has a direct connection to an ILOGIC/OLOGIC pair containing the input and output logic resources for data and 3-state control for the IOB (Xilinx, Inc 2015). The IO Standard has following feature;

1. **Input/Output:** In UltraScale, I/O pins are used for communicating with external components. I/Os are classified in high-range (HR), high-performance (HP) and high-density (HD). The HR I/Os support voltage from 1.0 V to 3.3 V, HP I/Os supports 1.0 V to 1.8 V. The I/O bank requires the particular design requirements before selecting the IO Standard (Aggarwal, 2014).
2. **I/O Electrical Characteristics:** The FPGA supports the single-ended and differential-ended outputs. Each single-ended or differential-ended are configured in such a way that single-ended uses CMOS push/pull, using VCCO and ground and differential ended uses a weak pull-up/down resistor for signal termination with a 100Ω internal resistor (Xilinx, Inc 2015).
3. **I/O Logic:** The I/O logic consist of IO delay, serializer and deserializer for IO Standard (Xilinx, Inc 2015).
    a. **Input and Output Delay**: In IO Standard, input and output logics can be configured as either combinatorial or registered logic with double data rate (DDR) operation and individually delay up to 1,270 ps at resolution of 5–16 ps called IDELAY and ODELAY. The volume can be doubled in single direction (Kadam & Sawarkar, 2014).
    b. **ISERDES and OSERDES**: IO standard coines high-speed, bit-serial I/O with slower parallel operation that requires a serializer and deserializer (SerDes) inside the I/O logic. Each I/O pin owns an IOSERDES (ISERDES and OSERDES) capable of performing serial-to-parallel or parallel-to-serial conversions with programmable widths of 2, 4, 8 or 16 bits. These I/O logic features enable high-performance interfaces such as Gigabit Ethernet/1000BaseX/SGMII from the transceivers to the SelectIO interface (Quinn, 2013).
    c. **SelectIO Resources**: In IO Standard the IO resources are configurable to achieve high-performance at low power consumption. SelectIO drivers and receivers, supporting a wide variety of standard interfaces. The robust feature set includes programmable control of output strength, slew rate and on-chip ability to generate a reference voltage (INTERNAL_VR) internally (Xilinx, Inc. 2014). Table 1 illustrates the types of IO Standards based on single ended and differential ended Select IO pins.

Table 1. Types of IO standard configure by select IO pins

Single-Ended I/O Standards	Differential I/O Standards
• Low Voltage Complementary Metal Oxide Semiconductor (LVCMOS) • Low Voltage Transistor-Transistor Logic (LVTTL) • High Speed Transceiver Logic (HSTL) • Stub-Series Terminated Logic(SSTL) • High Speed Unterminated Logic (HSUL) • Pseudo Open Drain (POD) • Low Voltage digitally controlled Impendence (LVDCI)	• Low Voltage Differential Signaling(LVDS) • Mini Low Voltage Differential Signaling (Mini-LVDS • Low Voltage CMOS (LVCMOS) • Subset LVDS (SUB-LVD) • Reduced Swing Differential Signaling) (RSDS) • Point-to-Point Differential Signaling (PPDS) • Transition Minimized Differential Signaling (TMDS) • High Speed Unterminated Logic (HSUL) • Low-voltage positive/pseudo emitter-coupled logic (LVPECL) • Differential HSTL

(Xilinx, Inc. 2014)

The IO standards require several parameters to be set before defining in VLSI based design. Such as voltages, attributes and etc. some of them are discussed as;

1. **VCCO**: I/O circuitry is powered using VCCO supply. The (Xilinx.Inc, 2014) shows the requirements of VCCO for each supported I/O standards for inputs, outputs termination circuit. VCCO for HP I/O bank must be same for board voltage. Inappropriate VCCO voltages can result in loss of functionality or damage to the device.
2. **VR**: VR is input reference voltage for Single-ended I/O standards. When VR is required within an I/O bank, the dedicated VR pin can be used as a VR supply input. UltraScale FPGA also have an internally generated reference voltage that is enabled by using the INTERNAL_VR constraint (Xilinx, Inc. 2014).
3. **VCCAUX**: The auxiliary (VCCAUX) supply provides power to the interconnect logics of various blocks inside the device that power's the input buffer circuits for some of the I/O standards in the range of 1.8 V to 2.2 (Xilinx, Inc. 2014).
4. **UltraScale FPGA SelectIO Attributes/Constraints**: UltraScale FPGA provide the access for I/O resource features such as location constraints, input delay, output drive strength, slew rate. using IO attributes/constraints (Xilinx, Inc. 2014).
5. **PACKAGE_PIN Constraint**: The PACKAGE_PIN constraint is used to stipulate the I/O location of an external port identifier. The values depend upon the package size. The syntax of defining PACKAGE_PIN attribute in the XDC file is (Xilinx, Inc. 2014); set_property PACKAGE_PIN pin_name [get_ports port_name]
6. **IOSTANDARD Attribute**: The IOSTANDARD attribute is used to select values for an I/O standard for I/O buffers. UltraScale FPGA is divided in HR and HP IO bank. The various IO standards support HR, HP, or both) by UltraScale device are discussed in (Xilinx, Inc. 2014). The syntax of defining IOSTANDARD attribute in the XDC file: set_property IOSTANDARD value [get_ports port_name]
7. **IBUF_LOW_PWR Attribute**: This attribute allows an optional trade-off between performance and power. This attribute is set to TRUE by default (Xilinx, Inc. 2014). The IBUF_LOW_PWR attribute is defined in the XDC file by syntax: set_property IBUF_LOW_PWR TRUE|FALSE [get_ports port_name]
8. **Output Slew Rate Attributes**: This attributes used to select the desired slew rate SLOW, MEDIUM (HP I/O banks only), or FAST for I/O output buffers (Xilinx, Inc. 2014). The syntax of defining the SLEW attribute in the XDC file is: set_property SLEW value [get_ports port_name]

The above terms are used in designing the low power design of VLSI based high speed communication systems implemented on UltraScale FPGA. There are several IO Standard are available in (Xilinx, Inc. 2014). The selection of IO Standard depends upon the requirements of VLSI systems. These IO Standards are specified in the Electronic Industry Alliance JEDEC (Cámara Nebreda, 2015). The IO Standard selected for case study VLSI based communication systems are as follows;

- LVDCI (Low-Voltage Digitally Controlled Impedance)
- LVCMOS (Low Voltage CMOS)
- HSTL (High-Speed Transceiver Logic)
- HSUL_12 (High Speed Unterminated Logic)

The reason of choosing these IO Standard is that these IO Standards falls in high performance IO bank at low power consumption. The main aim for this research is that to select a low power IO Standard for VLSI based high speed communication system. In the next section, the methodology for designing the high speed communication system using VLSI design is discussed.

METHODOLOGY FOR DESIGNING THE LOW POWER DESIGN FOR VLSI BASED HIGH SPEED COMMUNICATION SYSTEM

The high speed communication system is realized over 28 nm chip packed in UltraScale device. The VLSI based design for communication system is realized using flow chart discussed in Figure 3 It is illustrated that first the communication system is developed using Matlab Simulink model. After the successful development of Simulink model, the next step is developed the HDL design for communication system. The HDL design is imported in Xilinx suite 14.7 using vhdl language. This will create the test bench for the designed system. At this stage, the VLSI based high speed communication system is developed. After that proposed IO Standard technique is applied. The power analysis for VLSI based communication system is demonstrated. The power analysis yields the low power IO Standard for designed system. Finally, the VLSI based communication system is realized 28 nm chip packaged in UltraScale FPGA.

It is important to note that communication system is taken as case study for designing the low power VLSI system design using proposed technique. Like this several VLSI designed system can be converted in low power VLSI design systems.

Figure 3. Methodology for designing the low power design of VLSI based high speed communication system

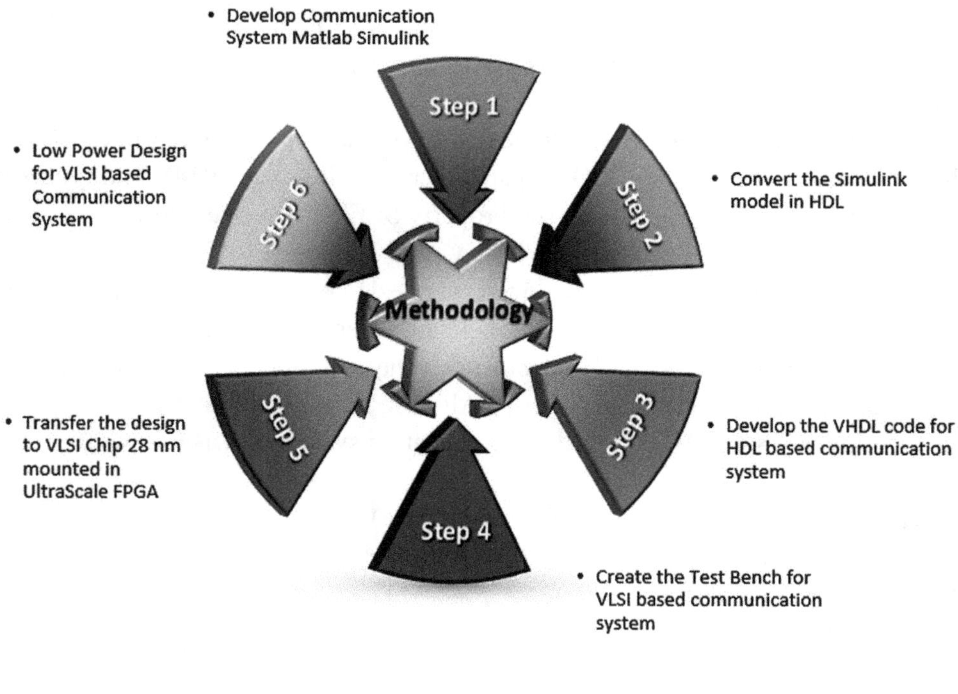

Development of Communication System Simulink Model

The high speed communication system is taken as case study to demonstrate the low power VLSI design system for designed system using proposed technique on 28 nm chip. The first step is to realize the communication system in Matlab using Simulink model is discussed here.

The 100Gb/s Differential Phase Shift Keying communication system is designed using different steps presented in Figure 4. In the first step, 100Gb/s DPSK signal is generated and in the second step, 100Gb/s DPSK signal is modulated using Non-Return-Zero (NRZ) scheme. In the third step, the 100Gb/s modulated signal is travelled over 100 Km of AWGN channel. In the last step signal is received using demodulation technique.

Figure 5 describes the Matlab Simulink model of 100Gb/s DSPK communication system. The system is designed by generating the pseudo random binary sequence (PRBS) signal using Bernoulli Binary Generator. This PRBS signal is converted in DPSK format by exclusive or operation. This DPSK signal is modulated by mixing the 1550 nm laser signal in it using NRZ scheme. This 100Gb/s modulated signal is transferred to AWGN channel for the transmission distance of 100 Km. The signal is received using demodulation technique. This Simulink model is further defines in different sub models such as 100Gb/s DPSK transmitter, NRZ modulator, AWGN channel and receiver.

Figure 4. Design steps of 100 Gb/s DPSK communication

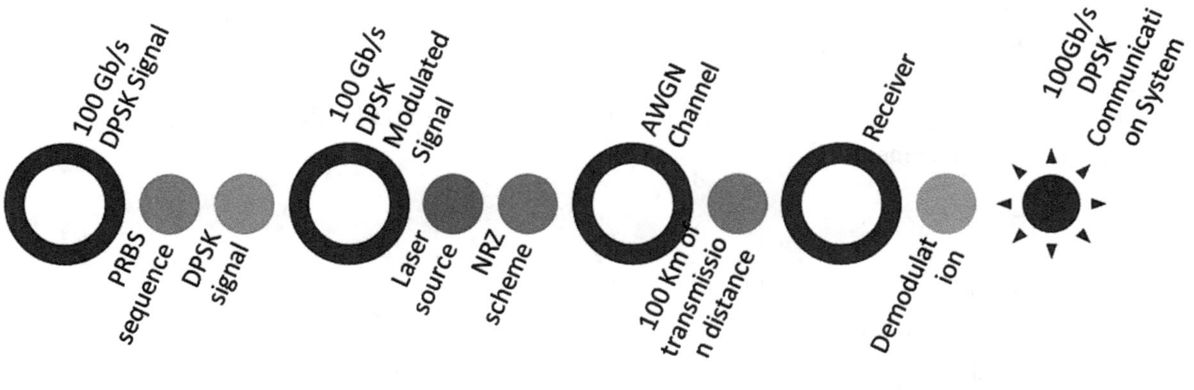

Figure 5. Simulink model of high speed 100Gb/s DPSK communication

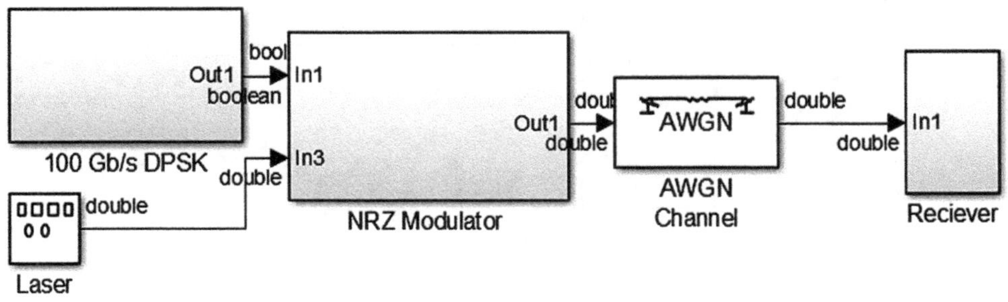

1. **100Gb/s DPSK Transmitter:** DPSK signal converts a binary signal information in terms of phase related information. Usually, binary "0"s produce phase shift of 0° and binary "1"s produce phase shift of 180° (Gnauck, & Winzer, 2005). DPSK signal principal uses one bit delay due to partial detection technique. DPSK signal is produced by applying differential encoding scheme to Phase Shift Keying (PSK) signal (Gnauck, & Winzer, 2005).. DPSK signal generator block realized in MATLAB as outlined in Figure 6.

PRBS sequence is XOR with one bit delay, which states that during transmission binary "0" produces no or 0° phase changes and for binary "1" π or 180° phase change is occurred for each single bit. The generated DPSK signal has 0.2 probability of Zero and the sample rate is of 100Gb/s. This 100GB/s DPSK signal is mixed with 1550 nm laser signal for NRZ modulator. The modulated signal is transferred to AWGN channel.

2. **NRZ Modulator:** The NRZ modulated signal is generated by integrating the 100Gb/s DPSK signal and laser signal using NRZ scheme. In this work, NRZ scheme is used as it delivers the noise resilience during transmission (Breuer & Petermann, 1997). The NRZ modulator provides the phase swing between optical carrier (laser signal) and 100Gb/s DPSK signal. This modulated signal is passed to phase swing block that produces the constructive interference for the signal (Breuer & Petermann, 1997). One input of modulator is 100Gb/s DPSK signal and the other is carrier signal. Figure 7 defines the MATLAB Simulink model for modulating the 1000Gb/s DPSK signal and carrier laser signal in which In1 represents the laser signal and In3 represents the 100Gb/s DPSK signal. NRZ technique modulates the phase of signal in conjunction with amplitude and phase of DPSK signal and optical carrier.

The main advantage of using the NRZ modulation is that it removes the noise from signal before transmitting it over channel.

Figure 6. DPSK signal from Bernoulli binary sequence

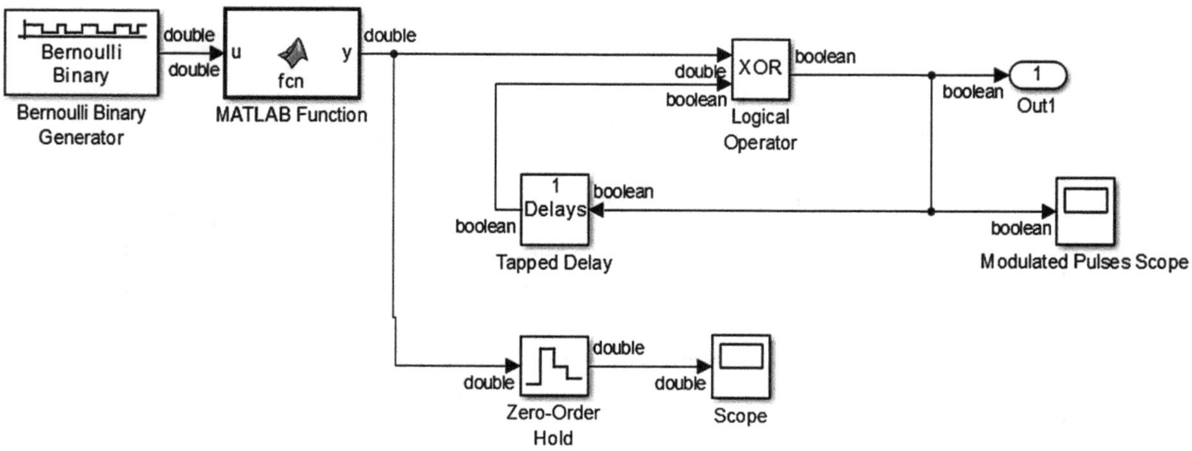

Figure 7. NRZ modulation for 100Gb/s DPSK signal

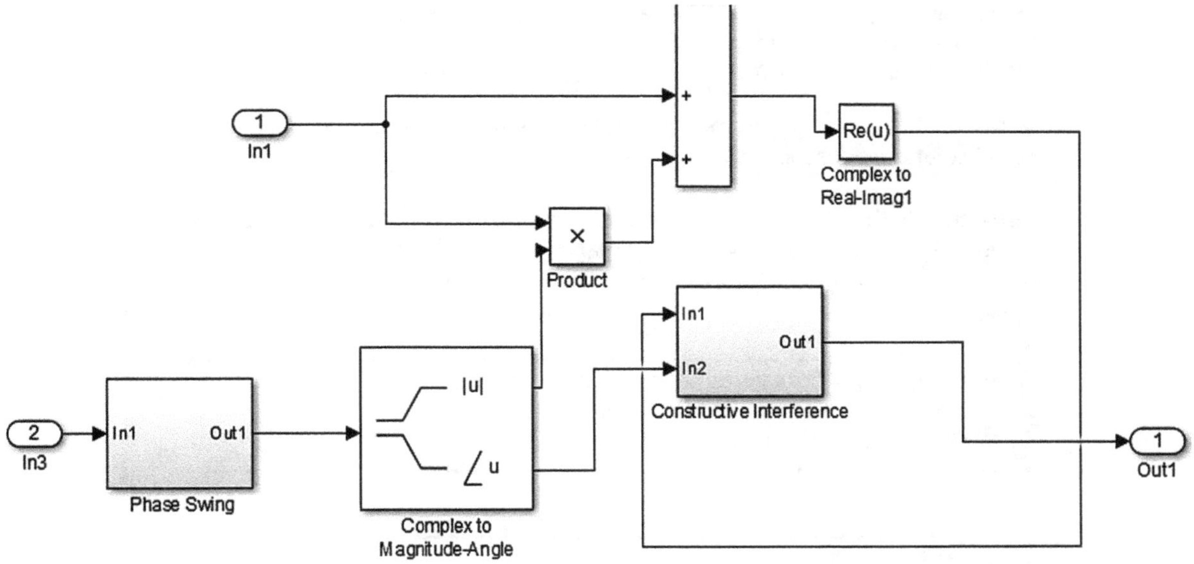

3. **Additive White Gaussian Noise Channel:** The AWGN Channel block adds white Gaussian noise to input signal. This block inherits its sample time from the input signal (Poltyrev, 1994). The channel provides the transmission of signal at specific distance, the channel also facilitates to transmit the signal with the inclusion of noise.

Figure 8 describes the parameters to be selected for configuring the AWGN channel for communication system. The response of DPSK transmitted signal over AWGN channel is analyzed using receiver model.

4. **DPSK Receiver:** DPSK receiver detects and demodulates the regenerated signal (Chinn, Boroson, & Livas, 1996). The receiver is modeled by mixing the local oscillator signal with information carrier that converts information signal to the intermediate frequency range (Chinn, Boroson, & Livas, 1996). Matlab Simulink based DPSK receiver model function is demonstrated in Figure 9. The local oscillator signal is mixed with the signal using optical coupler. The photodetector converts light wave signal in electronic signal and after that signal is demodulated.

In demodulation each tapped delay is simulated by a set of two "angles" blocks depending upon magnitude. The output from photodetector is passed through XOR with carrier signal to extract the carrier and receive the transmit signal. The delay block stores the phase of previous symbol. The XOR block extracts phase and amplitude of present and previous symbol. After that, signal is passed through low pass filter to remove the higher-order harmonics of the nonlinear photo detection process and other noises. The DPSK receiver is conventional asynchronous detection type, therefore the decoder is not necessary.

5. **Setting Up Simulink HDL Coder Configuration for 100Gb/s DPSK Communication System:** HDL Coder in Matlab creates synthesizable vhdl codes for Matlab Simulink models (Sbarcea & Nicula, 2004). The generated HDL code can be used for FPGA programming and prototype de-

Figure 8. AWGN channel in Matlab Simulink model

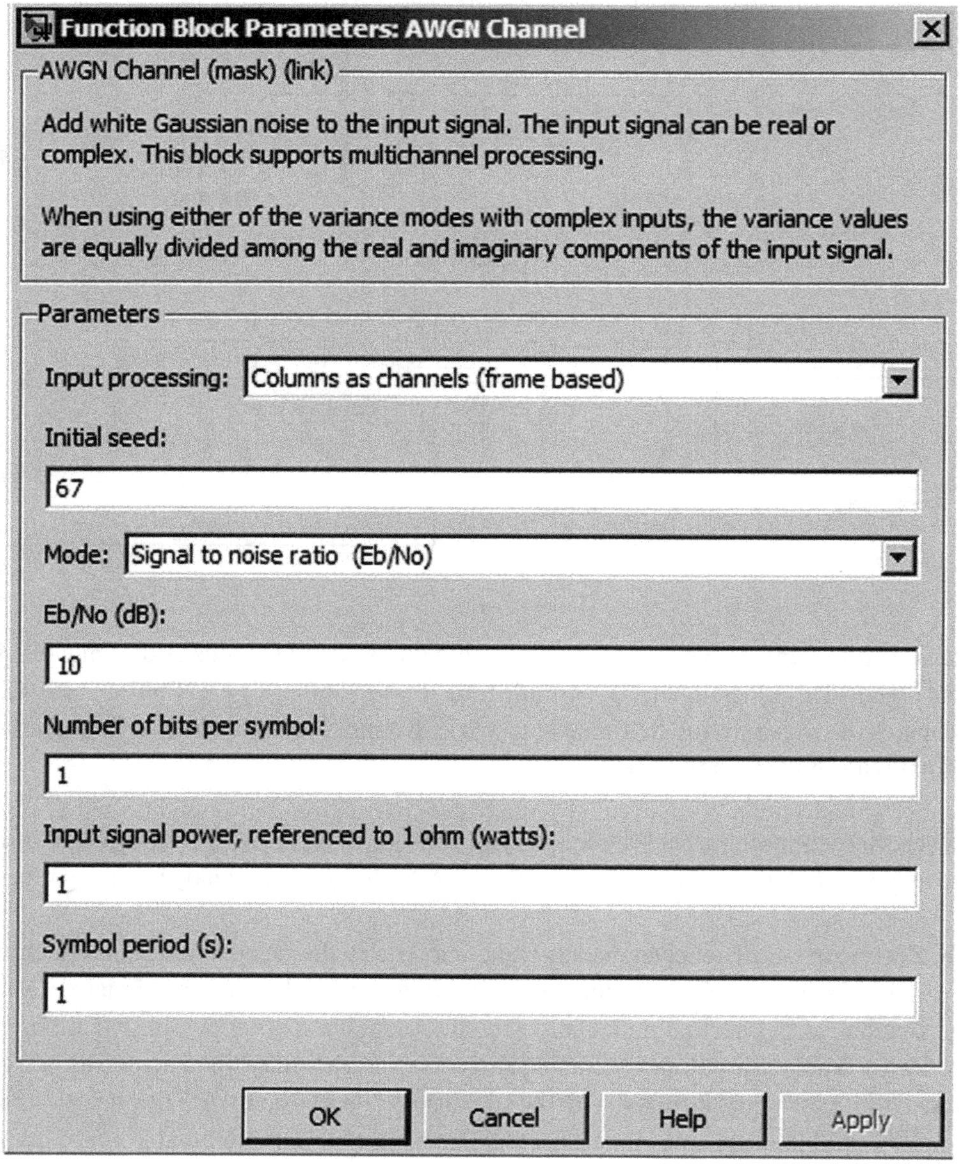

sign. It also offers a workflow advisor that presets the programming of Xilinx FPGAs. The HDL coder provides the control of hardware resource and traceability between Simulink model and the generated verilog and vhdl code (Sbarcea & Nicula, 2004). Figure 10 describes the method of converting Simulink in HDL code generation in vhdl for 100Gb/s DPSK communication system and also transfer of program in chip.

Figure 9. Receiver model

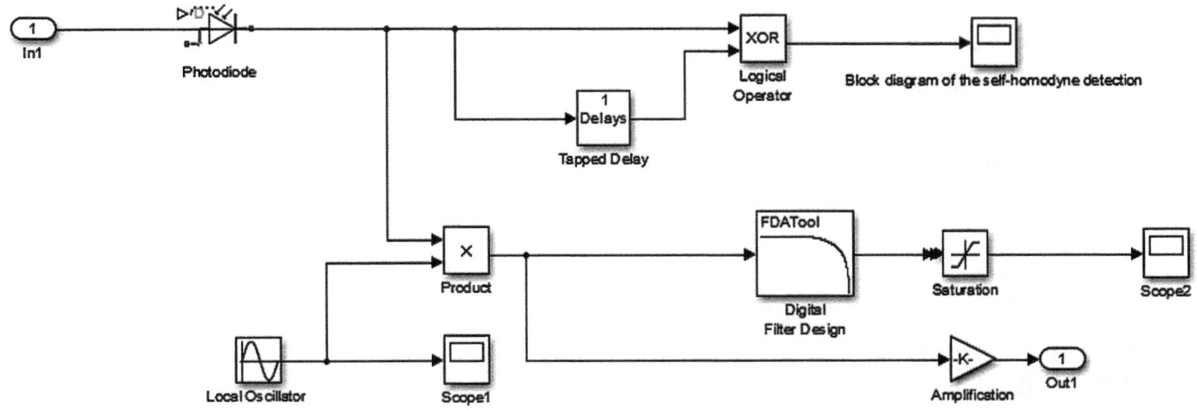

Figure 10. HDL code generation of Simulink model
Sbarcea and Nicula (2004)

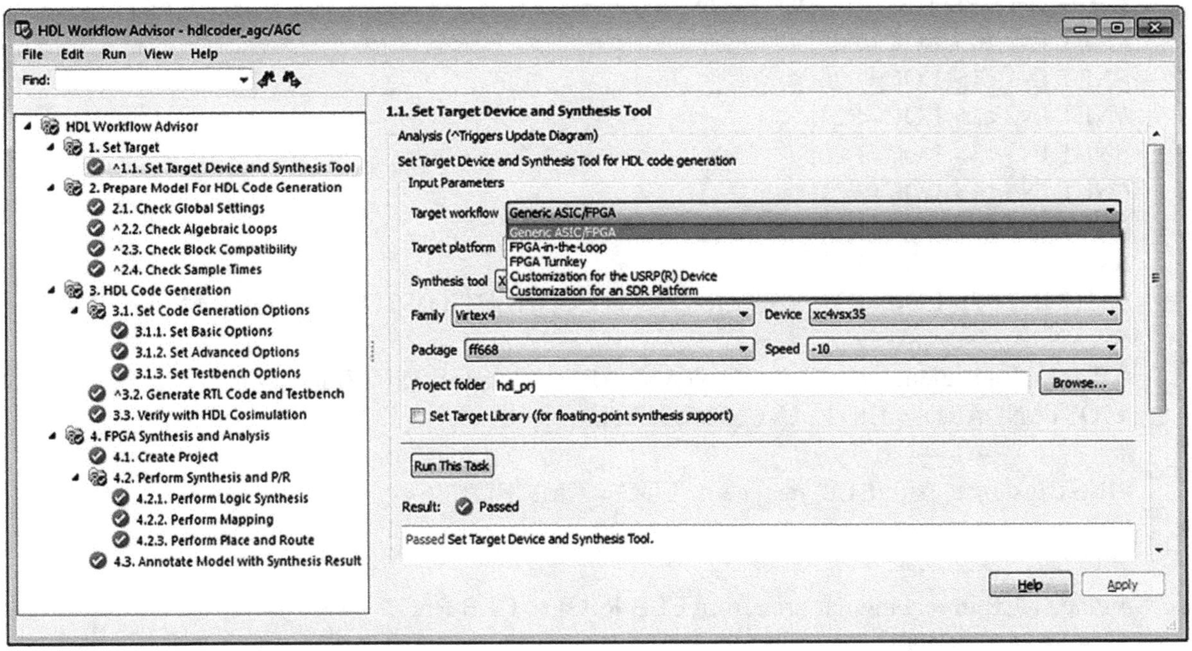

6. **VHDL Code for 100Gb/s DPSK Communication System:** The vhdl code is develop for designing the VLSI based communication system. The developed code is executed on Xilinx 14.7 suite for generating the test bench for VLSI based communication system. The VLSI based code for communication system is defined as;

```
#####################################################################
# ATTRIBUTE and LOCATION SPECIFICATIONS #
#####################################################################
# CLB and Pin location locking constraints #
```

```
##################################################################
#
# -----------------------
# IO pin number assign
# -----------------------
#INST io_buf_instance_name LOC = P11 ;
#NET io_net_name LOC = P110 ;
#
# -----------------------
# Assign a signal to a range of I/O pins
# -----------------------
#NET "signal_name" LOC=P62, P63, P64;
##################################################################
# CLOCK FROM EXTERNAL OSCILLATOR #
##################################################################
//NET CLK LOC=P13;
# ===== LED DRIVER OUTPUTS =====
//NET IN0<0> LOC=P25;
//NET IN1<1> LOC=P26;
//NET IN2<2> LOC=P24;
//NET IN3<3> LOC=P20;
//NET IN4<4> LOC=P23;
//NET IN5<5> LOC=P18;
//NET IN6<6> LOC=P19;
//NET IN7<7> LOC=P30; # THERE IS NO CONNECTION TO THE DECIMAL-POINT
# -----------------------
# Place a logic element(called a IO STANDARD) in a specific CLB location.
# IOSTANDARD = HSTL, LVCMOS, HSUL...
# -----------------------
#INST instance_path/BEL_inst_name LOC = CLB_F17C36 ;
#
# -----------------------
# Place CLB in rectangular area from CLB R1C1 to CLB R5C7
# -----------------------
#INST /U1/U2/reg<0> LOC=clb_r1c1:clb_r5c7;
#
# -----------------------
# Place hierarchical logic block in rectangular area from CLB R1C1 to CLB R5C7
# -----------------------
#INST /U1* LOC=clb_r1c1:clb_r5c7;
##################################################################
# Data Bits From The Pc (Via Parallel Port) #
##################################################################
//NET PC_D0 LOC=P41;
```

```
//NET PC_D1 LOC=P42;
//NET PC_D2 LOC=P44;
//NET PC_D3 LOC=P45;
//NET PC_D4 LOC=P46;
//NET PC_D5 LOC=P47;
//NET PC_D6 LOC=P38; # MUST USE SPECIAL-PURPOSE PINS FOR
//NET PC_D7 LOC=P39; # ACCESSING PC_D<6> AND PC_D<7>
# -----------------------
# Prohibit IO pin P26 or CLBR5C3 from being used:
# -----------------------
#CONFIG PROHIBIT = P25 ;
#CONFIG PROHIBIT = CLB_R6C4 ;
#Config Prohibit is very important for forcing the software to not use critical
# configuration pins like IN and OUT on the FPGA. The Mode pins and JTAG
# Pins require a special pad to constraint
#################################################################
# Data Bits From Pc (Parallel Port) #
#################################################################
//NET PAR<0> LOC=P71;
//NET PAR<1> LOC=P72;
//NET PAR<2> LOC=P64;
//NET PAR<3> LOC=P68;
#################################################################
# RAM….. pins #
#################################################################
//NET OE_ LOC=P81; # ACTIVE-LOW OUTPUT ENABLE
//NET CE_ LOC=P85; # ACTIVE-LOW CHIP ENABLE
//NET WE_ LOC=P82; # ACTIVE-LOW WRITE ENABLE
//NET RA<0> LOC=P88; # RAM'S ADDRESS PINS
//NET RA<1> LOC=P89;
//NET RA<2> LOC=P72;
//NET RA<3> LOC=P74;
//NET RA<4> LOC=P5;
//NET RA<5> LOC=P7;
//NET RA<6> LOC=P70;
//NET RA<7> LOC=P76;
//NET RA<8> LOC=P78;
//NET RA<9> LOC=P79;
//NET RA<10> LOC=P73;
//NET RA<11> LOC=P41;
//NET RA<12> LOC=P5;
//NET RA<13> LOC=P7;
```

```
//NET RA<14> LOC=P1;
//NET RD<0> LOC=P11; # RAM'S DATA PINS
//NET RD<1> LOC=P60;
//NET RD<2> LOC=P61;
//NET RD<3> LOC=P51;
//NET RD<4> LOC=P50;
//NET RD<5> LOC=P49;
//NET RD<6> LOC=P48;
//NET RD<7> LOC=P45;
# -----------------------
# Assign an OBUF to be FAST or SLOW:
# -----------------------
#INST obuf_instance_name FAST ;
#INST obuf_instance_name SLOW ;
#
# -----------------------
# FPGAs only: IOB input Flip-flop delay specification
# -----------------------
# Declare an IOB input FF delay (default = MAXDELAY).
# NOTE: MEDDELAY/NODELAY can be attached to a CLB FF that is pushed
# into an IOB by the "map -pr i" option.
#INST input_ff_instance_name MEDDELAY ;
#INST input_ff_instance_name NODELAY ;
#
# -----------------------
# Assign Global Clock Buffers Lower Left Right Side
# -----------------------
# INST gbuf1 LOC=SSW
NET "a" LOC=P30;
NET "b" LOC=P33;
NET "c_in" LOC=P34;
NET "sum" LOC=P76;
NET "c_out" LOC=P82;
################################################################
# IO Pins #
################################################################
//NET XTAL1 LOC=P27; # INPUT CLOCK
//NET RST LOC=P56; # ACTIVE-HIGH RESET
################################################################
```

7. **Test Bench of Simulink based High Speed DPSK Communication System:** The VHDL code for developed communication system is executed on Xilinx suite 14.7. The FPGA generic flow shown in Figure 11 describes the process of converting the Matlab Simulink model in to Xilinx design.

Figure 11. FPGA design flow for Matlab based communication design

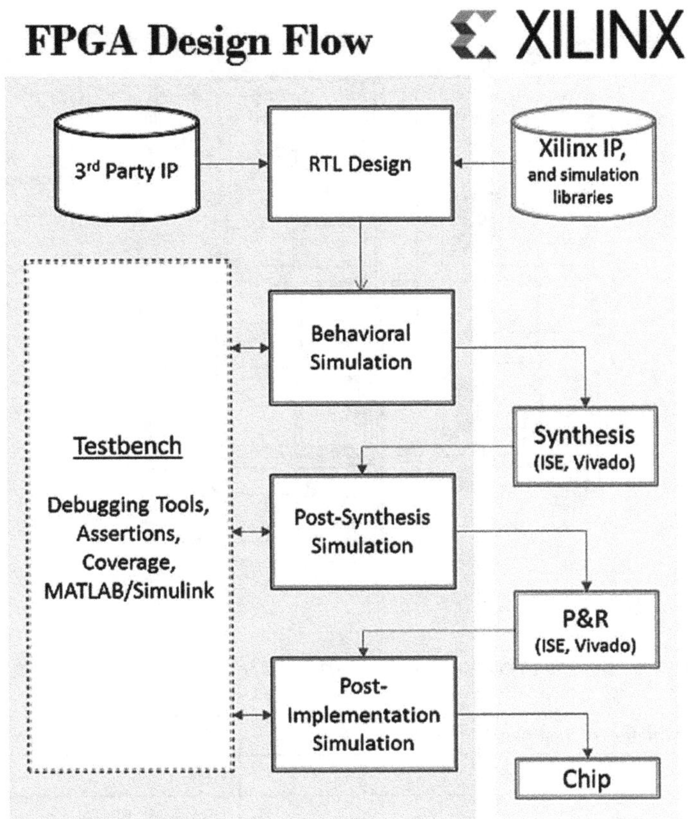

The test bench for designed VLSI based communication system requires the parameters for designing at UltraScale FPGA. Figure 12a shows the frame buffer of 100Gb/s DPSK system where, sys_clk is clock frequency, uart_tx and uart_rx are universal asynchronous transmitter and receiver respectively. Figure 12.b describes RTL schematic diagram for the VLSI based designed communication system. The LEDs are used for to monitor the 8 bit 100Gb/s transmitted signal, it also contains the Latches, input and output buffers and memory interfaces.

It is very important to configure the UltraScale device before implementing the VLSI design over device. This configuration requires the selection of several parameters such as device speed, programming language, simulator, voltage supply, environmental temperature and others. Table 2 describes the parameters for test bench of VLSI based designed system.

POWER REDUCTION AND ANALYSIS FOR 100Gb/s DPSK COMMUNICATION SYSTEM USING IO STANDARD TECHNIQUE

The UltraScale FPGA IO Standards and their basic characteristics are already discussed in basic terminologies section. In this section, the 100Gb/s DPSK communication system's test bench is tested over different IO Standards.

Figure 12. (a) Frame buffer for VLSI based 100Gb/s DPSK communication system; (b) RTL schematic diagram for VLSI based 100Gb/s DPSK communication system

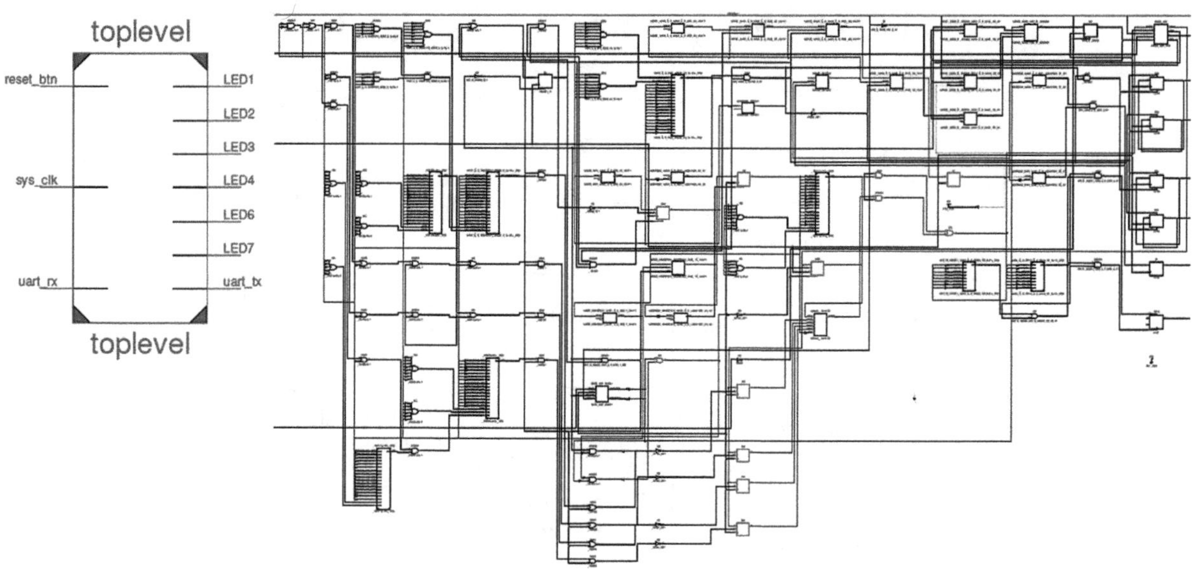

Table 2. Parameters for test bench creation of high speed DPSK communication system

Parameters of Test bench	Values
Top-level Source type	HDL
Evolution development board	UltraScale FPGA
Device	DS890
Speed	-3
Synthesis tool XST	VHDL
Simulator	Xilinx ISE
Heat sink	Medium profile
Supply voltage	1.5 v
Automatic FSM extraction	Yes
FSM encoding algorithm	Optimization algorithms
Clock	Buffer Clock of 1 GHz
IO standard	Support Various IO's
Ambient temperature	25°C

Power Reduction for 100Gb/S VLSI Based DPSK Communication System Using IO Standard Technique

These IO Standards are LVCMOS (Low Voltage CMOS), LVDCI (Low-Voltage Digitally Controlled Impedance), HSTL (High-Speed Transceiver Logic), HSUL (High-Speed Unterminated Logic) at voltage range of 1.8 V. These IO Standards are defined by (JEDEC, 2015) Standards.

The power reduction is achieved using IO Standard technique. The IO Standard technique is applied over designed system by defining the specific IO Standard syntax in VHDL code for designed VLSI based communication system. The power reduction is tested at different high carrier frequencies such as; 500 GHz (default carrier frequency), 900GHz, 10 THz and 17 THz. In the next section, the syntax for defining IO Standard in the VHDL code of designed is discussed. Furthermore, the power consumption is recorded for each IO Standard. This power consumption consist of design static power, device static power and device dynamic power is disused in (Xilinx, Inc. 2014), (Xilinx, Inc. 2014) for designed communication system.

1. **LVDCI (Low-Voltage Digitally Controlled Impedance) IO Standard:** LVDCI provides the controlled impedance for I/O buffers in 28 nm chip packed in UltraScale FPGA. The device supports the LVDCI_15, LVDCI_18 defined by (JEDEC, 2015). The 100Gb/s DPSK communication system test bench is tested over LVDCI_18 because this IO Standard support the 1.8 V, which is suitable for designed VLSI based communication system. The LVDCI syntax for VLSI based 100 Gb/s DPSK communication system is defined as;
   attribute IOSTANDARD: string;
   attribute IOSTANDARD of IDIOA0: label is- "LVDCI_18";

Table 3 describes the power ingesting for designed VLSI system using LVDCI_18 for different carrier frequencies of 500 GHz, 900GHz, 10 THz and 17 THz.

LVDCI IO standard is consuming total power of 6.1 W for 500 GHz, 7.05 W for 900 GHz, 9.2 W for 10 THz and 12.52 W for 17 THz. The chip power variation is recorded, when including the IO Standard in VHDL code of VLSI based communication. The detailed analysis is demonstrated in power analysis section.

2. **LVCMOS (Low Voltage CMOS) IO Standard:** LVCMOS IO Standard is specially used for high frequency and switching operation uisng 28 nm chip packed in UltraScale FPGA. LVCMOS supports LVCMOS12, LVCMOS15, LVCMOS18, LVCMOS25, and LVCMOS33 defined by (JEDEC, 2015). The designed VLSI based communication system is tested over LVCMOS18. The LVCMOS syntax for VLSI based system is shown as;
   attribute IOSTANDARD: string;
   attribute IOSTANDARD of IDIOA0: label is- "LVCMOS18";

Table 3. Power consumption of VLSI based communication system using LVDCI IO standard

Carrier Frequency→ Power Consumption ↓	500 GHz	900 GHz	10 THZ	17 THz
Device Static Power in W	1.2	1.4	1.95	2.7
Design Static Power in W	1.7	1.95	2.34	3.62
Device Dynamic Power in W	3.2	3.7	4.91	6.2
Total Power in W	6.1	7.05	9.2	12.52

Table 4. Power consumption of VLSI based communication system using LVCMOS IO standard

Carrier Frequency→ Power Consumption ↓	500 GHz	900 GHz	10 THZ	17 THz
Device Static Power in W	0.9	1.1	1.35	1.7
Design Static Power in W	1.1	1.2	1.74	2.6
Device Dynamic Power in W	2.5	2.94	3.81	4.5
Total Power in W	4.5	5.24	6.9	8.8

Table 4 defines the power consumption of VLSI based communication system at LVCMOS18 IO Standard. The power consumption is analyzed for for different carrier frequencies as discussed above. The power increased drastically, when frequency is increased because of Equation (5).

It is extracted that power consumption is reduced in comparison with LVDCI IO Standard. The power reduction is attained for LVCMOS IO Standard because it consumes less static and dynamic power than LVDCI IO Standard.

3. **HSTL_I (High-Speed Transceiver Logic) IO Standard:** The high-speed transceiver logic (HSTL) standard is used for a high-speed bus operation defined by (JEDEC, 2015). HSTL supports the high clocking, high-speed interfaces. UltraScale supports different class of HSTL for 1.2V (in HP I/O banks) and 1.5V, 1.8V (both HP and HR I/O banks). The VLSI based communication is demonstrated using 1.8 V i.e. HSTL_I_18. The syntax for defining the HSTL IO Standard in VHDL code of designed communication system is defined as;
attribute IOSTANDARD: string;
attribute IOSTANDARD of IDIOA0: label is- "HSTL_I_18";

The power consumption is recorded for designed VLSI based communication system as described in Table 5 at different carrier frequencies of 500 GHz, 900 GHz, 10 THz and 17 THz.

It is observed that for designed system using HSTL_I_18 consumes less power than LVCMOS and LVDCI for all carrier due to that HSTL supports the high-speed bus operation while LVCMOS and LVDCI IO Standard supports low bus operation. HSTL consumes less dynamic power than both LVCMOS and LVDCI IO Standard.

Table 5. Power consumption of VLSI based communication system using HSTL IO standard

Carrier Frequency→ Power Consumption ↓	500 GHz	900 GHz	10 THZ	17 THz
Device Static Power in W	0.4	0.6	0.7	1.0
Design Static Power in W	0.7	0.85	0.9	1.4
Device Dynamic Power in W	1.2	1.99	2.01	2.5

4. **High-Speed Unterminated Logic IO Standard:** The HSUL IO Standard is low power dynamic memory bus IO Standard distinct by (JEDEC, 2015). UltraScale support the HSUL and provides the optional untuned single termination feature offers a weak pull-up to VCCO. The syntax for defining the HSUL IO Standards for VLSI based system is described as;
attribute IOSTANDARD: string;
attribute IOSTANDARD of IDIOA0: label is- "HSUL_18";

The power consumption for designed VLSI based communication system is shown in Table 6. It is observed that HSUL IO Standard for designed system consumes less than rest of the IO Standards. This is because HSUL supports both HR and HP IO banks at high-speed bus operation with low power memory interface and furthermore, operates at low VCCO and low output impedance.

HSUL IO Standard provides the significant reduction in power consumption. For 500 GHz 87%, for 900 GHz 80%, for 10 THz 75.6% and for 17 THz 76% power reduction is achieved in comparison with LVDCI IO Standard that consumes more power.

Power Analysis for 100Gb/S VLSI Based DPSK Communication System Using IO Standard Technique

The power analysis is performed using XPA analyzer (Xilinx, Inc. 2012). XPower Analyzer (XPA) provides an analysis on real design data. XPA is precise tool that can read from the implemented design database the exact logic and routing resources used (Xilinx, Inc. 2012). Figure 13 defines the summary for power consumption report. XPA provides the facility to also allow to adjust environment settings to reduce power consumption. XPA is executed after designing the implementation in ISE, using Place & Route (PAR) (Xilinx, Inc. 2012). At the point in your design phase when XPA is used, your ISE project should have successfully completed Place & Route (PAR). Placed and routed design database contains all logic configuration and routing information such as physical constraint for all of the logic and I/Os design plus activity for specific nets such as clock networks (Xilinx, Inc. 2012).

The XPA analyzer provides detail analysis for on chip parameters such as; on chip power, temperature consumes by chip and power consumption for IO, leakage, logic and etc (Xilinx, Inc. 2012). The XPA analyzer is executed for test bench of designed VLSI based communication system using LVDCI, LVCMOS, HSTL and HSUL IO Standard.

Table 6. Power consumption of VLSI based communication system using HSUL IO standard

Carrier Frequency→ Power Consumption ↓	500 GHz	900 GHz	10 THZ	17 THz
Device Static Power in W	0.1	0.2	0.35	0.6
Design Static Power in W	0.2	0.36	0.75	0.95
Device Dynamic Power in W	0.5	0.89	1.21	1.45
Total Power in W	0.8	1.45	2.31	3

Figure 13. XPA analyzer
Xilinx, Inc. (2012)

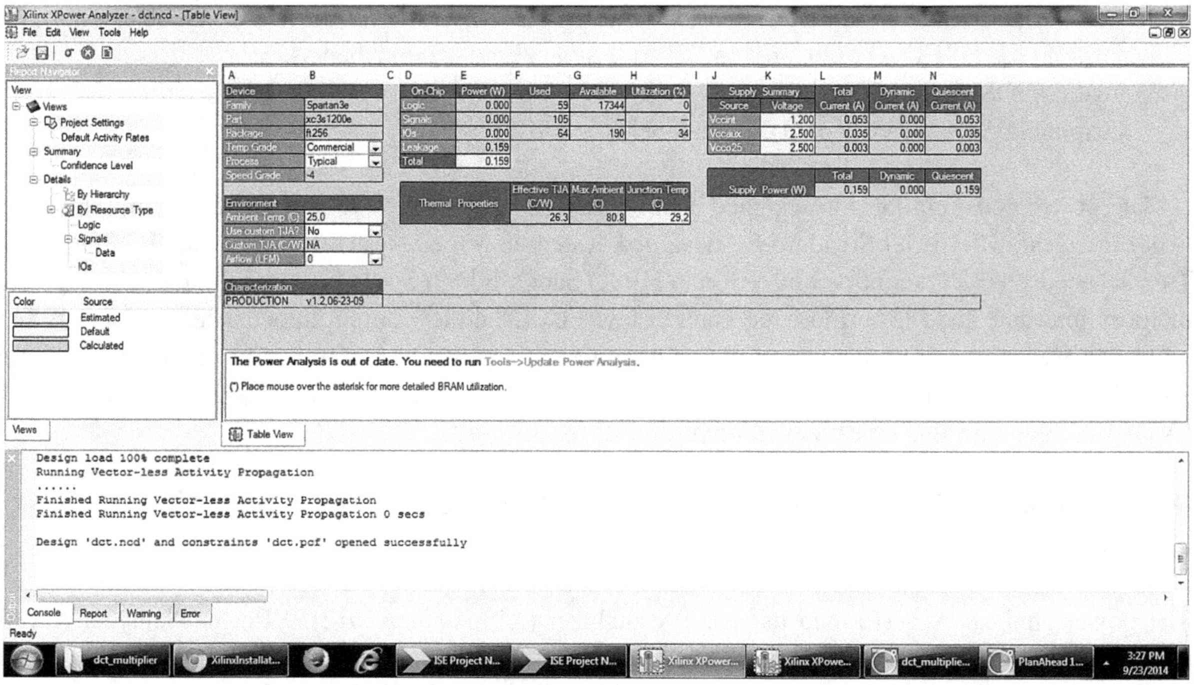

When the test bench of VLSI based communication is realized using LVDCI IO Standard, it is noted that leakage power is very high and temperature is around 36°C for operated voltage and maximum chip power is consumed is of 2.54 W as presented in Figure 14.

It is observed that execution of test bench at LVCMOS IO Standard for VLSI based communication has less leakage power than LVDCI IO Standard. Figure 15 states that chip temperature at operated IO Standard voltage is 30°C and maximum chip power is consumed is of 2.50 W. It is illustrated that suitable change in IO Standard produce reduction in power consumption of chip and also reducing the chip temperature.

Figure 14. XPA based power consumption analysis VLSI based designed system using LVDCI IO standard

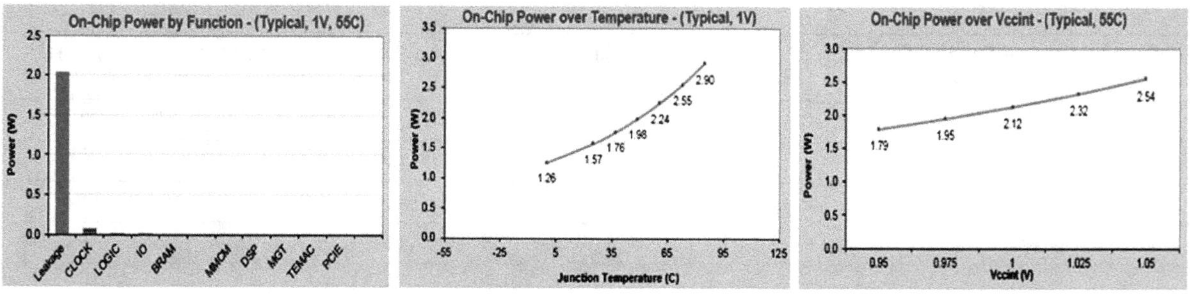

Figure 15. XPA based power consumption analysis VLSI based designed system using LVCMOS IO standard

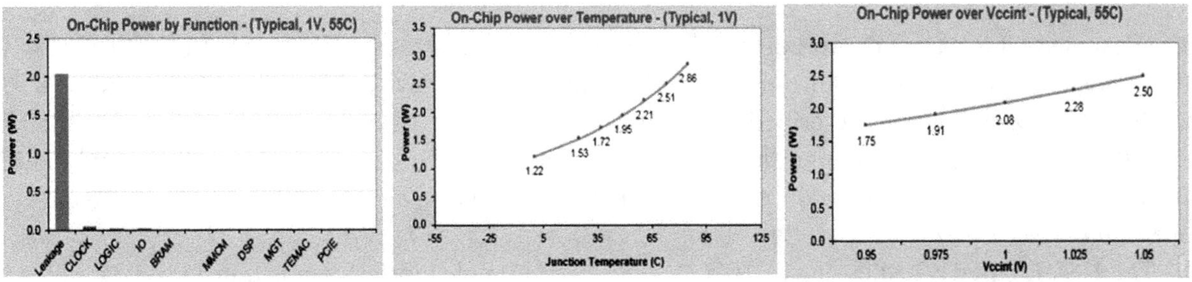

XPA analyzer execution for HSTL IO Standard based VLSI designed system yields less chip power consumption, leakage power and less chip temperature. It is described in Figure 16 that chip temperature is 29°C for operated IO Standard voltage and maximum chip power is recorded is of 2.34 W.

Finally, when designed VLSI system is demonstrated using HSUL IO standard the design consumes the chip power of 2.33 W at chip temperature of 28°C for operated IO Standard voltage as described in Figure 17. It is also noted that leakage power is also reduced up to 1.8 W. It is extracted that HSUL consuming very less chip than other IO Standard.

It is also determined that changing the suitable IO Standard is significantly reducing the power consumption for the designed VLSI system. In the next, hardware implementation for the designed system is realized over UltraScale FPGA.

Figure 16. XPA based power consumption analysis VLSI based designed system using HSTL IO standard

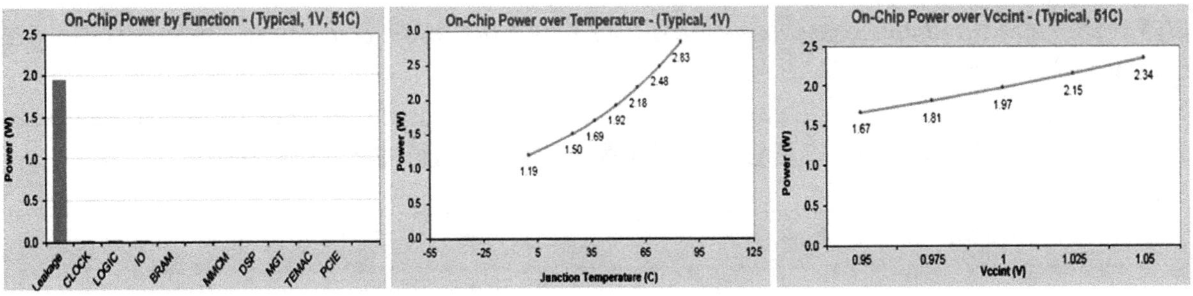

Figure 17. XPA based power consumption analysis VLSI based designed system using HSUL IO standard

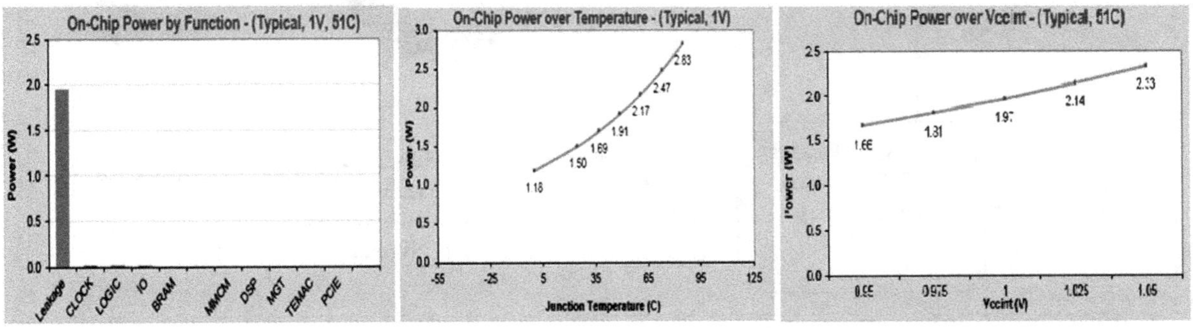

Hardware Implementation of VLSI Based 100GB/S DPSK Communication System over UltraScale FPGA Using Proposed Technique

The designed VLSI based DPSK communication system is realized over UltraScale FPGA. The HSUL IO Standard is selected for System on Chip (SoC) realization as shown in Figure xx. The design flow for hardware realization is also described in Figure 18. The 100Gb/s DPSK system is designed in Matlab after that using system C compiler the design is converted in HDL coding in Xilinx. In Xilinx, the IO standard technique is applied for developed system to reduce the power consumption for high frequency operation.

The HSUL based VLSI design of 100Gb/s DPSK communication system is transferred over 28 nm chip using programmer. The 100Gb/s DPSK system is also connected with hardware simulator using GCC complier. After that, 100Gb/s DPSK signal of 8 digit binary is transmitted over AWGN channel. The AWGN channel is configured for 100 km transmission distance. At the receiving end LED are connected as shown in frame buffer of RTL design of 100Gb/s DPSK communication system. The signal is received with low power consumption using UltraScale FPGA IO Standard technique.

RESULTS AND DISCUSSION

The Matlab based Simulink model of high speed PSK communication consumes 75 W. When the VLSI based design of communication system is realized using proposed IO Standard the significant power reduction is achieved. When the VLSI design is realized using IO Standards a significant power reduction is achieved comparing the power consumption of Simulink model. When the VLSI based design of

Figure 18. Hardware implementation of VLSI based 100Gb/s DPSK communication system using IO standard over UltraScale FPGA

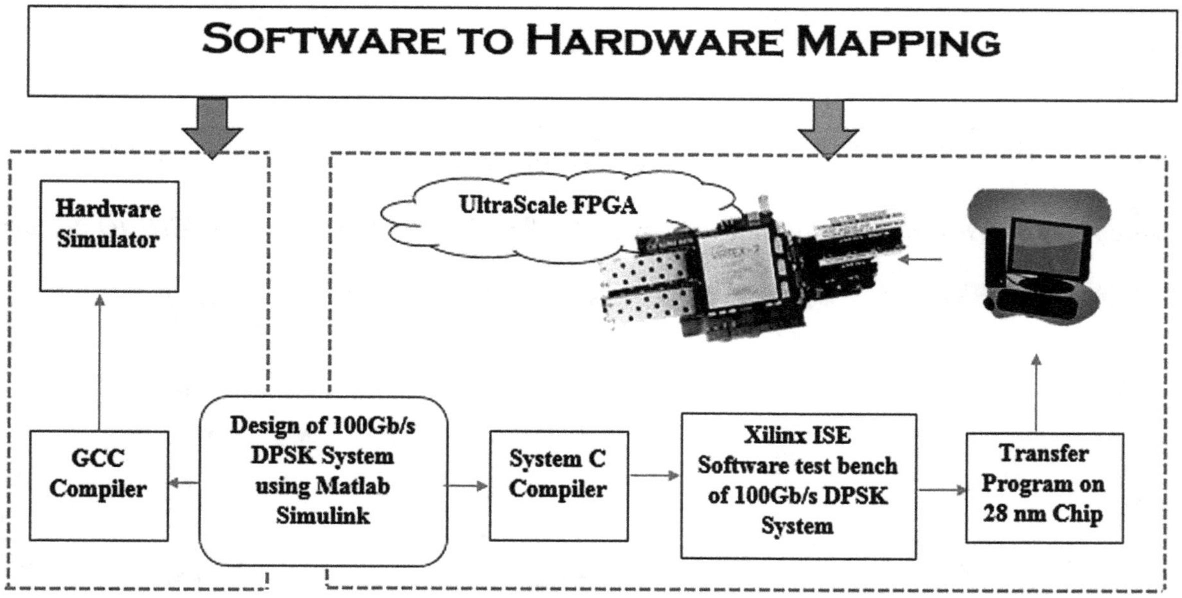

communication system is operated at LVCMOS, power consumption is reduced, for 500 GHz and 900 GHz 26%, for 10 THz 25% and for 17 THz 29% in comparison with LVDCI. Similarly, using HSTL IO Standard for 500 GHz, 900 GHz, 10 THz and 17 THz the 62%, 51%, 60% and 60% power reduction is achieved respectively on contrary to the LVDCI IO Standard. Finally, when VLSI based designed system is operated at HSUL IO Standard, the 87%, 80%, 75% and 76% power reduction is achieved respectively for 500 GHz, 900 GHz, 10 THz and 17 THz in comparison with LVDCI IO Standard. HSUL IO Standard consumes less power due to high speed bus operation, supports both HR and HP IO banks, low power memory interface and high performance with improved switching for signaling. Therefore, the rest of the IO Standards LVDCI, LVCMOS and HSTL are not advisable to be use for low power design of VLSI based communication. Figure 19 defines the total power reduction of VLSI based communication using different IO standards. It is extracted that HSUL IO Standard consumes less power at all carrier frequencies. After HSUL IO Standard, HSTL consumes less power but issue with the HSTL is that it consumes the equal amount of power for both static and dynamic. This increase the leakage power and reduce the switching efficiency during signaling. LVDCI consumes the highest power for the designed system. It is recommended to use HSUL IO Standard for low power deign of VLSI designed system.

The power analysis using XPA analyzer states that IO Standard based design for VLSI designed system is consuming less chip power. It can be seen in Figure 20 that HSUL consumes chip power of 2.33 W for VLSI based communication system.

Furthermore, it also less chip temperature and less leakage power of 1.8 W. It is determined that developed technique is reducing the power consumption for VLSI based system. The hardware implantation verified the output of low power design of VLSI based communication system.

Figure 19. Comparison of VLSI based communication system using IO standards

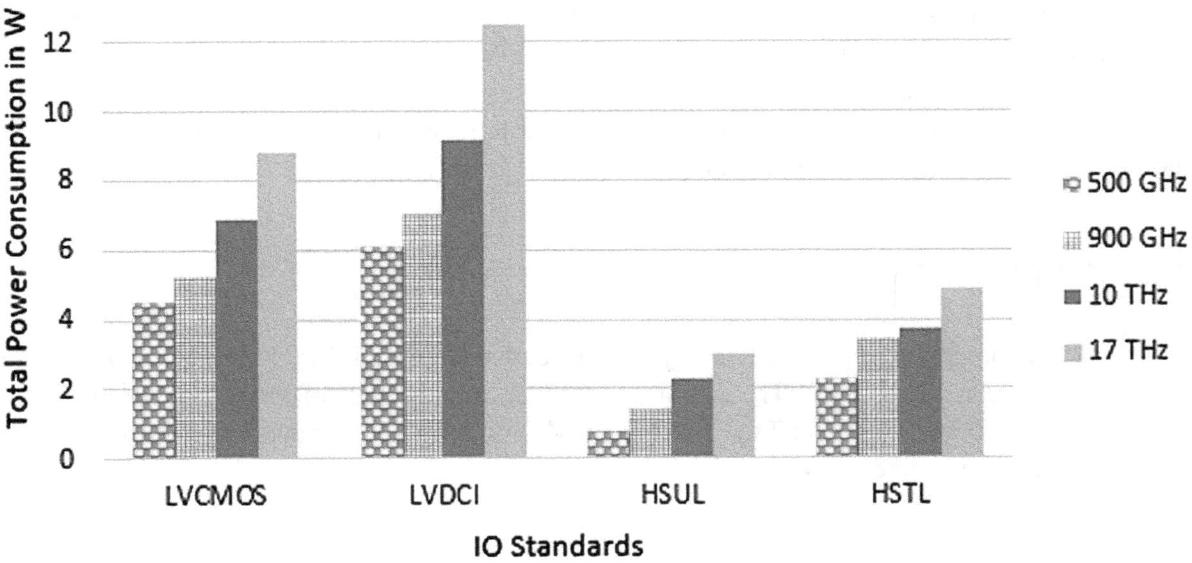

Figure 20. Chip Power reduction for VLSI based communication system using IO standards

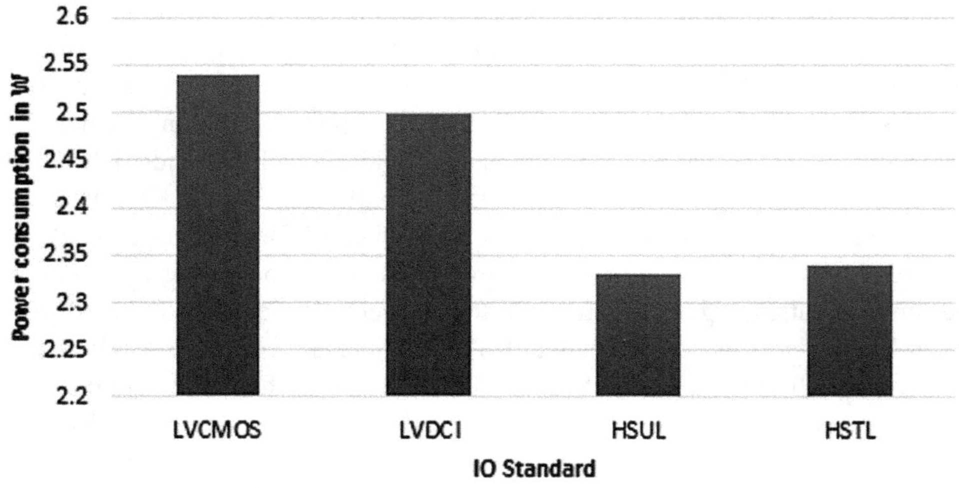

CONCLUSION AND FUTURE SCOPE

In this chapter, the low power deign of VLSI based communication system design is demonstrated. Using proposed technique the low power design is realized on 28 nm chip packed in UltraScale FPGA. Using IO standard the implementation of power optimized communication devices are possible. The power reduction is achieved for high-speed communication system using proposed technique and its real time implementation is demonstrated. The proposed technique yields power reduction in high frequency operation of VLSI based designed communication system. Overall, more than 80% power consumption is received for high frequency operation. Furthermore, the developed design consumes very low chip power of 2.33 W. It is also determined that more than 19% chip power reduction is achieved using proposed design. One more competent feature of this designed system using proposed technique is that it don't require any heat sink when operated at high frequency. The chip temperature is also maintained and controlled. The leakage power is also reduced for designed VLSI based designed system. The hardware realization for the designed system is low cost implementation using UltraScale FPGA SoC configuration. A great deal of work still needs to be done in order to reduce the consumption of the protocols, mobile phones and for other real time systems.

ACKNOWLEDGMENT

This work was supported in part by Postgraduates Incentive Grant (GIPS) (No.U168) Universiti Tun Hussein Onn Malaysia (UTHM). The authors would like to thank staff of Faculty of Electrical and Electronic Engineering, Universiti Tun Hussein Onn Malaysia (UTHM) for their support and efforts in handling article's experimental work.

REFERENCES

Abdullah, M., Das, B., & Mohd Shah, N. S. (2015). Energy Efficient Design of Hyper Transport Protocol based Laser Driver using Low-Voltage Differential Signaling. *International Journal of Control and Automation*, *8*(9), 131–138. doi:10.14257/ijca.2015.8.9.13

Abdullah, M., Das, B., & Mohd Shah, N. S., & Pandey. (2015). Energy-efficient Pseudo Noise generator based optical transmitter for Ethernet (IEEE 802.3 az). In *International Conference on in Computer, Communications, and Control Technology (I4CT)* (pp. 142-146), Kuching, Malaysia: IEEE press. doi:10.1109/I4CT.2015.7219554

Aggarwal, R. (2014). *FPGA place & route challenges.* Paper presented at the 2014 on International symposium on physical design. Petaluma, CA. doi:10.1145/2560519.2568050

Bacchus, R. M., Nguyen, V., & Benedict, M. K. (2014). *U.S. Patent No. 8,782,452.* Washington, DC: U.S. Patent and Trademark Office.

Bellaouar, A., & Elmasry, M. (2012). *Low-power digital VLSI design circuits and systems.* Springer Science & Business Media.

Breuer, D., & Petermann, K. (1997). Comparison of NRZ-and RZ-modulation format for 40-Gb/s TDM standard-fiber systems. *IEEE Photonics Technology Letters*, *9*(9), 398–400. doi:10.1109/68.556086

Cámara Nebreda, J. M. (2015). *Regulations on electronics.* Retrieved from http://riubu.ubu.es/bitstream/10259/3590/1/Regulations.pdf

Cetinkaya, O., & Akan, O. B. (2015). *A ZigBee based reliable and efficient power metering system for energy management and controlling.* Paper presented at the Computing, Networking and Communications (ICNC), 2015 International Conference on the green computing, Garden Grove, CA. doi:10.1109/ICCNC.2015.7069397

Chandrakasan, A. P., & Brodersen, R. W. (2012). *Low power digital CMOS design.* Springer Science & Business Media.

Chandrakasan, A. P., Sheng, S., & Brodersen, R. W. (1992). Low-power CMOS digital design. *IEICE Transactions on Electronics*, *75*(4), 371–382.

Chinn, S. R., Boroson, D. M., & Livas, J. C. (1996). Sensitivity of optically preamplified DPSK receivers with Fabry-Perot filters. *Journal of Lightwave Technology*, *14*(3), 370–376. doi:10.1109/50.485595

Cloetens, H. (2014). *Die temperature estimator: U.S. Patent No. 8,639,468.* Washington, DC: U.S. Patent and Trademark Office.

Dorsey, P. (2010). Xilinx stacked silicon interconnect technology delivers breakthrough FPGA capacity, bandwidth, and power efficiency. *Xilinx White Paper: Virtex-7 FPGAs*, 1-10.

Ghosh, A., & Ledwich, G. (2012). *Power quality enhancement using custom power devices.* London: Springer Science & Business Media.

Gnauck, A. H., & Winzer, P. J. (2005). Optical phase-shift-keyed transmission. *Journal of Lightwave Technology*, *23*(1), 115–130. doi:10.1109/JLT.2004.840357

Gonzalez, R., Gordon, B. M., & Horowitz, M. (1997). Supply and threshold voltage scaling for low power CMOS. *Solid-State Circuits. IEEE Journal of*, *32*(8), 1210–1216.

Hinton, K., Jalali, F., & Matin, A. (2015). Energy consumption modelling of optical networks. *Photonic Network Communications*, 1–13.

Horowitz, M., Indermaur, T., & Gonzalez, R. (1994). *Low-power digital design.* Paper presented at the Low Power Electronics. Digest of Technical Papers., IEEE Symposium, San Diego, CA.

Hussein, J., Klein, M., & Hart, M. (2011). Lowering power at 28 nm with Xilinx 7 series FPGAs. *Xilinx white paper WP389* (V.1.1). Xilinx Inc.

JEDEC. (2015). *E. I. A.* Retrieved from https://www.jedec.org/

Kadam, M., & Sawarkar, K. (2014). An Overview of Reconfigurable Hardware for Efficient Implementation of DSP Algorithms. *IOSR Journal of Engineering*, *4*(2), 34–43. doi:10.9790/3021-04253443

Leibson, S., & Mehta, N. (2013). Xilinx UltraScale: The Next-Generation Architecture for Your Next-Generation Architecture. *Xilinx White Paper WP435*. Xilinx Inc.

Li, J., Xie, Q., Wang, Y., Nazarian, S., & Pedram, M. (2015). *Leakage power reduction for deeply-scaled FinFET circuits operating in multiple voltage regimes using fine-grained gate-length biasing technique.* Paper presented at the 2015 Design, Automation & Test in Europe Conference & Exhibition, Grenoble, France. doi:10.7873/DATE.2015.0520

Mahoob, M. S., & Mehra, R. (2015). Layout Design of CMOS Buffer to Reduce Area and Power. *International Journal for Innovative Research in Science and Technology*, *2*(1), 22–25.

Maxfield, C. (2004). *The design warrior's guide to FPGAs: devices, tools and flows*. Elsevier.

Mendez, D. J., Riggins, M. D., Wagle, P., Bui, H. Q., Ng, M., Quinlan, S. M., … Aptekar-Strober, J. A. (2014). *System and method for globally and securely accessing unified information in a computer network: U.S. Patent No. 6,708,221*. Washington, DC: U.S. Patent and Trademark Office.

Poltyrev, G. (1994). On coding without restrictions for the AWGN channel. *IEEE Transactions on Information Theory*, *40*(2), 409–417. doi:10.1109/18.312163

Quinn, P. J. (2013). *FPGA based silicon innovation exploiting "More than Moore" technology.* Paper presented at the Ph. D. Research in Microelectronics and Electronics (PRIME) Conference, Glasgow, UK.

Sahni, K., Rawat, K., Pandey, S., & Ahmad, Z. (2015). *Power Optimization of Communication System Using Clock Gating Technique.* Paper presented at the Advanced Computing & Communication Technologies (ACCT), Fifth International Conference, Haryana, India. doi:10.1109/ACCT.2015.74

Santarini, M. (2013). Xilinx 20-nm Planar & 16-nm FinFET Go UltraScale. *Xcelljournal*, *84*(3), 8–15.

Sbarcea, B., & Nicula, D. (2004). *Automatic conversion of MATLAB/Simulink models to HDL models.* Paper presented at the International confernce of Optimization Electrical and Electronic Equipment, Brasov, Romania.

Seo, S., Lee, Y., Lee, J., & Kang, S. (2015). *A scan shifting method based on clock gating of multiple groups for low power scan testing.* Paper presented at the Quality Electronic Design (ISQED), 16th International Symposium, Santa Clara, CA. doi:10.1109/ISQED.2015.7085417

Shang, L., Kaviani, A. S., & Bathala, K. (2002). *Dynamic power consumption in Virtex™-II FPGA family.* Paper presented at the 2002 ACM/SIGDA tenth international symposium on Field-programmable gate arrays, Monterey, CA. doi:10.1145/503048.503072

Tessier, R., Pocek, K., & DeHon, A. (2015). Reconfigurable Computing Architectures. *IEE Proceedings, 103*(3), 332-354.

Trimberger, S. M. (2012). *Field-programmable gate array technology*. Springer Science & Business Media.

Tummala, R., Rymaszewski, E. J., & Klopfenstein, A. G. (2012). *Microelectronics packaging handbook: technology drivers*. Springer Science & Business Media.

Verhelst, M., Sylvester, D., Takamiya, M., Clinton, M., Wilcox, K., & Nose, K. (2015). *F4: Building the Internet of Everything (IoE): Low-power techniques at the circuit and system levels.* Paper presented at IEEE International the Solid-State Circuits Conference-(ISSCC).

Wenger, E. (2014). *Artificial intelligence and tutoring systems: computational and cognitive approaches to the communication of knowledge*. Morgan Kaufmann.

Whitfield, F. J., & Doyel, A. T. (1984). *Methods and means for conducting heat from electronic components and the like: U.S. Patent No. 4,466,483*. Washington, DC: U.S. Patent and Trademark Office.

Xilinx, Inc. (2012). *Xilinx Power Tools Tutorial*. Retrieved October 16, 2012, from http://www.xilinx.com/support/documentation/sw_manuals/xilinx14_3/ug733.pdf

Xilinx, Inc. (2014). *UltraScale Architecture SelectIO Resources*. Retrieved May 29, 2015, from http://www.xilinx.com/support/documentation/user_guides/ug571-ultrascale-selectio.pdf

Xilinx, Inc. (2015). *UltraScale Architecture and Product Overview*. Retrieved April 27, 2015, from http://www.xilinx.com/support/documentation/data_sheets/ds890-ultrascale-overview.pdf

Xu, Z., Gu, Q. J., Wu, Y.-C., & Chang, M.-C. F. (2015). Integrated D-band transmitter and receiver for wireless data communication in 65 nm CMOS. *Analog Integrated Circuits and Signal Processing, 82*(1), 171–179. doi:10.1007/s10470-014-0379-9

Yadav, P. S., & Pandey, V. K. (2015). Power Efficiency of Half Adder Design Using MTCMOS Technique in 35 Nanometre Regime. *International Journal for Innovative Research in Science and Technology, 1*(12), 111–117.

Yigit, M., Gungor, V. C., Tuna, G., Rangoussi, M., & Fadel, E. (2014). Power line communication technologies for smart grid applications: A review of advances and challenges. *Computer Networks, 70*, 366–383. doi:10.1016/j.comnet.2014.06.005

Compilation of References

Aalazaam, A., Roman, D., Nerguitzian, V., Stiharu, I., Bhat, R., Yasmeen, A., & Mosutafa, A. (2008). Real time continuous dielectrophoretic separation of malignant cells. *IEEE Conference on Microsystems and Nanoelectronics Research*. doi:10.1109/MNRC.2008.4683374

Abdullah, A., Fallah, F., & Pedram, M. (2004). Leakage current reduction in CMOS VLSI circuits by input vector control. *IEEE Trans. Very Large Scale Integration*, *12*(2), 140–154. doi:10.1109/TVLSI.2003.821546

Abdullah, M., Das, B., & Mohd Shah, N. S. (2015). Energy Efficient Design of Hyper Transport Protocol based Laser Driver using Low-Voltage Differential Signaling. *International Journal of Control and Automation*, *8*(9), 131–138. doi:10.14257/ijca.2015.8.9.13

Abdullah, M., Das, B., & Mohd Shah, N. S., & Pandey. (2015). Energy-efficient Pseudo Noise generator based optical transmitter for Ethernet (IEEE 802.3 az). In *International Conference on in Computer, Communications, and Control Technology (I4CT)* (pp. 142-146), Kuching, Malaysia: IEEE press. doi:10.1109/I4CT.2015.7219554

Abid, Z., El-Razouk, H., & El-Dib, D. A. (2008). Low power multipliers based on new hybrid full adders. *Microelectronics Journal*, *39*(12), 1509–1515. doi:10.1016/j.mejo.2008.04.002

Adiseno, H. M., & Olsson, H. (2003). *A 1.8-V Wide-Band CMOS LNA for Multiband Multistandard Front-End Receiver.* Radio Electronics-LECS/IMIT, Royal Institute of Technology (KTH). Retrieved from Sweden.www.imec.be/esscirc/essderc-esscirc-2003/papers/all/292.pdf

Aggarwal, R. (2014). *FPGA place & route challenges.* Paper presented at the 2014 on International symposium on physical design. Petaluma, CA. doi:10.1145/2560519.2568050

Akturk, A., & Goldsman, N. (2008). Electron transport and full-band electron–phonon interactions in graphene. *Journal of Applied Physics*, *103*(5), 053702. doi:10.1063/1.2890147

Alamo, J. A. (2010). Is nanometer-scale III–V CMOS cool enough to rejuvenate Moore's Law? *Compound Semiconductors*, *5*(16), 19–22.

Alioto, M., Consoli, E., & Palumbo, G. (2010). General strategies to design nanometer flip-flops in the energy-delay space. *IEEE Transactions on Circuits and Systems-Part I*, *57*(7), 1583–1596. doi:10.1109/TCSI.2009.2033538

Alioto, M., Consoli, E., & Palumbo, G. (2011a). Analysis and comparison in the energy- delay-area domain of nanometer CMOS flip-flops: Part I- Methodologies and design strategies. *IEEE Transactions on VLSI Systems*, *19*(5), 725–736. doi:10.1109/TVLSI.2010.2041376

Alioto, M., Consoli, E., & Palumbo, G. (2011b). Analysis and comparison in the energy-delay-area domain of nanometer CMOS flip-flops: Part II- Results and figures of merit. *IEEE Transactions on VLSI Systems*, *19*(5), 737–750. doi:10.1109/TVLSI.2010.2041377

Alioto, M., Consoli, E., & Palumbo, G. (2012). From energy-delay metrics to constraints on the design of digital circuits. *International Journal of Circuit Theory and Applications*, *40*(8), 815–834. doi:10.1002/cta.757

Alioto, M., Consoli, E., & Palumbo, G. (2015). Variations in Nanometer CMOS Flip-Flops: Part I—Impact of Process Variations on Timing. *IEEE Transactions on Circuits & Systems I*, *62*(8), 2035–2043. doi:10.1109/TCSI.2014.2366811

Alioto, M., & Palumbo, G. (2002). Analysis and Comparison on Full Adder Block in Submicron Technology. *IEEE Transaction on Very Large Scale Integration (VLSI) Systems*, *10*(6), 806–823.

Ali, S., Tanner, S., & Farine, P. A. (2011). A Robust, Low Power, High Speed Voltage Level Shifter with Built-in Short Circuit Current Reduction. *20th European Conference on Circuit Theory and Design (ECCTD)*, (pp. 142-145). doi:10.1109/ECCTD.2011.6043302

Allen, P. E., & Holberg, D. R. (2002). CMOS Analog Circuit Design (2nd ed.). Academic Press.

Alper, C., Lattanzio, L., Michielis, L. D., Palestri, P., Selmi, L., & Ionescu, A. M. (2013). Quantum mechanical study of the germanium electron–hole bilayer tunnel FET. *IEEE Transactions on Electron Devices*, *9*(60), 2754–2760. doi:10.1109/TED.2013.2274198

Anand, A., Das, A., Mandal, S.K., & Patro, B.S. (2013). FGMOS based low-voltage low-power high output impedance regulated cascode current mirror. *International Journal of VLSI Design & Communication Systems, 4*(2).

Anis, Mahmoud, & Elmasry. (2002). *Dynamic and Leakage Power Reduction in MTCMOS Circuits Using an Automated Efficient Gate Clustering Technique*. DAC 2002, New Orleans, LA.

Arora, N. D. (1993). MOSFETs models for VLSI Circuit Simulation, Theory and Practice. In Computational Microelectronics. Springer-Verlag.

Arreguit, X. (1989). *Compatible lateral bipolar transistors in CMOS technology: model and applications* (PhD Thesis N 817). EPFL, Lausanne.

Artan, N. S., Xu, X., Shi, W., & Chao, H. J. (2012). Optimizing Analog-To-Digital Converters for Sampling Extracellular Potentials. In *Proceedings of* the *34th Annual International Conference of the IEEE Engineering in Medicine and Biology Society (EMBC 2012), 1663-1666.* doi:10.1109/EMBC.2012.6346266

Ascoli, A., Corinto, F., Senger, V., & Tetzlaff, R. (2013). Memristor model comparison. *IEEE Circuits and Systems Magazine*, *13*(2), 89–105. doi:10.1109/MCAS.2013.2256272

Ashmila, E. M., Dlay, S. S., & Hinton, O. R. (2005). Adder methodology and design using probabilistic multiple carry estimates. *IEE Proc.-Comput. Digit. Tech.*, *152*(6), 697-703. doi:10.1049/ip-cdt:20045185

ASIC Design Flow Tutorial. (2010). Nano-Electronics and Computing Research Center, San Francisco State University.

Assaderaghi, F., Sinitsky, D., Parke, S., Boker, J., Ko, P. K., & Hu, C. (1994). A dynamic threshold voltage MOSFET (DTMOS) for very-low voltage operation. *IEDM Tech. Dig.*, 809-812.

Athas, W. C., Svensson, L. J., Koller, N., Tzartzanis, N., & Chou, E. (1994). Low power digital system based on adiabatic switching principles. *IEEE Trans. on VLSI Systems*, *2*(4), 398–407. doi:10.1109/92.335009

Athas, W. C., Svensson, L. J., & Tzartzanis, N. (1996). A resonant signal driver for two-phase, almost-non-overlapping clocks.*IEEE International Symposium on Circuits and Systems*. IEEE. doi:10.1109/ISCAS.1996.541916

Aufray, B., Kara, A., Vizzini, S. B., Oughaddou, H., Léandri, C., Ealet, B., & Le Lay, G. (2010). Graphene-like silicon nanoribbons on Ag(110): A possible formation of silicene. *Applied Physics Letters*, *96*(18), 183102. doi:10.1063/1.3419932

Avila, D., Alvarez, E., & Abusleme, A. (2014). Digital assistance for energy reduction in ADCs using a simple signal prediction algorithm. In *Proceedings of IEEE 5th Latin American Symposium on Circuits and Systems (LASCAS)*, 1-3. doi:10.1109/LASCAS.2014.6820264

Avital, M., Dagan, H., Levi, I., Keren, O., & Fish, A. (2015). DPA-Secured Quasi-Adiabatic Logic (SQAL) for Low-Power Passive RFID Tags Employing S-Boxes. IEEE Transactions on Circuits and Systems—I: Regular Papers, 62(1), 149-156.

Bacchus, R. M., Nguyen, V., & Benedict, M. K. (2014). *U.S. Patent No. 8,782,452*. Washington, DC: U.S. Patent and Trademark Office.

Balankuntty, A., Shih-An, Y., Yiping, F., & Kinget, P. R. (2010). A 0.6 V Zero-IF/Low-IF Receiver with Integrated Fractional-N Synthesizer for 2.4GHz ISM-Band Applications. *IEEE Journal of Solid-State Circuits*, *45*(3), 538–553. doi:10.1109/JSSC.2009.2039827

Basker, V., Standaert, T., Kawasaki, H., Yeh, C., Maitra, K., Yamashita, T., & Kumar, A. et al. (2010). A 0.063 μm2 FinFET SRAM cell demonstration with conventional lithography using a novel integration scheme with aggressively scaled fin and gate pitch*Proceedings International Symposium VLSI Technology, Systems Application.*

Bellaouar, A., & Elmasry, M. (2012). *Low-power digital VLSI design circuits and systems*. Springer Science & Business Media.

Bennet, C. (1982). The Thermodynamics of Computation – a Review. *International Journal of Theoretical Physics*, *12*(12), 905–940. doi:10.1007/BF02084158

Bennett, C. (1973). Logical reversibility of computation. *IBM Journal of Research and Development*, *17*, 525–532.

Bennett, C. H., & Landauer, R. (1985). The Fundamental Physical Limits of Computation. *Scientific American*, *253*(1), 38–46. doi:10.1038/scientificamerican0785-48

Berut, A., Arakelyan, A., Petrosyan, A., Ciliberto, S., Dillenschneider, R., & Lutz, E. (2012). Experimental verification of Landauer's principle linking information and thermodynamics. *Nature*, *483*(7388), 187–189. doi:10.1038/nature10872 PMID:22398556

Bhoj, A. N., Joshi, R. V., & Jha, N. K. (2013). *Efficient methodologies for 3D-TCAD modeling of emerging devices and circuits. IEEE Transanctions Computer-Aided Design.*

Bisdounis, L., Isdounis, L., Gouvetas, & Koufopavlou. (1998). A comparative study of CMOS circuit design styles for low-power high-speed VLSI circuits. *Int. J. Electronics*, *84*(6), 599-613.

Bishnoi, B., & Ghosh, B. (2014). Spin transport in buckled bilayer silicene. *Computational Materials Science*, *85*, 16–19. doi:10.1016/j.commatsci.2013.12.041

Black, C. T. (2005). Self-aligned self assembly of multi-nanowire silicon field effect transistors. *Applied Physics Letters*, *87*(16).

Blackrock. (2001). *Blackrock Microsystems Utah Array*. Retrieved August 31, 2015, from http://www.blackrockmicro.com/content.aspx?id=78

Blalock, B. J., Allen, V., & Rincon-Mora, G. A. (1998, July). Designing 1-V op-amps using standard digital CMOS technology. *IEEE Trns. Circuits and System-II. Analog and Digital Signal Processing*, *45*(7), 769–780.

Blotti, A., Pascoli, S.D., & Saletti. (2000) Simple model for positive-feed-back adiabatic logic power consumption estimation. Electronic Letters, 36(2).

Blotti, A., & Saletti, R. (2004). Ultralow-Power Adiabatic Circuit Semi-Custom Design. *IEEE Trans. on VLSI Systems*, *12*(11), 1248–1253. doi:10.1109/TVLSI.2004.836320

Bohr, M. (1996). A high performance 0.25m logic technology optimized for 1.8V operation. In *Proceedings of the International Electron Devices Meeting*, (pp. 847-850). Academic Press.

Bolomey, L., Meurville, E., & Ryser, P. (2009). Implantable ultra-low power DSP-based system for a miniature chemico-rheological biosensor. *Procedia Chemistry*, *1*(1), 1235–1238. doi:10.1016/j.proche.2009.07.308

Borkar. (2008). *Gigascale integration - challenges and opportunities*. Intel® Software Network.

Borkar, S. (1999). Design challenges of technology scaling. *IEEE Micro*, *19*(4), 23–29. doi:10.1109/40.782564

Borkar, S. (2006). Electronics beyond nano-scale CMOS. In *Proceedings of the 43rd annual Design Automation Conference*. ACM.

Boucart, K., & Adrian, M. I. (2007). Double-gate tunnel FET with high-κ gate dielectric. *Electron Devices, IEEE Transactions on*, *54*(7), 1725–1733. doi:10.1109/TED.2007.899389

Breuer, D., & Petermann, K. (1997). Comparison of NRZ-and RZ-modulation format for 40-Gb/s TDM standard-fiber systems. *IEEE Photonics Technology Letters*, *9*(9), 398–400. doi:10.1109/68.556086

Brews, J. (1990). *High Speed Semiconductor Devices*. New York: Wiley.

Brokaw, P. (1981). *Band-gap Voltage Reference with Curvature Correction*. United States Patent.

Bullis, K. (2013). Quantum Dots Produce More Colorful Sony TVs. *MIT Technology Review*.

Bult, K., & Geelen, J. G. M. (1992, December). An inherently linear and compact MOST-only current division technique. *IEEE Journal of Solid-State Circuits*, *27*(12), 1730–1735. doi:10.1109/4.173099

Cahangirov, S., Topsakal, M., Aktürk, E., Şahin, H., & Ciraci, S. (2009). Two- and One-Dimensional Honeycomb Structures of Silicon and Germanium. *Physical Review Letters*, *102*(23), 236804. doi:10.1103/PhysRevLett.102.236804 PMID:19658958

Cahay, M. (2012). Quantum Confinement VI: Nanostructured Materials and Devices. *Proceedings of the International Symposium*. The Electrochemical Society.

Cámara Nebreda, J. M. (2015). *Regulations on electronics*. Retrieved from http://riubu.ubu.es/bitstream/10259/3590/1/Regulations.pdf

Cao, Y., Sato, T., Sylvester, D., Orshansky, M., & Hu, C. (2000). New paradigm of predictive MOSFET and interconnect modeling for early circuit design. In *Proceedings of IEEE Custom Integrated Circuits Conference* (pp. 201-204). IEEExplore Digital Library. Retrieved Sep. 24, 2013, from http://ptm.asu.edu/

Carlson, A., Guo, Z., Balasubramanian, S., Zlatanovici, R., Liu, T.-J. K., & Nikolic, B. (2011). Modeling of width-quantization-induced variations in logic FinFETs for 22 nm and beyond.*Proceedings International Symposium VLSI Technology, Systems Applications*.

Castellanos-Gomez, A., Vicarelli, L., Prada, E., Island, J., Narasimha, A. K. L., Blanter, S. I., … van der Zant, H. S. J. (2014). Isolation and characterization of few-layer black phosphorus. *2D Materials, 1*.

Cetinkaya, O., & Akan, O. B. (2015). *A ZigBee based reliable and efficient power metering system for energy management and controlling*. Paper presented at the Computing, Networking and Communications (ICNC), 2015 International Conference on the green computing, Garden Grove, CA. doi:10.1109/ICCNC.2015.7069397

Chandrakasan, A. P., & Brodersen, R. W. (1995). Low-power CMOS digital design. Kluwer Academic.

Chandrakasan, A. P., Sheng, S., & Brodersen, R. W. (1999). Low Power CMOS Digital Design. *IEEE Journal of Solid-State Circuits*, *27*(4), 473–484.

Chandrakasan, A. P., & Brodersen, R. W. (2012). *Low power digital CMOS design*. Springer Science & Business Media.

Chandrakasan, A. P., & Robert, W. B. (2012). *Low power digital CMOS design*. Springer Science & Business Media.

Chandrakasan, A. P., Sheng, S., & Brodersen, R. W. (1992). Low-power CMOS digital design. *IEICE Transactions on Electronics*, *75*(4), 371–382.

Chandrakasan, A., Bowhill, W. J., & Fox, F. (2000). *Design of High- Performance Microprocessor Circuits*. Wiley-IEEE Press. doi:10.1109/9780470544365

Chang, F., Lee, Y., & Chiu, Ch. (2008). Multiple Electrodes Arrayed Dielectrophoretic Chip with Application on Micro-Bead Manipulation. *IEEE Proceedings of the Nano/Micro Engineered and Molecular Systems*. IEEE.

Chang, Gu, & Zhang. (2005). A Review of 0.18-μm Full Adder Performances for Tree Structured Arithmetic Circuits. *IEEE Transaction on Very Large Scale Integration (VLSI) Systems*, *13*(6), 686–695.

Chao, J., & Johnston, C. A. (1989). Behaviour analysis of CMOS D flip-flops. *IEEE Journal of Solid-State Circuits*, *24*(5), 1454–1458. doi:10.1109/JSSC.1989.572637

Chatelain, J. D. (1979). *Dispositifs à semiconductours. In Traité d'électricité* (Vol. 7). Presses Polytechniques Romandes.

Chen, C. X. (2008). A cost effective 32nm high effective k/metal –gate technology for low power application with single metal/gate –first process. Symposium on VLSI Technology Digest of Technical Paper, (pp. 88-99). Academic Press.

Chen, Z. (2004). Nanotubes for nanoelectronics. Encyclopedia of Nanoscience and Nanotechnology, 7, 919-942.

Chen, Z., Johnson, M., Wei, L., & Roy, K. (1998). Estimation of standby leakage power in CMOS circuits considering accurate modeling of transistor stacks. In *Proc. of Int. Symposium on LPED*, (pp. 239-244). Academic Press.

Cheng, K.-H., Cheng, S.-W., & Lee, W.-S. (2006). 64-bit pipeline carry lookahead adder using all-n-transistor TSPC logics. *Journal of Circuits, Systems, and Computers*, *15*(1), 13–27. doi:10.1142/S0218126606002915

Chen, M. J., Ho, J. S., Huang, T. H., Yang, C. H., Jou, Y. N., & Wu, T. (1996, June). Back-gate forward bias method for low-voltage CMOS Digital Circuits. *IEEE Transactions on Electron Devices*, *43*(6), 904–910. doi:10.1109/16.502122

Chernichenko, D., Kushnerov, A., & Ben-Yaakov, S. (2012). *Adiabatic Charging of Capacitors by Switched Capacitor Converters with Multiple Target Voltages*. 2012 IEEE 27th Convention of Electrical and Electronics Engineers, Israel.

Chiang, K. C., Artan, N. S., & Chao, H. J. (2013). A Signal-Specific Approach for Reducing SAR-ADC Power Consumption. In *Proceedings of* the *IEEE Biomedical Circuits and Systems Conference (BioCAS), 278-281*. doi:10.1109/BioCAS.2013.6679693

Chinn, S. R., Boroson, D. M., & Livas, J. C. (1996). Sensitivity of optically preamplified DPSK receivers with Fabry-Perot filters. *Journal of Lightwave Technology*, *14*(3), 370–376. doi:10.1109/50.485595

Choi, Y. S., Kim, Y. J., Im, M., Kim, B. G., Yun, K. S., & Yoon, E. (2006). Three Dimensional Electrode Structure Controlled by Dielectrophoresis for Flow-Through Micro Electroporation System. *IEEE International Conference on Micro Electro Mechanical Systems*. doi:10.1109/MEMSYS.2006.1627837

Chopra, K., & Vrudhula, B. K. (2004). Implicit pseudo boolean enumeration algorithms for input vector control. In *Proc. of IEEE-ACM Design Automation Conference*. doi:10.1145/996566.996774

Chua, L. O. (1971). Memristor—The Missing Circuit Element. *IEEE Transactions on Circuit Theory*, *CT-18*(5), 507–519. doi:10.1109/TCT.1971.1083337

Chun, J. W., Roger, C. Y., & Chen. (2010). A Novel Leakage reduction technique for CMOS circuit design. *Proceedings of ISOCC*.

Cloetens, H. (2014). *Die temperature estimator: U.S. Patent No. 8,639,468*. Washington, DC: U.S. Patent and Trademark Office.

Colinge, J-P., et al. (2010). Nanowire transistors without junctions. *Nature Nanotechnology*, *5*(3), 225-229.

Consoli, E., Palumbo, G., & Pennisi, M. (2012). Reconsidering high-speed design criteria for transmission-gate-based master-slave flip-flops. *IEEE Transactions on VLSI Systems*, *20*(2), 284–295. doi:10.1109/TVLSI.2010.2098426

Consoli, E., Palumbo, G., Rabaey, J. M., & Alioto, M. (2014). Novel class of energy-efficient very high speed conditional push-pull pulsed latches. *IEEE Transactions on VLSI Systems*, *22*(7), 1593–1605. doi:10.1109/TVLSI.2013.2276100

Cook, B. W. (2007). *Low energy RF transceiver design*. Berkeley, CA: University of California.

Cui, Y., Zhong, Z., Wang, D., Wang, W. U., & Lieber, C. M. (2003). High performance silicon nanowire field effect transistors. *Nano Letters*, *3*(2), 149–152. doi:10.1021/nl0258751

Danova, T. (2015). *The Wearables Report: Growth trends, consumer attitudes, and why smartwatches will dominate. Business Insider*. Retrieved August 31, 2015, from http://www.businessinsider.com/the-wearable-computing-market-report-2014-10

Dao, H. Q., Nowka, K., & Oklobdzija, V. G. (2001). Analysis of clocked timing elements for dynamic voltage scaling effects over process parameter variation. *InProceedings of IEEE International Symposium on Low Power Electronic Design* (pp. 56-59). IEEExplore Digital Library. doi:10.1145/383082.383094

DasGupta, S., McMorrow, D., Reed, R. A., Schrimpf, R. D., Boos, J. B., & Ramachandran, V. (2010). Process and contamination effects on the single-event response of AlSb/InAs HEMTs. *IEEE Transactions on Nuclear Science*, *57*, 3262–3266.

Davila, M. E., Xian, L., Cahangirov, S., Rubio, A., & Le Lay, G. (2014). Germanene: A novel two-dimensional germanium allotrope akin to graphene and silicene. *New Journal of Physics*, *16*(9), 095002. doi:10.1088/1367-2630/16/9/095002

Denis, P. A., & Iribarne, F. (2013). Comparative Study of Defect Reactivity in Graphene. *The Journal of Physical Chemistry C*, *117*(37), 19048–19055. doi:10.1021/jp4061945

Denkar, J. S. (1994). A Review of Adiabatic Computing. AT & T Bell laboratories. *1994 IEEE Symposium on Low Power Electronics*. doi:10.1109/LPE.1994.573218

Design Compiler User Guide. (2009). Synopsys, Version C-2009.06.

Dewey, G., et al. (2011). Fabrication, characterization, and physics of III–V heterojunction tunneling field effect transistors (H-TFET) for steep sub-threshold swing. *Proceedings IEEE IEDM*.

Dickinson, A. G., & Denker, J. S. (1995). Adiabatic dynamic logic. IEEE Solid State Circuits, 30(3), 311-315. doi:10.1109/4.364447

Do, A. T., Lam, C. K., Tan, Y. S., Yeo, K. S., Jia, H. C., Zou, X., & Je, M. et al. (2012). A 160 nW 25 kS/s 9-bit SAR ADC for neural signal recording applications. In *Proceedings IEEE 10th International New Circuits and Systems Conference (NEWCAS)*, (pp. 525-528). doi:10.1109/NEWCAS.2012.6329072

Do, A. V., Boon, C. C., Yeo, K. S., & Cabuk, A. (2008). A subthreshold low-noise amplifier optimized for ultra-low-power applications in the ISM band. *IEEE Transactions on Microwave Theory and Techniques*, *56*(2), 286–292. doi:10.1109/TMTT.2007.913366

Dorsey, P. (2010). Xilinx stacked silicon interconnect technology delivers breakthrough FPGA capacity, bandwidth, and power efficiency. *Xilinx White Paper: Virtex-7 FPGAs*, 1-10.

Drake, A. J., Zamdmer, N., Nowka, K. J., & Brown, R. B. (2003). Analysis of the impact of gate body signal phase on DTMOS Inverters in 0.13μm PD-SOI. IEEE International SOI Conference, (pp. 99-100). IEEE.

Dresselhaus, M. S., Dresselhaus, G., & Avouris, P. (2001). *Carbon nanotubes: synthesis, structure, properties, and applications*. Springer. doi:10.1007/3-540-39947-X

Dueck, G. W. (2012). Synthesis of Toffoli Networks: status and challenges. In *International Symposium on Electronic System Design (ISED)*. doi:10.1109/ISED.2012.26

Elakkumanan, P., Thondapu, C., & Sridhar, R. (2004). A gate leakage reduction strategy for sub-70 nm memory circuits. In *Implementation of High Performance Circuits. (DCAS-04),Proceedings of the IEEE Dallas/CAS Workshop*. IEEE. doi:10.1109/DCAS.2004.1360446

EntroMedics. (2015). *EntroMedics VBLOC Maestro® Rechargeable System*. Retrieved August 30, 2015, from http://www.enteromedics.com/usa/vbloc_maestro_system.asp

Enz, C. C. (1989). *High Precision CMOS Micro-power Amplifiers* (PhD Thesis N 802). EPFL, Lausanne.

Enz, C. C. (1996). The EKV model: a MOST model dedicated to low-current and lowvoltage analog circuit design and simulation in Low-power HF microelectronics. In IEEE Circuits. and Systems. IEEE Publications.

Enz, C. C., Krummenacher, & Vittoz, E. A. (1995, July). An Analytical MOS Transistor Model Valid in All Regions of Operation and Dedicated to Low-Voltage and Low- Current Applications. *Analog Integrated Circuits and Signal Processing Journal*, *8*, 83-114.

Eriksson, H., Larsson-Edefors, P., & Eckerbert, D. (2006). Toward Architecture-Based Test-Vector Generation for Timing Verification of Fast Parallel Multipliers. *IEEE Transaction on Very Large Scale Integration (VLSI) Systems*, *14*(4), 370–379.

Fallah, F., & Pedram, M. (2005). Standby and active leakage current control and minimization in CMOS VLSI circuits. *IEICE Transactions on Electronics*, *E88–C*(4), 509–519. doi:10.1093/ietele/e88-c.4.509

Fei, W., & Yu, H. (2011). Design exploration of hybrid cmos and memristor circuit by new modified nodal analysis. *IEEE Transactions on Very Large Scale Integration (VLSI) Systems*, *99*, 1–15.

Feynman, R. (1985). Quantum Mechanical Computers. *Optics News*, *11*(2), 11–20. doi:10.1364/ON.11.2.000011

Feynman, R. (1986). Quantum Mechanical Computers. *Foundations of Physics*, *12*(6).

Fitbit. (2015). *FitBit Activity Monitors*. Retrieved August 30, 2015, from http://www.fitbit.com

Gao, F., & Hayes, J. P. (2004). Exact and Heuristic approaches to input vector control for leakage power reduction. In *Proc. of IEEE Int. Conference on CAD*, (pp. 527-532). IEEE.

Gascoyne, P., & Vykonkal, J. (2004). Dielectrophoresis-based sample handling in general-purpose programmable diagnostic instrument. *IEEE Proceedings, 92*, 22-42. DOI: 10.1109/JPROC.2003.820535

Gascoyne, P., Mahidol, C., Ruchirawat, M., Satayavivad, J., Watcharasit, P., & Becker, F. (2002). Microsample preparation by dielectrophoresis: isolation of malaria. *Lab on Chip Magazine, 2*, 70-75. DOI: 10.1039/b1109990c

Geim, A. (2009). Graphene: Status and Prospects. *Science*, *324*(5934), 1530–1534. doi:10.1126/science.1158877 PMID:19541989

Gerosa, G., Gary, S., Dietz, C., Pham, D., Hoover, K., Alvarez, J., & Kahle, J. et al. (1994). A 2.2 W, 80 MHz superscalar RISC microprocessor. *IEEE Journal of Solid-State Circuits*, *29*(12), 1440–1454. doi:10.1109/4.340417

Ghosh, A., & Ledwich, G. (2012). *Power quality enhancement using custom power devices*. London: Springer Science & Business Media.

Ghosh, S., & Roy, K. (2008). Exploring high-speed low-power hybrid arithmetic units at scaled supply and adaptive clock-stretching. *ASPDAC*, *2008*, 635–640.

Giacomotto, C., Nedovic, N., & Oklobdzija, V. (2007). The effect of the system specification on the optimal selection of clocked storage elements. *IEEE Journal of Solid-State Circuits*, *42*(6), 1392–1404. doi:10.1109/JSSC.2007.896516

Gnani, E., Gnudi, A., Reggiani, S., & Baccarani, G. (2011). Theory of the junctionless nanowire FET. *Electron Devices, IEEE Transactions on*, *58*(9), 2903–2910. doi:10.1109/TED.2011.2159608

Gnauck, A. H., & Winzer, P. J. (2005). Optical phase-shift-keyed transmission. *Journal of Lightwave Technology*, *23*(1), 115–130. doi:10.1109/JLT.2004.840357

Goldberger, A. L., Amaral, L. A. N., Glass, L., Hausdorff, J. M., Ivanov, P. Ch., Mark, R. G., & Stanley, H. E. et al. (2000). PhysioBank, PhysioToolkit, and PhysioNet: Components of a New Research Resource for Complex Physiologic Signals. *Circulation*, *101*(23), e215–e220. doi:10.1161/01.CIR.101.23.e215 PMID:10851218

Gonzalez, R., Gordon, B. M., & Horowitz, M. (1997). Supply and threshold voltage scaling for low power CMOS. *Solid-State Circuits. IEEE Journal of*, *32*(8), 1210–1216.

Gray, P. R., Hurst, P. J., Lewis, S. H., & Meyer, R. G. (2001). *Analysis and Design of Analog Integrated Circuits* (4th ed.). New York: Wiley.

Grochowski & Annavaram. (Eds.). (2006). Energy per Instruction Trends in Intel Microprocessors. Technology Intel Magazine, 1-8.

Guang, Y., Geng, H., Lai, K-Y., Shi, Y., & Chang, S-C. (2014). Multibit Retention Registers for Power Gated Designs: Concept, Design, and Deployment. *IEEE Transaction on Computer Aided Design of Integrated Circuit and System*, *33*(4).

Guerra, D., Akis, R., Marino, F. A., Ferry, D. K., Goodnick, S. M., & Saraniti, M. (2010). Aspect ratio impact on RF and DC performance of state-of the- art short-channel GaN and InGaAs HEMTs. *IEEE Electron Device Letters*, *11*(31), 1217–1219.

Gustafsson, O., Dempster, A. G., & Wanhammar, L. (2004). Multiplier blocks using carry-save adders. *IEEE Proceeding, ISCAS*. doi:10.1109/ISCAS.2004.1329311

Gusynin, V. P., & Sharapov, S. G. (2005). Unconventional Integer Quantum Hall Effect in Graphene. *Physical Review Letters*, *95*(14), 146801. doi:10.1103/PhysRevLett.95.146801 PMID:16241680

Guyot, A., Hochet, A. B., & Muller, M. J. (1987). A way to build efficient carry-skip adders. *IEEE Transactions on Computers*, *C-36*(4), pp1144–pp1151. doi:10.1109/TC.1987.1676855

Ha, H., Lee, S.-K., Kim, B., Park, H.-J., & Sim, J.-Y. (2014). A 0.5-V, 1.47- μW 40-kS/s 13-bit SAR ADC With Capacitor Error Compensation. *IEEE Transactions on Circuits and Wystems. II, Express Briefs*, *61*(11), 840–844. doi:10.1109/TCSII.2014.2350378

Hamid, M.-M., & Afzali, K. A. (2001). Efficient power-clock generation for adiabatic logic. *The 2001 IEEE Int.Symp. on Circuits and Systems, ISCAS 2001.*

Hanchate, N., & Ranganathan, N. (2004). LECTOR:ATechnique for leakage reduction in CMOS circuits. *IEEE Trans. On VLSI Integration System, 22*(2), 196–205. doi:10.1109/TVLSI.2003.821547

Han, G. H., Jung, H. D., & Kim, W. T. (2015). A 2.88mW + 9.06dBm IIP3 Common-Gate LNA with Dual Cross Coupled Capacitive Feedback. *IEEE Transactions on Microwave Theory and Techniques, 63*(3), 1019–1025. doi:10.1109/TMTT.2014.2377036

Hänninen, I., Lu, H., Blair, E. P., Lent, C. S., & Snider, G. L. (2014). Reversible and Adiabatic Computing: Energy-Efficiency Maximized. *Field-Coupled Nanocomputing, LNCS, 8280*, 341–356.

Harrison, R. R., & Charles, C. (2003). A low-power low-noise CMOS amplifier for neural recording applications. *IEEE Journal of Solid-State Circuits, 38*(6), 958–965. doi:10.1109/JSSC.2003.811979

Ha, S., Kim, C., Chi, Y. M., Akinin, A., Maier, C., Ueno, A., & Cauwenberghs, G. (2014). Integrated Circuits and Electrode Interfaces for Noninvasive Physiological Monitoring. *IEEE Transactions on Bio-Medical Engineering, 61*(5), 1522–1537. doi:10.1109/TBME.2014.2308552 PMID:24759282

Haug, H., & Koch, S. W. (2012). *Quantum Theory of the Optical and Electronic Properties of Semiconductors.* World Scientific.

Hedayati. (2011). *A Study of Successive Approximation Registers and Implementation of an Ultra-Low Power 10-bit SAR ADC in 65nm CMOS Technology.* (Master's thesis). Linköping Institute of Technology.

Heo, S., & Asanovic, K. (2001). Load-sensitive flip-flop characterization. In *Proceedings of IEEE Computer Society Workshop on VLSI* (pp. 87–92). IEEExplore Digital Library.

Hinton, K., Jalali, F., & Matin, A. (2015). Energy consumption modelling of optical networks. *Photonic Network Communications*, 1–13.

Ho, Y., Chang, C., & Chauchin. (2012). Design of a subthrehold–supply bootstrapped CMOS inverter based on an active leakage current reduction technique. *IEEE Transaction on Circuit and System Express brief–II, 59*(1), 55-59.

Ho, Y.-T., & Hwang, T-T. (2004). Low Power Design Using Dual Threshold Voltage. In *IEEE ASP Design Automation Conference.*

Holler, M., Tam, S., Castro, H., & Benson, R. (1989). An electrically trainable artificial neural network with 10240 'floating gate' synapses.*Proceeding of the International Joint Conference on Neural Networks*, (vol. 2, pp. 191-196). doi:10.1109/IJCNN.1989.118698

Horowitz, M. (2005). *Scaling, power, and the future of CMOS.* Electron Devices Meeting. IEDM Technical Digest. IEEE. doi:10.1109/IEDM.2005.1609253

Horowitz, M., Indennaur, T., & Gonzalez, R. (1994, October). Low Power Digital Design. *Technical Digest IEEE Symposium Low Power Electronics.*

Horowitz, M., Indermaur, T., & Gonzalez, R. (1994). *Low-power digital design.* Paper presented at the Low Power Electronics. Digest of Technical Papers., IEEE Symposium, San Diego, CA.

Hossain, R., Wronski, L. D., & Albicki, A. (1994). Low power design using double edge triggered flip-flops. *IEEE Transactions on Very Large Scale Integration (VLSI) Systems, 2*(2), 261–265. doi:10.1109/92.285754

Hosseini, S. R., Saberi, M., & Lotfi, R. (2014). A Low-Power Subthreshold to Above-Threshold Voltage Level Shifter. *IEEE Transactions on Circuits and Wystems. II, Express Briefs*, *61*(10), 753–757. doi:10.1109/TCSII.2014.2345295

Ho, Y., Huang, G., & Li, P. (2011). Dynamical properties and design analysis for nonvolatile memristor memories. *IEEE Transactions on Circuits and Systems. I, Regular Papers*, *58*(4), 724–736. doi:10.1109/TCSI.2010.2078710

Hsiao, S.-F., Yeh, J.-S., & Chen, D.-Y. (2002). High-performance Multiplexer-based Logic Synthesis Using Pass-transistor Logic. *VLSI Design*, *15*(1), 417–426. doi:10.1080/1065514021000054736

Huang, G.-Y., Chang, S.-J., Liu, C.-C., & Lin, Y.-Z. (2012). A 1-μW 10-bit 200-kS/s SAR ADC With a Bypass Window for Biomedical Applications. *IEEE Journal of Solid-State Circuits*, *47*(11), 2783–2795. doi:10.1109/JSSC.2012.2217635

Huijsing, J. (2013). *Analog circuit design: low-noise, low-power, low-voltage; mixed-mode design with CAD tools; voltage, current and time references*. Springer Science & Business Media.

Hun, K., & Mooney. (2006). Sleepy Keeper: A New Approach to Low-leakage Power VLSI Design. In *Proc. IFIP International Conference on Very Large Scale Integration* (pp. 367-372). Academic Press.

Hussein, J., Klein, M., & Hart, M. (2011). Lowering power at 28 nm with Xilinx 7 series FPGAs. *Xilinx white paper WP389* (V.1.1). Xilinx Inc.

Ibrahim, M., Elsayed, F., Ghallab, Y., & Badawy, W. (2009). An Electric Field Array Microsystem for Lab-on-Chip and Biomedical Analysis. *IEEE Conference on Biomedical Circuits and Systems*. doi:10.1109/BIOCAS.2009.5372076

Iijima, S. (1991). Helical microtubules of graphitic carbon. *Nature, 354*(6348), 56-58.

Imran, A., Hasan, M., Islam, A., & Abbasi, S. A. (2012). Optimized design of a 32-nm CNFET-based low-power ultra wideband CCII. *Nanotechnology, IEEE Transactions on*, *11*(6), 1100–1109. doi:10.1109/TNANO.2012.2212248

Inose, H., Aoki, T., & Watanabe, K. (1966). Asynchronous delta-modulation system. *Electronics Letters*, *2*(3), 95–96. doi:10.1049/el:19660077

International Technology Roadmap for Semiconductor. (2009). Retrieved from http//www.itrs.net

Inukai, T., Hiromoto, T., & Sakurai, T. (2001). Variable threshold voltage CMOS (VTCMOS). *Proceedings of ISPLED.*

Islam, M. S., Nasrin, M. S., Mansur, N., & Tasneem, N. (2010). Dual Stack Method: A Novel Approach to Low Leakage and Speed Power Product VLSI Design. In *Proc. ICECE2010.*

ITRS. (2013). Available from http://www.itrs.net

Javey, A., et al. (2003). Ballistic carbon nanotube field-effect transistors. *Nature, 424*(6949), 654-657.

JEDEC. (2015). *E. I. A.* Retrieved from https://www.jedec.org/

Johnson, M. C., Somasekhar, D., Chiou, L. Y., & Roy. (2002). Leakage Control with Efficient Use of Transistor Stacks in Single Threshold CMOS. *IEEE Trans. on VLSI Systems*, *10*, 1-5.

Jonsson, B. E. (2011). An empirical approach to finding energy efficient ADC architectures. In *Proceedings 2011 International Workshop on ADC Modelling, Testing and Data Converter Analysis and Design and IEEE ADC Forum.*

Jo, S., Chang, T., Ebong, I., Bhadviya, B. B., Mazumder, P., & Lu, W. (2010). Nanoscale memristor device as synapse in neuromorphic systems. *Nano Letters*, *10*(4), 1297–1301. doi:10.1021/nl904092h PMID:20192230

Jose, D., & Datta, A. (2014). Structures and Chemical Properties of Silicene: Unlike Graphene. *Accounts of Chemical Research*, *47*(2), 593–602. doi:10.1021/ar400180e PMID:24215179

Jung, G., Sundarajan, V., & Sobelman, G. E. (2005). A robust self- Resetting CMOS 32-bit parallel adder. Department of Electrical and Computer Engineering, University of Minnesota.

Juniper Research. (2014). *Smart Wearables Market to Generate $53bn Hardware Revenues by 2019*. Retrieved August 25, 2015, from http://www.juniperresearch.com/press/press-releases/smart-wearables-market-to-generate-$53bn-hardware

Jupiter XT. (2006). *Synopsys, Top-Down Hierarchical Flow, User Guide, Version Y-2006*. Author.

Kadam, M., & Sawarkar, K. (2014). An Overview of Reconfigurable Hardware for Efficient Implementation of DSP Algorithms. *IOSR Journal of Engineering*, *4*(2), 34–43. doi:10.9790/3021-04253443

Kaeting, Flynn, Aitken, Gibbons, & Shi. (2007). *Low Power Methodology Manual for system-on-chip design*. Springer.

Kahng, A. B. (2007). Design challenges at 65nm and beyond. In *Proceedings of the Conference on Design, Automation and Test in Europe*, (pp. 1466-1467). Academic Press.

Kahng, D., & Sze, S. M. (1967). A floating-gate and its application to memory devices. *The Bell System Technical Journal*, *4*(46), 1288–1295. doi:10.1002/j.1538-7305.1967.tb01738.x

Kanchana Bhaaskaran V. S. (2011). Energy Recovery Performance of Quasi-Adiabatic Circuits using Lower Technology Nodes. *IICPE2010*.

Kanchana Bhaaskaran, V. S., & Raina, J. P. (2008). Differential Cascode Adiabatic Logic Structure for Low Power. *Journal of Low Power Electronics*, *4*(2), 178–191. doi:10.1166/jolpe.2008.264

Kanchana Bhaaskaran, V. S., & Raina, J. P. (2010). Two-Phase Sinusoidal Power Clocked Quasi-Adiabatic Logic Circuits. *Journal of Circuits, Systems, and Computers*, *19*(2), 335–347. doi:10.1142/S0218126610006098

Kanchana Bhaaskaran, V. S., & Raina, J. P. (2012). Pre-resolve and Sense Adiabatic Logic for 100 kHz to 500 MHz Frequency Classes. *Journal of Circuits, Systems, and Computers*, *21*(5).

Kang, S., & Leblebici, Y. (2003). CMOS Digital Integrated Circuits - Analysis and Design. McGraw-Hill.

Kang, S., & Lebleblici, Y. (2003). *CMOS Digital Integrated Circuits*. New York: McGraw Hill.

Kao, J. T., & Chandrakasan, A. P. (2000, July). Dual-threshold voltage techniques for low-power digital circuits. *IEEE Journal of Solid-State Circuits*, *35*(7), 1009–1018. doi:10.1109/4.848210

Kao, J., & Chandrakasan, A. (2001). MTCMOS sequential circuits. In *Proceedings of European Solid-State Circuits Conference*.

Kao, J., Narendra, S., & Chandrakasan, A. (2002). Subthreshold Leakage Modeling and Reduction Techniques. In *Proceedings of the International Conference on Computer Aided Design*.

Kaur, S. (2013). On chip network. *IETE Technical Review*, *30*(3), 168–172. doi:10.4103/0256-4602.113467

Kawaguchi, H., Nose, K., & Sakurai, T. (2002). A Super Cut-Off CMOS(SCCMOS) Scheme for 0.5-V Supply Voltage with Picoampere Stand-By Current. *IEEE Journal of Solid-State Circuits*, *35*(10), 498–1501.

Kawaguchi, H., & Sakurai, T. (1998). A reduced clock-swing flip-flop (RCSFF) for 63 percent power reduction. *IEEE Journal of Solid-State Circuits*, *33*(5), 807–811. doi:10.1109/4.668997

Keating, M., Flynn, D., Aitken, R., & Alan, G. K. S. (2007). *Low Power Methodology Manual For System-on-Chip Design*. Springer Publications.

Kenington, P. B., & Astier, L. (2000). Power consumption of A/D converters for software radio applications. *IEEE Transactions on Vehicular Technology*, *49*(2), 643–650. doi:10.1109/25.832996

Khatami, Y., & Kaustav, B. (2009). Steep subthreshold slope n-and p-type tunnel-FET devices for low-power and energy-efficient digital circuits. *Electron Devices, IEEE Transactions on*, *56*(11), 2752–2761. doi:10.1109/TED.2009.2030831

Khondaker, S. I., Kang, L., & Zhen, Y. (2010). The fabrication of single-electron transistors using dielectrophoretic trapping of individual gold nanoparticles. *Nanotechnology*, *21*(9).

Kiani, M., Jow, U. M., & Ghovanloo, M. (2011). Design and Optimization of a 3-Coil Inductive Link for Efficient Wireless Power Transmission. *IEEE Transactions on Biomedical Circuits and Systems*, *5*(6), 579–591. doi:10.1109/TBCAS.2011.2158431 PMID:21922034

Kim, Y. B., Yong-B, K., & Fabrizio, L. (2009). A novel design methodology to optimize the speed and power of the CNTFET circuits. *Circuits and Systems, 2009. MWSCAS'09. 52nd IEEE International Midwest Symposium on*. IEEE. doi:10.1109/MWSCAS.2009.5235967

Kim, D., & Alamo, J. A. (2010). Scalability of sub-100 nm InAs HEMTs on InP substrate for future logic applications. *IEEE Transactions on Electron Devices*, *7*(57), 1504–1511. doi:10.1109/TED.2010.2049075

Kim, K. K., Kim, Y. B., Choi, M., & Park, N. (2007). Leakage minimization technique for nanoscale CMOS VLSI. *IEEE Design & Test of Computers*, *24*(4), 322–330. doi:10.1109/MDT.2007.111

Kim, K.-H., Gaba, S., Wheeler, D., Cruz-Albrecht, J. M., Hussain, T., Srinivasa, N., & Lu, W. (2012). A functional hybrid memristor crossbar-array/CMOS system for data storage and neuromorphic applications. *Nano Letters*, *12*(1), 389–395. doi:10.1021/nl203687n PMID:22141918

Kim, S., Ziesler, C. H., & Papaefthymiou, M. C. (2005). Charge-recovery computing on silicon. *IEEE Transactions on Computers*, *54*(6), 651–659. doi:10.1109/TC.2005.91

Kim, T. H., Artan, N. S., Viventi, J., & Chao, H. J. (2012). Spatiotemporal compression for efficient storage and transmission of high-resolution electrocorticography data. In *Proceedings of Engineering in Medicine and Biology Society* (pp. 1012–1015). EMBC.

Kim, Y. B. (2010). Challenges for nanoscale MOSFETs and emerging nanoelectronics. *Transactions on Electrical and Electronic Materials*, *11*(3), 93–105. doi:10.4313/TEEM.2010.11.3.093

Kim, Y., Song, B.-S., Grosspietsch, J., & Gillig, S. F. (2001). A carry-free 54b x 54b Multiplier Using Equivalent Bit Conversion Algorithm. *IEEE Journal of Solid-State Circuits*, *36*(10), 1538–1545. doi:10.1109/4.953482

Klimach, H., Costa, A. L. T., Monteiro, M. F. C., & Bampi, S. (2013). Resistorless switched-capacitor bandgap voltage reference with low sensitivity to process variations. *Electronics Letters*, *49*(23), 1448–1449. doi:10.1049/el.2013.1761

Kong, K., Yun, S., & Ruqian, L. (2010). An optimized majority logic synthesis methodology for quantum-dot cellular automata. *Nanotechnology, IEEE Transactions on*, *9*(2), 170–183. doi:10.1109/TNANO.2009.2028609

Kramer, J. S., Denker, B., & Moroney. (1995). 2nd order adiabatic computation with 2N-2P and 2N-2N2P logic circuits. *Proc. Int. Symp. Low Power Design*. doi:10.1145/224081.224115

Kuhn, K. J. (2012). Considerations for ultimate CMOS scaling. *IEEE Transactions on Electron Devices*, *59*(7), 1813–1828. doi:10.1109/TED.2012.2193129

Kuo, Hwang, & Wu. (2006). A Power-Driven Multiplication Instruction-Set Design Method for ASIPs. *IEEE Transaction on Very Large Scale Integration (VLSI) Systems*, *14*(1), 81–85.

Kuroda & Sakurai. (1995). Overview of low-Power ULSI techniques. *ICICE Transaction of Electronics, E78-C*(4), 334-344.

Kuroda, T., & Sakurai, T. (1995, April). Overview of low-power ULSI circuit techniques. *Inst. Of Electronics, Information & Communication Engineers Tran. Electron*, *78*(4), 334–343.

La Rosa, C., Tilley, P., Fox, J., & Kaler, K. (2008). Microfluidic Device for Dielectrophoresis Manipulation and Electrodisruption of Respiratory Pathogen Bordetella pertussis. *IEEE Transactions on Biomedical Engineering*, 2426-2432. DOI: 10.1109/TBME.2008923148

Lam, Y.-H., & Ki, W.-H. (2010). CMOS Bnadgap References with Self-biased Symmetrically Matched Current-Voltage Mirror and Extension of Sub-1-V Design. *IEEE Transactions on Very Large Scale Integration (VLSI) Systems*, *18*(6), 857–865.

Landauer, R. (1961). Irreversibility and heat generation in the computing process, IBM Jl. *Research for Development*, *5*(3), 183–191.

Laszlo, V., Kovacs, F., & Hosszu, G. (2001). An Improved Pass-Gate Adiabatic Logic.*Proceedings, 14th Annual IEEE Int. ASIC/SOC Conference*.

Lattanzio, L., Michielis, L. D., & Ionescu, A. M. (2012). Complementary germanium electron–hole bilayer tunnel FET for sub-0.5-V operation. *IEEE Electron Device Letters*, *2*(33), 167–169. doi:10.1109/LED.2011.2175898

Lattanzio, L., Michielis, L. D., & Ionescu, A. M. (2012). The electron–hole bilayer tunnel-FET. *Solid-State Electronics*, *74*, 85–90. doi:10.1016/j.sse.2012.04.016

Lay, G. L. (2015). 2D materials: Silicene transistors. *Nature Nanotechnology*, *10*(3), 202–203. doi:10.1038/nnano.2015.10 PMID:25643254

Leatherdale, C. A., Woo, W.-K., Mikulec, F. V., & Bawendi, M. G. (2002). On the Absorption Cross Section of CdSe Nanocrystal Quantum Dots. *The Journal of Physical Chemistry B*, *106*(31), 7619–7622. doi:10.1021/jp025698c

Lee, D., Kwong, W., Blaauw, D., & Sylvester, D. (2003). Analysis and Minimization Techniques for Total Leakage Considering Gate Oxide Leakage. In *Proceedings of the 40th design automation conference*, (pp. 175-180).

Lee, B. G. (2015). Power and Bandwidth Scalable 10-b 30-MS/s SAR ADC. *IEEE Transactions on Very Large Scale Integration (VLSI) Systems*, *23*(6), 1103–1110.

Lee, D., Blaauw, D., & Sylvester, D. (2005). Static leakage reduction through simultaneous Vt/Tox and state assignment. *IEEE Transactions on Computer-Aided Design of Integrated Circuits and Systems*, *24*(7).

Lee, D., Seong, D. J., Jo, I., Xiang, F., Dong, R., Oh, S., & Hwang, H. (2007). Resistance switching of copper doped MoO(sub x) films for nonvolatile memory applications. *Applied Physics Letters*, *90*(12), 122104. doi:10.1063/1.2715002

Lee, S. W., Mao, C., Flynn, C. E., & Belcher, A. M. (2002). Ordering of quantum dots using genetically engineered viruses. *Science*, *296*(5569), 892–895. doi:10.1126/science.1068054 PMID:11988570

Lee, T. H. (1998). *The Design of CMOS Radio-Frequency Integrated Circuits* (1st ed.). New York, NY: Cambridge University Press.

Lehmann, T. (2014). Design of fast low-power floating high-voltage level-shifters. *Electronics Letters*, *50*(3), 202–204. doi:10.1049/el.2013.2270

Leibson, S., & Mehta, N. (2013). Xilinx UltraScale: The Next-Generation Architecture for Your Next-Generation Architecture. *Xilinx White Paper WP435*. Xilinx Inc.

Lent, C. S., Liu, M., & Lu, Y. (2006). Bennett clocking of quantum-dot cellular automata and the limits to binary logic scaling. *Nanotechnology*, *17*(16), 4240–4251. doi:10.1088/0957-4484/17/16/040 PMID:21727566

Li, J., Xie, Q., Wang, Y., Nazarian, S., & Pedram, M. (2015). *Leakage power reduction for deeply-scaled FinFET circuits operating in multiple voltage regimes using fine-grained gate-length biasing technique.* Paper presented at the 2015 Design, Automation & Test in Europe Conference & Exhibition, Grenoble, France. doi:10.7873/DATE.2015.0520

Library Data Preparation for IC Compiler. (2010). *User Guide, Synopsys, Version D-2010.03.* Author.

Li, H., Yanan, Z., & Akin, D. (2005). Characterization and modeling of a microfluidic dielectrophoresis filter for biological species. *IEEE Journal of Microelectromechanical Systems*, *14*(1), 103–112. doi:10.1109/JMEMS.2004.839124

Lim, J., Kim, & Chae. (2000). NMOS reversible energy recovery logic for ultra-low-energy applications. *IEEE J. of Solid-State Circuits, 35*(6), 865-875.

Lima Silva, V. M., Lisboa de Souza, A. A., & Cavalcanti Catunda, S. Y. (2014). Flexible A/D converter architecture targeting sparse signals. In *Proceedings, IEEE International Instrumentation and Measurement Technology Conference (I2MTC)*, (pp. 1496-1500). doi:10.1109/I2MTC.2014.6860995

Lin, C. H., Chang, J., Guillorn, M., Bryant, A., Oldiges, P., & Haensch, W. (2010). Non-planar device architecture for 15 nm node: FinFET or trigate?. In *Proceedings of International SOI Conference.*

Lin, S. Y., Wang, C. C., & Lee, J. H. (2014). A 9.96mW 3.24dB NF 1.9GHz Wideband Low Noise Amplifier Using 90nm CMOS Technology. In *Proceedings of the IEEE Radio and Wireless Symposium*, (vol. 14, pp. 208-210). IEEE.

Liu, H., Neal, A. T., Zhu, Z., Luo, Z., Xu, X., Tománek, D., & Ye, P. D. (2014). Phosphorene: An Unexplored 2D Semiconductor with a High Hole Mobility. *ACS Nano*, *8*(4), 4033–4041. doi:10.1021/nn501226z PMID:24655084

Liu, T., Bihr, U., Becker, J., Anders, J., & Ortmanns, M. (2014). In vivo verification of a 100 Mbps transcutaneous optical telemetric link. In *Proceedings of IEEE Biomedical Circuits and Systems Conference (BioCAS)*, (pp. 580-583). doi:10.1109/BioCAS.2014.6981792

Liu, Y. P., Yuan, C., & Hung, Y. L. Y. (2014). A capacitor constructed bypass window switching scheme for energy-efficient SAR ADC. In *Proceedings of IEEE International Symposium on Circuits and Systems (ISCAS)*, (pp. 1352-1355). doi:10.1109/ISCAS.2014.6865394

Li, Y., Mansano, A. L., Yuan, Y., Zhao, D., & Serdijn, W. A. (2014). An ECG Recording Front-End With Continuous-Time Level-Crossing Sampling. *IEEE Transactions on Biomedical Circuits and Systems.*, *8*(5), 626–635. doi:10.1109/TBCAS.2014.2359183 PMID:25330494

Li, Y., Zhao, D., & Serdijn, W. A. (2013). A 0.8V 8-bit low-power asynchronous level-crossing ADC with programmable comparison windows. In *Proceedings of IEEE Biomedical Circuits and Systems Conference (BioCAS)*, (pp. 138-141). doi:10.1109/BioCAS.2013.6679658

Lopez, M. S., Gerstlauer, A., Avila, A., & Martinez, S. O. (2011). A programmable and configurable multi-port System-on-Chip for stimulating electrokinetically-driven microfluidic devices. In *Proceedings of the Annual International Conference of the IEEE Engineering in Medicine and Biology Society, EMBS* (pp. 8361-8364). doi:10.1109/IEMBS.2011.6092062

Lopez, M. S., Moncada, H., Perez, V., Lapizco, B., & Martinez, S. (2013). An electric stimulation system for electrokinetic particle manipulation in microfluidic devices. *Review of Scientific Instruments, 84*(3). DOI: 10.1063/1.4793559

Lorenzo, R., & Chaudhury, S. (2013). A Novel all NMOS Leakage Feedback with Data Retention Technique. In *Proc. of CARE 2013, IEEE Int. Conference.*

Lorenzo, R., & Chaudhury, S. (2014). Analysis of leakage feedback technique. In *Electronics, Communication and Instrumentation (ICECI), 2014 International Conference on*. IEEE.

Lotze, N., & Manoli, Y. (2012). A 62mV 0.13 μm CMOS standard cell based design technique using Schmitt-trigger logic. *IEEE Journal of Solid-State Circuits*, *47*(1), 47–60. doi:10.1109/JSSC.2011.2167777

Lu, L., Yan, G., Zhao, K., & Xu, F. (2015). An Implantable Telemetry Platform System With ASIC for in Vivo Monitoring of Gastrointestinal Physiological Information. *IEEE Sensors Journal*, *15*(6), 3524–3534. doi:10.1109/JSEN.2015.2392105

Mahoob, M. S., & Mehra, R. (2015). Layout Design of CMOS Buffer to Reduce Area and Power. *International Journal for Innovative Research in Science and Technology*, *2*(1), 22–25.

Maitra, K., Khakifirooz, A., Kulkarni, P., Basker, V. S., Faltermeier, J., Jagannathan, H., & Bu, H. et al. (2011). Aggressively scaled strained-silicon-on-insulator undoped-body high-κ/metal-gate nFinFETs for high-performance logic applications. *IEEE Electron Device Letters*, *6*(32), 713–715. doi:10.1109/LED.2011.2126556

Mak, K. F., Lee, C., Hone, J., Shan, J., & Heinz, T. F. (2010). Atomically Thin MoS_2: A New Direct-Gap Semiconductor. *Physical Review Letters*, *105*(13), 136805. doi:10.1103/PhysRevLett.105.136805 PMID:21230799

Manaresi, N., Romani, A., Medoro, G., Altomare, L., Leonardi, A., Tartagni, M., & Guerrieri, R. (2003). A CMOS Chip for Individual Cell Manipulation and Detection. *IEEE Journal of Solid-State Circuits*, *38*(12), 2297–2305. doi:10.1109/JSSC.2003.819171

Mano, M. (1994). *Computer system architecture* (3rd ed.). Prentice Hall.

Markovic, D., Tschanz, J., & De, V. (2003). *Transmission-gate based flip-flop*. U.S. Patent 6642765.

Marković, D., Wang, C. C., Alarcon, L. P., Tsung-Te Liu, , & Rabaey, J. M. (2010). Ultralow-power design in near-threshold region. *Proceedings of the IEEE*, *98*(2), 237–252. doi:10.1109/JPROC.2009.2035453

Maxfield, C. (2004). *The design warrior's guide to FPGAs: devices, tools and flows*. Elsevier.

Mayer, F. (2011). Demonstration of MOSFET-like ON-current performance in arsenide/antimonide tunnel FETs with staggered hetero-junctions for 300 mV logic applications. In Proceedings IEEE IEDM.

Medtronic. (2015a). *Medtronic Activa® RC Neurostimulator.* Retrieved on August 22, 2015, from http://www.medtronic.eu/your-health/essential-tremor/device/dbs-therapy/ActivaRCNeurostimulator/index.htm

Medtronic. (2015b). *Medtronic MiniMed 530G with Enlite*. Retrieved on August 22, 2015, from http://www.medtronicdiabetes.com/home

Mehonic, A., Cueff, S. B., Wojdak, M., Hudziak, S., Jambois, O., Labbé, C., & Kenyon, A. J. et al. (2012). Resistive switching in silicon suboxide films. *Journal of Applied Physics*, *111*(7), 074507. doi:10.1063/1.3701581

Meindl, J. D. (1995). Low Power Microelectronics: Retrospect and Prospect. *Proceedings of the IEEE*, *83*(4), 619–635. doi:10.1109/5.371970

Mendez, D. J., Riggins, M. D., Wagle, P., Bui, H. Q., Ng, M., Quinlan, S. M., … Aptekar-Strober, J. A. (2014). *System and method for globally and securely accessing unified information in a computer network: U.S. Patent No. 6,708,221.* Washington, DC: U.S. Patent and Trademark Office.

Mesgarzadeh, B., Hansson, M., & Alvandpour, A. (2007). Jitter characteristic in charge recovery resonant clock distribution. *IEEE Journal of Solid-State Circuits*, *42*(7), 1618–1625. doi:10.1109/JSSC.2007.896691

Meuffels, P., & Soni, R. (2012). *Fundamental Issues and Problems in the Realization of Memristors*. Academic Press.

Min, L-S., Kawaguchi, H., & Sakurai, T. (2003). Zig-zag super cut-off CMOS (ZSCCMOS) block activation and self-adaptive voltage level controller: An alternative to clock–gating scheme in leakage dominant Era. *IEEE International Solid–State Circuits Conference*, (pp. 400-401). IEEE.

Ming, X., Ying-qian, M., Ze-kun, Z., & Zhang, B. (2010). A high precision compensated CMOS bandgap voltage reference without resistors. IEEE Transaction on Circuits and System, 57(10).

Mirhassani, M., Ahmadi, M., & Jullien, G. A. (2008). Low-Power Mixed-Signal CVNS-Based 64-Bit Adder for Media Signal Processing. *IEEE Transactions on Very Large Scale Integration (VLSI) Systems*, *16*(9), 1141–1150.

Miryala, S., Calimera, A., Macii, E., & Poncino, M. (2014). Ultra Low-Power Computation via Graphene-Based Adiabatic Logic Gates. *2014 17th Euromicro Conference on Digital System Design.*

Moaiyeri, M. H., Akbar, D., & Keivan, N. (2011). Design of energy-efficient and robust ternary circuits for nanotechnology. *IET Circuits, Devices & Systems*, *5*(4), 285–296. doi:10.1049/iet-cds.2010.0340

Montanaro, J., Witek, R. T., Anne, K., Black, A. J., Cooper, E. M., Dobberpuhl, D. W., & Thierauf, S. C. et al. (1996, November). A 160 MHz, 32-b 0.5-W CMOS RISC microprocessor. *IEEE Journal of Solid-State Circuits*, *31*(11), 1703–1714.

Monteiro, C. (2015). Yasuhiro Takahashi & Toshikazu Sekine. (2014). Low-power secure S-box circuit using charge-sharing symmetric adiabatic logic for advanced encryption standard hardware design. *IET Circuits Devices Syst.*, *9*(5), 362–369. doi:10.1049/iet-cds.2014.0150

Moody, G. B., & Mark, R. G. (2001). The impact of the MIT-BIH Arrhythmia Database. *IEEE Eng in Med and Biol*, *20*(3), 45–50. doi:10.1109/51.932724 PMID:11446209

Moon, Y., & Jeong, D. K. (1996). An Efficient Charge Recovery Logic Circuit. *IEEE Journal of Solid-State Circuits*, *31*(4), 514–522. doi:10.1109/4.499727

Morrison, M., & Ranganathan. (2013). Novel optimization method for reversible logic circuit minimization. *2013 IEEE Computer Society Annual Symposium on VLSI* (ISVLSI).

Mudassir, R., & Abid, Z. (2005). New parallel multipliers based on low power adders. 2005 IEEE CCECE/CCGEI, Saskatoon.

Mukundrajan, R. (2012). Ultra low power circuit design using Tunnel FETs. *VLSI (ISVLSI), 2012 IEEE Computer Society Annual Symposium on*. IEEE. doi:10.1109/ISVLSI.2012.70

Mutoh, S., Douseki, T., Matsuya, Y., Aoki, T., Shigematsu, S., & Yamada, J. (1995). 1-V Power Supply High-speed Digital Circuit Technology with Multithreshold-Voltage CMOS. *IEEE Journal of Solid-State Circuits*, *30*(8), 847–854. doi:10.1109/4.400426

Myjak, M. J., Frias, J. G. D., & Jeon, S. K. (2006). An energy-efficient differential flip-flop for deeply pipelined systems. In *Proceedings of 49th IEEE International Midwest Symposium on Circuits and Systems* (pp. 203-207). IEEExplore Digital Library. doi:10.1109/MWSCAS.2006.382032

Navarro-Botello, V., Montiel-Nelson, J. A., & Nooshabadi, S. (2007). High performance low power CMOS dynamic logic for arithmetic circuits. *Microelectronics Journal*, *38*(4-5), 482–488. doi:10.1016/j.mejo.2007.03.018

Navi, K., Farazkish, R., Sayedsalehi, S., & Rahimi Azghadi, M. (2010). A new quantum-dot cellular automata full-adder. *Microelectronics Journal*, *41*(12), 820–826. doi:10.1016/j.mejo.2010.07.003

Nikolic, B., Oklobdizija, V. G., Stojanovic, V., Jia, W., Chiu, J. K., & Leung, M. M. (2000). Improved sense-amplifier based flip-flop: Design and measurements. *IEEE Journal of Solid-State Circuits*, *35*(6), 876–884. doi:10.1109/4.845191

Ni, Z., Liu, Q., Tang, K., Zheng, J., Zhou, J., Qin, R., & Lu, J. et al. (2011). Tunable bandgap in silicene and germanene. *Nano Letters*, *12*(1), 113–118. doi:10.1021/nl203065e PMID:22050667

Ni, Z., Zhong, H., Jiang, X., Quhe, R., Luo, G., Wang, Y., & Lu, J. et al. (2014). Tunable band gap and doping type in silicene by surface adsorption: Towards tunneling transistors. *Nanoscale*, *6*(13), 7609–7618. doi:10.1039/C4NR00028E PMID:24896227

Nogawa, M., & Ohtomo, Y. (1998). A data-transition look-ahead DFF circuit for statistical reduction in power consumption. *IEEE Journal of Solid-State Circuits*, *33*(5), 702–706. doi:10.1109/4.668984

Norris, D. J. (1995). *Measurement and Assignment of the Size-Dependent Optical Spectrum in Cadmium Selenide (CdSe) Quantum Dots*. (PhD thesis). MIT, Cambridge, MA.

O'Driscoll, S., & Meng, T. H. (2010). Adaptive signal acquisition and wireless power transfer for an implantable prosthesis processor. In *Proceedings of IEEE International Symposium on Circuits and Systems (ISCAS)*, (pp. 3589-3592). doi:10.1109/ISCAS.2010.5537805

O'Driscoll, S., Shenoy, K., & Meng, T. (2011). Adaptive Resolution ADC Array for an Implantable Neural Sensor. *IEEE Transactions on Biomedical Circuits and Systems*, *5*(2), 120–130. doi:10.1109/TBCAS.2011.2145418 PMID:23851200

Oh, N., & Lee, S. (2005). Building a 2.4-GHz radio transceiver using IEEE 802.15.4. *IEEE Circuits Devices Magazine.*, *21*(6), 43–51. doi:10.1109/MCD.2005.1578587

Oklobdizija, V. G., Stojanovic, V., Markovic, D., & Nedovic, N. (2003). *Digital system clocking: High performance and low power aspects*. Wiley. doi:10.1002/0471723703

Oktyabrsky, S., & Ye, P. D. (2010). *Fundamentals of III-V Semiconductor MOSFETs*. New York: Springer. doi:10.1007/978-1-4419-1547-4

OpenCores. (2014). *OpenCores*. Retrieved in 2014, http://opencores.org/

Osaki, Y., Hirose, T., Kuroki, N., & Numa, M. (2012). A Low-Power Level Shifter with Logic Error Correction for Extremely Low-Voltage Digital CMOS LSIs. *IEEE Journal of Solid-State Circuits*, *47*(7), 1776–1783. doi:10.1109/JSSC.2012.2191320

Pal, A., Sachid, A. B., Gossner, H., & Rao, V. R. (2011). Insights into the design and optimization of tunnel-FET devices and circuits. *Electron Devices, IEEE Transactions on*, *58*(4), 1045–1053. doi:10.1109/TED.2011.2109002

Palumbo, G., & Pennisi, M. (2008). Design guidelines for high-speed transmission-gate latches: Analysis and comparison. In *Proceedings of IEEE International Conference on Electronics, Circuits and Systems* (pp. 145-148). IEEExplore Digital Library. doi:10.1109/ICECS.2008.4674812

Pao, H. C., & Sah, C. T. (1996). Effects of diffusion current on the characteristics of metal oxide (insulator)-semiconductor transistors. *Solid-State Electronics*, *10*, 927–937.

Parhami, B. (2006). Fault Tolerant Reversible Circuits.*Proceedings of 40th Asilomar Conf. Signals, Systems, and Computers.*

Park, J. (2005). *Sleepy Stack: a New Approach to Low Power VLSI and Memory*. (Ph.D. Dissertation). School of Electrical and Computer Engineering, Georgia Institute of Technology. Available http://etd.gatech.edu/theses

Park, J. Y., Kim, S.-H., Hong, S.-M., & Kim, K. R. (2015). Physical analysis and design of resonant plasma-wave transistors for terahertz emitters. *IEEE Transactions THz Science Technology*, *5*, 244-250.

Patra, P. (1995). *Approaches to Design of Circuits for Low-Power Computation.* (Ph.D dissertation). Univ. Texas at Austin.

Pierret, R. F. (1996). *Semiconductor Device Fundamentals*. Reading, MA: Addison-Wesley.

Piguet, C. (2007). Low power design in deep submicron 65 and 45nm technology. *14th IEEE Intl. Conf. on Electronics, Circuits & System 2007, ICECS, 2007*. doi:10.1109/ICECS.2007.4511140

Poltyrev, G. (1994). On coding without restrictions for the AWGN channel. *IEEE Transactions on Information Theory*, *40*(2), 409–417. doi:10.1109/18.312163

Possani, T., Lucas, C. S., & Alessandro, G. (2012). *Automatic Design of Micropower Carbon Nanotube Operational Transconductance Amplifiers*. SIM - South Symposium on Microelectronics.

Prati, E., De Michielis, M., Belli, M., Cocco, S., Fanciulli, M., Kotekar-Patil, D., & Sanquer, M. et al. (2012). Few electron limit of n-type metal oxide semiconductor single electron transistors. *Nanotechnology*, *23*(21), 215204. doi:10.1088/0957-4484/23/21/215204 PMID:22552118

Psilogeorgopoulos, Munteanu, Chuang, Ivey, & Seed. (1999). *Contemporary Techniques for Lower Power Circuit Design PREST Deliverable D2.1*. The Department of Electronic and Electrical Engineering The University of Sheffield Mappin Street Sheffield S1 3JD.

PTM. (2013). Retrieved from http://www.eas.asu.edu/~ptm

Quinn, P. J. (2013). *FPGA based silicon innovation exploiting "More than Moore" technology*. Paper presented at the Ph. D. Research in Microelectronics and Electronics (PRIME) Conference, Glasgow, UK.

Rabaey, J. M., & Pedram, M. (2002). Low Power Design Methodologies. Kluwer Academic Publishers.

Rabaey, J. (2009). *Low power design essentials*. Springer Science & Business Media. doi:10.1007/978-0-387-71713-5

Rabaey, J. M., Chandrakasan, A., & Nikolic, B. (2012). *Digital Integrated Circuits*. Prentice Hall.

Rabaey, J. M., & Massoud, P. E. (2012). *Low power design methodologies* (Vol. 336). Springer Science & Business Media.

Rahman, H., & Chakrabarti, C., (2005). An efficient control point insertion technique for leakage reduction of scaled CMOS circuits. *IEEE Transaction on Circuit And system Brief-II*, *52*(8), 496-500.

Rajendran, J., Rose, G., Karri, R., & Potkonjak, M. (2012). Nano-PPUF: A memristor-based security primitive. In *IEEE Computer Society Annual Symposium on VLSI (ISVLSI)*, (pp. 84–87). IEEE.

Rao, R. M. (2003) A heuristic to determine low leakage sleep state vectors for CMOS combinational circuits. In *Proc. of IEEE Int. Conf. CAD*, (pp. 689-692). IEEE.

Ravanshad, N., Rezaee-Dehsorkh, H., Lotfi, R., & Lia, Y. (2014). A Level-Crossing Based QRS-Detection Algorithm for Wearable ECG Sensors. *IEEE Journal of Biomedical and Health Informatics*, *18*(1), 183–192. doi:10.1109/JBHI.2013.2274809 PMID:24403416

Raychaudhuri, A. (2004). *Edge-triggered D-flip-flop circuit*. U. S. Patent 6762637.

Raychowdhury, A., & Kaushik, R. (2007). Carbon nanotube electronics: design of high-performance and low-power digital circuits. *Circuits and Systems I: Regular Papers, IEEE Transactions on*, *54*(11), 2391-2401. doi:10.1109/TCSI.2007.907799

Razak, Z., Erdogan, A., & Arslan, T. (2010). ASIC Design of an Adaptive Control Unit for Reconfigurable Analog-to-Digital Converters. In *Proceedings IEEE Computer Society Annual Symposium on VLSI (ISVLSI)*, (pp. 179-184). doi:10.1109/ISVLSI.2010.79

Razavipour, G., Afzali-Kusha, A., & Pedram, M. (2009). Design and analysis of two low-power SRAM cell structures. *IEEE Transactions on Very Large Scale Integration (VLSI) Systems*, *17*, 1551–1555.

Reed, M. A., Randall, J. N., Aggarwal, R. J., Matyi, R. J., Moore, T. M., & Wetsel, A. E. (1988). Observation of discrete electronic states in a zero-dimensional semiconductor nanostructure" (PDF). *Physical Review Letters, 60*(6), 535–537. doi:10.1103/PhysRevLett.60.535 PMID:10038575

Risch, L. (2005). Pushing CMOS Beyond the Roadmap.*Proceedings of ESSCIRC.*

Roche, B., et al. (2012). A tunable, dual mode field-effect or single electron transistor. *Applied Physics Letters, 100*(3).

Rose, G., McDonald, N., Yan, L.-K., Wysocki, B., & Xu, K. (2013). Foundations of memristor based PUF architectures. In *IEEE/ACM International Symposium on Nanoscale Architectures (NANOARCH)*, (pp. 52– 57). doi:10.1109/NanoArch.2013.6623044

Roy, K., & Prasad, S.C. (2000). *Low Power CMOS VLSI Circuit Design.* Wiley Interscience.

Roy, K., & Prasad, S. C. (1990). *Low-Power CMOS VLSI Circuit Design.* New York: Wiley Interscience.

Roy, K., & Sharat, C. P. (2009). *Low-power CMOS VLSI circuit design.* John Wiley & Sons.

Ruiz, G. A. (1998). Evaluation of three 32-bit CMOS adders in DCVS logic for self-timed circuits. *IEEE Journal of Solid-State Circuits, 33*(4), 604–613. doi:10.1109/4.663566

Ruiz, G. A., & Granda, M. (2004). An area-efficient static CMOS carry-select adder based on a compact carry look-ahead unit. *Microelectronics Journal, 35*(12), 939–944. doi:10.1016/j.mejo.2004.09.002

Rusu, A., Gonzalez, D. R. L., & Ismail, M. (2006). Reconfigurable ADCs enable smart radios for 4G wireless connectivity. *IEEE Circuits and Devices Magazine, 22*(3), 6–11. doi:10.1109/MCD.2006.1657844

Saberi, M., Lotfi, R., Mafinezhad, K., & Serdijn, W. A. (2011). Analysis of Power Consumption and Linearity in Capacitive Digital-to-Analog Converters Used in Successive Approximation ADCs. *IEEE Transactions on Circuits and Systems. I, Regular Papers, 58*(8), 1736–1748. doi:10.1109/TCSI.2011.2107214

Sahni, K., Rawat, K., Pandey, S., & Ahmad, Z. (2015). *Power Optimization of Communication System Using Clock Gating Technique.* Paper presented at the Advanced Computing & Communication Technologies (ACCT), Fifth International Conference, Haryana, India. doi:10.1109/ACCT.2015.74

Sakurai, T., & Newton, A. R. (1990, October). Alpha-Power Law MOSET Model and its Applications to CMOS Inverter Delay and other Formulas. *IEEE JSSC, 25*(2), 584-594.

Sakurai, T. (2004). Perspective of Low Power *VLSI. IEICE Transactions on Electronics, 87*(4), 429–436.

Santarini, M. (2013). Xilinx 20-nm Planar & 16-nm FinFET Go UltraScale. *Xcelljournal, 84*(3), 8–15.

Sathe, V. S., Chuch, J.-Y., & Papaefthymiou, M. C. (2007). Energy-efficient GHz-class charge-recovery logic. *IEEE Journal of Solid-State Circuits, 42*(6), 38–46. doi:10.1109/JSSC.2006.885053

Sbarcea, B., & Nicula, D. (2004). *Automatic conversion of MATLAB/Simulink models to HDL models.* Paper presented at the International confernce of Optimization Electrical and Electronic Equipment, Brasov, Romania.

Schedin, F., Geim, A. K., Morozov, S. V., Hill, E. W., Blake, P., Katsnelson, M. I., & Novoselov, K. S. (2007). Detection of individual gas molecules adsorbed on graphene. *Nature Materials, 6*(9), 652–655. doi:10.1038/nmat1967 PMID:17660825

Seabaugh, A. C., & Qin, Z. (2010). Low-voltage tunnel transistors for beyond CMOS logic. *Proceedings of the IEEE, 98*(12), 2095–2110. doi:10.1109/JPROC.2010.2070470

Second Sight. (2015). *Second Sight Argus II Retinal Prosthesis System.* Retrieved on August 31, 2015, from http://www.2-sight.com/g-the-argus-ii-prosthesis-system-pf-en.html

Seitz, C. L., Frey, A. H. Mattisson Sven, S. D., Rabin, D. A., & Snepscheut, J (1985). *Hot Clock nMOS.* California Institute of Technology. Retrieved from http://resolver.caltech.edu/CaltechCSTR:1985.5177-tr-85

Senthilpari, C., Mohamad, Z. I., & Kavitha, S. (2011). Proposed low power, high speed adder-based 65-nm Square root circuit, Vol41(2) pp 445-451. *Microelectronics Journal.*

Senthilpari, C., Singh, A. K., & Diwakar, K. (2008). Design of a low-power, high performance, 8× 8bit multiplier using a Shannon-based adder cell. *Microelectronics Journal, 39*(5), 812–821. doi:10.1016/j.mejo.2007.12.016

Seo, S., Lee, Y., Lee, J., & Kang, S. (2015). *A scan shifting method based on clock gating of multiple groups for low power scan testing.* Paper presented at the Quality Electronic Design (ISQED), 16th International Symposium, Santa Clara, CA. doi:10.1109/ISQED.2015.7085417

Sepehrian, H., Saberi, M., & Lotfi, R. (2011). A signal-specific successive-approximation analog-to-digital converter. *IEEE International Symposium on Circuits and Systems (ISCAS),* (pp. 1624-1627).

Shang, L., Kaviani, A. S., & Bathala, K. (2002). *Dynamic power consumption in Virtex™-II FPGA family.* Paper presented at the 2002 ACM/SIGDA tenth international symposium on Field-programmable gate arrays, Monterey, CA. doi:10.1145/503048.503072

Shang, Y., Fei, W., & Yu, H. (2012). Analysis and modeling of internal state variables for dynamic effects of nonvolatile memory devices. *Circuits and Systems I: Regular Papers, IEEE Transactions on, 99.* doi:10.1109/TCSI.2011.2180441

Shannon, C. E. (1950c). A Symmetrical Notation for Numbers. *The American Mathematical Monthly, 57*(2), 90–93. doi:10.2307/2304993

Shao, H., & Tsui, C. (2007). A Robust, input voltage adaptive and low energy consumption level converter for subthreshold logic. 33rd ESSCIRC, (pp. 312-315).

Sharma, J., Ansari, M. S., & Sharma, J. (2014). Current-Mode Electronically Tunable Resistor-less Universal Filter in ±0.5V 32 nm CNFET. *Devices, Circuits and Communications (ICDCCom), 2014 International Conference on.* IEEE.

Sharma, V.K., Pattanaik, M., & Raj, B. (2014). INDEP approach for leakage reduction in CMOS circuits. *International Journal of Electronics, 102*(1), 200-215.

Sharma, V.K., Pattanaik, M., & Raj. (2013). ONOFIC approach: Low power high speed nanoscale VLSI circuits design. *International Journal of Electronics, 101*(1), 61-73.

Sharuddin & Lee. (2014).An Ultra-low Power and Area Efficient 10 bit Digital to Analog Converter Architecture. *IEEE Int. Conference of Semiconductor Electronics 2014, IEEE-ICSE2014.*

Shih, T. C., Chu, K,. & Liu, C. (2007). A Programmable Biochip for the Applications of Adaptive Multisorting Using Dielectrophoresis Array. *IEEE Journal of Microelectromechanical Systems, 16*(4), 816-825. DOI 10.1109/JMEMS.2007.897558

Shin, S. J., et al. (2010). Si-based ultrasmall multiswitching single-electron transistor operating at room-temperature. *Applied Physics Letters, 97*(10).

Sinangil, M. E., Yip, M., Qazi, M., Rithe, R., Kwong, J., & Chandrakasan, A. P. (2012). Design of Low-Voltage Digital Building Blocks and ADCs for Energy-Efficient Systems. *IEEE Transactions on Circuits and Wystems. II, Express Briefs, 59*(9), 533–537. doi:10.1109/TCSII.2012.2208675

Singh, K., Tiwari, S. C., & Gupta, M. (2014). A modified implementation of tristate inverter based static master-slave flip-flop with improved power-delay-area product. The Scientific World Journal.

Singh, N., Agarwal, A., Bera, L. K., Liow, T. Y., Yang, R., Rustagi, S. C., & Kwong, D.-L. et al. (2006). High-Performance fully depleted Silicon Nanowire Gate-All-Around CMOS devices. *IEEE Electron Device Letters*, *27*(5), 383–386. doi:10.1109/LED.2006.873381

Sinha, A. K., & Caviglia, D. D. (2013). Designing high value resistive network using weak inversion region of PMOS device at the floating Gate of a sensor. *IETE Technical Review*, *30*(6), 473–482. doi:10.4103/0256-4602.125665

Skotnicki, T., Hutchby, J. A., King, T., Wong, H.-S. P., & Boeuf, F. (2005). The end of CMOS scaling: Toward the introduction of new materials and structural changes to improve MOSFET performance. *Circuits and Devices Magazine, IEEE*, *21*(1), 16–26. doi:10.1109/MCD.2005.1388765

Smoline, J., & DiVincenzo. (1996). Five two-qubit gates are sufficient to implement the quantum Fredkin gate. *Physics Review A*, *53*(4), 2855-2856.

Sobhy, E. A., Hemy, A. A., Hoyos, S., Entesari, K., & Sanchez-Sinencio, E. (2011). A 2.8mW Sub 2-dB Noise Figure Inductorless Wideband CMOS LNA Employing Multiple Feedback. *IEEE Transactions on Microwave Theory and Techniques*, *59*(12), 3154–3161. doi:10.1109/TMTT.2011.2169081

Splendiani, A., Sun, L., Zhang, Y., Li, T., Kim, J., Chim, C. Y., & Wang, F. et al. (2010). Emerging Photoluminescence in Monolayer MoS_2. *Nano Letters*, *10*(4), 1271–1275. doi:10.1021/nl903868w PMID:20229981

Srikanth, K., & Dhireesha, K. (2008). GALEOR: Leakage reduction for CMOS circuits. In *Proceedings of 15th IEEE International Conference on Electronics Circuits and Systems*. IEEE.

St. Jude Medical. (2015). *St Jude Medical Nanostim Leadless Pacing*. Retrieved on August 29, 2015, from http://sjm.com/leadlesspacing.aspx

Standard library 0.13um - 0.25 um ROM Generator from Artisan: User Manual. (2004). *revision ug_2004q3v1*.

Standard library SRAM Generator from Artisan. (2004). User Manual (revision ug_2004q1v0). Author.

Stojanovic, V., & Oklobdzija, V. (1999). Comparative analysis of master-slave latches and flip-flops for high-performance and low-power systems. *IEEE Journal of Solid-State Circuits*, *34*(4), 536–548. doi:10.1109/4.753687

Stojcevski, A., Veljanovski, R., Singh, J., Faulkner, M., & Zayegh, A. (2003). A Real-Time Reconfigurable Pipelined Architecture with Advanced Power Management for UTRA-FDD. In *Proceedings of the IEEE Personal Indoor and Mobile Radio Communications (PIMRC)*, (pp. 64-68). doi:10.1109/PIMRC.2003.1264233

Strickland, E. (2013). Teeny Tiny Pacemaker Fits Inside the Heart. *IEEE Spectrum*.

Strollo, A., & Caro, D. (2000). Low power flip-flop with clock gating on master and slave latches. *Electronics Letters*, *36*(4), 294–295. doi:10.1049/el:20000268

Strollo, A., Napoli, E., & Caro, D. (2001). Low power flip-flops with reliable clock gating. *Microelectronics Journal*, *32*(1), 21–28. doi:10.1016/S0026-2692(00)00072-0

Subramanian, V. (2010). Multiple gate field-effect transistors for future CMOS technologies. *IETE Technical Review*, *27*(6), 446–454. doi:10.4103/0256-4602.72582

Sultana, J., Mitra, S. K., & Chowdhury, A. R. (2015). On the Analysis of Reversible Booth's Multiplier. *28th International Conference on VLSI Design* (VLSID).

Supreeth, M.S., & Bhaaskaran, V.S. (2015). Self-gated resonant-clocked flip-flop optimized for power efficiency and signal integrity. *IET Circuits, Devices & Systems*.

Supreeth, M. S., Sagar, D., & Kanchana Bhaaskaran, V. S. (2015). Low Power Energy Recovery Clock Distribution. *International Conference on Communications and Signal Processing (ICCSP'14).*

Sutherland, I. E., Sproull, B. F., & Harris, D. L. (2004). *Logical Effort: Designing fast CMOS circuits* (1st ed.). Morgan Kaufmann Publishers.

Suzuki, Y., Odagava, K., & Abe, T. (1973). Clocked CMOS calculator circuitry. *IEEE Journal of Solid-State Circuits, 8*(6), 462–469. doi:10.1109/JSSC.1973.1050440

Svensson, L. J., & Koller, J. G. (1994). Adiabatic charging without inductors.*International Workshop on Low-Power Design.*

Tang, H., Sun, Z. C., Chew, K. W. R., & Siek, L. (2014). A 1.33 μW 8.02-ENOB 100 kS/s Successive Approximation ADC With Supply Reduction Technique for Implantable Retinal Prosthesis. *IEEE Transactions on Biomedical Circuits and Systems, 8*(6), 844–856. doi:10.1109/TBCAS.2014.2300186 PMID:25608284

Tang, W., Osman, A., Kim, D., Goldstein, B., Huang, C., Martini, B., & Culurciello, E. et al. (2013). Continuous Time Level Crossing Sampling ADC for Bio-Potential Recording Systems. *IEEE Transactions on Circuits and Systems. I, Regular Papers, 60*(6), 1407–1418. doi:10.1109/TCSI.2012.2220464 PMID:24163640

Tao, L., Cinquanta, E., Chiappe, D., Grazianetti, C., Fanciulli, M., Dubey, M., & Akinwande, D. et al. (2015). Silicene field-effect transistors operating at room temperature. *Nature Nanotechnology, 10*(3), 227–231. doi:10.1038/nnano.2014.325 PMID:25643256

Taur, Y., & Ning, T. H. (1998). *Fundamentals of Modern VLSI Devices.* Cambridge University Press.

Teherani, J. T., Agarwal, S., Yablonovitch, E., Hoyt, J. L., & Antoniadis, D. A. (2013). Impact of Quantization Energy and Gate Leakage in Bilayer Tunneling Transistors. *IEEE Electron Device Letters, 34*(2), 298–300. doi:10.1109/LED.2012.2229458

Tessier, R., Pocek, K., & DeHon, A. (2015). Reconfigurable Computing Architectures. *IEE Proceedings, 103*(3), 332-354.

Thapliyal & Srinivas. (2005). A Novel Reversible TSG Gate and Its Application for Designing Reversible Carry Look-Ahead and Other Adder Architectures.*Tenth Asia-Pacific Computer Systems Architecture Conference (ACSAC05).*

Thompson, S. E., & Srivatsan P. (2006). Moore's law: the future of Si microelectronics. *Materials Today, 9*(6), 20-25.

Thomsen, A., & Brooke, M. A. (1991). A floating-gate MOSFET with tunneling injector fabricated using a standard double-polysilicon CMOS process. *IEEE Electron Device Letters, 12*(12), 111–113. doi:10.1109/55.75728

Tierney, J., Rader, C., & Gold, B. (1971). A digital frequency synthesizer. *IEEE Transactions on Audio and Electroacoustics, 19*(1), 48–57. doi:10.1109/TAU.1971.1162151

Tomishima, S., Tsuji, T., Kawasaki, T., Ishikawa, M., Inokuchi, T., Kato, H., & Takahashi, K. et al. (2001, November). A 1.0-V 230-MHz column access embedded DRAM for portable MPEG applications. *IEEE Journal of Solid-State Circuits, 36*(11), 1728–1737.

Trakimas, M., & Sonkusale, S. R. (2011). An Adaptive Resolution Asynchronous ADC Architecture for Data Compression in Energy Constrained Sensing Applications. *IEEE Transaction on Circuits and Systems I:Regular Papers, 58*(5), 921–934.

Tran, T. T. N., Boon, C. C., Do, M. A., & Yeo, K. S. (2011). Ultra-low power sub-mA series input resonance differential common gate LNA. *IET Electronic Letters, 47*(12), 703-704.

Tran, T. T. N., Boon, C. C., Do, M. A., & Yeo, K. S. (2012). *Ultra-low power Low noise amplifier designs for 2.4GHz ISM band applications.* (Doctoral dissertation). Nanyang Technological University, Singapore.

Trimberger, S. M. (2012). *Field-programmable gate array technology*. Springer Science & Business Media.

Tripathi, S. K., & Ansari, M. S. (2014). Voltage-mode universal filter for ZigBee using ±0.9 V 32nm CNFET ICC-II. (2014). *Confluence The Next Generation Information Technology Summit (Confluence),5th International Conference*. IEEE.

Trung-Kien, N., Krizhanovskii, V., Jeongseon, L., Seok-Kyun, H., Sang-Gug, L., Nae-Soo, K., & Cheol-Sig, P. (2006). A Low Power RF Direct-Conversion Receiver/Transmitter for 2.4GHz Band IEEE 802.15.4 Standard in 0.18um CMOS Technology. *IEEE Transactions on Microwave Theory and Techniques*, *54*(12), 4062–4071. doi:10.1109/TMTT.2006.885556

Tsai, J., Wang, J., & Su, Y. (2013). Piezoelectric rubber films for human physiological monitoring and energy harvesting. In *Proceedings of IEEE 26th International Conference on Micro Electro Mechanical Systems (MEMS)*, (pp. 841-844). doi:10.1109/MEMSYS.2013.6474374

Tsividis, Y. P. (1987). *Operation and modelling of the MOS Transistor*. Mc Graw-Hill.

Tsividis, Y., & Suyama, K. (1993, May). MOSFET Modeling for analog circuit CAD: problems and prospects.*Proc. IEEE Custom Integrated Circuits Conf.* doi:10.1109/CICC.1993.590710

Tummala, R., Rymaszewski, E. J., & Klopfenstein, A. G. (2012). *Microelectronics packaging handbook: technology drivers*. Springer Science & Business Media.

Tzartzanis, N., & Athas, W. C. (1997). Clock-powered logic for a 50 MHz low-power RISC Data path. In *Proc. of ISSCC*.

Usmani, F. A., & Mohammad, H. (2009). *Design and parametric analysis of 32nm op-amp in cmos and cnfet technologies for optimum performance. In Micro-Nanoelectronics, Technology and Applications. EAMTA 2009*. IEEE.

Usmani, F. A., & Mohammad, H. (2010). Carbon nanotube field effect transistors for high performance analog applications: An optimum design approach. *Microelectronics Journal*, *41*(7), 395–402. doi:10.1016/j.mejo.2010.04.011

Uyemura. (1999). *CMOS logic circuit design*. Kluwer Academic Publishers.

Van Rethy, J., De Smedt, M., Verhelst, M., & Gielen, G. (2013). Predictive sensing in analog-to-digital converters for biomedical applications. In *Proceedings of the International Symposium on Signals, Circuits and Systems (ISSCS)*, (pp. 11-12). doi:10.1109/ISSCS.2013.6651263

Vankamamidi, V., Marco, O., & Fabrizio, L. (2008). A serial memory by quantum-dot cellular automata (QCA). *Computers, IEEE Transactions on*, *57*(5), 606–618. doi:10.1109/TC.2007.70831

Varma, P. (2004). *Double edge-triggered flip-flops*. U.S. Patent 6794916.

Vasefi & Abid. (2005). *Low Power N-bit Adders and Multiplier using Lowest-Number of transistor 1 bit adders*. Saskatoon: Department of Electrical and Computer Engineering IEEE CCECE/CCGEI.

Veendrick, H. J. M. (1984, August). Short-circuit Dissipation of Static CMOS Circuitry and its Impact on the Design of Buffer Circuits. *IEEE JSSC*, 468-473.

Verhelst, M., Sylvester, D., Takamiya, M., Clinton, M., Wilcox, K., & Nose, K. (2015). *F4: Building the Internet of Everything (IoE): Low-power techniques at the circuit and system levels*. Paper presented at IEEE International the Solid-State Circuits Conference-(ISSCC).

Vesterbacka, M. (1999). A 14-transistor CMOS full adder with full voltage swing nodes. In *Proc. IEEE Workshop Signal Processing Systems*, (pp. 713–722). doi:10.1109/SIPS.1999.822379

Vetuli, A., Pascoli, S. D., & Reyneri, L. M. (1996). Positive Feedback in Adiabatic Logic. *Electronics Letters*, *32*(20), 1867–1869. doi:10.1049/el:19961272

Vieri, C. J. (1999). *Reversible Computer Engineering and Architecture*. (Thesis). Massachusetts Institute of Technology.

Vieru, C. (1995). *Pendulum: A Reversible Computer Architecture*. (SM thesis). Massachusetts Inst. of Technology.

Villemejane, J., Mottet, G., Francais, O., Pioufle, B., Woytasik, M., & Dufour, E. (2010). Nanomanipulation of Living Cells on a Chip Using Electric Field. *IEEE International Symposium in Electronic Design, Test and Application*. DOI doi:10.1109/DELTA.2010.26

Vittoz, E. A. (1985, June). The Design of High-Performance Analog Circuits on Digital CMOS Chips. *IEEE Journal of Solid-State Circuits*, *20*(3), 657–665. doi:10.1109/JSSC.1985.1052365

Vittoz, E. A., & Neyroud, O. (1979, June). A low-voltage CMOS bandgap reference. *IEEE Journal of Solid-State Circuits*, *14*(3), 573–577. doi:10.1109/JSSC.1979.1051218

Vojin,, G., Oklobdˇzija, F., Maksimovi'c, D., & Fengcheng. (1997). Pass-transistor adiabatic logic using single power-clock supply. *IEEE Transactions on Circuits and Systems. 2, Analog and Digital Signal Processing*, *44*(10).

Walden, R. H. (1999). Analog-to-digital converter survey and analysis. *IEEE Journal on Selected Areas in Communications*, *17*(4), 539–550. doi:10.1109/49.761034

Wallace, P.R. (1947). The Band Theory of Graphite. *Physical Review, 71*(9), 622–634.

Wang, M., Zhang, M., Fan, X., & Liu, Y. (2014). A Power Reduction Technique for Wideband Common Gate Low Noise Amplifiers. In *Proceedings of the IEEE 57th International Midwest Symposium on Circuits and Systems,* (vol. 10, pp. 969-972). IEEE.

Wang, W.Y., & Lau, K.T. (1005). Adiabatic pseudo-domino logic. *Electronics Letters, 31*(23), 1982-83.

Wang, B., & Law, M. K. (2015). A Precision CMOS Voltage Reference Exploiting Silicon Bandgap Narrowing Effect. *IEEE Transactions on Electron Devices*, *62*(7), 2128–2135. doi:10.1109/TED.2015.2434495

Wan, J., Royer, C. L., Zaslavsky, A., & Cristoloveanu, S. (2011). Tunneling FETs on SOI: Suppression of ambipolar leakage, low-frequency noise behavior, and modeling. *Solid-State Electronics*, *65-66*, 226–233. doi:10.1016/j.sse.2011.06.012

Watkins, N., Vankatesan, B., Toner, M., Rodriguez, W., & Bashir, R. (2009). A Robust Electrical Microcytometer with 3-Dimensional Hydrofocusing. *La bon a Chip, 9*(22), 3177-3184. DOI .10.1039/b91224a

Wei, L., Chen, Z., & Roy, K. (1999). Mixed-Vth (MVT) CMOS circuit design methodology for low power applications. In *Proc. of IEEE-ACM Design Automation Conference*. doi:10.1145/309847.309974

Wei, G. Y., & Horowitz, M. (1999, April). A fully digital, energy-efficient, adaptive power supply regulator. *IEEE Journal of Solid-State Circuits*, *34*(4), 520–528. doi:10.1109/4.753685

Wenger, E. (2014). *Artificial intelligence and tutoring systems: computational and cognitive approaches to the communication of knowledge*. Morgan Kaufmann.

Weste, N. H. E., & Eshraghian, K. (1993). *Principles of CMOS VLSI Design: A systems perspective* (2nd ed.). Addison-Wesley Publishing Company.

Weste, N. H. E., & Harris, D. M. (2011). *CMOS VLSI Design: A circuits and systems perspective* (4th ed.). Addison-Wesley.

Weste, N. H. E., Harris, D., & Banerjee, A. (2008). *CMOS VLSI Design: A circuits and system perspective* (3rd ed.). Pearson education.

Whitfield, F. J., & Doyel, A. T. (1984). *Methods and means for conducting heat from electronic components and the like: U.S. Patent No. 4,466,483*. Washington, DC: U.S. Patent and Trademark Office.

Wong, H.-S., Chan, K., & Taur, Y. (1997). Self-Aligned (Top and Bottom) Double-Gate MOSFET with a 25 nm Thick Silicon Channel. IEDM, 427.

Woo, S., Kim, W., Lee, C., & Laskar, J. (2012). A Wideband low-power CMOS LNA with positive-negative feedback for noise, gain, and linearity optimization. *IEEE Transactions on Microwave Theory and Techniques*, *60*(10), 3169–3178. doi:10.1109/TMTT.2012.2211379

Wu, J. (2008). Interactions of Electrical Fields and Fluids: Laboratory on a Chip Applications, IET. *Journal of Nanobiotechnology*, *2*(1), 14–27. doi:10.1049/iet-nbt:20070023 PMID:18298196

Xia, L., Cheng, J., Glover, N. E., & Chiang, P. (2014). 0.56 V, –20 dBm RF-Powered, Multi-Node Wireless Body Area Network System-on-a-Chip With Harvesting-Efficiency Tracking Loop. *IEEE Journal of Solid-State Circuits*, *49*(6), 1345–1355. doi:10.1109/JSSC.2014.2305074

Xilinx, Inc. (2012). *Xilinx Power Tools Tutorial*. Retrieved October 16, 2012, from http://www.xilinx.com/support/documentation/sw_manuals/xilinx14_3/ug733.pdf

Xilinx, Inc. (2014). *UltraScale Architecture SelectIO Resources*. Retrieved May 29, 2015, from http://www.xilinx.com/support/documentation/user_guides/ug571-ultrascale-selectio.pdf

Xilinx, Inc. (2015). *UltraScale Architecture and Product Overview*. Retrieved April 27, 2015, from http://www.xilinx.com/support/documentation/data_sheets/ds890-ultrascale-overview.pdf

Xu, Y., Yan, B., Zhang, H. J., Wang, J., Xu, G., Tang, P., & Zhang, S. C. et al. (2013). Large-Gap Quantum Spin Hall Insulators in Tin Films. *Physical Review Letters*, *111*(13), 136804. doi:10.1103/PhysRevLett.111.136804 PMID:24116803

Xu, Z., Gu, Q. J., Wu, Y.-C., & Chang, M.-C. F. (2015). Integrated D-band transmitter and receiver for wireless data communication in 65 nm CMOS. *Analog Integrated Circuits and Signal Processing*, *82*(1), 171–179. doi:10.1007/s10470-014-0379-9

Yadav, P. S., & Pandey, V. K. (2015). Power Efficiency of Half Adder Design Using MTCMOS Technique in 35 Nanometre Regime. *International Journal for Innovative Research in Science and Technology*, *1*(12), 111–117.

Yang, X., Maitra, K., Yeh, C., Zeitzoff, P., Raymond, M., Kulkarni, P., & Miller, R. J. et al. (2011) Analysis of parasitic resistance in double gate FinFETs with different fin length. In *Proceedings of International SOI Conference*. doi:10.1109/SOI.2011.6081799

Yaul, F. M., & Chandrakasan, A. P. (2014). A 10 bit SAR ADC With Data-Dependent Energy Reduction Using LSB-First Successive Approximation. *IEEE Journal of Solid-State Circuits*, *49*(12), 2825–2834. doi:10.1109/JSSC.2014.2352304

Ye & Roy. (2001). QSERL: quasi-static energy recovery logic. *IEEE Jl. of Solid-State Circuits, 36*(2), 239–248.

Yeap, G. K. (1998). *Practical Low power Digital VLSI Design* (1st ed.). Springer. doi:10.1007/978-1-4615-6065-4

Yeo, K. S., Rofail, S. S., & Goh, W. L. (2002). *CMOS/BiCMOS ULSI-Low Voltage, Low Power*. Prentice Hall PTR.

Ye, R., Horng, T., & Wu, J. (2013). Two CMOS Dual Feedback Common Gate Low-Noise Amplifiers with wideband Input and Noise Matching. *IEEE Transactions on Microwave Theory and Techniques*, *61*(10), 3690–3699. doi:10.1109/TMTT.2013.2277997

Ye, X. S., Zhi-G., S., Hongbo, Z., Lei, Y., & Cang-L., W. (2014). Intrinsic carrier mobility of germanene is larger than graphene's: First-principle calculations. *RSC Advances*, *4*(41), 21216–21220. doi:10.1039/C4RA01802H

Ye, Y., & Kaushik, R. (1996). Energy Recovery Circuits Using Reversible and Partially Reversible Logic. *IEEE Trans. on Circuits and Systems-I: Fund Theory and Applications*, *43*, 769–778.

Yigit, M., Gungor, V. C., Tuna, G., Rangoussi, M., & Fadel, E. (2014). Power line communication technologies for smart grid applications: A review of advances and challenges. *Computer Networks*, *70*, 366–383. doi:10.1016/j.comnet.2014.06.005

Yingchieh, H., & Su. (2012). A 0.1–0.3 V 40–123 fJ/bit/ch On-Chip Data Link With ISI-Suppressed Bootstrapped Repeaters. *IEEE Journal of Solid-State Circuits*, *47*(5).

Yip, M., & Chandrakasan, A.P. (2013). A Resolution-Reconfigurable 5-to-10-Bit 0.4-to-1 V Power Scalable SAR ADC for Sensor Applications. *IEEE Journal of Solid-State Circuits, 48*(6), 1453-1464.

Youngjoon, S., Chanho, L., & Yong, M. (2004). A low power 16-bit RISC microprocessor using ECRL circuits. ETRI Journal, 26(6).

Younis, S., & Knight, T. (1994). *Asymptotically zero energy computing using split-level charge recovery logic*. Technical Report AITR-1500. MIT AI Laboratory.

Younis, S. G., & Knight, T. F. (1993). Practical implementation of charge recovering asymptotically zero power CMOS. *Proc. of the 1993 Symposium on Integrated Systems*.

Yuce, M. R., Keong, H. C., & Chae, M. S. (2009). Wideband Communication for Implantable and Wearable Systems. *IEEE Transactions on Microwave Theory and Techniques*, *57*(10), 2597–2604. doi:10.1109/TMTT.2009.2029958

Yuk, K., McConaghy, C., Gascoyne, P., Schwartz, J. A., Vykoukal, J., & Andrews, C. (2007). A High-Voltage SOI CMOS Exciter Chip for a Programmable Fluidic Processor System Current, K.W.; Biomedical Circuits and Systems. *IEEE Transactions on Biomedical Circuits and Systems*, *1*(2), 105–115. doi:10.1109/TBCAS.2007.908110 PMID:23851665

Zaare, M., & Maymandi-Nejad, M. (2013). A new power efficient wireless non-uniform adaptive-sampling ECG recording system. In *Proceedings 21st Iranian Conference onElectrical Engineering (ICEE)*. doi:10.1109/IranianCEE.2013.6599768

Zhang, L., Wu, C., Mao, L.-F., & Zheng, J. (2012). Integrated SRAM compiler with clamping diode to reduce leakage and dynamic power in nano-CMOS process. *Micro & Nano Letters*, *7*(2), 171–173. doi:10.1049/mnl.2011.0680

Zhang, X., & Lian, Y. (2014). A 300-mV 220-nW Event-Driven ADC With Real-Time QRS Detection for Wearable ECG Sensors. *IEEE Transactions on Biomedical Circuits and Systems*, *8*(6), 834–843. doi:10.1109/TBCAS.2013.2296942 PMID:25608283

Zhao, X., Yi, J., Tong, D., & Cheng, X. (2006). Leakage power reduction for CMOS combinational circuits. In *Proc. of ICSICT*. doi:10.1109/ICSICT.2006.306330

Zheng, L., Leene, L. B., Liu, Y., & Constandinou, T. G. (2014). An adaptive 16/64 kHz, 9-bit SAR ADC with peak-aligned sampling for neural spike recording. In *Proceedings of IEEE International Symposium on Circuits and Systems (ISCAS)*. doi:10.1109/ISCAS.2014.6865652

Zhou, J., Wang, C., Liu, X., Zhang, X., & Je, M. (2015). An Ultra-low Voltage Level Shifter Using Revised Wilson Current Mirror for Fast and Energy-Efficient Wide-Range Voltage Conversion from Sub-threshold to I/O Voltage. *IEEE Transactions on Circuits and Systems I*, *62*(3), 697–706. doi:10.1109/TCSI.2014.2380691

Zhu, H., & Kursan, V. (2014). Novel low leakage and high speed triple threshold voltage buffers with skewed inputs and outputs. *IEEE Transactions on Circuits and Systems*, (61), 7.

Zhu, J., Bai, N., & Wu, J. (2013). A review of sense amplifier for static random access memory. *IETE Technical Review*, *30*(1), 72–81. doi:10.4103/0256-4602.107343

Zhuo, W., Shekhar, S., Embabi, S. H. K., Gyvez, J. P. D., Allstot, D. J., & Sanchez-Sinencio, E. (2005). A capacitor cross-coupled common-gate low-noise amplifier. *IEEE Transactions on Circuits and Wystems. II, Express Briefs*, *52*(12), 875–879. doi:10.1109/TCSII.2005.853966

Zhu, X., & Chen, Y. (2012). A 9-bit 100MS/s tri-level charge redistribution SAR ADC with asymmetric CDAC array. *2012 International Symposium onVLSI Design, Automation, and Test (VLSI-DAT)*.

About the Contributors

Manoj Sharma, SM- IEEE, Executive Committee Member IEEE India SSCS/CASS, has 11 plus years of experience in project implementation, research, development and academics in govt and private sector . He has successfully communicated over 20 research papers in International Journals and IEEE /Conferences. He continues to serve the research community in various volunteer activities including as Session Chair, PC Member and Reviewer for International Journal and Conferences. He has 2 Patent applications pending in IPO. His interest areas are circuit design, low power, timing analysis, optimization, ASIC design, VLSI, SOC, Radio Frequency Identification systems, CAD systems, communication system, NFC, Algorithms for VLSI systems and tools.

Ruchi Gautam is working as Director (Technical) in MyResearch Labs, Gr. Noida, India with interest in embedded and VLSI systems. She is also Managing Editor of International Journal of Circuits, Architecture and Design, Inderscience. She is contributing to the research community by various volunteer activities.

Mohammad Ayoub Khan is working with Department of Computer Science, Sharda University, Gr. NOIDA, INDIA with interests in Radio Frequency Identification, microcircuit design, and signal processing, NFC, front end VLSI(Electronic Design Automation, Circuit optimization, Timing Analysis), Placement and Routing in Network-on-Chip etc. He has more than 15 years' experience in his research area. He has published more than 70 publications in the reputed journals, proceedings and patents. He is contributing to the research community by various volunteer activities. He has served as Conference chair in various reputed IEEE/Springer international conferences. He is member of professional bodies of IEEE, ACM, ISTE and EURASIP society. He is also editorial member of technical committee at various journals of IEEE, Springer and Elsevier. Recently, he has also co-authored/edited three books.

* * *

Mohammad Faiz Liew Abdullah received BSc (Hons) in Electrical Engineering (Communication) in 1997, Dip Education in 1999 and MEng by research in Optical Fiber Communication in 2000 from University of Technology Malaysia (UTM). He completed his PhD in August 2007 from The University of Warwick, United Kingdom in Wireless Optical Communication Engineering. He started his career as a lecturer at Polytechnic Seberang Prai (PSP) in 1999 and was transferred to UTHM in 2000 (formerly known as PLSP). At present he is an Associate Professor and the Deputy Dean (Research and Development), Faculty of Electrical & Electronic Engineering, University Tun Hussein Onn Malaysia (UTHM).

He had 15 years' experience of teaching in higher education, which involved the subject Optical Fiber Communication, Advanced Optical Communication, Advanced Digital Signal Processing and etc. His research area of interest are wireless and optical communication, photonics and robotic in communication (Email: faiz@uthm.edu.my).

Mohd. Samar Ansari was born in Bareilly, India. He received B.Tech., M. Tech. and Ph.D. degrees in Electronics Engineering from Aligarh Muslim University, Aligarh, India, in 2001, 2007 and 2012, respectively. He is currently Assistant Professor in the Department of Electronics Engineering, Aligarh Muslim University, Aligarh, India, on leave from NIT, Jaipur. His research interests include analog VLSI, analog signal processing and nonlinear feedback neural networks. He has published around 65 research papers in reputed international journals & conferences and authored 2 books and contributed 5 book chapters.

Nabi Sertac Artan is an Assistant Professor of Electrical and Computer Engineering at New York Institute of Technology (NYIT). He got his Ph.D. degree in Electrical Engineering from New York University (formerly Polytechnic University). Prior to joining NYIT, Dr. Artan was on the faculty of the New York University School of Engineering. He also worked as an ASIC (Application Specific Integrated Circuit) Design Engineer and designed integrated circuits for commercial, academic and military applications. Dr. Artan served in the organizing committees of the ACM Conference on Security and Privacy in Wireless and Mobile Networks (WiSec), Workshop on RFID Security (RFIDSec), IEEE Sarnoff Symposium, and ACM/IEEE Symposium on Architectures for Networking and Communications Systems (ANCS). Dr. Artan's current research interests include low-power embedded systems and VLSI circuits for implantable medical devices and high-speed circuits for network security.

V. S. Kanchana Bhaaskaran is a Professor and Dean of the School of Electronics Engineering at VIT University Chennai. She obtained her undergraduation degree in Electronics and Communication Engineering from Institution of Engineers (India), Calcutta and her M.S. degree in Systems and Information from Birla Institute of Technology and Sciences, Pilani and Ph D from V I T University. She has more than 35 years of industry, research and teaching experience by serving the Department of Employment and Training, Government of Tamil Nadu, IIT Madras, Salem Cooperative Sugar Mills' Polytechnic College, SSN College of Engineering and VIT University Chennai. Her specializations include Low Power VLSI Circuit Design, Microprocessor architectures and Linear Integrated Circuits. She has published around 75 papers in International Journals and conferences, and has one patent published and one filed. She is a reviewer for peer reviewed international journals and conferences. She is the Fellow of the Institution of Engineers (India), Fellow of the Institution of Electronics and Telecommunication Engineers, Life Member of the Indian Society for Technical Education and Member of the Institute of Electrical and Electronics Engineers Inc., USA.

Satish Chandra received the M.Tech degree in VLSI design from Indraprastha University, New Delhi, in 2009. He is currently pursuing the Ph.D. degree in electronics engineering at the Division of ECE, NSIT, University of Delhi. He has authored and co-authored over 11 research papers in the low power VLSI domain in various international/national journals and conferences.

Saurabh Chaudhury graduated from REC Silchar (now NIT Silchar) with Honours in Electrical Engineering in 1994. He joined as a lecturer in REC Silchar in 1997, after a brief experience in industry (one year) and teaching (one year). Subsequently he has obtained his M.Tech and Ph.D degree from IIT Kharagpur in the field of Microelectronics & VLSI design in 2001 and 2009 respectively, through QIP program, Govt. of India. He has always strived for knowledge and research and dreams big. He is working in the domain of low power VLSI design, synthesis, leakage minimization, CNTs and image processing. He has published more than 35 papers in Int. journals and Conferences. He has reviewed many papers in technical journals including IEEE, IET and other international conferences.Currently he is holding the position of Associate Professor, in Electrical Engineering.

Bhagwan Das received his Bachelor of Electronic Engineering Degree in 2008 from MUET, Pakistan. He has received his Masters of Engineering Degree (communication engineering) from QUEST, Pakistan in 2013. He is in services with Quiad-e-Awam University of Engineering, Science and Technology since Jan 2011 as Lecturer in Electronic department as a regular employee. He is also professional member of IEEE. He is Registered Engineer in Pakistan Engineering council and member in Pakistan Engineering Congress. He is currently doing PhD in Electrical Engineering from Universiti Tun Hussein Onn Malaysia (UTHM) under supervision of Assoc. Prof. Dr. Mohammad Faiz Liew bin Abdullah. His research fields of interest are optical communication and signal processing and FPGAs based Energy Efficient design. He successfully published many research articles in SCOPUS, ISI, EI compendex indexed journals. He has presented more than 15 papers in IEEE index conferences held in Vietnam, Thailand, India, Malaysia, and Pakistan. He is nominated international program committee member in International Conference on Computing and Informatics (ICOCI) (ISI index). He reviewed many paper in International Conference on Soft Computing in Data Science 2015 (SCDS2015) (Springer index). Recently, he has reviewed the several articles of "Wireless Personal Communication" Impact factor Journal 0.998 (Thomas Reuters) and "Radio Engineering Journal" Impact factor Journal 0.7 (Thomas Reuters). Recently, won the Gold medal in Research and Innovation Festival 2015 held in Malaysia.

Martha Salome Lopez de la Fuente works at Universidad de Monterrey, Mexico, as a research professor for the intelligent systems and robotics programs. Her contributions in the research and education areas relate to digital systems design, application specific integrated circuits, embedded systems applications, intelligent systems, and service robots. Since 1994 she has been teaching courses on microprocessors, digital electronics, embedded systems, integrated circuit design, and hardware architectures for service robots. Martha has presented her research work in forums such as Engineering in Medicine and Biology Conference, CERMA Electronics Robotics and Automotive Mechanics Conference, Eurasian Multidisciplinary Forum, and World Forum on Internet of Things. She has also published scientific articles in the IEEE Xplore digital library, the Review of Scientific Instruments journal, and the European Scientific Journal.

Maneesha Gupta was born in 1960. She received her B.E. in Electronics & Communication Engineering from Government Engineering College, Jabalpur in 1981, M.E. in Electronics & Communication Engineering from Government Engineering College, Jabalpur in 1983 and PhD. in Electronics Engg. (Analysis, Synthesis & Applications of Switched Capacitor Circuits) from Indian Institute of Technology, Delhi in 1990. Dr. Gupta held the positions of Lecturer in Electronics & Communication Engineering Department at Government Engineering College, Jabalpur from 1981 to 1982, Kota Engg. College, Kota from 1986 to 1988, YMCA Institute of Engg., Faridabad in 1998 and Netaji Subhas Institute of Technology, New Delhi from 1998 to 2000. She joined as Assistant Professor in Electronics & Communication Engineering (ECE) Department of the Netaji Subhas Institute of Technology in 2000, and is currently working as Professor since 2011 in the same institute. Her teaching and research interests are Switched Capacitors Circuits and Analog Signal processing. She has co-authored over 100 research papers in the above areas in various international/national journals and conferences. She got best paper award for her paper in IETE journal of Education in 2001.

Lini Lee was born in Malaysia in 1977. She received the B. E., M. Sc and Ph.D. degrees from Universiti Putra Malaysia, Malaysia in 1999, 2001 and 2008 respectively. She joined the industry working on supporting Synopsys EDA tools in 2001. Then, she came back to university to pursue her Ph.D. In 2008, she joined Freescale working as engineer. In 2010, she left the industry to join Multimedia University (MMU) and since then, she has been with Faculty of Engineering, MMU where she's currently a Senior Lecturer. Her main areas of research interest are integrated circuit (IC) design in both digital and analog, especially at low power applications. Dr Lini is an active member of IEEE and a Chartered Engineer (CEng) with the Institution of Engineering and Technology.

Rohit Lorenzo received the B.Tech degree in electronics and communication engineering from U.P Technical University, Lucknow,India, in 2006 and the M.E degree in VLSI design from Karunya University, Coimbatore, India, in 2008 .He is currently working toward the Ph.D degree under the guidance of Dr. S.Chaudhury in the department of Electrical engineering, NIT Silchar,India. During 2008-2010, he was a Assistant professor at InvertisUniversity, Barielly, India and from 2011-2012. Healso work as a Assistant professor on contract at NIT Kurukshetra, India. His research interest include low power VLSI design, synthesis, leakage power minimization, low power SRAM circuits and digital VLSI design.He is reviewer of international journal of electronicsr (taylor and francis).He is a student member of IEEE and IEICE.

Kavyashree.P obtained her M.Tech in VLSI Design and Embedded Systems from Visvesvaraya Technological University Extension Center, UTL Technologies Limited, Bangalore, India. Her area of research is on VLSI Design, analog and mixed signal design, RF Integrated circuits and sensor design. She has published an international journal paper and four IEEE conference papers in these research areas.

Senthil C Pari had post-graduated in Applied Electronics and Material Science at National Institute of Technology, Tiruchirappalli. He had joined as a lecturer in AVC College, Mayiladuthurai and He elevated as a Head of the department in Adaikalamatha College, Vallam Thanjavur. He was join as a lecturer in Multimedia University in 2005 and he obtained Ph.D. Degree from Multimedia University in 2009. He promoted as a senior lecturer in 2010 at multimedia university. Dr. C.Senthilpari had a proposed a new adder circuit in VLSI Design research. His work on the VLSI Design and simulation analysis of new hardware circuits. Dr. C.Senthilpari invited committee member in various international conference. He immersed himself in conducting research in chosen area of specialization, guiding Ph.D. projects and Master students and teaching at undergraduate and post graduate level. Over 150 B.Eng students, 1 Master students, 4 Doctorates stand testimony for his productivity in VLSI Design research. Dr. C.Senthilpari has so far published more than 47 papers in national and international journals and conferences. He has attended many international conferences. Dr. Senthipari is a Senior Member in IEEE (USA) and Member in IET (U.K). Dr.Senthilpari obtained Charted Engineer from IET (U.K).

B Shivalal Patro has completed his B.Tech from Trident Academy of Technology in Electronics & Telecommunication Engineering in 2010, Odisha. He then completed M.Tech in Communication Systems from KIIT University in 2012. Currently, he is a Ph.D scholar at KIIT University, Bhubaneswar, Odisha. His area of research includes high speed, low power analog and mixed signal IC design, modeling and optimization.

Vandana B Rao has completed her B.Tech from JNTU (Hyderabad) and M.Tech from PCU (Puducherry). She is presently pursuing PhD in KIIT University (Bhubaneswar).Her research area is Device modeling in VLSI.

Kunwar Singh received his B.Tech in Electronics and Communication Engineering from GGSIP University, New Delhi and M.Tech in VLSI Design from CDAC, Noida in 2006 and 2009 respectively. He served as an intern in Cadence Design Systems during Feb. 2009 – Jul. 2009 where he worked as a product validation engineer in SPB group. He is currently serving as an Assistant Professor in the Division of ECE, NSIT wef Sep. 2013 and is working towards his PhD degree in the University of Delhi. Earlier, he has also served as Assistant Professor in the Department of Electrical Engineering, Delhi Technological University from Jul. 2010 – Sep. 2013. His research interests include Low power high performance digital CMOS circuits, CAD for VLSI Design, applications of artificial intelligence techniques in power-delay-area product optimization of digital CMOS integrated circuits, automated design and optimization of analog and mixed-signal integrated circuits, integrated circuit design using memristors. He has authored and co-authored over 14 research papers in the above areas in various international/national journals and conferences. He has filed one Indian and one US patent application.

Shailendra Kumar Tripathi was born in Kanpur, India. He has completed B.Sc. and M.Sc.(Electronics) degrees from C.S.J.M. University Kanpur, India, in 2002 and 2004 respectively. He received M.Tech., degree in Electronics Engineering from Aligarh Muslim University, Aligarh, India, in 2008. Presently, he is research scholar in the Department of Electronics & Communication Engineering, Malaviya National Institute of Technology, Jaipur, India. His research interests are analog signal processing, nanoelectronic circuits and VLSI Design. He has published 5 research papers in reputed international conferences.

Siva Yellampalli is currently working as a Professor with VTU Extension Centre, UTL Technologies Ltd. He obtained his MS & Ph.D from Louisiana State University. His area of research is system level design for power optimization. His area of research encompasses different research fields such as Very Large Scale Integration (VLSI), mixed signal circuits/systems development, micro-electromechanical systems (MEMS), and integrated carbon nanotube based sensors. He has published a book in the area of mixed signal design, and edited two books on carbon nano tubes. He also published about 100 International Journal papers & IEEE Conference papers in these areas of research. In addition he has delivered keynote speeches at International conferences held in Canada, Spain & Dubai. He has given many professional presentations including invited talks. He has been a consultant to a variety of industries and acts as a reviewer for technical journals and book publishers. Currently he is the Student Activities Chair, Executive Committee IEEE Bangalore Section. He is also a life member of Indian Society for Technical Education (ISTE), senior member, International Association of Computer Science and Information Technology (IACSIT), VLSI Society of India (VSI) and National Society of Collegiate Scholars (NSCS).

Index

www.ingramcontent.com/pod-product-compliance
Lightning Source LLC
LaVergne TN
LVHW060403300926
849162LV00006B/1337
* 9 7 8 1 5 2 2 5 0 1 9 0 9 *